Lecture Notes in Computer Scie

Commenced Publication in 1973
Founding and Former Series Editors:
Gerhard Goos, Juris Hartmanis, and Jan van Leeuwen

Mario A. Nascimento   Timos Sellis
Reynold Cheng   Jörg Sander   Yu Zheng
Hans-Peter Kriegel   Matthias Renz
Christian Sengstock (Eds.)

Advances in Spatial and Temporal Databases

13th International Symposium, SSTD 2013
Munich, Germany, August 21-23, 2013
Proceedings

 Springer

Volume Editors

Mario A. Nascimento
University of Alberta, Edmonton, AB, Canada, E-mail: mario.nascimento@ualberta.ca

Timos Sellis
RMIT University, Melbourne, VIC, Australia, E-mail: timos.sellis@rmit.edu.au

Reynold Cheng
The Unversity of Hong Kong, China, E-mail: ckcheng@cs.hku.hk

Jörg Sander
University of Alberta, Edmonton, AB, Canada, E-mail: jsander@ualberta.ca

Yu Zheng
Microsoft Research Asia, Beijing, China, E-mail: yuzheng@microsoft.com

Hans-Peter Kriegel
Matthias Renz
Ludwig Maximilians University, Munich, Germany
E-mail: {kriegel, renz}@dbs.ifi.lmu.de

Christian Sengstock
Ruprecht Karls University, Heidelberg, Germany
E-mail: sengstock@informatik.uni-heidelberg.de

ISSN 0302-9743   e-ISSN 1611-3349
ISBN 978-3-642-40234-0   e-ISBN 978-3-642-40235-7
DOI 10.1007/978-3-642-40235-7
Springer Heidelberg Dordrecht London New York

Library of Congress Control Number: 2013944578

CR Subject Classification (1998): H.2, H.3, I.2, G.2

LNCS Sublibrary: SL 3 – Information Systems and Application,
incl. Internet/Web and HCI

Typesetting: Camera-ready by author, data conversion by Scientific Publishing Services, Chennai, India

Printed on acid-free paper

Springer is part of Springer Science+Business Media (www.springer.com)

Preface

This volume contains the proceedings of the 13th International Symposium on Spatial and Temporal Databases (SSTD). Included are research contributions in the area of spatial and temporal data management and related technologies presented at SSTD 2013 at the Ludwig-Maximilians-Universität, just beside the beautiful and world-renowned English Garden, in the wonderful city of Munich, the capital of the Free State of Bavaria, Germany. The symposium brings together, for three days, researchers, practitioners, and developers for the presentation and discussion of current research on concepts, tools, and techniques related to spatial and temporal databases.

SSTD 2013 was the 13th in a series of biannual events. Previous symposia were successfully held in Santa Barbara (1989), Zurich (1991), Singapore (1993), Portland (1995), Berlin (1997), Hong Kong (1999), Los Angeles (2001), Santorini, Greece (2003), Angra dos Reis (2005), Boston (2007), Aalborg (2009), and Minneapolis (2011). Before 2001, the series was devoted solely to spatial database management, and called The International Symposium on Spatial Databases. Starting in 2001, the scope was extended in order to also accommodate temporal database management, in part due to the increasing importance of research that considers spatial and temporal dimensions of data as complementary.

This year we received 58 submissions from authors in 25 different countries, which were reviewed by at least three of the 43 Program Committee members, helped by 44 external reviewers. At the end of a thorough process of reviews and discussions, 24 submissions were accepted for presentation at the symposium. After careful evaluation of the top-ranked submissions, the paper "Stream-Mode FPGA Acceleration of Complex Pattern Trajectory," co-authored by Roger Moussalli, Marcos Vieira, Walid Najjar, and Vassilis Tsotras, was selected as SSTD 2013 best paper.

SSTD 2013 also continued several innovative aspects that have been successfully introduced in previous events. In addition to the research paper track, the conference hosted a demonstrations track and included two vision and challenge papers. Demonstrations and vision/challenge papers were solicited by separate calls for papers. While demonstrations proposals had to illustrate running systems that showcase the applicability of interesting and solid research, the vision/challenge submissions had to discuss novel ideas that are likely to guide research in the near future and/or challenge prevailing assumptions. The submissions to the demo and vision/challenge track (9 demonstration submissions and the 4 vision/challenge papers submissions) were evaluated by dedicated Program Committees, recruited by the demonstrations co-chairs.

We are very fortunate to have had two well-accomplished researchers from academia and industry as keynote speakers opening the first two days of the conference: Prof. Cyrus Shahabi (University of Southern California) made a presentation on "A Big-Data Framework for Decision-Making in Transportation Systems" and Ralf-Peter Schäfer (VP TomTom Traffic Product Unit & Fellow, Berlin) talked about "Car-Centric Traffic Monitoring and Management Using Probe-Based Community Data." Both are very attractive and timely topics, from the academic and industrial points of view.

The success of SSTD 2013 was the result of a team effort. Special thanks go to many people for their dedication and hard work, in particular to the Local Organizers, Publicity Chairs, Proceedings Chair, and Webmasters. Naturally, we owe our gratitude to a larger range of people, in particular we would like to thank the authors, irrespectively of whether their submissions were accepted or not, for supporting the symposium series and for sustaining the high quality of the submissions. Last but most definitely not least, we are very grateful to the members of the Program Committees (and the external reviewers) for their thorough and timely reviews.

Finally, these proceedings reflect the state-of-the-art in the domain of spatio-temporal data management, and as such we believe they form a strong contribution to the related body of research and literature.

June 2013

Mario A. Nascimento
Timos Sellis
Reynold Cheng
Jörg Sander
Yu Zheng
Hans-Peter Kriegel
Matthias Renz

Organization

Steering Committee

The SSTD Endowment

General Co-chairs

Hans-Peter Kriegel   Ludwig Maximilians University Munich, Germany

Matthias Renz   Ludwig Maximilians University Munich, Germany

Program Co-chairs

Mario A. Nascimento   University of Alberta, Canada

Timos Sellis   RMIT University, Australia

Reynold Cheng   University of Hong Kong

Demo Co-chairs

Jörg Sander   University of Alberta, Canada

Yu Zheng   Microsoft Research, Asia

Publicity Chair

Thomas Seidl   RWTH Aachen University, Germany

Michael Gertz   Ruprecht Karls University Heidelberg, Germany

Sponsorship Chair

Agnès Voisard   Freie Universität Berlin, Germany

Erik Hoel   ESRI, USA

Proceedings Chair

Christian Sengstock   Ruprecht Karls University Heidelberg, Germany

Local Arrangements

Peer Kröger   Ludwig Maximilians University Munich,
                             Germany
Matthias Schubert   Ludwig Maximilians University Munich,
                             Germany

Webmaster

Tobias Emrich   Ludwig Maximilians University Munich,
                             Germany
Johannes Niedermayer   Ludwig Maximilians University Munich,
                             Germany

Program Committee

Walid Aref	Dan Lin	Shashi Shekhar
Spiridon Bakiras	Xuemin Lin	Kian-Lee Tan
Michela Bertolotto	Hua Lu	Yufei Tao
Claudio Bettini	Nikos Mamoulis	Yannis Theodoridis
Michael Böhlen	Yannis Manolopoulos	Anthony K.H. Tung
Lei Chen	Mohamed Mokbel	Carola Wenk
Chi-Yin Chow	Kyriakos Mouratidis	Ouri Wolfson
Maria Luisa Damiani	Mirco Nanni	Michael Worboys
Ralf Hartmut Güting	Enrico Nardelli	Xiaokui Xiao
Christian S. Jensen	Dimitris Papadias	Xing Xie
Panagiotis Karras	Spiros Papadimitriou	Man Lung Yiu
George Kollios	Dieter Pfoser	Rui Zhang
Bart Kuijpers	Chiara Renso	Yu Zheng
Feifei Li	Dimitris Sacharidis	Xiaofang Zhou
Jianzhong Li	Bernhard Seeger	

Vision/Challenge Program Committee

Walid Aref	Dimitris Papadias	Timos Sellis
Spiridon Bakiras	Spiros Papadimitriou	Richard Snodgrass
Erik Hoel	Apostolos Papadopoulos	Goce Trajcevski
George Kollios	Dimitris Sacharidis	Vassilis Tsotras
Nikos Mamoulis	Simonas Saltenis	Karine Zeitouni
Mirco Nanni	Bernhard Seeger	

Demo Program Committee

Bertino Elisa
Sanjay Chawla
Fosca Giannotti
Ralf Hartmut Güting
Yan Huang

Christian S. Jensen
Nikos Mamoulis
Mohamed Mokbel
Peter Scheuermann
Cyrus Shahabi

Shashi Shekhar
Goce Trajcevski
X. Sean Wang
Ouri Wolfson
Xiaofang Zhou

External Reviewers

Achakeev, Daniar
Armenantzoglou, Nikos
Ayala, Daniel
Behr, Thomas
Bertolaja, Letizia
Bikakis, Nikos
Bouros, Panagiotis
Bozanis, Panayiotis
Camossi, Elena
Cintia, Paolo
Corral, Antonio
Dimokas, Nikos
Efstathiades,
    Christodoulos
Evans, Michael
Gounaris, Anastasios

Huang, Jin
Jiang, Jinling
Jiang, Zhe
Kellaris, George
Khodaei, Ali
Kvochko, Andrey
Liagouris, John
Lin, Yimin
Liu, Bin
Ma, Sol
Monreale, Anna
Papadopoulos, Apostolos
Parchas, Panos
Pelekis, Nikos
Qi, Jianzhong
Qi, Shuyao

Riboni, Daniele
Seidemann, Marc
Shang, Shuo
Skoutas, Dimitrios
Stenneth, Leon
Varriale, Roland
Vassilakopoulos, Michael
Wei, Ling-Yin
Xie, Xike
Xu, Bo
Xue, Andy Yuan
Yang, Kwangsoo
Zhang, Chengyuan
Zhang, Jilian
Zhou, Xun
Zhu, Haohan

TransDec: A Big-Data Framework for Decision-Making in Transportation Systems

Cyrus Shahabi

University of Southern California
shahabi@usc.edu

Abstract. The vast amounts of transportation datasets (traffic flow, incidents, etc.) collected by various federal and state agencies are extremely valuable in real-time decision-making, planning, and management of the transportation systems. In this talk, I will argue that considering the large volume of the transportation data, variety of the data (different modalities and resolutions), and the velocity of the data arrival, developing a scalable system that allows for effective querying and analysis of both archived and real-time data is an intrinsically challenging BigData problem. Subsequently, I will present our end-to-end prototype system, dubbed TransDec (short for Transportation Decision-Making), which enables real-time integration, visualization, querying, and analysis of these dynamic and archived transportation datasets. I will then discuss a GPS navigation application enabled by such a system and demonstrate its commercialization as a product called ClearPath (see http://myfastestpath.com). Motivated by ClearPath, we will look under the hood and focus on a route-planning problem where the weights on the road-network edges vary as a function of time due to the variability of traffic congestion. I will show that naïve approaches to address this problem are either inaccurate or slow, leading to our new approach to this problem: A time-dependent A* algorithm.

Cyrus Shahabi is a Professor of Computer Science and Electrical Engineering and the Director of the Information Laboratory (InfoLAB) at the Computer Science Department and also the Director of the NSF's Integrated Media Systems Center (IMSC) at the University of Southern California. He was also the CTO and co-founder of a USC spin-off and an In-Q-Tel portfolio company, Geosemble Technologies, which was acquired in July 2012. He received his B.S. in Computer Engineering from Sharif University of Technology in 1989 and then his M.S. and Ph.D. Degrees in Computer Science from the University of Southern California in May 1993 and August 1996, respectively. He authored two books and more than two hundred research papers in the areas of databases, GIS and multimedia.

Dr. Shahabi has received funding from several agencies such as NSF, NIJ, NASA, NIH, DARPA, AFRL, and DHS as well as several industries such as Chevron, Google, HP, Intel, Microsoft, NCR and NGC. He was an Associate Editor of IEEE Transactions on Parallel and Distributed Systems (TPDS) from

2004 to 2009. He is currently on the editorial board of the VLDB Journal, IEEE Transactions on Knowledge and Data Engineering (TKDE), ACM Computers in Entertainment and Journal of Spatial Information Science. He is the founding chair of IEEE NetDB workshop and also the general co-chair of ACM GIS 2007, 2008 and 2009. He chaired the nomination committee of ACM SIGSPATIAL for the 2011-2014 terms. He is a PC co-Chair of MDM 2013 and regularly serves on the program committee of major conferences such as VLDB, ACM SIGMOD, IEEE ICDE, ACM SIGKDD, and ACM Multimedia.

Dr. Shahabi is a recipient of the ACM Distinguished Scientist award in 2009, the 2003 U.S. Presidential Early Career Awards for Scientists and Engineers (PECASE), the NSF CAREER award in 2002, and the 2001 Okawa Foundation Research Grant for Information and Telecommunications. He was the recipient of US Vietnam Education Foundation (VEF) faculty fellowship award in 2011 and 2012, an organizer of the 2011 National Academy of Engineering "Japan-America Frontiers of Engineering" program, an invited speaker in the 2010 National Research Council (of the National Academies) Committee on New Research Directions for the National Geospatial-Intelligence Agency, and a participant in the 2005 National Academy of Engineering "Frontiers of Engineering" program.

Car-Centric Traffic Monitoring and Management Using Probe-Based Community Data

Ralf-Peter Schäfer

TomTom
ralf-peter.schaefer@tomtom.com

Abstract. With the introduction of TomTom's historic and real-time traffic technologies IQ Routes and HD Traffic in 2007 the portfolio has been implemented in many European countries, North America, South Africa, Australia and New Zealand.

The backbone of the technology is community data from GPS enabled navigation devices, fleet management solutions as well as GSM cell phone operation. Today, the entire TomTom Traffic Community consists of more than 140 Mio. users which bring TomTom in a position to get precise travel time data for the entire road network in the underlying markets. All the traffic content has been fully integrated into navigation software of TomTom and is also licensed to external parties for use in enterprise and government application of traffic information, traffic planning and traffic management.

The increasing community of connected navigation devices using precise traffic information is also contributing for a better utilization of the road network when drivers follow dynamic route guidance. In a Traffic Manifesto the TomTom vision has been published making TomTom's traffic information widely available. The availability of traffic information and use in dynamic navigation helps to improve the road network utilization and can contribute to travel behavior changes with positive impact for the network load and the environment. Even when only 10% of the entire driver population uses dynamic routes guidance with precise traffic as HD Traffic they can gain travel time wins of up to 15% toward a planned destination but also help non-informed users to decrease the travel time up to 5%.

Connected navigation can help traffic management authorities to make the road network utilization more efficient at lower costs as installing more expensive local detection systems. In the presentation the data collection, processing, storage and distribution system will be introduced. The data fusion engine to provide travel time and jam delay information is fully based on analytics of billions of real-time GPS probe data collected in 26 markets globally. Real-time and historic traffic information can be used in a variety of classical traffic management and standard routing/navigation applications generated in high quality and more cost efficient as infrastructure based systems using local detection system as inductive loops. Infrastructure performance measures are available in the entire road network and give traffic planners and consultants great tools at hand to put the investments at the right spots. TomTom discloses every 3 months the Traffic Congestion Report where over 100 cities globally are analyzed. Traffic delays in rush hour and over the days are computed against the free flow condition measure over the night hours.

In the most recent report Moscow was analyzed as the most congested city of all cities in the report. The presentation will also cover how the index is built.

Since January 2012 Ralf-Peter Schäfer is leading the TomTom Traffic Product Unit. He is responsible for the global product development for traffic in the TomTom Group incl. engineering, product and program management as well as sales and marketing support.

He joined TomTom in August 2006 and started to work in a position of the Research Director of TomTom's Mobility Solution Department and the Head of the TomTom R&D Centre in Berlin. His major scope from 2006 was the development of the TomTom traffic portfolio incl. algorithms and software for the realtime traffic information service HD Traffic and the historic speed profile product IQ Routes. With his team he also developed the dynamic location referencing technology OpenLR to allow cost efficient transmission and excahnge of map location.

Ralf-Peter Schäfer studied electrical engineering the the Technical University of Ilmenau and worked in several research organisations as the German Academy of Sciences, the German Centre for Computer Science, the German Aerospce Center (DLR) before he joined TomTom. Main areas of activities in the past and today include static and dynamic content generation for traffic products, probe technology development from GPS and GSM sources as well as modeling and data processing techniques for traffic systems.

Table of Contents

Session 1: Joins and Algorithms

Session 2: Mining and Discovery

Session 3: Indexing

Session 4: Trajectories and Road Network Data 1

Session 5: Nearest Neighbours Queries

Session 6: Trajectories and Road Network Data 2

Session 7: Uncertainty

Demonstrations

Efficient Top-k Spatial Distance Joins[*]

Shuyao Qi[1], Panagiotis Bouros[2,**], and Nikos Mamoulis[1]

[1] Department of Computer Science
The University of Hong Kong
{syqi2,nikos}@cs.hku.hk
[2] Department of Computer Science
Humboldt-Universität zu Berlin, Germany
bourospa@informatik.hu-berlin.de

Abstract. Consider two sets of spatial objects R and S, where each object is assigned a score (e.g., ranking). Given a spatial distance threshold ϵ and an integer k, the top-k spatial distance join (k-SDJ) returns the k pairs of objects, which have the highest combined score (based on an aggregate function γ) among all object pairs in $R \times S$ which have spatial distance at most ϵ. Despite the practical application value of this query, it has not received adequate attention in the past. In this paper, we fill this gap by proposing methods that utilize both location and score information from the objects, enabling top-k join computation by accessing a limited number of objects. Extensive experiments demonstrate that a technique which accesses blocks of data from R and S ordered by the object scores and then joins them using an aR-tree based module performs best in practice and outperforms alternative solutions by a wide margin.

1 Introduction

The spatial join operator retrieves pairs of objects that satisfy a spatial predicate. Spatial joins have been extensively studied [3,6,17,7,15] due to their applicability and potentially high execution cost. Still, this query type only focuses on spatial attributes, while in many applications spatial objects have additional attributes. For instance, restaurants shown in websites like Foursquare and Yelp are assigned user-generated ratings. As another example, consider collections of spatial objects created in the context of emerging scientific fields like atmospheric, oceanographic, and environmental sciences with an expertise that ranges, from the modeling of global climatic change to the analysis of earth's tectonics. The objects in such collections are associated with measurements of several attributes varying from temperature and pressure to earth's gravity and seismic activity.

These attributes can be used to derive a ranking for the objects. Ranking has been considered by the database community in the context of top-k queries [5,9] and top-k joins [13,8,12,16] where ranked inputs are joined to derive objects or tuple pairs which maximize an aggregate function on scoring attributes. Consider, for example, the following top-k join query expressed in SQL:

---

[*] Work supported by grant HKU 714212E from Hong Kong RGC.
[**] This work was conducted while the author was with the University of Hong Kong.

M.A. Nascimento et al. (Eds.): SSTD 2013, LNCS 8098, pp. 1–18, 2013.

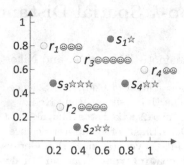

Fig. 1. Example of a top-k spatial distance join

```
SELECT R.id, S.id
FROM R, S
WHERE R.att = S.att
ORDER BY R.score + S.score DESC
STOP AFTER k;
```

The result of this query is the k pairs of objects (r, s) with $r \in R$ and $s \in S$ that qualify the equality predicate on their common attribute `att`, having the highest SUM of their `score` attributes.

Despite the vast availability of spatial objects associated with scoring attributes, to our knowledge, there exists no join operator that considers both spatial and score attributes at the same time.[1] On an attempt to fill this gap, we introduce the *top-k spatial distance join* (k-SDJ) query. Given two collections of spatial objects R and S that also carry a score attribute, the k-SDJ query retrieves a k-subset J of $R \times S$ such that for every pair of objects $(r, s) \in J$, r is spatially close to s based on a *distance threshold* ϵ (i.e., $dist(r, s) \leq \epsilon$, where $dist$ denotes the distance between the spatial locations of r and s), and for every $(r', s') \in R \times S - J$ such that $dist(r', s') \leq \epsilon$, it holds $\gamma(r, s) \geq \gamma(r', s')$, where γ is a *monotone aggregate function* (e.g., SUM) which combines the scores of two objects. k-SDJ finds application in tasks like recommending to the visitors of a city the k best pairs of restaurants and hotels within short distance that have the top combined ratings, or investigating the correlation between scientific attributes, e.g., identifying locations where earthquakes of high magnitude take place on a very large depth. For instance, Figure 1 illustrates a set R of four restaurants and a set S of four hotels. The objects carry a score shown next to every point. Assuming that the qualifying pairs should have Euclidean distance at most $\epsilon = 0.3$ and $\gamma = SUM$, the result of 2-SDJ contains pairs (r_2, s_3) with aggregate score 7 and (r_2, s_2) with aggregate score 6. Notice that, although $dist(r_4, s_4) < \epsilon$, pair (r_4, s_4) is not included in the query result because $\gamma(r_4, s_4) < \gamma(r_2, s_2) < \gamma(r_2, s_3)$. Further, while being the restaurant with the highest score, r_3 is not included in any result pair, as there is no hotel at a

---

[1] An exception is the work of [11] which, however, is restricted to a specific type of attributes (probabilities) and a specific aggregation function (product).

distance to r_3 smaller than 0.3. Note that k-SDJ is very similar to the top-k join problem in relational databases (see the example SQL query above); the only difference is that in k-SDJ the equality join predicate is replaced by a distance bounding predicate between the spatial locations of objects in R and S.

Contributions. In this paper, we study the efficient evaluation of the k-SDJ query. In brief, the key contributions of our work are summarized as follows:

- We introduce k-SDJ over two collections of spatial objects with scoring attributes. The k-SDJ query can be used either as a standalone operator or participate in complex query evaluation plans. For this purpose, we assume that the input collections are not indexed in advance.
- We present three algorithms, which access and process the data in different order; (i) the *Score-First algorithm* (SFA), which accesses the objects from R and S in decreasing order of their scores, (ii) the *Distance-First Algorithm*, which gives higher priority to the spatial distance join component of the query, and (iii) the *Block-based Algorithm* (BA), which performs block-wise evaluation, combining the benefits of SFA and DFA, without sharing their disadvantages. All techniques employ aR-trees [14] (albeit in different fashions) in order to combine spatial search with score-based pruning.
- We conduct extensive experiments to verify the effectiveness and efficiency of our proposed methods.

Outline. The rest of the paper is organized as follows. Section 2 reviews the related work. Section 3 presents algorithms for k-SDJ evaluation. Comprehensive experiments and our findings are reported in Section 4. Finally, Section 5 concludes the paper and discusses directions for future work.

2   Related Work

Our work is related to spatial joins, top-k queries and top-k joins, and spatial top-k joins. Sections 2.1 to 2.4 summarize related work done in these areas.

2.1   Spatial Joins

There exist two types of spatial distance join queries: the ϵ-*distance* and the k-*closest pairs* join. Given two collections of spatial objects R and S the ϵ-distance join identifies the object pairs (r, s) with $r \in R$, $s \in S$, such that $dist(r, s) \leq \epsilon$. An ϵ-distance join can be processed similarly to a spatial intersection join [2]. Specifically, assuming that each of the R and S collections are indexed by an R-tree, the two R-trees are concurrently traversed by recursively visiting pairs of entries (e_R, e_S) for which their MBRs have minimum distance at most ϵ. Minimizing the cost of computing the distance between an MBR and an object was studied in [3]. For non-indexed inputs, alternative spatial join algorithms can be applied (e.g., the algorithm of [1] based on external sorting and plane sweep). The k-closest pairs join computes, from two collections R and S, the k object pairs (r, s), $r \in R$, $s \in S$, with the minimum spatial distance $dist(r, s)$. Two different approaches exist for k-closest pairs. In the incremental approach [6,17] the results are reported one-by-one in ascending order of their spatial distance. For non-incremental computation of closest pairs, [4] extends the nearest neighbor algorithm of [15] achieving in this way, minimum memory requirements and better access locality for tree nodes.

2.2   Top-k Queries

Fagin et al. [5] present an analytical study of various methods for top-k aggregation of ranked inputs by monotone aggregate functions. Consider a set of objects (e.g., restaurants) which have scores (i.e., rankings) at two or more different sources (e.g., different ranking websites). Given an aggregate function γ (e.g., SUM) the top-k query returns the k restaurants with the highest aggregated scores (from the different sources). Each source is assumed to provide a sorted list of the objects according to their atomic scores there; requests for random accesses of scores based on object identifiers may also be possible. For the case where both sorted and random accesses are possible, a *threshold algorithm* (TA) retrieves objects from the ranked inputs (e.g., in a round-robin fashion) and a priority queue is used to organize the best k objects seen so far. Let l_i be the last score seen in source S_i; $T = \gamma(l_1, ..., l_m)$ defines a lower bound for the aggregate score of objects never seen in any S_i yet. If the kth highest aggregate score found so far is no less than T, the algorithm is guaranteed to have found the top-k objects and terminates. For the case where only sorted accesses are possible, [12] presents an optimized approach.

2.3   Top-k Joins

The top-k query is a special case of a top-k join query, which performs rank aggregation on top of relational join results; recall the SQL query example in the Introduction, where two tables R and S are joined (based on their common attribute `att`), but only the top-k join results according to the `score` attributes are required to be output.

Ilyas et al. [8] proposed a binary operator, called hash-based rank-join (HRJN) for top-k joins, which produces results incrementally and therefore can be used multiple times in an multi-way join evaluation plan. Assume that the tuples of R and S are accessed incrementally based on their values in the `score` attribute. HRJN accesses tuples from R (or S) and joins them using the join key `att` with the buffered tuples of S (or R), which have previously accessed (these tuples are buffered and indexed by a hash-table). Join results are organized in a priority queue based on their aggregate scores. Let l_R, h_R (l_S, h_S) be the lowest and highest scores seen in R (S) so far; all join results currently in the queue having aggregate scores larger than $T = \max\{\gamma(h_R, l_S), \gamma(l_R, h_S)\}$ are guaranteed to have higher aggregate score than any join result not found so far and therefore can be output (or pipelined to the operator than follows HRJN). A follow-up work [16] identifies optimal strategies for pulling tuples from the inputs in a multi-way top-k join and joining them with the buffered results of other inputs. An early work on multi-way top-k join evaluation is done by Natsev et al. [13].

The binary top-k join operator (i.e., HRJN) can be adapted to solve k-SDJ; the only difference is that the equality join predicate in HRJN is replaced by a spatial distance predicate. In Section 3, we describe our *Score-First Algorithm* (SFA), which is based on this idea.

2.4   Spatial Top-k Joins

The term "top-k spatial join" is defined in [19]; however, this problem definition is very different from what we study in this paper; given two spatial datasets R and S, the query of [19] retrieves k objects in R intersecting the maximum number of objects from S. Therefore the ranking criterion based on the number of spatial intersections and not based on the aggregations of (non-spatial) scores from the two inputs. The only work to our knowledge, closely related to k-SDJ is [11], which studies a spatial join between two datasets R and S, containing spatial data associated with probabilistic values; in this case each object o (e.g., a biological cell) is defined by a set of probabilistic locations and it is also assigned a confidence p_o to belong to a specific cell class. Given two objects r and s from datasets R and S, respectively, a score of the (r, s) pair is defined by multiplying their confidence probabilities p_r and p_s, and also considering the distance $dist(r, s)$ between their uncertain locations. Then, the top-k probabilistic join between R and S returns the top-k object pairs in order of their scores. Compared to k-SDJ, the problem definition in [11] is different. The aggregate score function for k-SDJ does not involve the distance of the objects, but the distance is used in the join predicate. Further, the solution proposed in [11] is of limited applicability as it is bound to a specific aggregation function and can efficiently work only with L_1 distance.

3   Algorithms

In this section, we study evaluation techniques for k-SDJ. According to the query definition, the results are object pairs with (i) large aggregate scores and (ii) nearby spatial locations. First, we discuss two solutions, which extend work on top-k joins [13,8,12,16] and spatial joins [3,6,17,7,15], respectively, to prioritize either of the two query components; i.e., they either consider object scores or spatial distances first, respectively. We present these methods in Sections 3.1 and 3.2; they are optimized to employ aR-trees [14] (in a different fashion) in order to prune the search space during query evaluation. In Section 3.3, we present a framework which processes the objects in order of their scores, in a block-wise fashion and spatially joins the blocks, using bounds to early terminate accessing of blocks. Figure 2 illustrates the running example that we use to demonstrate our algorithms; two spatial datasets $R = \{r_1 \ldots r_8\}$ and $S = \{s_1 \ldots s_8\}$ with 8 points each. The point coordinates are shown on the left of the figure, while the two tables on the right show the objects in each collection in descending order of their scores. Table 1 shows the notation frequently used.

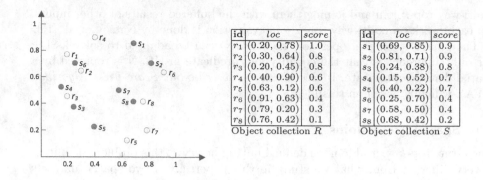

id	loc	score
r_1	(0.20, 0.78)	1.0
r_2	(0.30, 0.64)	0.8
r_3	(0.20, 0.45)	0.8
r_4	(0.40, 0.90)	0.6
r_5	(0.63, 0.12)	0.6
r_6	(0.91, 0.63)	0.4
r_7	(0.79, 0.20)	0.3
r_8	(0.76, 0.42)	0.1

Object collection R

id	loc	score
s_1	(0.69, 0.85)	0.9
s_2	(0.81, 0.71)	0.9
s_3	(0.24, 0.38)	0.8
s_4	(0.15, 0.52)	0.7
s_5	(0.40, 0.22)	0.7
s_6	(0.25, 0.70)	0.4
s_7	(0.58, 0.50)	0.4
s_8	(0.68, 0.42)	0.2

Object collection S

Fig. 2. Example of two datasets R and S with 8 points each

Table 1. Table of Symbols

Notation	Description
R/S	Collections with scored spatial objects
k	The number of required results
ϵ	The spatial distance threshold
γ	A monotone aggregate function
C	Candidate/result set of k-SDJ
θ	k-th smallest aggregate score in C

3.1   The Score-First Algorithm (SFA)

This method employs the framework of top-k join algorithms [8,16] to compute k-SDJ. In particular, a variant of the HRJN algorithm is applied using the spatial distance predicate instead of the equality predicate used in the original algorithm. The *Score-First Algorithm* (SFA) presumes that both R and S are ordered based on the object scores (e.g., as shown in Figure 2). This can be the case if they stem from underlying operators which produce such interesting orders; otherwise R and S need to be sorted before the application of SFA. SFA incrementally accesses objects either from R or from S. For each collection (e.g., R), it maintains an aR-tree [14] (e.g., A_R), which spatially organizes the buffered objects accessed so far.[2] In addition, SFA keeps track of the set C of distance join pairs found so far with the k highest aggregate scores and uses the lowest score θ in C as a bound for pruning and termination.

Algorithm 1 is a high-level pseudocode of SFA. After initializing C and the aR-trees, SFA incrementally accesses objects from R or S (we will shortly discuss about the access order). Assume that the current object is r accessed from R (i.e., $i = R$ and $o = r$ at Lines 5 and 7 of the pseudocode); the other case is symmetric. SFA performs the following steps:

---

[2] The aR-tree has identical structure and update algorithms as the R-tree, however, each non-leaf entry is augmented with the maximum score of all objects in the subtree pointed by it. Figure 3 illustrates the structure of two aR-trees for the data of Figure 2.

Algorithm 1. Score-First Algorithm (SFA)

Input: k, ϵ, γ, R, S
Output: C
1: initialize a min-heap $C:=\emptyset$ of candidate results; initialize $\theta:=-\infty$
2: sort R and S based on score, if not already sorted
3: initialize aR-trees $A_R:=\emptyset$ and $A_S:=\emptyset$
4: **while** more objects exist in R and S **do**
5: 　　$i :=$ next input to be accessed //either R or S
6: 　　$j :=$ other input //either S or R
7: 　　$o := get\_next(i)$ //get next object from input i
8: 　　$T:=\max\{\gamma(h_R, l_S), \gamma(l_R, h_S)\}$ //HRJN termination threshold
9: 　　**while** $(o' := get\_next\_pair(o, A_j, \epsilon, \theta)) \neq null$ and $T > \theta$ **do**
10: 　　　　update C and θ using (o, o')
11: 　　　if $T \leq \theta$ **then**
12: 　　　　**break** //result secured; no need to access more objects
13: 　　Insert o to A_i
14: **return** C

---

1. It updates T(Line 8)
2. It probes r against the aR-tree A_S for S to incrementally retrieve objects s from A_S such that $dist(r, s) \leq \epsilon$ and $\gamma(r, s) > \theta$ (function $get\_next\_pair()$ in Line 9 retrieves such objects in decreasing order of $\gamma(r, s)$);
3. For each qualifying pair (r, s) found, it updates C and θ (Line 10);
4. It checks whether the algorithm can terminate (Lines 11–12);
5. It inserts r to the aR-tree A_R for R (Line 13)

We now elaborate on the steps above. After each object access, SFA firstly updates the termination threshold $T = \max\{\gamma(h_R, l_S), \gamma(l_R, h_S)\}$ (Line 8). From the description of HRJN in Section 2.3, recall that l_R, h_R (l_S, h_S) are lowest and highest scores seen in R (S) so far (initially they are set as the maximum score in R (S)). In Step 2, aR-tree search on A_S is performed as a *score-based incremental ϵ-distance range query* centered at r (function $get\_next\_pair()$ in Line 9); during search, entries whose MBRs are further than ϵ from r are pruned and the remaining ones are prioritized based on their aggregate scores (i.e., the maximum score of any object indexed under them). Specifically, for each entry e, $\gamma(r, e.score)$ is computed (where $e.score$ is the aggregate score for e in the aR-tree) and if it is found not larger than θ, then the entry is pruned (as it would not be possible to find an object $s \in S$ in the subtree pointed by e, such that $\gamma(r, s) > \theta$). Otherwise, the entry is inserted in a *priority queue* which guides the aR-tree search to retrieve the spatial join pairs (r, s) in decreasing order of $s.score$ (note that this results in retrieving pairs in decreasing order of $\gamma(r, s)$, since r is fixed).

Whenever a new pair (r, s) is found, C and θ are updated immediately in order to tighten the θ bound and potentially prune additional aR-tree nodes during search: if $|C| < k$, (r, s) is inserted into C regardless of its aggregation score; otherwise, (r, s) is inserted into C only if $\gamma(r, s) > \theta$; in this case (r, s) *replaces* the k-th pair in C, such that C always keeps the best k pairs found so far. In either case, θ is updated to the k-th aggregate score in C. During the computation of new join results, as soon as $T \leq \theta$, SFA terminates reporting C as the k-SDJ result. Finally, r is inserted to the aR-tree A_R for R, which is used to probe objects from S.

We note that at Line 5 SFA chooses as input (R or S) to read the next object from the collection with the higher last seen score, according to the rationale of HRJN [8]; this increases the chances that the threshold T drops and that SFA terminates earlier.

To illustrate SFA, consider the example of Figure 2 and a k-SDJ query with $k = 1$, $\epsilon = 0.1$, and $\gamma = SUM$. After the first access (r_1 from R), there is no join result (since A_S is currently empty), so r_1 is just inserted to A_R. Then s_1 is accessed from S; s_1 is probed against A_R and no match is found (i.e., $dist(r_1, s_1) > \epsilon$). Then, since $l_R = 1.0 > l_S = 0.9$, r_2 is accessed and joined (unsuccessfully) with A_S; then s_2 and s_3 are accessed in turn, still without producing any distance join results. When r_3 is accessed, it is joined with A_S (now containing $\{s_1, s_2, s_3\}$) and produces the first join result (r_3, s_3), which is added to C; now $\theta = \gamma(r_3, s_3) = 1.6$. Note that T is currently $\max\{\gamma(1.0, 0.8), \gamma(0.8, 0.9)\} = 1.8 > \theta$, which means that a possibly better pair can be found and SFA cannot terminate yet. The next accessed object is s_4, which is joined with A_R, finding result (r_3, s_4), which is not better than the current top pair (r_3, s_3) (Note that in this case $get\_next\_pair(s_4, A_R, \epsilon, \theta)$ will not return r_3 but just $null$, because it uses θ for pruning). Next, both r_4 and s_5 give no new join pairs. Then s_6 is retrieved but still cannot produce a join pair with score better than θ; therefore SFA, after having updated T to 1.5, terminates reporting $C = \{(r_3, s_3)\}$.

SFA is expected to be fast, if the join results are found only after few accesses over the sorted collections R and S. If the best pairs include objects deep in the sorted collections, the overhead to maintain and probe the aR-trees can be high.

3.2   The Distance-First Algorithm (DFA)

The second evaluation technique extends a spatial distance join algorithm to compute the object pairs from R and S, which qualify the spatial threshold θ, *incrementally* in decreasing order of their aggregate scores. The algorithm terminates as soon as k pairs have been generated.

For implementing the spatial distance join, we could apply algorithms like the R-tree join [2], assuming that R and S are already indexed by R-trees, or methods that spatially join non-indexed inputs, like the (external memory) plane sweep algorithm [1], which first sorts R and S based on one of their coordinates and then sweeps a line along the sort axis to compute the results. The above approaches, however, do not have a way of prioritizing the join result computation according to the aggregate scores of qualifying distance join pairs.

We now present our optimized approach, which also employs aR-tree indices, however, it computes k-SDJ in a different way compared to SFA. The *Distance-First Algorithm* (DFA) (Algorithm 2) assumes that both R and S are indexed by two aR-trees A_R and A_S (if the trees do not exist, DFA bulk-loads them in a pre-processing phase, using the algorithm of [10]). DFA spatially joins the two trees by adapting the classic algorithm of [2] to traverse them not in a depth-first, but in a *best-first* order, which (1) still prunes entry pairs (e_R, e_S), $e_R \in A_R, e_S \in A_S$ for which $dist(e_R, e_S) > \epsilon$ (*dist* here denotes the minimum distance between the MBRs of the two entries), but (2) prioritizes the entry pairs

---

Algorithm 2. Distance-First Algorithm (DFA)

Input: k, ϵ, γ, R, S
Output: k-SDJ results incrementally
1: build aR-trees A_R on R and A_S on S, if not already indexed
2: initialize a max-heap H_e of aR-tree entry pairs (e_R, e_S) organized by $\gamma(e_R, e_S)$
3: **for** each pair (e_R, e_S) in $A_R.root \times A_S.root$ **do**
4:      **if** $dist(e_R, e_S) \leq \epsilon$ **then**
5:          push (e_R, e_S) into H_e
6: **while** $H_e \neq \emptyset$ **do**
7:      $(e_R, e_S) = H_e$.dequeue()
8:      **if** e_R and e_S are non-leaf node entries **then**
9:          n_R := node of A_R pointed by e_R; n_S := node of A_S pointed by e_S
10:          **for** each entry $e'_R \in n_R$ and each entry $e'_S \in n_S$ **do**
11:              **if** $dist(e'_R, e'_S) < \epsilon$ **then**
12:                  push (e'_R, e'_S) into H_e
13:      **else**
14:          output (e_R, e_S) as next k-SDJ result

---

to be examined based on $\gamma(e_R, e_S)$ (here, γ is applied on the aggregate scores stored at the entries). In other words, the entry pairs which have the maximum aggregate score are examined first during the join; this order guarantees that the qualifying object pairs will be computed incrementally in decreasing order of their aggregate scores. For this purpose, DFA initially puts in a priority queue (i.e., max heap) H_e all pairs of root entries within distance ϵ from each other in the two trees (Line 3 of Algorithm 2). Pairs of entries from H_e are de-heaped in priority of their aggregate scores $\gamma(e_R, e_S)$. Then, the spatial distance join is evaluated for the corresponding aR-tree nodes and the results are inserted to H_e if they are non-leaf entries (branching condition at Line 8). Otherwise, if a leaf node entry pair (r, s) (i.e., object pair) is de-heaped, it is guaranteed that (r, s) has higher aggregate score than any other object pair to be found later, since entry and object pairs are accessed in decreasing order of their γ-scores from H_e. Therefore the object pair is output as the next result of the k-SDJ query (Line 14). DFA terminates after k results have been computed.

As an example of DFA, consider the two aR-trees for the R and S datasets of Figure 2, as shown in Figure 3. Note that the aR-tree entries are augmented with the maximum scores of any objects in the subtrees indexed by them (e.g., R_2 has score 0.6). Assume that $k = 1$, $\epsilon = 0.1$, and $\gamma = SUM$. DFA begins by joining the two root nodes of A_R and A_S, which adds two entry pairs to H_e; (R_1, S_1) and (R_2, S_2); the other two combinations (R_1, S_2) and (R_2, S_1) are pruned by the ϵ-distance join predicate. The next pair to be examined is (R_1, S_1) because $\gamma(R_1, S_1) = 1.8 > \gamma(R_2, S_2)$. Thus, the nodes pointed by R_1 and S_1 are synchronously visited and their ϵ-distance join adds to H_e pairs (R_3, S_4), (R_4, S_4), and (R_4, S_3). The next entry pair to be de-heaped is (R_3, S_4) with $\gamma(R_3, S_4) = 1.7$; this results in object pair (r_1, s_6) being found and added to H_e. Next, (R_4, S_3) is de-heaped and (r_3, s_3) is added to H_e. The next pair to be popped from H_e is the object pair (r_3, s_3); note that this is guaranteed to be the top ϵ-distance join pair, since it is the first object pair to be extracted from H_e, and thus, the algorithm terminates.

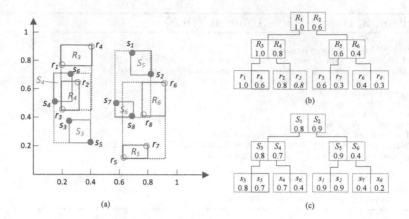

Fig. 3. Two aR-trees for R and S

DFA shares some common elements with the probabilistic spatial join algorithm of [11]. Adapting the solution proposed in [11] for k-SDJ would involve extending the classic plane-sweep approach to operate on spatially grouped objects, where group MBRs are annotated by aggregate probability values (used for pruning group pairs during evaluation). However, our approach is more optimized in the sense that it is applied on hierarchically multiple levels of grouping (instead of a single grouping level in [11]). Thus, DFA represents the best possible implementation of the framework of [11] for k-SDJ queries, which as we show in Section 4 is inferior to our best approach.

DFA is expected to be fast if the locations of objects are correlated with their atomic scores; in this case, the objects with the highest scores that are close to each other are identified fast; otherwise, many tree node pairs may have to be accessed and joined until DFA can terminate. In other words, since the data are primarily clustered by their locations in the aR-trees, pruning pairs of non-leaf node entries is mostly due to the spatial predicate while the aggregate scores may be too uniform to facilitate in this task.

3.3 The Block-based Algorithm (BA)

SFA and DFA both have certain shortcomings; on one hand, SFA fails to exploit the spatial domain to locate fast spatial join results; on the other hand, DFA does not necessarily find pairs of objects with high aggregate scores fast. Our third solution is a *Block-based Algorithm* (BA), which alleviates these shortcomings. Like SFA, BA examines the data by prioritizing objects with high scores first, however, it does not incrementally build and join the aR-trees for them (which is costly due to the repeated insertion and range query operations); instead it bulk-loads aR-trees for blocks of objects from R (or S) and spatially joins them with the blocks from S (or R) read so far, avoiding unnecessary block or node pair joins based on their score and distance bounds.

Algorithm 3. Block-based Algorithm (BA)

Input: k, ϵ, γ, R, S
Output: C
```
 1: initialize a min-heap C:=∅ of candidate results; initialize θ:=−∞
 2: sort R and S based on score, if not already sorted
 3: while more blocks of objects exist in R and S do
 4:     i := next input to be accessed //either R or S
 5:     j := other input //either S or R
 6:     b := get_next_block(i) //get next block of objects from input i
 7:     A_b := bulk-load aR-tree for b
 8:     for each block b' of j do
 9:         if γ(b^u, b'^u) > θ then
10:             apply DFA to join A_b with A'_b
11:             retrieve results and update C and θ
12:     T:=max{γ(b^u_{R1}, b^l_{Slast}), γ(b^l_{Rlast}, b^u_{S1})}
13:     if T ≤ θ then
14:         break //result secured; no need to access more objects
15: return C
```

More specifically, BA (like SFA) considers R and S sorted by score. It then divides them into blocks of λ objects each. At each step a block of objects b_R (or b_S) is accessed from R (or S) and then, joined with the blocks of S (or R), which have already been accessed. Thus, BA can be considered as an adaptation of SFA that operates at the block-level. For example, Figure 4 illustrates the blocks for collections R and S if $\lambda = 2$. BA would, for instance, first read b_{R1}, then read b_{S1} and join it with b_{R1}, then read b_{S2} and join it with b_{R1}, then read b_{R2} and join it with b_{S1} and b_{S2}, etc. Since the objects in R and S are sorted in decreasing order of their scores, for each block b, there is an upper score bound b^u corresponding to the score of the first object in b. Therefore, when considering the join between two blocks b_R and b_S, $\gamma(b_R^u, b_S^u)$ represents an upper score bound for all ϵ-distance join pairs in $b_R \times b_S$. If we have found at least k distance join pairs so far, then we know that joining b_R with b_S is pointless when $\gamma(b_R^u, b_S^u) \leq \theta$, where θ is the k-th best aggregate score. In other words, the current block, e.g., b_R, is only joined with the blocks b_S of S for which $\gamma(b_R^u, b_S^u) > \theta$. For the block-level join, we employ the DFA algorithm; an aR-tree is constructed for the current block and joined with the aR-trees of blocks read from the other input using Algorithm 2.

Algorithm 3 is a high-level pseudocode for BA. Line 4 chooses the input (i.e., R or S) from which the next block will be retrieved using the same policy as SFA (i.e., the input with the highest last seen score). Without loss of generality, assume that $i = R$. Then, for the next block b accessed from R, an aR-tree is created, and b is joined with all blocks b' from S which have already been examined using DFA (for these blocks, the corresponding aR-trees have already been constructed before). The blocks b' are joined in decreasing order of their score ranges (e.g., first b_{S1}, then b_{S2}, etc.). Each new ϵ-distance join pair is used for updating C and θ (in fact the current θ is used to terminate the block-wise DFA as soon as the top pair on the heap H^e has aggregate score at most θ). After handling b, the termination threshold T is updated (Line 12). Note that this threshold is the same as in the SFA algorithm: the maximum value between $\gamma(b_{R1}^u, b_{Slast}^l)$ and $\gamma(b_{Rlast}^l, b_{S1}^u)$, where b_{R1}^u (b_{S1}^u) is the highest score in the first

id		loc	score
b_{R1}	r_1	(0.20, 0.78)	1.0
	r_2	(0.30, 0.64)	0.8
b_{R2}	r_3	(0.20, 0.45)	0.8
	r_4	(0.40, 0.90)	0.6
b_{R3}	r_5	(0.63, 0.12)	0.6
	r_6	(0.91, 0.63)	0.4
b_{R4}	r_7	(0.79, 0.20)	0.3
	r_8	(0.76, 0.42)	0.1

(a) collection R

id		loc	score
b_{S1}	s_1	(0.69, 0.85)	0.9
	s_2	(0.81, 0.71)	0.9
b_{S2}	s_3	(0.24, 0.38)	0.8
	s_4	(0.15, 0.52)	0.7
b_{S3}	s_5	(0.40, 0.22)	0.7
	s_6	(0.25, 0.70)	0.4
b_{S4}	s_7	(0.58, 0.50)	0.4
	s_8	(0.68, 0.42)	0.2

(b) collection S

Fig. 4. Example of BA

block of R (S) (i.e., the score of the very first object in the collection) and b^l_{Rlast} (b^l_{Slast}) is the lowest score in the last block of R (S) (i.e., the score of the last object in it).

As an example of BA, consider the k-SDJ ($k = 1$, $\epsilon = 0.1$, $\gamma = SUM$) between collections R and S of Figure 4, which are partitioned into blocks. Initially, b_{R1} is read and an aR-tree A_{R1} is created for it. Then, b_{S1} is accessed, bulk-loaded to an A_{S1}, and joined using A_{R1} (Line 10 of BA), producing no spatial join results. Next, b_{S2} is accessed and (unsuccessfully) joined with b_{R1}. After reading b_{R2} the block is joined with b_{S1} and b_{S2} (in this order), to generate $C = \{(r_3, s_3)\}$ and set $\theta = 1.6$. The next block b_{S3} is joined with b_{R1}, but not b_{R2}, because $\gamma(b^u_{R2}, b^u_{S3}) = \gamma(0.8, 0.7) \leq \theta$; this means that in the best case a spatial distance join between b_{R2} and b_{S3} will produce a pair with score 1.5, which is not better than the current top pair's (r_3, s_3) score. The join between b_{S3} and b_{R1} does not improve the current k-SDJ result. At this stage, BA terminates because $T = 1.5$, which is not higher than $\theta = 1.6$.

Although BA is reminiscent to SFA in that it examines the records in order of their scores, BA has two main advantages compared to SFA. First, performing the joins at the block level is more efficient, because the aR-trees for the blocks are created just once efficiently by bulk-loading (instead of iterative insertions as in SFA) and could be used for multiple block joins (e.g., b_{R1} is joined with blocks b_{S1}–b_{S3} in our example). The joins between aR-trees are much faster compared to the record-by-record probing (i.e., index nested loops) approach of SFA. Second, in BA, the currently processed block is joined only with a small number of blocks of the other input (with the help of the upper score bounds of the blocks), while in SFA the current object is probed against the entire set of objects buffered from the other input.

4   Experimental Evaluation

In this section, we present an experimental evaluation of our techniques for k-SDJ. Section 4.1 details the setup of our analysis. Sections 4.2 and 4.3 experimentally prove the superiority of SFA and DFA compared to alternative implementations that follow the score-first, the distance-first evaluation paradigm, respectively. Section 4.4 carries out a comparison between the SFA, DFA and

Table 2. Experimental parameters

Parameter	Values	Default value
ϵ	0.001, 0.005, 0.01, 0.05	0.01
k	1, 5, 10, 50, 100	10
$\|P\|$	5, 10, 50, 100	10
$\lambda/\|R\|$	0.0005, 0.001, 0.005, 0.01, 0.02	0.005

BA algorithms. All algorithms involved in this study are implemented in C++ and the experiments were conducted on a 2.3 Ghz Intel Core i7 CPU with 8GB of RAM running OS X.

4.1 Setup

Our experimental analysis involves two collections of real spatial objects: (1) FLICKR that contains 1.68M locations associated with photographs taken from the city of London, UK over a period of 2 years and hosted on the Flickr photosharing website, and (2) ISLES that contains 20M POIs in the area of the British isles drawn from the OpenStreetMap project dump. To perform the experiments, every collection is split into two equally sized parts denoted by R and S. This is to avoid performing a self-join which will produce result pairs involving exactly the same object. Since real scores for objects were not available, we generated them as follows. For each part, we first randomly created $|P|$ seed points (simulating POIs) and generate the object scores with the following formula: $o.score = 1 - \min_{i=1}^{|P|}[dist(o, P_i)]$. Intuitively, real life objects are more likely to have high scores if they are close to a POI [18]. For example, the hotels close to the city center have higher potential to be highly rated for their convenience. The generated scores are normalized to take values in $[0, 1]$. In our study, we vary the total number of seed points created for each of R and S, from 5 to 100. Larger values of $|P|$ tend to generate more uniform score distributions and higher average scores. Finally, note that we consider SUM as the score aggregation function γ for our experiments.

   To assess the performance of each join method, we measure their response time that also includes index building cost and all sorting costs, varying (1) the distance threshold ϵ, (2) the number of returning pairs k, and (3) the number of seed points $|P|$. Further, in case of BA we also vary the ratio $\lambda/|R|$ of the block size λ over the size of the collection $|R|$ ($|S|$). Table 2 summarizes all the parameters involved in this study. On each experiment, we vary one of ϵ, k, $|P|$ and $\lambda/|R|$ while we keep the remaining parameters fixed to their default values. Finally, note that both the collections and the indexing structures used by the evaluation methods are stored in main memory (like relational top-k join studies, we consider k-SDJ evaluation in main memory), and that we set the page size for both the R-trees and the aR-trees to 4KB.

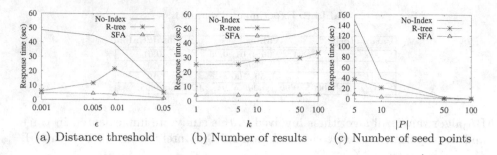

(a) Distance threshold   (b) Number of results   (c) Number of seed points

Fig. 5. Comparison of the score-first algorithms on FLICKR

4.2 Score-First Algorithms

The first set of experiments demonstrates that our SFA algorithm, as described in Section 3.1, is an efficient adaptation of the state-of-the-art top-k relational join algorithm HRJN. Recall that, in SFA, we use aR-trees to index the objects seen so far from R and S and use an efficient aR-tree search algorithm, which exploits the aggregate scores stored at intermediate entries to prune the search space while searching for join pairs for the current object o. In fact, the score-first paradigm could also be applied without using aR-trees for indexing (in this case, a scan would be applied against the buffered objects to find the spatial join pairs), or one could use R-trees instead of aR-trees. Figure 5 compares the response times of these two alternatives with SFA (denoted by No-Index and R-tree, respectively), while varying ϵ, k and $|P|$. As expected, SFA outperforms the other two methods since it is able to prune object pairs in terms of both their spatial distance and their aggregate scores. We also observe that increasing ϵ makes the k-SDJ evaluation faster for SFA and No-Index, but not for R-tree. The reason is the following. As ϵ increases, more object pairs qualify the spatial predicate. Since the objects are sorted in descending order of their scores, a smaller number of pairs needs to be examined. Although this holds also for R-tree, its time initially increases due to the increasing cost of the range queries involved. Another important observation is that the response time of SFA is less affected by the increase of k compared to the other methods. Particularly, for $k = 100$, SFA needs 10% more time to compute k-SDJ compared to $k = 1$, while in case R-trees or No-Index, this increase is 33% and 45%, respectively. Finally, with the usage of more seed nodes for score generation, the response time of all methods decreases since more object pairs have high aggregate scores.

4.3 Distance-First Algorithms

We perform a similar evaluation for DFA, by comparing it against two algorithms that adopt alternative techniques for a distance-first k-SDJ processing (instead of building and joining two aR-trees). Thus, one option is to generate and join two R-trees (on each of R and S), and another option is to sort R and S an apply

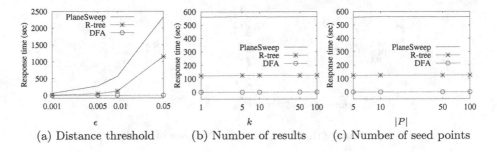

(a) Distance threshold   (b) Number of results   (c) Number of seed points

Fig. 6. Comparison of the distance-first algorithms on FLICKR

the plane sweep spatial join technique. Figure 6 compares the three approaches, showing that their response time is affected only by the increase of the distance ϵ and not by k or $|P|$; this is due to the fact that all methods primarily focus on the spatial predicate of the k-SDJ, which is independent of the scores. Due to its ability to use score aggregation bounds, DFA not only outperforms the other methods (in some cases for more than two orders of magnitude), but it is also very little affected by the increase of ϵ.

We also implemented BA with different versions of its block-based join module; i.e., the aR-tree based join as described in Section 3.3, and alternatives based on plane sweep and R-tree join. The results (not included due to space constraints) confirm that the aR-tree based module for joining blocks is superior to the other alternatives (the trends are similar to the experiments of Figure 6).

4.4 Comparison of the Evaluation Paradigms

We have already shown that SFA, DFA, and BA, as described in Section 3 are efficient implementations for the corresponding search paradigms (search-first, distance-first, and block-based, respectively). We now compare these three algorithms to identify the best evaluation paradigm and algorithm for k-SDJ. Figure 7 and 8 report on their response time for FLICKR and ISLES[3], respectively, while varying ϵ, k, $|P|$ and $\lambda/|R|$. The figures show that BA outperforms SFA and DFA in all cases. It is important to also notice that BA is very robust to the variation of the parameters. Specifically, its response time is always around 350msec for FLICKR and 3.5sec for ISLES. In contrast, SFA is (1) positively affected by the increase of ϵ since more object pairs qualify the spatial predicate and thus, the top-k results can be quickly identified, and (2) negatively by the increase of k as it needs to examine more object pairs. Further, DFA becomes slower with ϵ as it primarily focuses on the spatial predicate of the k-SDJ. BA manages to combine the above advantages of the score-first paradigm and SFA, and the distance-first paradigm and DFA, as it examines blocks of objects ordered by score and applies the spatial predicate at the block level instead on the whole collections.

---

[3] In this experiment we used only half of the ISLES collection; i.e., $|R| = |S| = 5M$.

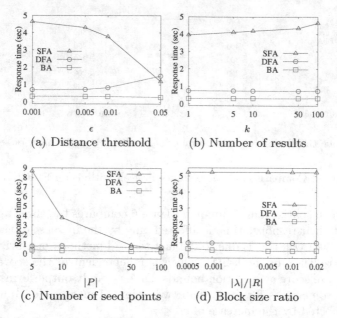

Fig. 7. Comparison of SFA, DFA and BA algorithms on FLICKR

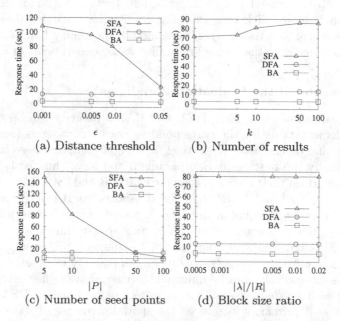

Fig. 8. Comparison of SFA, DFA and BA algorithms on ISLES

We also performed a scalability experiment, by joining samples R and S of the ISLES dataset of different sizes $|R|$ and $|S|$, while setting ϵ, k, $|P|$ and $\lambda/|R|$ to their default values. Figure 9(a) shows the results of the scalability test. We observe that all methods scale similarly, with BA being 1-2 orders of magnitude faster than the other methods. Note that even on the full ISLES collection, BA needs less than 7 seconds to evaluate a k-SDJ query.

In the last experiment, we evaluate the performance of the methods on datasets of different cardinalities. R and S are samples of the ISLES dataset of varying cardinality ratio $|R| : |S|$, while $|R| + |S|$ is fixed to 10M. The result is shown in Figure 9(b). We observe that the response times of all methods only grow slightly with the increase of $|R| : |S|$ (the effect is more obvious on SFA). This is because although the cardinality ratio is changing, the percentage of traversed records on both size does not change much. Still, maintaining and updating a larger aR-tree on R costs a little more than what is saved on a smaller aR-tree for S, leading to the slight cost growth as the ratio increases.

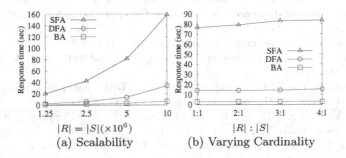

(a) Scalability   (b) Varying Cardinality

Fig. 9. Scalability and Cardinality test on ISLES

5   Conclusion

In this paper, we proposed and studied top-k spatial distance join (k-SDJ) queries. Although this operator finds practical application, it has not been studied in the past, therefore no efficient solutions exist for it, so far. We proposed three algorithms for k-SDJ queries. SFA accesses the objects from the joined collections incrementally and spatially joins them to find new results, using bounds to terminate early. DFA incrementally computes the distance join results ordered by score, by extending a spatial join algorithm to apply on aggregate R-trees that index the two inputs. BA is a hybrid approach, which considers blocks of objects from the two inputs, ordered by score, and joins them as necessary, until a termination condition is reached. Our experimental findings on large datasets show that BA performs best in practice, while the optimized use of aR-trees in all methods greatly improves their performance compared to baseline alternatives which rely on simpler indexing schemes. A direction for future work is to extend our algorithms to compute top-k object pairs according to a combined distance-based and score-based aggregate function.

References

1. Arge, L., Procopiuc, O., Ramaswamy, S., Suel, T., Vitter, J.S.: Scalable sweeping-based spatial join. In: VLDB, pp. 570–581 (1998)
2. Brinkhoff, T., Kriegel, H.P., Seeger, B.: Efficient processing of spatial joins using R-trees. In: SIGMOD Conference (1993)
3. Chan, E.P.F.: Buffer queries. IEEE Trans. Knowl. Data Eng. 15(4), 895–910 (2003)
4. Corral, A., Manolopoulos, Y., Theodoridis, Y., Vassilakopoulos, M.: Closest pair queries in spatial databases. In: SIGMOD Conference (2000)
5. Fagin, R., Lotem, A., Naor, M.: Optimal aggregation algorithms for middleware. In: PODS, pp. 102–113 (2001)
6. Hjaltason, G.R., Samet, H.: Incremental distance join algorithms for spatial databases. In: SIGMOD Conference (1998)
7. Hjaltason, G.R., Samet, H.: Distance browsing in spatial databases. ACM Trans. Database Syst. 24(2), 265–318 (1999)
8. Ilyas, I.F., Aref, W.G., Elmagarmid, A.K.: Supporting top-k join queries in relational databases. In: VLDB, pp. 754–765 (2003)
9. Ilyas, I.F., Beskales, G., Soliman, M.A.: A survey of top-k query processing techniques in relational database systems. ACM Comput. Surv. 40(4) (2008)
10. Leutenegger, S.T., Edgington, J.M., Lopez, M.A.: STR: A simple and efficient algorithm for R-tree packing. In: ICDE, pp. 497–506 (1997)
11. Ljosa, V., Singh, A.K.: Top-k spatial joins of probabilistic objects. In: ICDE, pp. 566–575 (2008)
12. Mamoulis, N., Yiu, M.L., Cheng, K.H., Cheung, D.W.: Efficient top-k aggregation of ranked inputs. ACM Trans. Database Syst. 32(3) (2007)
13. Natsev, A., Chang, Y.C., Smith, J.R., Li, C.S., Vitter, J.S.: Supporting incremental join queries on ranked inputs. In: VLDB, pp. 281–290 (2001)
14. Papadias, D., Kalnis, P., Zhang, J., Tao, Y.: Efficient OLAP operations in spatial data warehouses. In: Jensen, C.S., Schneider, M., Seeger, B., Tsotras, V.J. (eds.) SSTD 2001. LNCS, vol. 2121, pp. 443–459. Springer, Heidelberg (2001)
15. Roussopoulos, N., Kelley, S., Vincent, F.: Nearest neighbor queries. In: SIGMOD Conference (1995)
16. Schnaitter, K., Polyzotis, N.: Optimal algorithms for evaluating rank joins in database systems. ACM Trans. Database Syst. 35(1) (2010)
17. Shin, H., Moon, B., Lee, S.: Adaptive multi-stage distance join processing. In: SIGMOD Conference (2000)
18. Yiu, M.L., Lu, H., Mamoulis, N., Vaitis, M.: Ranking spatial data by quality preferences. IEEE Trans. Knowl. Data Eng. 23(3), 433–446 (2011)
19. Zhu, M., Papadias, D., Lee, D.L., Zhang, J.: Top-k spatial joins. IEEE Trans. Knowl. Data Eng. 17(4), 567–579 (2005)

Regional Co-locations of Arbitrary Shapes

Song Wang[1], Yan Huang[2], and Xiaoyang Sean Wang[3]

[1] Department of Computer Science, University of Vermont, Burlington, USA
swang2@uvm.edu
[2] Department of Computer Science, University of North Texas, Denton, USA
huangyan@unt.edu
[3] School of Computer Science, Fudan University, Shanghai, China
xywangCS@fudan.edu.cn

Abstract. In many application domains, occurrences of related spatial features may exhibit co-location pattern. For example, some disease may be in spatial proximity of certain type of pollution. This paper studies the problem of regional co-locations with arbitrary shapes. Regional co-locations represent regions in which two spatial features exhibit stronger or weaker co-location than that in other regions. Finding regional co-locations of arbitrary shapes is very challenging because: (1) statistical frameworks for mining regional co-location do not exist; and (2) testing all possible arbitrarily shaped regions is computational prohibitive even for very small dataset. In this paper, we propose frequentist and Bayesian frameworks for mining regional co-locations and develop a probabilistic expansion heuristic to find arbitrary shaped regions. Experimental results on synthetic and real world data show that both frequentist method and Bayesian statistical approach can recover the region with arbitrary shapes. Our approaches outperform baseline algorithms in terms of F measure. Bayesian statistical approach is approximately three orders of magnitude faster than the frequentist approach.

1 Introduction

In Epidemiology, different but related diseases occur in different places. These disease may exhibit co-location patterns where one type of disease tends to occur in spatial proximity of another. In Ecology, different types of animals can be observed in different locations. There exist patterns such as symbiotic relationship and predator-prey relationship. In transportation systems, trip demands and taxi supplies tend to co-locate. Different types of crimes committed and different types of road accidents may also exhibit co-location. In short, co-location is a common application scenario in spatial data sets.

In many of these applications, the co-location pattern may be dissimilar at different regions. *Regional co-location* refers to regions where co-location pattern is stronger or weaker than expected. This is possibly due to environmental factors or provincial social interaction structures. For example, related contagious respiratory diseases may exhibit stronger regional co-location in more interactive communities. As another example, trip requests and roaming taxis may

M.A. Nascimento et al. (Eds.): SSTD 2013, LNCS 8098, pp. 19–37, 2013.

show weaker co-location in over-served or under-served regions. In this paper, we study the problem of regional co-locations.

Problem Definition. In the regional co-location setting, we are interested in the interaction of two spatial features a and b given spatial proximity distance $Dist$. At each time snapshot, we have a dataset D. In D we have spatial feature a occurring at a set of discrete spatial locations L^a and spatial feature b occurring at another set of discrete spatial locations L^b (L^a and L^b may overlap). We also have two baseline location sets B^a and B^b which represent the possible locations where these two features can occur, respectively. B^a or B^b may correspond to the underlying locations that can host the occurrence of a feature. For example, if a is one type of disease, then B^a will be the base population that may be infected by the disease. For any region S, we use L^a_S to denote the occurrence of a that happen inside S and B^a_S to denote the baseline locations/population located inside S. L^b_S and B^b_S are defined similarly. $L^a, L^b, L^a_S, L^b_S, B^a_S$, and B^b_S will be used in defining our spatial statistics shortly.

We are interested in finding regional co-locations in a two-dimensional (2D for short) space and the framework can be extended to 3D easily. The 2D space is partitioned into an $n \times n$ grid G, where n is grid size. Each location $l \in L^a \cup L^b$ is hashed into a grid cell c. Given a user-specified proximity distance $Dist$, we want to find regions $S \subseteq G$ where features a and b tend to locate within distance $Dist$ more often (stronger co-location) or less often (weaker co-location) than those regions outside S based on pairs of occurrence of features. In the language of statistics, the null hypothesis H_0 is that the two spatial features may or may not exhibit co-location pattern but the co-location pattern is consistent across the whole 2D space. The alternative hypotheses $H_1(S)$ represent a higher or lower level of co-location inside S comparing to that outside S. We are interested in finding S of arbitrary shapes and not confined to rectangular (including squared) shapes. Because we can handle stronger and weaker regional co-location similarly in the same framework, we will focus on discussing stronger co-location hereafter and the discussion of the opposite is straightforward.

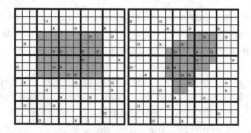

Fig. 1. Regional Co-locations of a Rectangle and an Arbitrary Shape: An Example

Figure 1 illustrates a regional co-location of a rectangle on the left and a regional co-location of an arbitrary shape on the right. The space consists of 15×15 locations and is partitioned into 5×5 grid cells. Note that a grid cell consists of 3×3 locations. A region is a set of connected grid cells. We assume that the baseline location sets B^a and B^b contain all locations and spatial proximity $Dist$ is a Manhattan distance of less than 3. Two spatial features are represented by symbols $\circ$ and $\times$. In both cases, the shaded (green) region has a higher level of co-location than outside. A observant reader may have noticed that while it may be possible to enumerate all the rectangle regions ($\sum_{i=0}^{n} \sum_{j=0}^{n} (n-i)(n-j) = O(n^4)$) [11], it is very challenging to enumerate all possible arbitrarily shaped regions. Welsh [15] states that the problem of counting the number of connected sub-graphs is #P-complete even in the very restricted case (a planar bipartite graph). It is therefore challenging to find regional co-location with arbitrary shapes.

Contributions. In this paper, we propose a *principled statistical framework* to study the *arbitrarily shaped regional co-location* problem. We develop a frequentist method and a Bayesian statistical method to identify regional co-locations with arbitrary shapes. This paper makes the following contributions:

- We propose a new spatial statistic for frequentist method (in section 2) and a Bayesian method (in section 3) to find arbitrarily shaped co-location regions. To the best of our knowledge, this is the first work that allows finding regional co-locations with arbitrary shapes without requiring extensive domain knowledge and inputs;
- We propose an effective heuristic region expansion algorithm (in section 4) to identify arbitrarily shaped regions with stronger (or weaker) co-location. The expansion heuristic applies to both frequentist method and Bayesian statistical method;
- Experimental results (in section 5) on both synthetic data sets and real world taxi data show that both frequentist method and Bayesian statistical method can recover regions of arbitrarily shaped co-location. Our approaches outperform the state-of-art algorithms in terms of F measure. The running time of Bayesian statistical approach is approximately three orders of magnitude faster than the frequentist approach. The real data contains locations for $17,139$ taxis with 48.1 million GPS records serving $468,000$ trips.

2   Frequentist Method

We first present an overview of the proposed frequentist method. Frequentist method searches over all possible region $S \subseteq G$. For each region S, it calculates a likelihood ratio statistic (defined shortly). The likelihood ratio statistic compares the "co-location strength" inside S with that of outside S, i.e. $G - S$. It then compares the likelihood ratio statistic of all regions and finds the region(s) which maximize the statistic. For a dataset and the regions with largest ratio statistic, it then performs a significance testing (detailed in section 2.2). If the test turns out to be insignificant, we decide that the data may be generated by the null hypothesis H_0 of uniform co-location across space. If not, the data with these regions are considered to be more likely to be generated by the alternative hypothesis of regional stronger co-location in those regions. This section focuses on defining the likelihood ratio statistic for a given region and significance testing used by frequentist method. We will present the identification of arbitrarily shaped regions in Section 4.

2.1   Likelihood Ratio Statistic

Frequentist method has commonly been used to identify spatial clusters with certain properties in spatial scan statistics [9]. For clusters, likelihood ratio statistic can be conveniently defined based on counts in a region. For regional co-locations, based on the concept of *participation index* proposed in [6], we propose the following likelihood ratio statistics to be used in our frequentist method for any region $S \subseteq G$, namely *participation probability ratio statistic*.

Participation Probability Ratio Statistic. The participation probability $P_S^{a \to b}$ of spatial feature a to b within spatial proximity distance $Dist$ of a region S is:

$$P_S^{a \to b} = [\text{probability of a random event of } a \text{ having an event} \tag{1}$$
$$\text{of } b \text{ within distance } Dist]$$

The participation probability ratio statistic $\mathcal{P}_S^{a \to b}$ is defined as $\mathcal{P}_S^{a \to b} = \frac{P_S^{a \to b}}{P_{G-S}^{a \to b}}$. The estimate $\hat{P}_S^{a \to b}$ can be obtained by:

$$\hat{P}_S^{a \to b} = \frac{|\{l | l \in L_S^a \wedge (\exists e, e \in L^b \wedge e \in Dist(l))\}|}{|L_S^a|}, \tag{2}$$

where $Dist(l)$ is the circle with radius $Dist$ around a location l. In words, the denominator $|L_S^a|$ is the total number of occurrences of a inside S and the nominator is the number of occurrences of a that are inside S and have an occurrence of b in their neighborhood as well. $\hat{P}_{G-S}^{a \to b}$ can be estimated in a similar manner to obtain $\hat{\mathcal{P}}_S^{a \to b}$.

For example, in the left side of Figure 1, $\hat{P}_{green}^{x \to o} = \frac{4}{4} = 1$ since all $\times$ has a o within $Dist$, $\hat{P}_{G-green}^{x \to o} = \frac{4}{7}$, and $\mathcal{P}_{green}^{x \to o} = \frac{P_S^{x \to o}}{P_{G-green}^{x \to o}} = \frac{7}{4}$. From hereafter, we use P statistic to represent participation probability ratio statistic.

2.2   Significance Testing

In order to verify the statistical significance of the statistic obtained for a given region (or a set of regions) S, we perform significance testing through Monte Carlo simulation. Specifically, given a dataset D, we first learn the occurrence rates of a and b, as well as rate of a, b together within $Dist$ with Expectation Maximization (EM) algorithm described in section 3.3. An occurrence rate refers to the percentage of a (b or a, b together e.g. b occurs in proximity distance $Dist$ of a) among the total population. We then generate Rep replica data sets based on these rates. For each replica C, we enumerate all possible regions and obtain the region S_C that maximizes $\mathcal{P}_{S_C}^{a \to b}$. For the given dataset D, let the p-value of S_D be $\frac{Rep_{beat}+1}{Rep+1}$, where Rep_{beat} is the number of replicas with statistic higher than that of the region found in D. If this p-value is less than some threshold (e.g. 5%), we conclude that the discovered region S is unlikely to happen by chance and reject the null hypothesis of uniform co-location level across the space.

3 Bayesian Statistical Method

3.1 Bayesian Statistic

For a given data set D, Bayesian statistic approach compares the null hypothesis H_0 that spatial features a and b follow the same independent statistical distribution uniformly across the whole space against the alternative set of hypothesis $H_1(S)$, each representing a higher co-location of features a and b in a region S. We will need to calculate the posterior probabilities $P(H_0|D)$ and $P(H_1|D)$ for a given dataset D. Using Bayesian rule, we have:

$$P(H_0|D) = \frac{P(D|H_0)P(H_0)}{P(D)},\tag{3}$$

and

$$P(H_1|D) = \frac{P(D|H_1)P(H_1)}{P(D)},\tag{4}$$

where $P(D) = P(D|H_0)P(H_0) + \sum_S P(D|H_1)P(H_1)$. To calculate posterior probabilities $P(H_0|D)$ and $P(H_1|D)$, we need to know the prior probabilities $P(H_1)$ and $P(H_0)$ as well as those conditional probabilities $P(D|H_0)$ and $P(D|H_1)$. To calculate $P(D|H_0)$ and $P(D|H_1)$, the data set is partitioned into cells with a grid G. For each cell c_i, we have a count c_i^a for occurrence of feature a and count c_i^b for occurrence of feature b. We use a Bivariate Poisson (BP) distribution [8] $BP((q^a - \delta)|B^a|, q^b|B^b| - \delta|B^a|), \delta|B^a|)$ to describe the data, where q^a, q^b are the occurrence rates of spatial feature a and b, respectively; B^a and B^b are baseline population for a and b, respectively, and δ is the occurrence rate of a, b outside the region S. The reason that we use BP distribution is because it is a popular distribution in modelling count data, which fits our problem well. q^a, q^b and δ can be learned by an EM algorithm [8]. We use q_{in}^a and q_{out}^a to denote occurrence rate of feature a inside and outside S, respectively. For a dataset D, given the counts of a and b in each cell and the BP distribution, conditional probability $P(D|H_0)$ can be computed in equation (5) and $P(D|H_1)$ in equation (6). $P(D|H_0)$ assumes all cells c_i in G follows the same BP distribution and the

$$P(D|H_0) = \prod_{c_i \in G} P((c_i^a, c_i^b) \sim BP((q^a - \delta)|B_{c_i}^a|, q^b|B_{c_i}^b| - \delta|B_{c_i}^a|, \delta|B_{c_i}^a|))\tag{5}$$

$$P(D|H_1) = \prod_{c_i \in S} P((c_i^a, c_i^b) \sim BP((q_{in}^a - \delta_{in})|B_{c_i}^a|, q_{in}^b|B_{c_i}^b| - \delta_{in}|B_{c_i}^a|, \delta_{in}|B_{c_i}^a|))$$
$$\times \prod_{c_i \in G-S} P((c_i^a, c_i^b) \sim BP((q_{out}^a - \delta_{out})|B_{c_i}^a|, q_{out}^b|B_{c_i}^b| - \delta_{out}|B_{c_i}^a|, \delta_{out}|B_{c_i}^a|))\tag{6}$$

probability is the product of the probability of all grid cells. $P(D|H_1)$ assumes cells in S follows a stronger co-location and cells in $G - S$ follows the given BP distribution. Choosing prior probability $P(H_1)$ and $P(H_0)$ is detailed in Section 3.2 and learning q^a, q^b using EM algorithm is described in 3.3.

3.2 Estimating Parameters and Choosing Prior

To choose prior, we follow the framework proposed in [11]. We assume that we know the prior probability of a co-location outbreak p. Then $P(H_0) = 1 - p$. We also assume the probability of the outbreak is equally distributed to all regions. So, $P(H_1(S)) = \frac{p}{N_s}$ where N_s is the number of possible arbitrarily shaped regions. Since we don't know N_s, we use the number of rectangular regions as an approximation. For any given region S, we assume that δ_{in} is the occurrence rate of a, b together inside S. We assume the outbreak will not change q^a or q^b inside S but it will increase δ_{in}. Therefore, $q_{in}^a = q_{out}^a = q^a$, $q_{in}^b = q_{out}^b = q^b$, $\delta_{out} = \delta$ and $\delta_{in} = \alpha\delta$, where δ is the occurrence rate of a, b outside S. Since we do not know the exact value of α, we use a discretized uniform distribution for α, ranging from $\alpha = [1, 3]$ with increment equals 0.2. The posterior probabilities can be calculated by averaging likelihoods over the distribution of α.

3.3 Learning Bivariate Poisson Distribution Using EM Algorithm

We apply the EM algorithm to learn BP distribution proposed in [8]. BP deals with random variable $\mathbf{X} = [X_1, X_2]$, where $X_1 = Y_0 + Y_1, X_2 = Y_0 + Y_2, Y_i, i \in [0, 2]$ are independent Poisson distribution with mean θ_i. In our context, $q^a = \theta_1, q^b = \theta_2$ and $\delta = \theta_0$. We have observations for X_1, X_2 but not for Y_0, Y_1 and Y_2. Y_0 represents the counts of feature a and b occurs in spatial proximity, Y_1 and Y_2 represent the counts of feature a and b, independently. Our purpose is to use EM to find out $\theta_i, i \in [0, 2]$. Given the probability density function for BP as follows:

$$P(\mathbf{X}) = P(X_1 = x_1, X_2 = x_2) = exp^{-(\sum_{i=0}^{2} \theta_i)} \frac{\theta_1^{x_1}}{x_1} \frac{\theta_2^{x_2}}{x_2} \sum_{j=0}^{min(x_1,x_2)} \binom{x_1}{i} \binom{x_2}{i} i! (\frac{\theta_0}{\theta_1 * \theta_2})^i$$

(7)

We are given N samples with observation for X_1 and X_2. In the E-Step, we compute the expectations of Y_0 based on the observations. At the k-th iteration, we compute $s_i = E(Y_{i0}|X_i, t_i, \theta^{(k)})$, where

- t_i is the base population for the i-th observation.
- $\theta^{(k)}$ is the vector of parameters $\langle \theta_0, \theta_1, \theta_2 \rangle$ for iteration k.

s_i is computed as follows:

$$s_i = \theta_0 * t_i * \frac{P(X_{i1} = x_{i1} - 1, X_{i2} = x_{i2} - 1)}{P(\mathbf{X}_i)}$$

(8)

$P(X_{i1} = x_{i1} - 1, X_{i2} = x_{i2} - 1)$ and $P(\mathbf{X}_i)$ (each $\mathbf{X}_i = (X_{i1}, X_{i2})$) are computed using equation (7). In the M-Step, we update those θs as follows: For θ_0:

$$\theta_0^{(k+1)} = \frac{\sum_{i=1}^{N} s_i}{\sum_{i=1}^{N} t_i}$$

(9)

where $\theta_0^{(k+1)}$ is value of θ_0 at iteration $k + 1$; $\sum_{i=1}^{N} s_i$ is the sum of all s_i for N observations computed from E-Step; $\sum_{i=1}^{N} t_i$ is the sum of populations. For θ_1 and θ_2, we update as follows:

$$\theta_i^{(k+1)} = \frac{\overline{x}_i}{\overline{t}} - \theta_0^{(k+1)} \tag{10}$$

where $\theta_i^{(k+1)}$ is the value of θ_1 and θ_2 at iteration $k + 1$; $\overline{x}_i$ is average number of x_i from N observations; $\overline{t}$ is the average of all population.

4   Finding Arbitrarily Shaped Regional Co-location

We now detail the region expansion heuristic to find regional co-location with arbitrary shapes. Our region expansion heuristic starts from a rectangular region $S \subseteq G$. We compute the statistics of S. After that, during each iteration of region expansion, we try to expand S by adding grid cells around S into it such that the statistic is maximized. Here, the statistic could be the aforementioned P statistic or Bayesian posterior probability. Since for any given S, we can add different number of cells towards different directions. It is impossible to enumerate all of them [15]. To make sure that the expansion process has statistical significance, at each iteration, we add K cells, we fix K at 30 in our current implementation since it is the smallest size to achieve statistical significance. We also generate M different groups of K cells and always expanding S by adding the group that maximizes the statistic score, where M is a user-specified parameter. For a given rectangular region S, this process repeats until the statistic score of S does not increase significantly based on user-specified threshold ϵ. We repeat this process for all possible rectangular regions and return the expanded region with maximized statistic as the result of region expansion. An overview of finding regional co-location with arbitrary shape is described in Algorithm 1 and pseudo code of the expansion process is presented in Algorithm 2 .

The input to Algorithm 1 is the grid G, grid size n and the spatial proximity distance $Dist$. The output is an arbitrarily shaped region with maximum statistic. Line 7 expand the current rectangular region and returns a score for the arbitrarily shaped region as expansion result. Following the work in [11], we only expand rectangular regions with size from 36 cells up to size $(\frac{n}{2})^2$ cells. When all rectangular regions with size in this range have been expanded, the region with the largest statistic score is found.

The input to Algorithm 2 is the rectangular region S, represented as a quadruple of integers, the grid G built from the data set D, number of candidate sets M, as well as the statistical significance threshold value ϵ. Line 7 to Line 22 generate one candidate grid cells set. For each candidate grid cells set $R_{candidate}$ generated during the expansion process, we compute its score. Once we have generated M candidate grid cell sets, we keep the candidate that maximize the statistic (Line 19). We then check whether the expanded region $R_{candidate}$ has statistic score higher than ϵ percent of the region currently found R_{found},

---

Algorithm 1. Expansion Method Overview
___

Input: grid G, grid size n, spatial distance $Dist$
Output: Arbitrarily shaped region G_{found}
Method:

 1: $maxScore = 0.0$;
 2: **for** $x_{min} = 0$ to $n/2$ **do**
 3:   **for** $x_{max} = x_{min} + 5$ to $n/2$ **do**
 4:     **for** $y_{min} = 0$ to $n/2$ **do**
 5:       **for** $y_{max} = y_{min} + 5$ to $n/2$ **do**
 6:         $S = \{x_{min}, x_{max}, y_{min}, y_{max}\}$;
 7:         $R_{found} = expand(S, G, Dist)$ region S with P statistic or Bayesian
            method, as detailed in Algorithm 2
 8:         **if** $(R_{found} > maxScore)$ **then**
 9:           $maxScore = R_{score}$ and record maximum scored region R_{found}
10:         **end if**
11:       **end for**
12:     **end for**
13:   **end for**
14: **end for**

___

if the gain of the statistical score is significant (i.e., higher than ϵ percent), we will upgrade R_{found} to $R_{candidate}$ until we cannot find such $R_{candidate}$ (Line 25).

5   Evaluation and Analysis

In this section, we compare our approaches with some baseline approaches that we propose. The proposed baseline approaches are simply to apply the same framework to find rectangular regions without arbitrary shape expansion and return the rectangular region with maximum statistic. Our approaches are termed $P, B, PBase$, and $BBase$. We use P to represent P statistic based method, B to represent Bayesian statistical method and $PBase$, and $BBase$ to represent their corresponding baseline methods.

The purpose of our experiments is to demonstrate the effectiveness of two approaches in discovering the arbitrarily shaped region with co-location. We show that for synthetic data, the proposed approaches recover the injected arbitrary shaped region with high accuracy. In addition, Bayesian method is approximately 1,000 times faster than frequentist method for both synthetic and real data. For real world data, frequentist method can find the region with high accuracy but is computationally expensive. We also show how our approaches react to arbitrariness of regions. From these experiments, we conclude that frequentist method works well in both synthetic and real world data sets. However, if computational power is not available, it is advisable to use Bayesian method to find the region with high precision.

Performance Metrics. We adopt the well-known *precision and* recall framework as performance metric. Formally, denote the injected region as $R_{true} =$

Algorithm 2. Expansion of Region S

Input: Initial region $S = \{x_{min}, x_{max}, y_{min}, y_{max}\}$, grid G, grid size n, number of candidate sets to expand M, candidate cell size K, threshold value ϵ

Output: Arbitrarily shaped region G_{found}

Method:

1: Compute initial score R_{score} (i.e., P statistics and Bayesian posterior probability) for S;
2: initialize $R_{found} = S$ and find initial set of neighbors V for R_{found}
3: **while** (true) **do**
4:    $i = 0$;
5:    **while** $i < M$ **do**
6:      $V' = V$ and initialize sampled cells set $S_N = \phi$;
7:      **while** (true) **do**
8:        sample one grid cell c from V'
9:        **if** $c \notin S_N$ **then**
10:          $S_N = S_N \cup c$
11:          update V by adding valid neighbors of c into V
12:        **else**
13:          goto Line 8;
14:        **end if**
15:        **if** $|S_N| < K$ **then**
16:          goto Line 7;
17:        **end if**
18:        compute score of current extended region $R' = R \cup S_N$,denoted as R'_{score}
19:        **if** $R'_{score} \geq R_{score} * (1 + \epsilon)$ **then**
20:          $R_{score} = R'_{score}$; $R_{candidate} = R'$; $R_{candidate}.score = R'.score$
21:        **end if**
22:      **end while**
23:      $i = i + 1$;
24:    **end while**
25:    **if** $(R_{canddiate}.score \geq R_{found}.score * (1 + \epsilon))$ **then**
26:      $R_{found} = R_{candidate}$; $R_{found}.score = R_{canddiate}.score$
27:      goto line 3
28:    **else**
29:      break;
30:    **end if**
31: **end while**
32: return R_{found} and $R_{found}.score$

$\{c_1, c_2, ..., c_n\}$ and the found region from our methods as $R_{found} = \{r_1, r_2, ...r_m\}$, where $c_i, i \in [1, n]$ and $r_j, j \in [1, m]$ are grid cells inside G.

Precision is defined as: $precision = \frac{|R_{true} \cap R_{found}|}{|R_{found}|}$. Recall is defined as: $recall = \frac{|R_{true} \cap R_{found}|}{|R_{true}|}$. After computing precision and recall, we compute the F measure in equation (11):

$$F_{measure} = 2 * \frac{precision * recall}{precision + recall} \tag{11}$$

We also measure the running time for $P, B, PBase$ and $BBase$. We now detail a metric to measure the arbitrariness of a region.

Measure of Region Arbitrariness. Formally, assume that R_{true} has bounding cells indices for x and y as x_{min}, x_{max} and y_{min}, y_{max}, respectively. The size of the bounding region for R_{true} can be defined as: $BR_{size} = (x_{max} - x_{min} + 1) * (y_{max} - y_{min} + 1)$. Then the arbitrariness of R_{true} is defined as follows: $ArbRatio_{R_{true}} = \frac{BR_{size} - |R_{true}|}{BR_{size}}$. Intuitively, the arbitrariness of a region is the ratio of number of grid cells that are not in R_{true} to the total number of grid cells of the bounding rectangular region.

5.1 Experiment Set-Up

Synthetic Data Experiment Set-Up Parameters and their values used in synthetic data experiments are listed in Table 1. Default values are in bold. We

Table 1. Synthetic data experiments parameters

n	32,64,128
q_{out}^a	0.02, 0.04,**0.06**,0.08,0.10
q_{out}^b	0.02, 0.04,**0.06**,0.08,0.10
q_{out}^{ab} (a.k.a δ)	0.01, 0.02, **0.03**,0.04,0.05
q_{in}^a	0.01, 0.02, **0.03**,0.04,0.05
q_{in}^b	0.01, 0.02, **0.03**,0.04,0.05
q_{in}^{ab}	0.03,0.06,**0.09**,0.12,0.15
number of candidate set M	4,8,**12**,16,20
arbitrary region size	66,96,**126**,156, 186
distance $Dist$	80, 160, **240**, 320,400

randomly generate 5 different arbitrary shaped regions with different sizes shown in Table 1. Occurrence rate of feature a, b outside the injected region are denoted as q_{out}^a, q_{out}^b and q_{out}^{ab}. Each group of q_{out}^a, q_{out}^b, q_{out}^{ab}, q_{in}^a, q_{in}^b and q_{in}^{ab} is defined as an occurrence rate combination.

For each fixed occurrence rate combination, we pick default region size (126 grid cells as shown in Table 1) and generate five different synthetic data sets. Similarly, we fix the occurrence rate at default combination, i.e., rate values in bold in Table 1 and vary the size of arbitrary region, we generate another five different synthetic data sets. Synthetic data is generate using Algorithm 3. For all data sets, we fix the population size as $30k$. We first generate the coordinates for each entity (i.e., person in context of Epidemiology) of the whole population. We then assign for each entity, spatial features a, b based on the input occurrence rate combination by random sampling. Spatial features a, b can be different types of disease in the context of Epidemiology and can be different types of accidents in crash data, etc. Algorithm 3 is straightforward. When those data sets have been generated, we apply our proposed approaches to find the arbitrary sized region. For each data set and each fixed parameter setting, we repeat the experiments 5 times. All results reported are based on those 5 independent runs.

Algorithm 3. Synthetic data generation

Input: occurrence rate $q_{out}^a, q_{out}^b, q_{out}^{ab}, q_{in}^a, q_{in}^b, q_{in}^{ab}$, total population TP ,range of region L, spatial proximity distance $Dist$, G_{true}
Output: data set D
Method:

1: initialize a location array $loc[TP]$;
2: **for** $i = 1$ to TP **do**
3:     generate x_i and y_i uniformly from $[1, L]$ and put $\langle x_i, y_i \rangle$ into $loc[i]$
4: **end for**
5: **for** each person p_i **do**
6:     generate a random number $r \in [0, 1]$
7:     **if** $p_i.\langle x_i, y_i \rangle \in G_{true}$ **then**
8:         **if** $r \le q_{in}^a$ **then**
9:             label p_i with event a
10:         **end if**
11:     **else if** $r \le q_{out}^a$ **then**
12:         label p_i with event a
13:     **end if**
14: **end for**
15: repeat Line 5 to Line 14 by replacing q_{in}^a with q_{in}^b and q_{out}^a with q_{out}^b, and label p_i with event b.
16: **for** each person p_i with no label **do**
17:     generate a random number $r \in [0, 1]$
18:     **if** $p_i.\langle x_i, y_i \rangle \in G_{true}$ **then**
19:         **if** $r \le q_{in}^{a,b}$ **then**
20:             label p_i with event a and find p_j inside radius $Dist$ of p_i, label as b
21:         **end if**
22:     **else if** $r \le q_{out}^{a,b}$ **then**
23:         label p_i with event a and find p_j inside $Dist$ of p_i, label as b
24:     **end if**
25: **end for**

For frequentist method, we also need to generate replicas in order to do the Monte Carlo simulation. For this purpose, for each given synthetic data set, we first apply EM algorithm to learn the overall rate of spatial features: $q_{overall}^a, q_{overall}^b$ and $q_{overall}^{ab}$, we then generate 1000 replica data by replacing $q_{in}^a = q_{out}^a = q_{overall}^a$, $q_{in}^b = q_{out}^b = q_{overall}^b$ and $q_{in}^{ab} = q_{out}^{ab} = q_{overall}^{ab}$. Meanwhile, we don't do region injection for replica data generation, i.e., Line 16 to Line 25 are excluded for replica generation. After that, we apply the frequentist method on each of those replica data, compute the statistics values for each arbitrarily shaped region returned. Finally, we compute the p-value of the frequentist method.

Real Data Experiment Set-Up. For real data experiments, we use the taxi data in Shanghai, China([12]). The data is collected with a frequency of 300 seconds from 12am, May 29, 2009 to 6pm, May 30,2009. It contains locations for $17,139$ taxis from 3 different taxi companies, which forms over 48.1 million

GPS records of 468,000 trips. It contains location data of taxi and pick-ups for 215 different time intervals. We randomly select 5 data sets from 5 different time intervals. In this taxi data context, we assume that all the requests have been met. To model the co-location problem, we assume that spatial feature a is request, which are *pick-up*; spatial feature b is *empty taxi*. Those two features can be directly read from the taxi data. Our goal is to find a region such that the co-location of request and empty taxi is the largest in a given spatial proximity. In other words, we want to find out regions such that the number of taxis is much larger than the number of requests. Parameters and their values used in real data experiments are listed in Table 2. Since q_{out}^a, q_{out}^b, and q_{out}^{ab} are learned from a given data set by applying EM algorithm as detailed in section 3.3, we do not need to provide them. Default values are in bold. Since we model over-serve colocation, we inject an arbitrarily shaped region into a given snapshot data, we first pick an arbitrary region and a removal percentage (the maximum percentage of requests that should be removed), we then remove those requests based on the removal percentage inside the arbitrarily shaped region randomly. After that, we run our proposed approaches to recover the injected region. For each data set and each fixed parameter setting, we repeated our experiments 5 times, all results reported are averaged over those 5 independent runs.

Table 2. Real data experiments parameters

Grid Size n	32,64,128
number of candidate set M	4,8,**12**,16,20
arbitrary region size	66,96,**126**,156, 186
distance $Dist$	40,60,**80**,100,120
removal percentage	15%, 20%, **25%**, 30%, 35%

5.2   Experiment Results and Analysis

We record the total number of rectangular regions to expand based on different grid size n, the result is presented in Table 3. We expect that the running time of those methods will increase dramatically with increment in grid size n. It is intuitive to see that the total number of regions to expands increase with n.

Table 3. Number of Rectangular Regions to Expand

Grid Size n	32	64	128
Total Number of Rect. Regions	8281	164836	3348900

Arbitrariness of Region. The arbitrariness of these 5 regions used in experiments is reported in Figure 2(a). We will show in other result figures that the higher the arbitrariness of a region is, the worse the performance of our approaches will be.

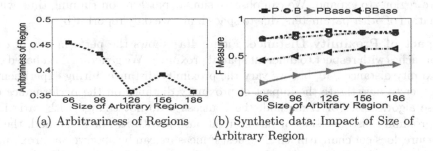

(a) Arbitrariness of Regions   (b) Synthetic data: Impact of Size of
Arbitrary Region

Fig. 2. Arbitrariness of Regions

Synthetic Data Results. In this section, we analyze our results on synthetic data with respect to various parameters including arbitrariness and size of region,grid size, distance proximity, as well as M.

Impact of Arbitrariness/Size of Region. We present the impact of size of arbitrary region in Figure 2(b). In general, when the arbitrariness of region is large, our approaches do not work well. One exception is when region size equals 156 and arbitrariness increases. However, our method still works very well, the reason is that when the region is reasonably large, it can tolerate more false positives. This experiments provide an insight that we need to make the approaches more robust when the arbitrariness is high.

Impact of Grid Size. Impact of grid size is presented in Figure 3. We fix grid size equals 32 for all the data sets we generated. We can see that P and B methods work pretty well in identifying the input arbitrary shaped region. PBase and BBase do not work very well since the input region is arbitrarily shaped instead of rectangular. We can observe that all methods do not work well when the grid size is 64 and 128 . The reason is that the data is partitioned using grid size equals 32. This result provides an insight that we need to apply those methods with the same grid size as the data is being partitioned. Figure 3(b) shows the running time of our methods as well as baseline approaches. We can observe that the running time for frequentist methods(P,PBase) is approximately 1000 times larger than that of Bayesian method. When grid size is larger, all methods run longer since there are

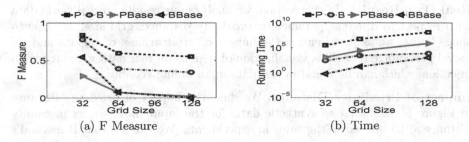

(a) F Measure   (b) Time

Fig. 3. Impact of Grid Size Synthetic Data

more regions to expand. We can observe similar patterns on running time with variation of other parameters, due to space limit, we only report 3(b).

Impact of Proximity Distance. Figure 4(a) shows the performance of our approaches with respect to different spatial proximity. We generate data based on proximity distance 240, but we vary the proximity distance during experiments in order to investigate the impact of proximity distance on the performance of those algorithms. For B method, the F measure do not change with variation on proximity distance since it has no relationship with it. For P method, the F measure does not change much. The best F measure can be observed at proximity distance equals 240 since it is the true spatial proximity distance. For baseline algorithms, they do not work well. This experiment provides an insight that our approaches are not very sensitive to spatial proximity distance.

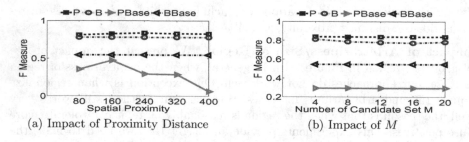

(a) Impact of Proximity Distance   (b) Impact of M

Fig. 4. Experiment Results: Synthetic Data

Impact of M. Performance of our approaches regarding different M values is shown in Figure 4(b). We can see that P, and B perform well in recovering the injected arbitrary shaped region. However, for baseline algorithms, they do not work very well. We can also observe that all approaches are relatively stable regarding changes of M generated during expansion. This is due to the fact that our expansion heuristic always pick the group of candidate grid cells that maximize the score. Meanwhile, the input arbitrarily shaped region is relatively small and varieties of the candidate sets is then relatively small. Therefore, with a small number of candidate sets, we can reach similar performance.

Real Data Results. In this section, we analyze our results on real world data with respect to different parameters: proximity distance, grid size, M. Due to space limits, we skip presenting the impact of arbitrariness of regions and removal percentage as well as visualization of regions on real data without region injection, which can be found in real data in an online version[1].

Impact of Proximity Distance. We show the impact of proximity distance in Figure 5(a). Similar to synthetic data, for the input data, we fix proximity distance at 80 but varies the value in experiments. We can see that B method's F measure does not change since it has nothing to do with proximity distance.

---

[1] https://www.dropbox.com/s/gnluegcq2f36ip6/colocationFullVersion.pdf

P method has high F measure. However, B does not work as well as P. The reason is that the real data does not necessarily obey bivariate Poisson distribution. We can see from Figure 5(b) that B method actually achieves the highest precision, but its recall is not good enough. Therefore, F measure is not good enough. We can observe similar performance of B method for other parameters. We omit them due to space limit. In all real data experiments, we can see that P works better than PBase and B outperforms BBase.

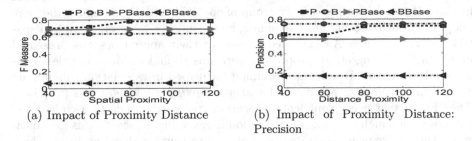

(a) Impact of Proximity Distance

(b) Impact of Proximity Distance: Precision

Fig. 5. Experiment Results: Real Data

Impact of Grid Size. We show the impact of grid size in Figure 6(a). Our injected data are based on grid size 32. We can see that our approaches only work for grid size 32. This is intuitive since the data is generated with grid size equals 32. Therefore, we can observe that F measure decreases with increment of grid size.

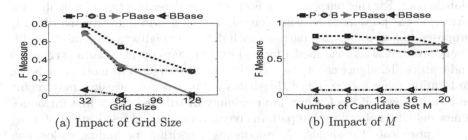

(a) Impact of Grid Size

(b) Impact of M

Fig. 6. Experiment Results: Real Data

Impact of M. Performance of our approaches with respect to M is shown in Figure 6(b). We can observe similar performance as that of grid size that B method does not work as well as P due to underlying distribution of data. The reason is the same as summarized in grid size experiments.

6 Related Works

Our work is related to previous works from two main categories: spatial co-location pattern mining and Bayesian spatial scan statistic.

Spatial Co-location Pattern Mining. Co-location patterns have been studied extensively in literature ([3], [5],[6],[7], [13],[14],[16]). Given a collection of
boolean spatial features, general spatial co-location pattern mining methods try
to find the subsets of features that are frequently located in spatial proximity.
Finding spatial co-location pattern is an important problem in ecology, epidemics, transportation system and others.

An initial summary of results on general spatial co-location mining was proposed in [14]. The authors proposed the notion of user-specified neighborhoods
in place of transactions to specify group of spatial items. By doing so, they can
adopt traditional association rule mining algorithms, i.e., Apriori [1] to find spatial co-location rules. An extended version of their approaches was presented
in [6]. In [7], Huang et. al. proposed algorithms to find co-location rules without using a support threshold, a common term used in association rule mining
algorithms. A novel measure called *maximal participation index* was proposed.
They found that every confident co-location rule corresponds to a co-location
pattern with a high maximal participation index value. Based on this new measure, Apriori-like pruning strategies were used to prune co-location rules. These
works on spatial co-location pattern discovery focus on finding global spatial co-
location patterns with a fixed interest measure threshold. Methods for mining
co-location patterns with rare spatial features were studied in [5].

Spatial co-location patterns with dynamic neighborhood constraint was proposed in [13]. The motivation of work in [13] is that existing work for finding
co-location patterns uses static neighborhood threshold. They argue that identifying the dependence relationship of spatial features and computing the prevalence measure of features have different distributions in different areas of the
global space. For this purpose, they postpone the determination of neighbor relations to the prevalence measure computation step and a greedy algorithm is
proposed to find co-location patterns with different neighbors constraints in different areas. A statistical model for co-location that considers auto-correlation
and feature-abundance effect was recently discussed in [2]. The motivation is the
co-location using user specified thresholds for prevalence measures may report
co-locations even if the features are randomly distributed. They first introduce
a new definition of co-location patterns based on statistical test instead of using
global prevalence thresholds. Corresponding algorithm for finding co-location
patterns based on this new statistical test is also proposed.

However, all these co-location methods focus on global co-location patterns,
which are not directly applicable to find regional co-locations. In [3], zonal co-
location pattern find co-location in a subset of the space, i.e.zone or region.
They used repeated specification of zone and interest measure values according
to user preference instead of discovering global spatial co-location patterns with
a fixed interest measure threshold. They propose an algorithm,namely, Zoloc-
Miner to discover regional co-location patterns with dynamic parameters. For
this purpose, a Quadtree index structure is proposed to store co-location patterns to handle dynamic parameters. They assume that the regions and interest
measure for those regions are given beforehand, which requires sophisticated

domain knowledge. However, domain knowledge is hardly obtainable in real world applications. Meanwhile, they do not find arbitrary shaped regions.

Those works mentioned before focused on spatial features with categorized value, i.e., the location of spatial features are not continuous. A framework for finding regional co-location patterns in continuous valued spatial variables was proposed in [4]. Their motivation is to find regions in a spatial dataset such that certain continuous quantities have high concentration (e.g., concentrations of different chemicals in sets of wells) inside the region than that of outside areas. It views regional co-location mining as a clustering problem in which an externally given fitness function has to be maximized. For this purpose, they propose a framework named CLEVER that uses randomized hill climbing to discover regional co-location. This work can find arbitrary shaped regions. However, they also require that extensive domain knowledge be available to find the initial representative regions for clustering and the co-location pattern is also available. This is usually not true in real world applications.

To our knowledge, no prior work deals with finding regional patterns with arbitrary shape without domain knowledge, which is the focus of our work.

Bayesian Spatial Scan Statistic. Spatial scan statistics have been studied extensively. The purpose of spatial scan statistics was to find spatial clusters where certain quantity of interest occurs significantly higher than expected. The state-of-art is based on Kulldorff's spatial scan statistic ([9]). An extended discussion of spatial scan statistic was presented in [10]. In [9], Kulldorff defined a general model for the multidimensional spatial scan statistic. There are three basic properties of the scan statistic: the geometry of the area being scanned, the underlying probability distribution generating the observed data under the null hypothesis and shapes of the scanning window. Calculation of the spatial scan statistic is based on rigid mathematical inductions based on different underlying probability models. The main idea of the spatial scan statistic was to calculate the defined statistic from the data being scanned, then a hypothesis testing against the null hypothesis was conducted by using Monte Carlo simulation. However, the computational cost of the spatial scan statistic is high. Furthermore, they only deal with clusters not co-locations.

Recently, a Bayesian spatial scan statistic was proposed to discover spatial clusters [11]. In disease surveillance systems, it is useful to find spatial regions with high frequency of certain disease and report an emergent disease outbreak to save lives. For this purpose, given the number of occurrences of certain diseases in a spatial region, a Bayesian spatial scan statistic was proposed to examine the posterior probability of every possible rectangular region under the null hypothesis and the alternative hypothesises. Bayesian spatial scan statistic can find significant spatial clusters with less running time and higher detection accuracy.

Closely related with epidemics and Bayesian spatial scan statistic are the Poisson distribution and bivariate Poisson distribution. Poisson distribution are often used as prior probability model in Bayesian spatial scan statistic. Bivariate Poisson distribution was widely used to model the count of disease in epidemics. In this work, our Bayesian statistics based approach uses the bivariate Poisson distribution and Bayesian spatial scan statistic to discover arbitrarily shaped regions with co-location patterns.

7 Conclusion

In this paper, we studied the problem of finding regional co-location with arbitrary shapes. For this purpose, we proposed two approaches: frequentist method and Bayesian statistics. We evaluated our approaches in both synthetic and real world data. Experimental results demonstrate that our two approaches work effectively. Bayesian method runs approximately three orders of magnitude faster than frequentist method. When computational power is not available, we can use Bayesian method to recover the region with high precision; otherwise, it is better to apply frequentist methods. However, in this work, the expansion process is stochastic, it will be interesting to study about deterministic expansion approaches in the future. Meanwhile, it will be interesting to investigate how to speed up the expansion process with more sophisticated heuristics.

References

1. Agrawal, R., Srikant, R.: Fast algorithms for mining association rules in large databases. In: Proceedings of the 20th International Conference on Very Large Data Bases, VLDB 1994, pp. 487–499. Morgan Kaufmann Publishers Inc., San Francisco (1994)
2. Barua, S., Sander, J.: SSCP: Mining statistically significant co-location patterns. In: Pfoser, D., Tao, Y., Mouratidis, K., Nascimento, M.A., Mokbel, M., Shekhar, S., Huang, Y. (eds.) SSTD 2011. LNCS, vol. 6849, pp. 2–20. Springer, Heidelberg (2011)
3. Celik, M., Kang, J.M., Shekhar, S.: Zonal co-location pattern Discovery with dynamic parameters. In: ICDM 2007, pp. 433–438. IEEE Computer Society, Washington, DC (2007)
4. Eick, C.F., Parmar, R., Ding, W., Stepinski, T.F., Nicot, J.-P.: Finding regional co-location patterns for sets of continuous variables in spatial datasets. In: Proceedings of the 16th ACM SIGSPATIAL International Conference on Advances in Geographic Information Systems, GIS 2008, pp. 30:1–30:10. ACM, New York (2008)
5. Huang, Y., Pei, J., Xiong, H.: Mining co-location patterns with rare events from spatial data sets. Geoinformatica 10(3), 239–260 (2006)
6. Huang, Y., Shekhar, S., Xiong, H.: Discovering colocation patterns from spatial data sets: A general approach. IEEE Trans. on Knowl. and Data Eng. 16(12), 1472–1485 (2004)
7. Huang, Y., Xiong, H., Shekhar, S., Pei, J.: Mining confident co-location rules without a support threshold. In: Proceedings of the 2003 ACM Symposium on Applied Computing, SAC 2003, pp. 497–501. ACM, New York (2003)
8. Karlis, D.: An em algorithm for multivariate poisson distribution and related models. Journal of Applied Statistics 30(1), 63–77 (2003)
9. Kulldorff, M.: A spatial scan statistic. Communications in Statistics - Theory and Methods 26(6), 1481–1496 (1997)

10. Kulldorff, M.: Spatial scan statistics: Models,calculations, and applications. In: Glaz, J., Balakrishnan, N. (eds.) Scan Statistics and Applications, Statistics for Industry and Technology, pp. 303–322. Birkhuser, Boston (1999)
11. Neill, D.B., Moore, A.W., Cooper, G.F.: A bayesian spatial scan statistic. In: NIPS (2005)
12. Powell, J.W., Huang, Y., Bastani, F., Ji, M.: Towards reducing taxicab cruising time using spatio-temporal profitability maps. In: Pfoser, D., Tao, Y., Mouratidis, K., Nascimento, M.A., Mokbel, M., Shekhar, S., Huang, Y. (eds.) SSTD 2011. LNCS, vol. 6849, pp. 242–260. Springer, Heidelberg (2011)
13. Qian, F., He, Q., He, J.: Mining spatial co-location patterns with dynamic neighborhood constraint. In: Buntine, W., Grobelnik, M., Mladenić, D., Shawe-Taylor, J. (eds.) ECML PKDD 2009, Part II. LNCS, vol. 5782, pp. 238–253. Springer, Heidelberg (2009)
14. Shekhar, S., Huang, Y.: Discovering spatial co-location patterns: A summary of results. In: Jensen, C.S., Schneider, M., Seeger, B., Tsotras, V.J. (eds.) SSTD 2001. LNCS, vol. 2121, pp. 236–256. Springer, Heidelberg (2001)
15. Welsh, D.: Approximate Counting. Cambridge University Press (2007)
16. Zhang, X., Mamoulis, N., Cheung, D.W., Shou, Y.: Fast mining of spatial collocations. In: Proceedings of the Tenth ACM SIGKDD International Conference on Knowledge Discovery and Data Mining, pp. 384–393 (2004)

MNTG: An Extensible Web-Based Traffic Generator

Mohamed F. Mokbel[1], Louai Alarabi[1], Jie Bao[1], Ahmed Eldawy[1], Amr Magdy[1],
Mohamed Sarwat[1], Ethan Waytas[1], and Steven Yackel[2]

[1] University of Minnesota, Minneapolis, MN 55455, USA
[2] Microsoft

{mokbel,louai,baojie,eldawy,amr,sarwat}@cs.umn.edu,
wayt0012@umn.edu, spazard1@live.com

Abstract. Road network traffic datasets have attracted significant attention in the
past decade. For instance, in spatio-temporal databases area, researchers harness
road network traffic data to evaluate and validate their research. Collecting real
traffic datasets is tedious as it usually takes a significant amount of time and effort. Alternatively, many researchers opt to generate synthetic traffic data using
existing traffic generation tools, e.g., Brinkhoff and BerlinMOD. Unfortunately,
existing road network traffic generators require significant amount of time and
effort to install, configure, and run. Moreover, it is not trivial to generate traffic
data in arbitrary spatial regions using existing traffic generators. In this paper,
we propose Minnesota Traffic Generator (MNTG); an extensible web-based road
network traffic generator that overcomes the hurdles of using existing traffic generators. MNTG does not provide a new way to simulate traffic data. Instead, it
serves as a wrapper over existing traffic generators, making them easy to use,
configure, and run for any arbitrary spatial road region. To generate traffic data,
MNTG users just need to use its user-friendly web interface to specify an arbitrary spatial range on the map, select a traffic generator method, and submit the
traffic generation request to the server. MNTG dedicated server will receive and
process the submitted traffic generation request, and notify the user via email
when finished. MNTG users can then download their generated data and/or visualize it on MNTG map interface. MNTG is extensible in two frontiers: (1) It can
be easily extended to support various traffic generators. It is already shipped with
the two most common traffic generators, Brinkhoff and BerlinMOD, yet, it also
has the interface that can be used to add new traffic generators. (2) It can be easily
extended to support various road network sources. It is shipped with U.S. Tiger
files and Open Street Map, yet, it also has the interface that can be used to add
other sources. MNTG is launched as a web service for public use; a prototype can
be accessed via http://mntg.cs.umn.edu.

1 Introduction

Road network traffic data consist of a set of spatial locations reported by a set of objects
moving over a road network. Traffic data have been already leveraged by researchers
in different areas, e.g., spatial-temporal databases, transportation, urban computing and
data mining. The process of extracting real traffic data requires installing and configuring many GPS-enabled devices and continuously monitoring the locations of such
devices, which is a cumbersome task. For instance, GeoLife project [1] took more than

M.A. Nascimento et al. (Eds.): SSTD 2013, LNCS 8098, pp. 38–55, 2013.
ⓒ Springer-Verlag Berlin Heidelberg 2013

four years to collect 17,621 trajectories dataset with the involvement of 182 volunteers in Beijing. Alternatively, many researchers opt to generate synthetic road network traffic data. As a consequence, several efforts have been dedicated to develop road network traffic generators, e.g., Brinkhoff [2] and BerlinMOD [3].

Even though existing traffic generators are quite useful, nonetheless, most of them suffer from the following: (1) It may take the user significant amount of effort to install and configure the traffic generation tool. For example, in order to run BerlinMOD, the user needs to first install a moving object database, i.e., SECONDO [4], and then get familiar with the script commands used to install it. After the installation, users still need to understand an extensive list of configuration parameters for each traffic generator. (2) It is not trivial to generate traffic data in arbitrary spatial regions using existing traffic generators. For example, to be able to use Brinkhoff or BerlinMOD generators for a different city than the default shipped one (Oldenburg and Berlin for Brinkhoff and BerlinMOD generators, respectively), the user needs to first obtain the road network information for the city of interest, which is a tedious task by itself. For example, to get the road network information for the city of Munich, a user may need to understand the format of OpenStreetMap [5], and then write a program that extracts the road network of Munich from OpenStreetMap. After obtaining the new road network data, the user will then need to understand how to modify the obtained format to match the required one by either Brinkhoff or BerlinMOD. Such set of tedious operations made it hard for casual users to use these traffic generators for arbitrary spatial areas. As a testimony, one can observe that almost all the literature that have used these generators for traffic data have used it for their default cities.

In this paper, we propose Minnesota Traffic Generator (MNTG); an extensible web-based road network traffic generator that overcomes the hurdles of using existing traffic generators. MNTG is basically a wrapper around existing traffic generators, with the mission of enabling an easy usage of all existing traffic generators, and hence help all researchers worldwide in validating and benchmarking their research techniques against various workloads of movings objects over real road networks.

MNTG has three main features that significantly help in achieving its goals: (1) MNTG is a web service with an easy-to-use user-friendly map interface. Behind the scenes, MNTG carries the burden of configuring and running existing traffic generators. Thus, MNTG users do not need to install or configure anything on their local machines. This is in contrast to the traditional usage of Brinkhoff and BerlinMOD generators that require various installations as mentioned above. (2) MNTG can be used for any arbitrary city or spatial area worldwide. Users can just navigate through the map interface, and mark their area of interest with a rectangular area. Once the traffic generation request is submitted, MNTG is responsible for extracting the road network information for the requested area and generating the traffic on the extracted area using one of the existing traffic generators. This is in contrast to the traditional usage of existing traffic generators that is hard to be tailored for arbitrary cities. (3) MNTG users do not need to worry about the processing time or computing resources, where MNTG has its own dedicated server machine that (a) receives a traffic request from the user, (b) internally processes the request in a multi-core multi-threaded program, and (c) emails the user back when the requested data is generated. The notifying email includes a link to download the data as well as an option to visualize the generated data. This is in contrast to the traditional usage of existing traffic generators that may take significant portion of the user time and computing resources.

Minnesota Traffic Generator (MNTG) is extensible in two frontiers: (1) It can be easily extended to support various traffic generators through few deterministic functions. Currently, MNTG is shipped with the two most common traffic generators, Brinkhoff and BerlinMOD, yet, it also has the interface that can be used to add new traffic generators. As a proof of concept, we extend it with a random generator which generates some kind of random walks over the road network. (2) It can be easily extended to support various road network sources. It is currently shipped with the support for U.S. Tiger files [6] and OpenStreetMap [5], yet, it also has the interface that can be used to add other sources for road network data.

MNTG is equipped with three main components, listed as follows: (1) *Road Network Converter*: that extracts the road network for the area of interest from either US. Tiger files or OpenStreetMap, and converts it to match the format of the underlying traffic generator. (2) *Traffic Processor*: that schedules and executes the received traffic generation requests. The traffic processor processes the incoming requests in parallel using a multi-threading paradigm to increase the overall system throughput. Both the road network converter and traffic processor provide an interface for the system users to incorporate a newly developed traffic generator. To plug-in a newly developed traffic generator, MNTG defines a set of abstract functions that users need to implement. The implemented functions deal with converting/extracting the road network data, executing the traffic generator, and preparing the generated traffic output. Once a traffic generator is plugged-in, users may leverage it to generate traffic data. (3) *System Front-End*: which contains a web interface for users to submit traffic generation requests, an email notifier to send messages or notifications to the user, and a set of tools for the user to download and visualize the generated traffic data.

A preliminary version of MNTG is launched as a web service for public use; a prototype can be accessed via http://mntg.cs.umn.edu. The preliminary version supports Brinkhoff, BerlinMOD and the random traffic generators on both U.S TIGER files and OpenStreetMap data. The extensibility interface for adding more generators or other road network sources is currently working internally under our support. Yet, these functionalities will be released to public use in our next version. Since its launch last month, MNTG has received more than 1000 traffic generation requests from researchers world wide. All requests have been efficiently satisfied, and results were sent back to the requesting users. We envision that MNTG will be the de facto standard for generating road network data for researchers in spatial and spatio-temporal databases worldwide.

The rest of this paper is organized as follows: Section 2 highlights related work. Section 3 gives an overview of MNTG. Sections 4 and 5 describe the two main components in the system back-end; (1) road network converter and (2) traffic models, respectively. Section 6 provides the description of the system front-end with detailed usage guide. Finally, Section 7 concludes the paper with pointers to future work.

Table 1. Existing Moving Objects Generators

Environment	Generators
Free Movement	Pfoser and Theodoridis [7], Oporto [8], GSDT [9], G-TERD [10]
Road Network	Brinkhoff [2] , BerlinMOD [3] , ST-ACTS [11] , GAMMA [12], SUMO [13] , Micro Simulators [14]
Multi Environments	MWGen [15]

2 Related Work

Road network traffic data (i.e., moving objects data) have been widely used by re-searchers to test and validate their techniques in various spatio-temporal data man-agement problems. This includes objects tracking [16], predictive queries [17], range queries [18], kNN queries [19], continuous queries [20], data uncertainty [21], and lo-cation privacy [22]. As a consequence, several efforts have focused on creating standard benchmarks for evaluating research on moving objects data [3, 23–27]. As part of cre-ating such benchmarks, generating synthetic moving objects data gained considerable attention in the literature.

Table 1 gives a summary of existing moving objects data generators. Based on the spatial environment where the objects move on, existing moving objects data generators can be classified into the following three main categories:

1. *Free movement* [7–10]. This category assumes that objects can move freely in a two-dimensional Euclidean space. The GSTD generator [9] generates data for ei-ther moving points or rectangular regions, where it allows its users to control the lifetime of each moving object. The GSTD generator is later extended to incorpo-rate real-life behaviors like group and obstructed movements [7]. G-TERD [10] has introduced new features to traffic generators, where users can generate arbitrary-shaped objects with tuning parameters that control object speed, color, and rotation over time. Unlike all other moving objects generators, Oporto [8] is particularly concerned with generating the movement of ships, which depends on fishing sce-narios where ships head to fish shoals and avoid storms.

2. *Road networks [2, 3, 28, 11–13, 29, 14]*. This category is mainly concerned with generating moving objects data in a road network environment. Constrained by the predefined road network paths, they basically generate traffic data based on real-life trip planning scenarios that simulate the human behavior. Brinkhoff [2], SUMO [13] and micro simulators [14] depend on short-term observations, where representative human behavior is observed for short discrete trips. On the other hand, BerlinMOD [3] and ST-ACTS [11] rely on long-term observations where human behavior is observed for several consecutive days.

3. *Multi-Environments* [15]. This category considers moving objects in multi-environments, e.g., Indoor $\rightarrow$ walk $\rightarrow$ Bus $\rightarrow$ walk. MWGen [15] surpassed the typical functionality of generating data only for a single environment and extends it to support multiple environments, i.e., indoors and outdoors, at the same time.

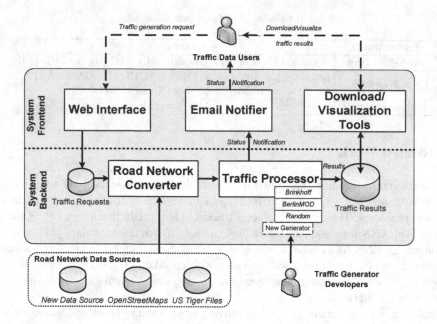

Fig. 1. MNTG System Overview

GMOBench simulates a scenario like an employee moving in her work building, then walks to the parking lot to drive her car all the way to home, then walks again to enter home and move indoors.

Minnesota Traffic Generator (MNTG) distinguishes itself from all the work mentioned above as it does not provide yet another technique for traffic data generation. Instead, it is an extensible wrapper built around any of the existing traffic generators in the second category mentioned above (road network movement). As an extensible easy-to-use wrapper, MNTG enables a practical use of all road network traffic generators developed over the last two decades. Due to its simplicity, MNTG is expected to give a boost to existing traffic generators by gaining wider user community.

3   System Overview

Figure 1 gives an overview of the MNTG system architecture. A user interacts with MNTG through its system front-end that includes three main components: (1) *Web Interface*, which allows users to submit traffic generation requests by selecting a geographical area on the map and setting the corresponding parameters in a very intuitive way, (2) *Email Notifier*, which retrieves the status updates from the back-end and keeps users posted on their traffic generation request progress, and (3) *Download and Visualization tools*, which allow users to download their generated traffic data as a text file and/or visualize the generated traffic data on the map. The details of the system front-end, with its three components, will be discussed in Section 6.

Internally, the system back-end of MNTG processes incoming traffic generation requests and generates traffic data for the system users. The system back-end consists of the following two main components:

1. *Road Network Converter*, which is responsible for extracting the road network data from the traffic generation request. It receives a rectangular spatial area with two corner coordinates, each represented as a *<latitude, longitude>*. Then, the *road network converter* exploits its underlying road network data source, US Tiger files or OpenStreetMaps, to extract the information of the selected area, and convert it into an appropriate format understood by the requested traffic generator. The *road network converter* is extensible to support other road network data sources beyond its default ones, US tiger files and OpenStreetMaps. Details of the *road network converter* will be discussed in Section 4.

2. *Traffic Processor*, which takes the road network data from the *road network converter* and feed it to the requested traffic generator, which is currently either Brinkhoff or BerlinMOD. The traffic processor is implemented in a multi-threading paradigm to: (a) allow multiple requests to be served concurrently, and (b) avoid the starving of small traffic generation requests waiting for large requests to finish. The *traffic processor* is highly extensible as it is equipped with modules that allow the traffic model developers to easily plug-in a new traffic generator. Details of the *traffic processor* will be discussed in Section 5.

4   Road Network Converter

Despite the abundance of road network data sources, such as US Tiger files and OpenStreetMap, none of them provides an intuitive way to extract road network paths, i.e., nodes and edges, for an arbitrary geographical area. MNTG, on the other hand, needs to generate road network traffic data for any geographical area selected by the user. To achieve that, MNTG employs a *road network converter* that is responsible for extracting the road network data for each incoming traffic generation request. Moreover, The road network converter is designed to support a wide variety of road network data sources.

In this section, we first discuss the main idea behind the road network converter. Then, we provide two detailed case studies featuring two road network data sources, which are currently implemented in MNTG: US Tiger Files and OpenStreetMap. Finally, we discuss the extensibility of the road network converter to support new road network data sources.

4.1   Main Idea

The road network converter is responsible for extracting road network nodes and edges from different sources and transforming the extracted data into a standard format that can be utilized by different traffic generators. The functionality of the road network converter does *not* depend on the underlying traffic model (e.g., Brinkhoff or BerlinMOD). Instead, it heavily depends on the underlying data source (e.g., US Tiger files or OpenStreetMap). To achieve its goals, the road network converter performs the following two steps for each traffic generation request:

1. *Step 1. Extracting Road Network.* The input to this step is :(a) a rectangular spatial area, defined by two corner *<latitude, longitude>* coordinates, and (b) the road network data source (e.g., US Tiger files or OpenStreetMap). The output of this step is the road network information of the selected area, based on the selected road network data source. This is done through an abstract function, called ExtractRoadNetwork, that exploits the underlying road network data source to: (a) prune all information that are outside the selected rectangular spatial area, and (b) prune the non road network information from the selected rectangular area. We do so because each road network data source provides data in different formats; for example, US Tiger Files are stored in a binary format with extra information about zip codes, rivers, demographics, etc, while OpenStreetMap stores data in an XML format with extra information about buildings, parks, traffic lights, etc.

2. *Step 2. Preparing Standard Output.* The input to this step is: (a) The road network information of the selected rectangular spatial area, i.e., the output of the ExtractRoadNetwork abstract function, and (b) the road network data source. The output of this step is two standard text road network data files: node.txt and edge.txt, which contain the final set of nodes and edges in the selected spatial area, respectively. This is done through an abstract function, called PrepareStandardOutput, that is aware of the data format of the underlying data source and converts it to our standard output format.

We opt to transform the road network information into a standard text format, to make it portable to various traffic generators. Each traffic generator uses its own different input file format and perhaps different spatial coordinate system. For example, Brinkhoff generator uses binary files to store nodes and edges, whereas BerlinMOD expects one text file with two bracketed locations to represent a road segment. Moreover, Brinkhoff generator uses its own spatial coordinates system, where the *latitude* and *longitude* of a location are the offsets instead of absolute values.

An example of the standard format of our generated node.txt file is as follows:

```
Node_ID             Lat                     Lng
54956254019183      44.85923581362268       -92.989281234375
19567871005131      45.032414105220745      -93.2028993984375
27380416518383      44.99418225712112       -93.4431044765625
```

Node_ID is a unique identifier for the node on the road networks, whereas Lat and Lng are the latitude and longitude coordinates that represent the geographical location of the node, respectively.

Similarly, an example of our generated edge.txt file is as follows:

```
Edge_ID   Node_1            Node_2            Tags
0         33352568523324    33481417542144    highway
1         35667555893384    38510824242033    oneway
2         34881576878577    35839354585144
```

Edge_ID is a unique identifier for the edge on the road networks, whereas Node_1 and Node_2 are node IDs contained in the node.txt file. It means that the two nodes (Node_1 and Node_2) are connected by the edge Edge_ID. Tags attach extra information to road edges which can be used by some generators.

4.2   Case Study 1: US Tiger File

US Topologically Integrated Geographic Encoding and Referencing (Tiger) Files [6] are published by US census bureau on a yearly basis to provide the most recent information about US geographical data, which include city boundaries, road networks, address information, water features, and many more. In MNTG, we focus on extracting the road network information from the Road directory of Tiger files.

A very unique feature of US Tiger Files is that the files are partitioned and organized based on US counties. In other words, all roads in a county are packed in a compressed file with a unique file identifier, e.g., tl_2010_01001_roads.zip, where tl means tiger line, 2010 indicates the publishing year of the data, 01001 is a unique identifier for the county (in this case is *Autauga, Alabama*), and roads represents the type of data.

The tricky part of the road network conversion with US tiger files is that a user may select a geographical area that covers multiple counties. This means that the road network converter needs to access road network data that spans multiple files. Hence, the most important step here is to find the corresponding counties covered by the user-selected area. To this end, we extract a minimum bounding box (UpperLat, UpperLng and LowerLat, LowerLng) for each US county. Then, we create a database table to store the bounding box corresponding to each county ID.

When a traffic generation request is received, we retrieve the road network data from US Tiger Files by first selecting all counties that overlap with the spatial region selected by the user. Then, we load the road network files for all overlapped counties and filter out the nodes and edges based on the user specified area. Finally, we write the qualified nodes and edges in the standard output format.

4.3   Case Study 2: OpenStreetMap

OpenStreetMap is a project that aims at digitizing geographical data for the whole world by providing geographical data that is free to use, distribute, and manipulate. Since it is maintained by volunteers, data in OpenStreetMap is updated frequently, whereas the data quality may not be as good as the data extracted from other commercial/official data sources. OpenStreetMap maintains a very large file, i.e., Planet.osm, to record all spatial objects in the whole world, e.g., road networks, buildings, rivers, etc. Essentially, planet.osm is one large XML (Extensible Markup Language) file that consists of the following four primitive data types:

- *Node*, that represents a spatial point by its latitude and longitude coordinates.
- *Way*, that consists of a sequence of *nodes* which, connected together, form a line or a polygon.
- *Relation*, that specifies the relation between *ways* and *nodes*. For example, two *ways* are connected together.
- *Tags*, that provides description for any of the other data types, *node*, *way*, or *relation*, using a key-value pair.

We carry out the extraction of OSM data in three phases, namely, *parsing*, *indexing* and *querying* where the first two phases are offline and the third phase is online. In the parsing phase, the XML `planet.osm` file is processed to extract node and edge files for the whole world in the format discussed in Section 4.1. All nodes are extracted and stored in one `node` file. Ways are filtered on the fly based on the associated tags and only those associated to the road network are extracted. Each way is stored as a sequence of edges in one `edge` file. The indexing phase preprocesses the `node` and `edge` files (56GB and 98GB, respectively) to speedup range queries that selects a particular area. We initially tried to load them in a PostGIS database with an R-tree index but the loading process did not finish in a reasonable time (we terminated the process when it took more than a week). As an alternative solution, we used SpatialHadoop [30], a MapReduce framework for spatial data, to build an R-tree index in a cluster of 20 machines which took around four hours. The first step is to join the node and edge files to project coordinates of both ends of an edge, and then build an index over the edges based on their minimal bounding rectangles (MBRs). Once the R-tree index is constructed, it is extracted out of the cluster in the format of a *master* and *data* files where the master file stores the region occupied by each data file as an MBR while the data files store the data records. The `node-edge` joined file ended in 10,000 data files with an average file size of 12MB. The final phase is the querying phase in which range queries are processed on the R-tree to extract node and edge files in a particular area based on a user traffic request. First, the master file is examined to select the data files that need to be processed. Next, these data files are processed to generate the node and edge files that are then processed by the selected generator.

4.4   Extensibility with Other Road Network Data Sources

As was pointed out in Figure 1, MNTG is extensible to support new road network data sources. Thanks to the modular design of the *road network converter*, extending MNTG with another road network data source is as simple as providing the contents of two abstract functions. Assume a service provider that has a new road network data source, termed *MyRoadNetwork*. To include this data into MNTG, we provide a template java file, where the service provider needs to fill the contents of: (a) the `ExtractRoadNetwork` abstract function which will basically select the information of the selected area form *MyRoadNetwork*, as discussed in Section 4.1, and (b) the `PrepareStandardOutput` abstract function that outputs the nodes and edges information of the selected area in the standard output format, as discussed in Section 4.1.

It is important to note that filling the contents of the abstract functions in the template does not really have to be by the service provider of the new data source. Instead, third parties or volunteers can provide this functionality. In other words, crowd sourcing can play a major role here in extending MNTG to support various road network data sources. We have started this by providing the abstraction of US Tiger files and Open-StreetMap, as described above in Sections 4.2 and 4.3, respectively. Yet, we call for the efforts of research community and volunteers to support more data sources within MNTG.

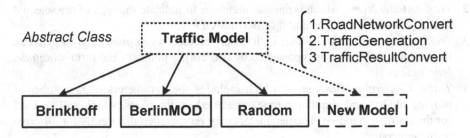

Fig. 2. Traffic Model Class

5   Traffic Processor

The *Traffic Processor* in MNTG is responsible for generating the requested traffic data based on the selected traffic generator. It takes the extracted road network from the *road network converter* component (Section 4) and feeds it to the selected traffic generator. The challenge here is on how to accommodate the various input formats, parameters, and running environments for different traffic generators. To this end, the *Traffic Processor* component in MNTG provides an abstract way to accommodate various traffic generators. It currently includes two famous ones, Brinkhoff and BerlinMOD, however, its abstract design makes it highly extensible to support more traffic generators. In this section, we first discuss the main idea behind the traffic processor. Then, we provide three detailed case studies featuring Brinkhoff, BerlinMOD and random walk traffic generators. Finally, we discuss the extensibility of the traffic processor to support other traffic generators.

5.1   Main Idea

To generate the traffic data based on a particular traffic generator, MNTG basically aims to run the traffic generator as is. However, this is hindered by the fact that different traffic generators: (a) employ different execution methods and (b) require different configuration files and/or parameters. For example, Brinkhoff model is executed with a java jar file, while BerlinMOD runs with a script file. As the purpose of MNTG is to enclose various traffic generators, it builds a wrapper around each traffic generator to make them all look the same when it comes to receiving a traffic generation request and producing the final result.

The main idea is to create an abstract class Traffic Model in MNTG, as depicted in Figure 2. Then, all definitions and functions for each traffic generator has to be incorporated inside this abstract class. In general, there are four key data structures that should be inherited by all traffic generators in MNTG:

1. *Traffic Request ID*, as the traffic request identifier, which is automatically generated for each submitted request. It is used to link the input/output traffic data to the corresponding traffic data requester and to send the traffic result as well as the status notifications to the submitting user.

2. *Traffic Model Name*, which is another identifier to indicate the type of the selected traffic generators, e.g., Brinkhoff or BerlinMOD.
3. *Traffic Generation Area*, which is the user selected rectangular area to generate traffic data in. The area is represented by two corner points of the form <*latitude*, *longitude*>.
4. *Traffic Generation Parameters*, which includes the parameters (e.g., number of moving objects and simulation time), specified by the users, which will be used for the traffic generation. The parameters may be specified differently for different traffic generators.

Additionally, there are three main abstract functions that need to be implemented for each traffic generator to be included in MNTG:

1. `RoadNetworkConvert`: this function converts the standard road network format received from the Road Network Converter (Section 4) to the specific format used by the traffic generator.
2. `TrafficGeneration`: this function produces the traffic data based on the various parameters specified by the user request. MNTG runs the traffic generator with its own scripts or commands.
3. `TrafficResultConvert`: this function converts the output of the traffic generation process into a standard simple output format. The main reason behind this function is that different traffic generators produce different formats of traffic data, while users may want to use the same program to analyze them.

An example of the standard output format of the traffic processor is as follows:

```
OID TS Type      Lat               Lng
0   0  newpoint  44.986362410452   -93.2982044219971
1   0  newpoint  44.998948892253   -93.1812858581543
2   0  newpoint  44.966607085432   -93.2727378845215
0   1  move      45.031348772862   -93.2991374040413
1   1  move      44.953949943361   -93.3676484298706
```

where `OID` is a unique identifier for the moving object. `Lat` and `Lng` are latitude and longitude coordinates that represent the spatial location of the object. `TS` represents the time unit at which object `OID` was at (`Lat`,`Lng`) spatial location. `Type` determines whether the generated point is a new object or an existing object that has just moved to a new location.

5.2 Case Study 1: Brinkhoff Model

Brinkhoff traffic generator is one of most widely used traffic generators [2] (cited 650+ per Google Scholar). The general idea behind Brinkhoff generator is to simulate the object movements from two random locations using the shortest path. To realize Brinkhoff generator inside MNTG, we have implemented the three abstract functions (introduced in Section 5.1), as follows:

1. `RoadNetworkConvert`. In this function, we convert the output of the *Road Net-work Converter* into two binary files based on the descriptions in Brinkhoff documentation[1], and rename them as `request_ID.node` and `request_ID.edge`.
2. `TrafficGeneration`. In this function, we prepare Brinkhoff configuration file, i.e., `property.txt`, where we update the corresponding path for the input files (the two generated binary road network files) and the output path. Then, we assemble the command using the parameters specified by the user in this request. Finally, we make the following external call:

```
java -classpath generator.jar generator2.DefaultDataGenerator RequestID
```

where the only modification for the original generator is that it now takes the `RequestID`, and produces the traffic result accordingly.
3. `TrafficResultConvert`. In this function, we convert the traffic data produced by Brinkhoff generator into our standard output format. An example of the Brinkhoff output is as follows:

```
Type      OID Seq Class TS X        Y        Speed  Next_X Next_Y
newpoint 0  1   0   0  14839.0 10262.0 1093.0 14782 10765
newpoint 1  1   2   0  26319.0 1430.0  922.9  26317 1260
newpoint 2  1   0   0  11443.0 10983.0 1093.0 11431 15703
```

where`Type` determines whether the point is a new object or an existing object. `OID` is a unique identifier for the moving object. `Seq` is the sequence number for the moving object, and `Class` determines the type of the moving object. `TS` represents the time unit during the simulation time. `X` and `Y` show the location of the object, as Brinkhoff employs a different coordinating system that uses the offsets to represent the location. `Speed` is the current moving speed of the object. `Next_X` and `Next_Y` are the locations for the node in the road networks, where the moving object will pass for the next movement.

As a result, we write a program to: (1) extract only the `OID`, `Type`, `TS` from the original output, and (2) convert the `X` and `Y` to be the latitude and longitude coordinates. After that, MNTG is able to generate traffic with any road networks using Brinkhoff model.

5.3   Case Study 2: BerlinMOD

BerlinMOD is another very popular traffic generator [3], where it simulates human movements during the weekdays and weekends. Users can specify their work and home areas in the road networks, then the generator simulates the users movements based on two rules: (1) during the weekdays, a user leaves Home in the morning (at 8 a.m.+ *T1*), drives to Work, stays there until 4 p.m.+ *T2* in the afternoon, and then returns back Home, (2) during the weekends, a user has an 0.4 probability to do an additional trip which may have 1-3 intermediate stops and ends at home.

---

[1] http://iapg.jade-hs.de/personen/brinkhoff/generator/
FormatNetworkFiles.pdf

To run the BerlinMOD traffic generator, a user would need to set up a SECONDO database [4], and uses a set of script instructions to query it. To realize BerlinMOD generator inside MNTG, we have implemented the three abstract functions (introduced in Section 5.1), as follows:

1. `RoadNetworkConvert`. In this function, we read the standard road network files and transform it to the format used in BerlinMOD. Ultimately, we produce a data file named `street.data` with the following information:

```
(OBJECT streets ()
    (rel (tuple ((Vmax real)(geoData line))))
        ((50.0(
            ( -93.276029  45.035464  -93.275936  45.035877 )
            ( -93.275936  45.035877  -93.275764  45.037752 )
```

As a result, BerlinMOD requires us to represent the road segments with a pair of locations bounded by a set of brackets.

2. `TrafficGeneration`. In this function, we prepare the script based on the generation parameters specified by the user, i.e., `BerlinMOD_DataGenerator_RequestID.SEC`, to query the underlying SECONDO database. In MNTG, we prepare a generic script for BerlinMOD and replace its parameters based on the user's request. Then, we run the following command line to execute the BerlinMOD generator:

`SecondoTTYNT -i BerlinMOD_DataGenerator_RequestID.SEC`

3. `TrafficResultConvert`. In this function, we build a program that converts the traffic data produced by BerlinMOD into a standard format. An example of the traffic data generated by BerlinMOD is as follows:

```
Mid Tid Tstart               Tend     Xstart    Ystart  Xend     Yend
1   2   2007-05-26 10:34:40 10:34:42 -93.1767 45.0449 -93.1767 45.0448
1   2   2007-05-26 10:34:42 10:34:44 -93.1767 45.0448 -93.1766 45.0446
1   2   2007-05-26 10:34:44 10:34:46 -93.1766 45.0446 -93.1765 45.0444
```

where `Mid` is the unique identifer of the moving object, `Tid` is the trip identifer, `Tstart` and `Tend` represent the start and end timestamps for the record, while `Xstart`, `Ystart`, `Xend`, and `Yend` are the corresponding locations when the object starts and ends during that time period.

As a result, we write a program to: (1) extract only the `Mid`, `Tid`, `Tstart` from the original output to identify the moving objects, and (2) convert the `Xstart` and `Ystart` to be the latitude and longitude. Then, MNTG is able to generate traffic with any road networks using BerlinMOD.

5.4   Cast Study 3: Random Generator

As a proof of the concept of generation model extensibility, we implement a random generator that generates random walks over the road network. The simplicity of the model used in this generator allows it to handle requests with large areas and hundreds of thousands of objects in a reasonable time. In addition to the road map of the selected area, the random generator takes as input two user parameters, number of moving objects and total simulation time. The generator starts by assigning an initial position for each object by selecting a random node in the road network. At each time step, each object advances one step by selecting a random edge from the edges adjacent to current node. To avoid going back and forth between two nodes, the last visited node is stored and is removed from possible choices of next nodes. If an object cannot find a possible next node (i.e., the only next node is the last visited node) or if the next node falls off the grid, the object is removed from the map and a new object is placed in a new random position. This simulates the event of a vehicle ending its trip and a new vehicle starting a new trip. This also ensures that the total number of objects in the map is fixed at the user defined parameter. Although this generation model does not accurately simulate real life, it is very useful for generating huge traffic data is a very short time which allows end users to test the scalability of their systems.

5.5 Extensibility with Other Traffic Generators

As was pointed out in Figure 1, MNTG is extensible to support various traffic generators. Extending MNTG with another traffic generators is as simple as providing the contents of the three abstract functions, defined in Section 5.1. Assume that a traffic generator developer has invented a new traffic generator, termed *RandomGenerator*. To include the *RandomGenerator* into MNTG, we provide a template java file, where the traffic generator developer needs to fill the contents of the three abstract functions: RoadNetworkConvert, TrafficGeneration, and TrafficResultConvert, as described in Section 5.1.

Similar to extending MNTG for new data sources, filling the contents of the abstract functions of a new traffic generator may be done by third parties or volunteers. Again, crowd sourcing can play a major role here in extending MNTG to support new traffic generators. We envision that MNTG will act as a vehicle that gives existing and forthcoming traffic generators a boost to gain wide users community. Thus, it is to the benefit of the traffic generator developers and to the research community in large to incorporate new data generation tools within MNTG.

6 System Front-End

The system front-end provides a set of tools for users to generate and visualize their requested traffic data. As MNTG is deployed as a web service, the system front-end represents a web interface that users can access over the internet. The web interface is designed for simplicity where users may generate, download, and visualize traffic data with few interactive, rather intuitive, steps.

The system front-end consists of three main modules: (1) *Web interface*, which allows users to easily interact with MNTG in terms of submitting traffic generation requests (Section 6.1), (2) *Email Notifier*, which acknowledges the receipt of the traffic request as well as notifies the user back when the request is finished with links to

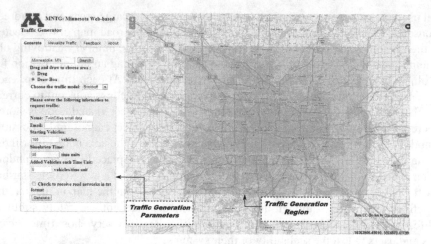

Fig. 3. MNTG Web GUI: Traffic Generation

download and visualize the generated data (Section 6.2), and (3) *Download & Visualization tools*, which allows the user to download its traffic data in a plain text format and/or visualize the generated data in an OpenLayers map interface (Section 6.3).

6.1 Web Interface

Figure 3 depicts MNTG web interface. To generate road network traffic data, a user would perform the following four easy steps:

1. Either drag/zoom the map or write an address in the search field to get the surroundings of the geographical area of interest.
2. Draw a rectangle around the area that you want to generate traffic within. This is done by two left mouse clicks for rectangle corners.
3. From the drop down menu, select the traffic generator that you want to use as either *Brinkhoff* or *BerlinMOD* traffic generators.
4. Click on the *Generate* button, and enter the traffic simulation parameters.

6.2 Email Notifier

MNTG may take a while to process a traffic generation request for two main reasons: (1) Depending on the size of the submitted traffic generation request (e.g., large number of moving objects or large simulation time), the underlying traffic generator (e.g., Brinkhoff or BerlinMOD) may spend significant time in simulating the requested traffic parameters, (2) Even though MNTG employs a multi-threading paradigm where several traffic requests can be processed concurrently, the system may be overloaded when the number of concurrent requests is more than the number of available threads. In that case, MNTG employs a waiting queue, where incoming requests have to be enqueued waiting for a system thread to be available.

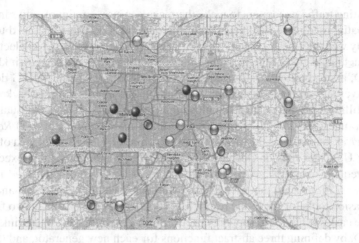

Fig. 4. MNTG Traffic Visualization

To this end, MNTG *email notifier* has two functionalities: (1) when the user submits a traffic generation request, the email notifier sends an email back to the user acknowledging the receipt of the request, and (2) Once MNTG finishes processing the user's traffic generation request, the email notifier sends a notification message that contains two links; the first one is where the user can download the generated traffic data as a text file while the second one is where the user can visualize the generated traffic data on the map.

6.3   Download and Visualization Tools

As mentioned earlier, MNTG produces its output generated data in a uniform text format. Users may download the generated traffic data, including object ids, timestamps, latitude, and longitude coordinates and/or visualize the generated traffic data on the map using MNTG Map interface. MNTG stores the generated traffic data in the unified format mentioned above inside a MySQL database. Traffic visualization in OpenStreetMap is performed using OpenLayers v2.12 API for displaying overlays in HTML. The data is loaded via Javascript into the web page which then creates an overlay for each time stamp of the traffic results. Overlays are an OpenLayers concept and can consist of many different types of data, as shown in Figure 4. In this case, document fragments are created for each object at a time stamp, which is then added to the overlay for that time stamp. When the data is being animated, it simply consists of displaying the corresponding overlay to the time stamp and hiding the remaining overlays. Overlays are used instead of traditional markers because of the speed at which they can be loaded in comparison to the maps built-in markers.

7   Conclusion and Future Work

This paper has proposed Minnesota Traffic Generator (MNTG); an extensible web-based road network traffic generator. MNTG is basically a wrapper that can be built around existing traffic generators to make them easy-to-use, configure, and run for any arbitrary spatial road region. To generate traffic data, MNTG users just need to use its user-friendly web interface to specify an arbitrary spatial area on the map, select a traffic generator method as one of the two most highly used traffic generators, Brinkhoff and BerlinMOD, and submit the traffic generation request to the server. MNTG dedicated server receives and processes the submitted request, and emails the user back once the request is fulfilled. Users can then download their generated data and/or visualize it on MNTG map interface. MNTG is composed of three main components: (1) *Road Network Converter* that extracts the road network information of the spatial area of interest from either US Tiger files or OpenStreetMap, (2) *Traffic Processor* that executes the submitted request using the selected traffic generator on the extracted road network, and (3) *System Front-End*, that includes the web interface, email notifier, and download/visualasion tools for the traffic result. MNTG is highly extensible in two frontiers: (1) It can be easily extended to support various traffic generators, beyond Brinkhoff and BerlinMOD, by defining three abstract functions for each new generator, and (2) It can be easily extended to support various road network sources, beyond US Tiger files and OpenStreetMap, by defining two abstract functions for each new data source.

MNTG is still an undergoing project in data management lab at the University of Minnesota. Its first release is already available for a public use at http://mntg.cs.umn.edu, where it has received and fulfilled over 1000 traffic generation requests since its release. Future work of MNTG includes: (a) supporting more traffic generators beyond the two we have for now, Brinkhoff and BerlinMOD, and (b) supporting more new data sources beyond US Tiger files and OpenStreetMap. A distinguishing feature in MNTG is that its future plans can be fulfilled via crowd sourcing, where interested developers and researchers world wide can enrich the infrastructure of MNTG by their contributions of new traffic generators and data sources. Plug-in functions are available for that purpose. With the increase of volume for traffic generation requests, we plan to move our server to a more powerful server machine with GPU cards to support large-volume traffic visualization.

References

1. Zheng, Y., Chen, Y., Xie, X., Ma, W.-Y.: GeoLife2.0: A Location-Based Social Networking Service. In: MDM, pp. 357–358 (2009)
2. Brinkhoff, T.: A Framework for Generating Network-based Moving Objects. GeoInformatica 6(2), 153–180 (2002)
3. Düntgen, C., Behr, T., Güting, R.H.: BerlinMOD: a Benchmark for Moving Object Databases. VLDB Journal 18(6), 1335–1368 (2009)
4. Güting, R.H., Behr, T., Düntgen, C.: Secondo: A platform for moving objects database research and for publishing and integrating research implementations. IEEE Data Engineering Bulletin 33(2), 56–63 (2010)
5. OpenStreetMaps, http://www.openstreetmap.org/
6. US TIGER LINES, http://www.census.gov/geo/maps-data/data/tiger-line.html
7. Pfoser, D., Theodoridis, Y.: Generating Semantics-based Trajectories of Moving Objects. Computers, Environment and Urban Systems 27(3), 243–263 (2003)

8. Saglio, J.-M., Moreira, J.: Oporto: A realistic scenario generator for moving objects. GeoInformatica 5(1), 71–93 (2001)
9. Theodoridis, Y., Silva, J.R.O., Nascimento, M.A.: On the Generation of Spatiotemporal Datasets. In: Güting, R.H., Papadias, D., Lochovsky, F.H. (eds.) SSD 1999. LNCS, vol. 1651, pp. 147–164. Springer, Heidelberg (1999)
10. Tzouramanis, T., Vassilakopoulos, M., Manolopoulos, Y.: On the Generation of Time-Evolving Regional Data. GeoInformatica 6(3), 207–231 (2002)
11. Gidófalvi, G., Pedersen, T.B.: ST-ACTS: A Spatio-temporal Activity Simulator. In: GIS, pp. 155–162 (2006)
12. Hu, H., Lee, D.-L.: GAMMA: A Framework for Moving Object Simulation. In: Medeiros, C.B., Egenhofer, M., Bertino, E. (eds.) SSTD 2005. LNCS, vol. 3633, pp. 37–54. Springer, Heidelberg (2005)
13. Krajzewicz, D., Hertkorn, G., Rössel, C., Wagner, P.: SUMO (Simulation of Urban MObility): An Open-Source Traffic Simulation. In: Proceedings of the 4th Middle East Symposium on Simulation and Modelling, pp. 183–187 (2002)
14. SMARTEST: Simulation Modelling Applied to Road Transport European Scheme Tests, http://www.its.leeds.ac.uk/projects/smartest/
15. Xu, J., Güting, R.H.: MWGen: A Mini World Generator. In: MDM, pp. 258–267 (2012)
16. Tsai, H.-P., Yang, D.-N., Chen, M.-S.: Mining Group Movement Patterns for Tracking Moving Objects Efficiently. IEEE TKDE 23(2), 266–281 (2011)
17. Jeung, H., Liu, Q., Shen, H.T., Zhou, X.: A Hybrid Prediction Model for Moving Objects. In: ICDE, pp. 70–79 (2008)
18. Mokbel, M.F., Xiong, X., Aref, W.G.: SINA: Scalable Incremental Processing of Continuous Queries in Spatio-temporal Databases. In: SIGMOD, pp. 623–634 (2004)
19. Wu, W., Guo, W., Tan, K.-L.: Distributed Processing of Moving K-Nearest-Neighbor Query on Moving Objects. In: ICDE, pp. 1116–1125 (2007)
20. Mokbel, M.F., Aref, W.G.: SOLE: Scalable On-line Execution of Continuous Queries on Spatio-temporal Data Sreams. VLDB Journal 17(5), 971–995 (2008)
21. Chung, B.S.E., Lee, W.-C., Chen, A.L.P.: Processing Probabilistic Spatio-temporal Range Queries Over Moving Objects with Uncertainty. In: EDBT, pp. 60–71 (2009)
22. Hu, H., Xu, J., Lee, D.L.: PAM: An Efficient and Privacy-Aware Monitoring Framework for Continuously Moving Objects. IEEE TKDE 22(3), 404–419 (2010)
23. Chen, S., Jensen, C.S., Lin, D.: A Benchmark for Evaluating Moving Object Indexes. VLDB Journal 1(2), 1574–1585 (2008)
24. Laender, A.H.F., Borges, K.A.V., Carvalho, J.C.P., Medeiros, C.B., da Silva, A.S., Davis, C.A.: Integrating Web Data and Geographic Knowledge into Spatial Databases. In: Spatial Databases, pp. 23–47 (2005)
25. Shen, C., Huang, Y., Powell, J.W.: The Design of a Benchmark for Geo-stream Management Systems. In: GIS, pp. 409–412 (2011)
26. Tzouramanis, T.: Benchmarking and Data Generation in Moving Objects Databases. In: Encyclopedia of Database Technologies and Applications, pp. 23–28 (2005)
27. Xu, J., Güting, R.H.: GMOBench: A Benchmark for Generic Moving Objects. In: GIS, pp. 410–413 (2012)
28. Güting, R.H., de Almeida, V.T., Ding, Z.: Modeling and Querying Moving Objects in Networks. VLDB Journal 15(2), 165–190 (2006)
29. Vazirgiannis, M., Wolfson, O.: A Spatiotemporal Model and Language for Moving Objects on Road Networks. In: Jensen, C.S., Schneider, M., Seeger, B., Tsotras, V.J. (eds.) SSTD 2001. LNCS, vol. 2121, pp. 20–35. Springer, Heidelberg (2001)
30. Eldawy, A., Mokbel, M.F.: A Demonstration of SpatialHadoop: An Efficient MapReduce Framework for Spatial Data. In: VLDB (2013)

Capacity-Constrained Network-Voronoi Diagram: A Summary of Results

KwangSoo Yang, Apurv Hirsh Shekhar, Dev Oliver, and Shashi Shekhar

Department of Computer Science, University of Minnesota, Minneapolis, MN 55455
ksyang@cs.umn.edu, shekh020@umn.edu, {oliver,shekhar}@cs.umn.edu

Abstract. Given a graph and a set of service centers, a Capacity Constrained Network-Voronoi Diagram (CCNVD) partitions the graph into a set of contiguous service areas that meet service center capacities and minimize the sum of the distances (min-sum) from graph-nodes to allotted service centers. The CCNVD problem is important for critical societal applications such as assigning evacuees to shelters and assigning patients to hospitals. This problem is NP-hard; it is computationally challenging because of the large size of the transportation network and the constraint that Service Areas (SAs) must be contiguous in the graph to simplify communication of allotments. Previous work has focused on honoring either service center capacity constraints (e.g., min-cost flow) or service area contiguity (e.g., Network Voronoi Diagrams), but not both. We propose a novel Pressure Equalizer (PE) approach for CCNVD to meet the capacity constraints of service centers while maintaining the contiguity of service areas. Experiments and a case study using post-hurricane Sandy scenarios demonstrate that the proposed algorithm has comparable solution quality to min-cost flow in terms of min-sum; furthermore it creates contiguous service areas, and significantly reduces computational cost.

Keywords: Capacity Constrained Network Voronoi Diagram, Pressure Equalization, Spatial Network Partitioning.

1 Introduction

Given a graph and a set of service centers (e.g., gas stations) with capacity constraints (e.g., amount of gasoline, size of parking lot, etc.), a Capacity Constrained Network-Voronoi Diagram (CCNVD) partitions the graph into a set of contiguous service areas (SAs) that honor service center capacities and minimize the sum of the distances (min-sum) from graph-nodes to allotted service centers. Figure 1(a) shows an example input of CCNVD consisting of a graph with 15 graph-nodes $(A, B, \ldots, O)$ and three service centers $(X, Y, \text{ and } Z)$ with capacities of 5 each. Figure 1(b) shows an example output of CCNVD where the graph is partitioned such that 5 graph-nodes are allotted to each service center, as shown by the distinct colors.

M.A. Nascimento et al. (Eds.): SSTD 2013, LNCS 8098, pp. 56–73, 2013.
© Springer-Verlag Berlin Heidelberg 2013

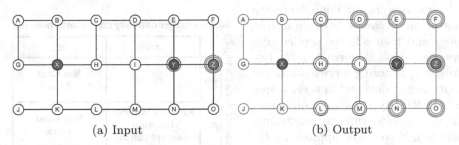

(a) Input   (b) Output

Fig. 1. Example of the Input and Output of CCNVD (Colors show service center allotment)

The CCNVD problem is important for critical applications such as assigning consumers to gas stations in the aftermath of a disaster, assigning evacuees to shelters, assigning patients to hospitals, and assigning students to school districts. For example, during hurricane Sandy, New Jersey residents had to wait for hours to fill up car tanks and containers for home generators [1]. Figure 2(a) shows vehicles waiting in line for fuel at a gas station and Figure 2(b) shows people waiting at a gas station, both in the New York-New Jersey area. Such fuel shortages serve as a reminder of the importance of resource allotment amid natural or man-made disasters such as floods, hurricanes, tsunamis, fires, terrorist acts, and industrial accidents.

(a) Vehicles wait in line for fuel at gas station in the New York-New Jersey area, on Nov. 1, 2012.
(Courtesy: www.bloomberg.com)

(b) Waiting line at a gas station on Nov. 1, 2012 in the New York-New Jersey area. (Courtesy: Andrew Burton/Getty)

Fig. 2. Long lines at gas station after Hurricane Sandy

CCNVD may also be used for shelter allocation where contiguous service areas reduce movement conflicts (which raise risk of congestion, stampede, etc.) across people heading to different shelters.

The CCNVD problem is NP-Hard and its proof is provided in Section 2.1. Intuitively, this problem is computationally challenging because of the large size of the transportation network and the constraint that service areas must be contiguous in the graph to simplify communication of allotments.

Previous work on minimizing the sum of the distances between graph-nodes and their allotted service centers can be categorized into two groups: 1) honoring service center capacity constraints and 2) service area (SA) contiguity. Prior work on honoring service center capacity constraints include min-cost flow approaches [2–18]. However, such approaches do not always preserve service area contiguity. Previous work on SA contiguity include the creation of Network Voronoi diagrams (NVDs) by assigning each node to the nearest service

Service center capacity constraints honored

		no	yes
Service Area Contiguity	no		Min-Cost Flow
	yes	Network Voronoi Diagram (nearest center)	Proposed Work

Fig. 3. Approaches to minimizing the sum of the distances between nodes and their allotted service centers

center [19–22]. However, previous NVDs have not been designed to account for capacity constraints of service centers. By contrast, this paper proposes a novel approach for creating Capacity Constrained Network Voronoi Diagrams (CCNVD) that honors both the capacity constraints of service centers and SA contiguity while reducing the sum of the distances from graph-nodes to their allotted service centers.

Our Contributions: In this paper, we propose a novel algorithm for creating a CCNVD based on the idea of pressure equalization (PE). Pressure equalization follows three main steps: 1) an initial solution that assigns graph-nodes to the nearest service centers (e.g., using traditional Network Voronoi Diagrams); 2) construction of a new data structure called the PE-graph that has PE-nodes, which represent service centers with a pressure attribute and PE-edges, which connect adjacent service areas (PE-nodes); 3) re-allotment of graph-nodes from overloaded (excess) service centers to underloaded (deficit) service centers. Specifically, our contributions are as follows:

- We prove that the CCNVD problem is NP-hard.
- We propose a Pressure Equalizer (PE) algorithm that creates CCNVD.
- We provide a cost model for our proposed approach.
- We experimentally evaluate our proposed algorithm using post-hurricane Sandy scenarios. Experimental results and a case study demonstrate that the proposed algorithm has comparable solution quality (in terms of min-sum) to min-cost flow, creates contiguous service areas, and significantly reduces the computational cost.

Scope and Outline: In constructing our novel algorithm for CCNVD, we assume undirected edges, unit demand at each non-service center node (graph-node), and no edge-capacity constraints. Directed edges, non-uniform node demand, and honoring capacity constraints at edges in the given transportation network are beyond the scope of this paper and may be addressed in future work. The rest of the paper is organized as follows: Section 2 provides the problem definition. Section 3 presents our proposed approach. In Section 4, we give a cost model of our proposed approach. Section 5 describes the experiment design and presents the experimental observations and results. Section 6 reports a case study using post-hurricane Sandy scenarios. Finally, Section 7 concludes the paper.

2 Problem Definition

In our problem formulation, a transportation network is represented and analyzed as an undirected graph composed of nodes and edges. Each node represents a spatial location in geographic space (e.g., road intersections). Each edge represents a connection between two nodes and has a travel time. Service centers have a capacity (supply). The $CCNVD(N, E, S, C, D)$ problem may be formalized as follows:

Input: A transportation network G with
- a set of graph-nodes N and a set of edges E,
- a set of service centers $S \subset N$,
- a set of positive integer capacities of service centers $C : S \rightarrow Z^+$, and
- a set of non-negative real distances of edges $D : E \rightarrow R_0^+$
Output: Capacity Constrained Network Voronoi Diagram (CCNVD)
Objective:
- Min-Sum: Minimize the sum of the distances from graph-nodes to their allotted service centers.
Constraints:
- Adequate total capacity across all service centers to accommodate all graph-nodes.
- Service Area (SA) contiguity: If $SA(u) = s$, then there is a path $p(u, s)$, such that $SA(u) = SA(v) = s$ for all nodes v in $p(u, s)$, where $SA(n)$ is the allotted service center for a graph-node n
- G is a k-node-connected graph, where $k = |S|$ to ensure the existence of a solution [23–25].

Figure 4(a) illustrates the input with a transportation network (15 graph-nodes $(A, B, \ldots, O)$ and three service centers $(X, Y,$ and $Z)$). Every edge is associated with a distance (e.g., travel time), as indicated by the number displayed above it. For simplicity, every graph-node has one unit of demand. Every service center has a capacity to serve 5 units of graph-nodes. Figure 4(b) shows a Network Voronoi Diagram (NVD), allotting every graph-node to the nearest (e.g., shortest path) service center. NVD assigns 8 graph-nodes (color=blue) to service center X, 5 graph-nodes (color=red) to service center Y, and 2 graph-nodes (color=green) to service center Z. The dotted lines represent the boundary edge between two adjacent SAs. Although NVD allots every graph-node to the nearest service center, it does not account for load balance, which may lead to congestion and delay at service center X while service center Z may have few graph-nodes. Figure 4(c) shows one possible example of the min-cost flow approach that minimizes the sum of the distances from graph-nodes to their allotted service centers. Although it achieves its goal (min-sum), service areas for Y and Z violate SA contiguity. Figure 4(d) shows an example of the proposed Capacity Constrained Network Voronoi Diagram (CCNVD). As can be seen, the load is balanced, 5 nodes are allotted to every service center as shown by distinct colors, and service areas are contiguous for all service centers.

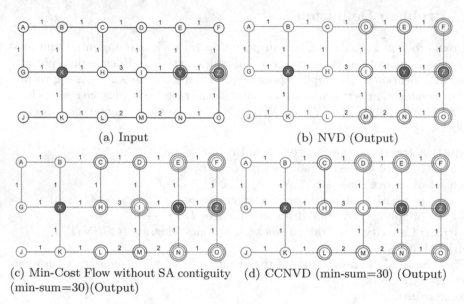

(a) Input   (b) NVD (Output)

(c) Min-Cost Flow without SA contiguity   (d) CCNVD (min-sum=30) (Output)
(min-sum=30)(Output)

Fig. 4. Example of the Input and Output of NVD, Min-Cost Flow, and CCNVD (Colors show service center allotment)

Solution Existence: The CCNVD problem is related to the connected k-partition problem [23–27], which is known to be NP-hard. In addition, it is also known that a solution exists for k-connected graphs, formally stated as follows: Let $G = (N, E)$ be a k-node-connected graph. Let $s_1, s_2, \ldots, s_k \in N$, $c = |N|$, and let $c_1, c_2, \ldots, c_k$ be positive integers such that $c_1 + c_2 + \ldots + c_k = c$. Then there exists a k-partition of $N(G)$ such that this partition separates the nodes $s_1, s_2, \ldots, s_k$, and the partition p_i containing s_i is a connected sub-graph consisting of c_i vertices, for $i = 1, 2, \ldots, k$ [23–25].

2.1   Problem Hardness

That CCNVD is NP-hard follows from a well-known result about the NP-hardness of the connected k-partition problem [26–28] that partitions a graph into k connected sub-graphs where $k \geq 3$.

Theorem 1. *The CCNVD problem is NP-hard.*

Proof. Given a graph $G = (N, E)$, service center nodes $s_1, s_2, \ldots, s_k \in N$, and positive integer capacities for service centers $c_1, c_2, \ldots, c_k$, where $c_1 + c_2 + \ldots + c_k =$

$|N|$ (e.g., equal sized sub-graphs), the connected-k-partition ($k-CP(N, E, S, C)$) problem separates the service center nodes $s_1, s_2, \ldots, s_k$ and the partition p_i containing s_i is a connected sub-graph consisting of c_i nodes for $i = 1, 2, \ldots, k$. It has been proved that a connected-k-partition of N is a NP-hard problem [26, 27]. The CCNVD problem clearly belongs to NP since, given an instance of CCNVD and a maximum bound T, we can take a set of connected sub-graphs such that the sum of the distances from graph-nodes (N) to their allotted service centers (S) is lower than T as a valid certification. Let $A = (N, E, S, C)$ be an instance of a connected-k-partition problem, where N is a set of nodes, E is a set of edges, S is a set of service center nodes, and C is a set of capacities for service centers. Let $B = (N, E, S, C, D, T)$ be an instance of the CCNVD problem, where D is a set of distances of E and T is a maximum bound of the sum of the distances from graph-nodes (N) to their allotted service centers (S). Then it is easy to show that a connected-k-partition is a special case of CCNVD, where k is the number of service centers S, $d \in D$ has zero value of distance, and T is unbounded. Since A is constructed from B in polynomial time, the proof is complete.

3   Proposed Approach for CCNVD

In this section, we introduce our Pressure Equalizer approach to the CCNVD problem.

3.1   Pressure Equalizer Algorithm

The Pressure Equalizer (PE) algorithm starts with the partitions of a Network Voronoi Diagram (NVD) and iteratively adjusts the shelter allotments until capacity constraints are met for all service centers. A core idea in PE is the Pressure Equalization Graph (PE-Graph), where pressure for a shelter s refers to the difference between the capacity of s and the number of nodes allotted to s. Positive values of pressure indicate overload and negative values indicate slack or available capacity. The PE-nodes of PE-Graph are service centers S in the transportation network. The PE-nodes are of three types: excess, deficit, and balanced. Let $capacity(s)$ be the capacity of a PE-node s and $allotment(s)$ be the number of graph-nodes allotted to s. If $allotment(s) > capacity(s)$, we refer to s as an excess PE-node and $allotment(s) - capacity(s)$ as the excess of the PE-node s. On the other hand, if $allotment(s) < capacity(s)$, we refer to s as a deficit PE-node and $capacity(s) - allotment(s)$ as the deficit of the PE-node s. We refer to a PE-node s with $allotment(s) = capacity(s)$ as balanced. A PE-edge from PE-node $s1$ to PE-node $s2$ is inserted if any allotted graph-node on $s1$ is connected to any allotted graph-node on $s2$. The collection of allotted graph-nodes for a PE-node s represents the service area (SA) for s. We refer to $SA(s)$ as the service area for s.

Figure 5(b) shows the PE-Graph for the Network Voronoi Diagram (NVD) of Figure 5(a) (reproduced in Figure 4(a)). It has three PE-nodes for service centers X, Y, and Z. PE-node X has an excess of 3 and PE-node Z has a deficit

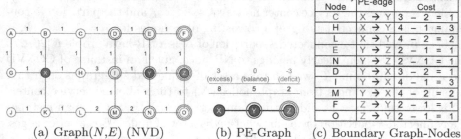

Boundary Node	PE-edge	Re-allotment Cost				
C	X → Y	3	–	2	=	1
H	X → Y	4	–	1	=	3
L	X → Y	4	–	2	=	2
E	Y → Z	2	–	1	=	1
N	Y → Z	2	–	1	=	1
D	Y → X	3	–	2	=	1
I	Y → X	4	–	1	=	3
M	Y → X	4	–	2	=	2
F	Z → Y	2	–	1	=	1
O	Z → Y	2	–	1	=	1

(a) Graph(N,E) (NVD)   (b) PE-Graph   (c) Boundary Graph-Nodes

Fig. 5. PE algorithm: Iteration 1 (Colors show service center allotment)

of 3. There are two PE-edges showing that the service area of Y is adjacent to the service areas of X and Z.

The PE algorithm tries to satisfy capacity constraints for every service center, maintain service area contiguity constraints, and reduce the sum of the distances from graph-nodes to their allotted service centers (PE-nodes). At each step, the algorithm re-allots a graph-node from an excess PE-node to fulfill the capacity constraint. The effect of re-allotting a graph-node n from $s1$ to $s2$ on an objective function (min-sum) can be defined using the following cost function:

$$re{-}allotmentCost(n, s1, s2) = shortestDist(n, s2) - shortestDist(n, s1), \quad (1)$$

where $shortestDist(n, s)$ is the length of the shortest path from n to s.

Given a Network Voronoi Diagram (NVD) and two service centers $s1$ and $s2$, the cost for $re{-}allotting(n \in SA(s1), s1, s2)$ can be minimized if n is a boundary graph-node of $SA(s1)$ and is connected to a boundary node of $SA(s2)$. A key idea behind the PE algorithm is to first choose the best boundary graph-node that minimizes the cost of re-allotment and then re-allot this graph-node to fulfill capacity constraints. Figure 5(c) shows boundary nodes for the NVD of Figure 5(a). In this example, the best boundary nodes in terms of minimizing the re-allotment cost are node C for re-allotment from X to Y, node E (or N) for re-allotment from Y to Z, node D for re-allotment from Y to X, and node F (or O) for re-allotment from Z to Y.

Challenges during this re-allotment include maintaining contiguity for SAs. The PE algorithm uses the PE-path that traverses from one excess PE-node, s_i to one deficit PE-node, s_j, on the PE-Graph and re-allots all the best boundary graph-nodes on the PE-path ($s_i \rightarrow s_j$). This approach smoothly expands (or shrinks) SAs while minimizing the violation of SA contiguity.

It also terminates in at most $O(N)$ iterations, where N is the number of graph-nodes since the sum of the excess at all service nodes is bounded by N. In this example, there is one PE-path ($X \rightarrow Y \rightarrow Z$) to traverse from the excess PE-node (X) to the deficit PE-node (Z). Thus, the algorithm may re-allot node C from X to Y and node E from Y to Z to reduce $excess(X)$ by 1 in the first iteration, This process may be repeated two more times to meet capacity constraints by moving node (L,H) from X to Y and node (N,D) from Y to Z.

A large sized network may have many possible PE-paths. The best way to minimize the cost for re-allotment is to find the PE-path that has minimum re-allotment cost across all pairs of excess and deficit PE-nodes. We refer to this as the best PE-path for the current iteration. A naive approach may use one shortest path (or cost) algorithm for each pair of excess and deficit PE-nodes and choose the lowest re-allotment cost pair and PE-path among all pairs. However, this becomes a bottleneck operation due to the need to apply multiple shortest path algorithms in every iteration. We can further reduce the computational cost to a single invocation of a shortest path algorithm by introducing a super-source and a super-sink node. We first connect the super-source node with all excess PE-nodes and the super-sink node with all deficit PE-nodes. These new connections become PE-edges with a re-allotment cost of 0. Finally, one shortest path algorithm on this transformed PE-Graph can identify the pair of excess and deficit PE-nodes with the lowest re-allotment cost.

Algorithm 1. Pressure Equalizer (PE) Algorithm (Pseudo-code)

Inputs:
 - A transportation network ($Graph(N, E)$) with a set of graph-nodes N and edges E.
 - A set of PE-nodes (service centers) $S \subset N$ with their capacity C
 - Every edge has a distance $d(e)$

Outputs: Capacity Constrained Network Voronoi Diagram ($CCNVD$)

Steps:
1: Compute all shortest distances from N to S.
2: Create Network Voronoi Diagram (NVD) and allot $n \in N$ to the nearest service
   center $s \in S$.
3: **while** Any PE-node $s \in S$ has excess graph-nodes **do**
4:     Create $PE-Graph(S, E_{pe})$ where PE-edge $e_{pe} \in E_{pe}$ connects two adjacent SAs.
5:     Find all boundary graph-nodes $N_{bdy} \subset N$ and compute re-allotment cost for
       N_{bdy}.
6:     Find the best boundary graph-nodes $N_{best-bdy} \subset N_{bdy}$.
7:     Group all excess PE-nodes $S_{ex} \subset S$ with a super-source src_{ex} and group all
       deficit PE-nodes $S_{df} \subset S$ with a super-sink $sink_{df}$.
8:     Find the best PE-path p in terms of minimizing the sum of re-allotment costs
       from src_{ex} to $sink_{df}$ without fragmenting any service area (e.g., path and exit).
9:     Re-allot the best boundary graph-nodes ($n_{best-bdy}$) on the best path p.
10: **end while**
11: return CCNVD. i.e, final allotment of graph-nodes to their service centers.

Algorithm 1 presents the pseudo-code for PE. First, PE initializes the CC-NVD with the given NVD (lines 1-2). It then creates a PE-Graph and finds all boundary graph nodes, as well as the best boundary graph nodes (lines 4-6). After that, it groups excess PE-nodes using a super-source and deficit PE-nodes using a super-sink (line 7). Next it searches the PE-graph and finds the best PE-path (line 8). PE then re-allots the best boundary graph-nodes on the PE-path (line 9). This process continues until the allotment is in line with the capacity of the service centers (line 3). Finally, the updated CCNVD with balanced service centers is returned (line 11).

Figures 5–8 show the execution of the proposed PE algorithm on the input network given in Figure 4(a). PE starts with NVD as an initial solution (Figure 5(a)) and creates a PE-Graph (Figure 5(b)). In this example, the service center with an excess is X and the service center with a deficit is Z. PE finds a PE-path to traverse from X to Z (e.g., $X \rightarrow Y \rightarrow Z$) as well as the best boundary graph-nodes adjacent to other SAs (e.g., C and E).

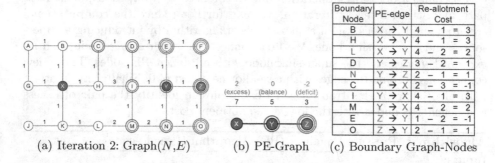

Boundary Node	PE-edge	Re-allotment Cost
B	X → Y	4 – 1 = 3
H	X → Y	4 – 1 = 3
L	X → Y	4 – 2 = 2
D	Y → Z	3 – 2 = 1
N	Y → Z	2 – 1 = 1
C	Y → X	2 – 3 = -1
I	Y → X	4 – 1 = 3
M	Y → X	4 – 2 = 2
E	Z → Y	1 – 2 = -1
O	Z → Y	2 – 1 = 1

(a) Iteration 2: Graph(N,E)   (b) PE-Graph   (c) Boundary Graph-Nodes

Fig. 6. PE algorithm: Iteration 2 (Colors show service center allotment)

Next, PE re-allots the best boundary graph-nodes (C and E) to their adjacent SAs (Figure 6(a)) and updates the PE-Graph (Figure 6(b)). After three iterations, PE achieves a balanced allotment (Figure 8(b)) by re-allotting nodes L and H from X to Y and nodes N and D from Y to Z and the algorithm terminates. Figure 8(a) shows the resulting Capacity Constrained Network Voronoi Diagram.

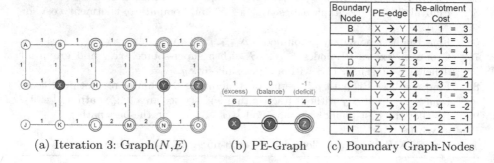

Boundary Node	PE-edge	Re-allotment Cost
B	X → Y	4 – 1 = 3
H	X → Y	4 – 1 = 3
K	X → Y	5 – 1 = 4
D	Y → Z	3 – 2 = 1
M	Y → Z	4 – 2 = 2
C	Y → X	2 – 3 = -1
I	Y → X	4 – 1 = 3
L	Y → X	2 – 4 = -2
E	Z → Y	1 – 2 = -1
N	Z → Y	1 – 2 = -1

(a) Iteration 3: Graph(N,E)   (b) PE-Graph   (c) Boundary Graph-Nodes

Fig. 7. PE algorithm: Iteration 3 (Colors show service center allotment)

4   Analysis of the PE Algorithm

The following lemma proves the correctness of the Pressure Equalizer algorithm.

Lemma 1. *The PE algorithm terminates after at most $n/2$ iterations, where n is the number of graph-nodes*

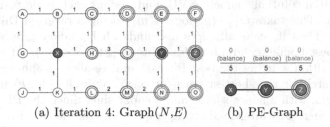

(a) Iteration 4: Graph(N,E)   (b) PE-Graph

Fig. 8. PE algorithm: Iteration 4 (Colors show service center allotment)

Proof. Each iteration reduces the number of allotted graph-nodes for excess service centers by one and increases the number of allotted graph-nodes for deficit service centers by one. The maximum possible number of allotted graph-nodes for excess service centers is n. Therefore, the maximum iteration is at most $n/2$.

Lemma 2. *Given a graph node v and two service centers $s1$ and $s2$, the cost for re-allotting v from $s1$ to $s2$ is defined by $re-allotmentCost(v, s1, s2) = shortestDist(v, s2) - shortestDist(v, s1)$, where $shortestDist(v, s)$ is the length of the shortest path from v to s.*

Proof. Let the service center for v be $s1$. After re-allotting v from $s1$ to $s2$, v is removed from service center $s1$ and added into service center $s2$. Therefore, the increased distance (cost) is defined by $shortestDist(v, s2) - shortestDist(v, s1)$.

Lemma 3. *Given a Network Voronoi Diagram (NVD) and two service centers $s1$ and $s2$, the cost for $re-allotting(v \in NVR(s1), s1, s2)$ can be minimized if v is a boundary graph-node of $SA(s1)$.*

Proof. By the definition of NVD, the farthest graph-node in $SA(s1)$ from a service center $s1$ should be the boundary graph-node. When re-allotting a graph node v from $s1$ to $s2$, the cost can be minimized when $shortestDist(v, s1)$ is maximum.

Lemma 4. *The normal termination of PE meets service center capacity constraints and preserves contiguity of service area.*

Proof. At termination, there are no excess or deficit PE-nodes on the PE-Graph, by Lemma 1. Since each re-allotment satisfies SA contiguity constraints, PE meets service center capacity constraints and preserves contiguity of service area at termination.

4.1   Algebraic Cost Model of the PE Algorithm

We present cost model for the PE algorithm to estimate its computational cost. Assume that n is the number of graph-nodes, k is the number of service centers, m is the number of edges, and n_{maxdeg} is the maximum node degree. First, PE creates an initial solution, namely NVD, and assigns each node to its nearest service center. This takes $O(k \cdot (n \cdot \log n + m))$ using a reversed Dijkstra's algorithm [29]. Then it scans all edges and finds all boundary graph-nodes, as well as the best boundary graph-nodes. This step takes $O(m)$. At each iteration, the algorithm searches the best PE-path and re-allots graph-nodes on the path. The search for the shortest path (or cost) takes $O(k \cdot \log k)$ using Dijkstra's algorithm with a super source and a super sink. Since the PE algorithm preserves SA contiguity after re-allotting graph-nodes on the best PE-path, each relaxation for Dijkstra's algorithm needs to check the SA contiguity by examining all edges. Therefore, the search for the best PE-path takes $O(m \cdot k \cdot \log k)$. The re-allotment takes at most $O(k)$ since the length of a path is bounded by the number of service centers. In the next iteration, we do not need to scan all edges to find all boundary graph-nodes. Rather, we can simply update the changed boundary graph-nodes and the best boundary graph-nodes. Since the number of re-allotted graph-nodes is bounded by $O(k)$ and the maximum number of incidents is bounded by n_{maxdeg}, it takes $O(k \cdot n_{maxdeg})$. The number of iterations is bounded by $n/2$. Therefore the cost model of the PE algorithm is $O(k \cdot (n \log n + m) + m + n \cdot (m \cdot k \cdot \log k + k \cdot n_{maxdeg}))$.

Table 1. Algebraic Comparison Computational Cost

Algorithm	Computational Cost	Capacity Constraint	Minimize $odDistances$	NVR Contiguity
NVD [7, 20]	$k \cdot (n \cdot \log n + m)$	No	Yes	No
Min-Cost Flow [7, 9]	$n \cdot (n \log n + k \cdot n)$ $+ k \cdot (n \cdot \log n + m)$	Yes	Yes	No
PE	$n \cdot (m \cdot k \cdot \log k + k \cdot n_{maxdeg})$ $m + k \cdot (n \cdot \log n + m)$	Yes	Yes	Yes

5   Experimental Evaluation

In this section, we present the experiment design and an analysis of the experiment results.

5.1   Experiment Layout

Figure 9 shows our experimental setup. For the transportation network, we used a Brooklyn, NY road map consisting of $7,450$ nodes and $22,377$ edges, and a Monmouth County, NJ road map consisting of $23,014$ nodes and $61,196$ edges, taken from OpenStreetMap [30]. For the service centers, we used the locations of gas stations and created a Capacity Constrained Network Voronoi Diagram (CCNVD). For simplicity, we assumed that all gas stations have the same capacity and that the gas stations together can serve all people allowed during time intervals of interest. We tested two different approaches: 1) a min-cost flow approach and 2) Pressure Equalizer (PE). The algorithms were implemented in

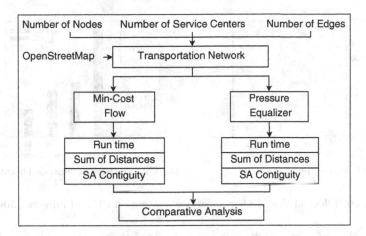

Fig. 9. Experiment Layout

Java 1.7 with a 1 GB memory run-time environment. All experiments were performed on an Intel Core i7-2670QM CPU machine running MS Windows 7 with 8 GB of RAM.

5.2   Experiment Results and Analysis

We experimentally evaluated min-cost flow and PE by comparing the impact on performance of 1) number of service centers and 2) size of the network (i.e., number of graph-nodes). Performance measurements were execution time and the sum of the distances from graph-nodes to their allotted service centers.

Effect of the Number of Service Centers: We used a Brooklyn, NY road map consisting of 7, 450 nodes and 22, 377 edges, and varied the number of service centers. The locations of service centers were chosen based on size favoring larger capacities. Figure 10(a) gives the execution time. As can be seen, PE outperforms the min-cost flow approach. As the number of service centers increases, the performance gap also increases. This is because the effect of the number of service centers in the cost model for the min-cost approach is higher than for the PE algorithm. When comparing the sum of the distances from graph-nodes to their allotted service centers (Figure 10(b)), we see that two algorithms performs almost identically, although the min-cost flow approach performs slighter better. As the number of service centers increases, the sum of the distances from the graph-nodes to their allotted service centers decreases.

Effect of Network Size: We used a Monmouth, NJ road map and created three road networks with $7.4K$, $14.3K$, and $23.0K$ nodes, respectively. We fixed the number of service centers to 10 and incrementally increased the number of graph-nodes from 7, 402 to 23, 014. Service center locations were chosen randomly and execution time were averaged over 50 test runs for each road networks. Figure 11(a) shows that the PE algorithm outperforms the min-cost flow approach.

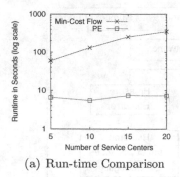
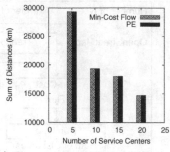

(a) Run-time Comparison   (b) Comparison of Sum of Distances

Fig. 10. Effect of the number of service centers on PE and min-cost flow

As the number of graph-nodes increases, so does the performance gap. This is because the effect of the number of graph-nodes in the cost model for the min-cost approach is higher than that of the PE algorithm. Figure 11(b) shows that the min-cost flow approach performs slightly better than PE. As the number of graph-nodes increases, the sum of the distances from graph-nodes to their allotted service centers also increases. The results of the experiments were that PE was faster than min-cost flow and had similar, albeit slightly lower performance in terms of sum of distances.

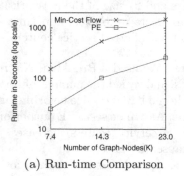
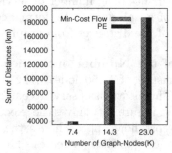

(a) Run-time Comparison   (b) Comparison of Sum of Distances

Fig. 11. Effect of the number of graph-nodes

6   Case Study with Brooklyn, NY Road Network

In our case study, we imagined a scenario in which victims of hurricane Sandy who needed to fuel up their cars were guided to the best gas station during the chaotic aftermath of the storm. For the transportation network, we used a Brooklyn, NY road map consisting of 7,450 nodes and 22,377 edges. We chose three different sets of gas stations (with 5, 7, and 10, respectively) and created a Network Voronoi Diagram (NVD), gas station allotment with the min-cost flow approach, and a CCNVD with the PE-algorithm. The number of allotted nodes for every gas station is represented by the size of the circles in the figures. For simplicity, we gave all gas stations equal capacity.

6.1 Case Study Results and Analysis

In Section 5, we showed that the PE algorithm incurred much lower computational cost and created an almost identical solution (min-sum) to the min-cost flow algorithm. In this section, the goal was to investigate the following: (1) Are the algorithms able to handle the balanced gas allotment for affected regions? (2) How does the number of service centers affect the SA contiguity of the algorithms?

Case Study 1: We chose five gas stations and put three of them to the west side of Brooklyn and two of them to the north of Brooklyn (Figure 12(a)). In our analysis, the NVD (Figure 12(b)) shows unbalanced allotments, which may lead to longer wait times for larger regions (e.g., gas station 3 (green), gas station 4 (purple), and gas station 5 (red)). Figure 12(c) shows that min-cost flow violates service area contiguity (the three green circles show areas of discontiguity). Figure 12(d) shows the CCNVD produced by our algorithm. We can see that the CCNVD is able to remove excesses and deficits so that all service centers are balanced in terms of the number of allotted nodes.

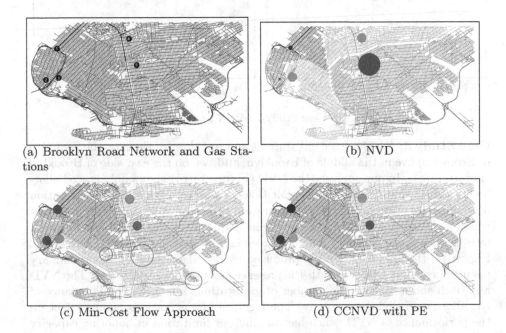

(a) Brooklyn Road Network and Gas Stations

(b) NVD

(c) Min-Cost Flow Approach

(d) CCNVD with PE

Fig. 12. Case Study 1 with 5 gas stations

Case Study 2: We chose seven gas stations and put three of them on the west side of Brooklyn and four in the middle of Brooklyn (Figure 13(a)). Again, in our analysis, the NVD (Figure 13(b)) shows unbalanced allotments, lead to longer wait times for larger regions (e.g., gas station 4 (purple), gas station 5 (red), gas station 7(rose)). Figure 13(c) shows that min-cost flow violates service area contiguity (the two green circles show areas of dis-contiguity). Figure 13(d) shows the CCNVD produced by our algorithm. As can be seen, all service centers on the CCNVD are balanced in terms of the number of allotted nodes.

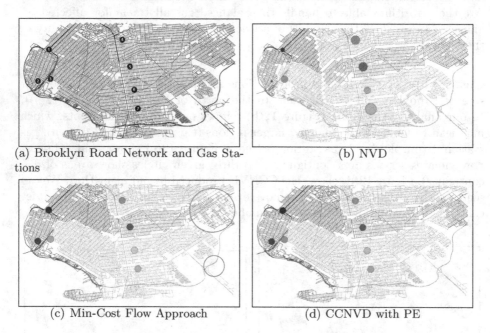

(a) Brooklyn Road Network and Gas Stations

(b) NVD

(c) Min-Cost Flow Approach

(d) CCNVD with PE

Fig. 13. Case Study 2 with 7 gas stations

Case Study 3: We chose ten gas stations and put three of them on the west side of Brooklyn, five in the middle of Brooklyn, and two on the east side of Brooklyn (Figure 14(a)). In our analysis, the NVD (Figure 14(b)) shows unbalanced allotments, which may lead to longer wait times for larger regions (e.g., gas station 3 (green), gas station 4 (purple), gas station 8(dark purple), gas station 9(light yellow), gas station 10(turquoise)). Figure 14(c) shows that min-cost flow violates service area contiguity (the five green circles show areas of dis-contiguity). Figure 14(d) shows the CCNVD produced by our algorithm. These preliminary results show that CCNVD with PE preserves service area contiguity. The NVD approach shows that as the number of gas stations increases, the imbalance of gas allotments slightly decreases. The location of gas stations has an effect on the performance of NVD which has an inherent limitation in honoring capacity constraint resource allotments. The min-cost flow results show SA discontinuity is increasing with increasing number of gas stations. In general, min-cost flow approaches are limited in preserving SA contiguity for CCNVD.

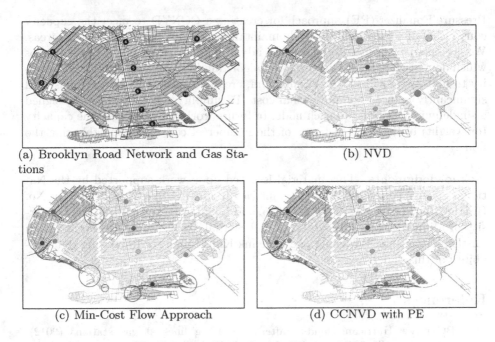

(a) Brooklyn Road Network and Gas Stations

(b) NVD

(c) Min-Cost Flow Approach

(d) CCNVD with PE

Fig. 14. Case Study 3 with 10 gas stations

Discussion: The proposed Pressure Equalizer algorithm advances the state of the art in computational techniques for CCNVD. The algorithm achieves a significant computational performance gain over current techniques. This improvement was obtained using three key features: (1) Pressure Equalization, (2) a PE-graph, and (3) a re-allotment cost function. Given unbalanced allotments (e.g., NVD), our approach smoothly expands (or shrinks) service areas to meet capacity constraints. Since we can reduce the sum of distances based on the re-allotment cost function, the PE algorithm can minimize both min-sum and service area dis-contiguity. The computational cost is dramatically reduced because the iterations in the PE algorithm are bounded by the number of graph-nodes and the size of the PE-graph is smaller than the min-cost flow graph.

Due to time limitations, we have only used three geographic areas for preliminary evaluation of the proposed algorithm. In the future, we plan to test our algorithm on a larger number of geographic areas.

7 Conclusion and Future Work

We presented the problem of creating a Capacity Constrained Network Voronoi Diagram (CCNVD), which is important for assigning evacuees to shelters, assigning patients to hospitals, assigning students to school districts, etc. Creating a CCNVD is challenging because of the large size of the transportation network and the constraint that service areas must be contiguous in the graph to simplify communication of allotments. In this paper, we introduced a novel

Pressure Equalizer (PE) approach for creating a CCNVD to meet the capacity constraints of service centers while maintaining the contiguity of service areas. We presented experiments and a case study using post-hurricane Sandy scenarios which demonstrated that our proposed algorithm had comparable solution quality to min-cost flow in terms of min-sum, created contiguous service areas, and significantly reduced computational cost. To simplify the analysis, we assigned a single unit of demand to each node. In future work, we will study the capacity (or weight) of the node in terms of the number of consumers, neighboring the node.

Acknowledgments. This material is based upon work supported by the National Science Foundation under Grant No. 1029711, USDOD under Grant No. HM1582-08-1-0017, HM1582-07-1-2035, and U-Spatial. We thank to the University of Minnesota Spatial Databases and Spatial Data Mining Research Group for their comments. We would like to thank Kim Koffolt for improving the readability of this paper.

References

1. ABC News: Hurricane sandy's aftermath: Long lines at gas stations (2012), http://goo.gl/omQ62 (retrieved March 2013)
2. Edmonds, J., Karp, R.M.: Theoretical improvements in algorithmic efficiency for network flow problems. Journal of the ACM 19(17), 248–264 (1972)
3. Tomizawa, N.: On some techniques useful for solution of transportation network problems. Networks 2(17), 173–194 (1971)
4. Daskin, M.: Network and discrete location: models, algorithms, and applications. Wiley-Interscience series in discrete mathematics and optimization. Wiley (1995)
5. Orlin, J.B.: A faster strongly polynomial minimum cost flow algorithm. Operations Research 41(2), 338–350 (1993)
6. Goldberg, A.V.: An efficient implementation of a scaling minimum-cost flow algorithm. Operations Research 22(1), 1–29 (1997)
7. Ahuja, R., Magnanti, T., Orlin, J.: Network flows: theory, algorithms, and applications. Prentice Hall (1993)
8. Schrijver, A.: Combinatorial Optimization: Polyhedra and Efficiency, vol. A. Springer (2003)
9. Korte, B., Vygen, J.: Combinatorial Optimization. Algorithms and Combinatorics. Springer, Heidelberg (2012)
10. Kuhn, H.W.: The hungarian method for the assignment problem. Naval Research Logistics Quarterly 2(1-2), 83–97 (1955)
11. Munkres, J.: Algorithms for the assignment and transportation problems. Journal of the Society for Industrial & Applied Mathematics 5(1), 32–38 (1957)
12. Goldberg, A., Tarjan, R.: Solving minimum-cost flow problems by successive approximation. In: Proceedings of the Nineteenth Annual ACM Symposium on Theory of Computing, pp. 7–18. ACM (1987)
13. Frank, A.: Connections in combinatorial optimization, vol. 38. Oxford Univ. Pr. (2011)
14. Cook, W., Cunningham, W., Pulleyblank, W., Schrijver, A.: Combinatorial Optimization. Wiley Series in Discrete Mathematics and Optimization. Wiley (2011)

15. Johnson, D.S., McGeoch, C.C.: Network flows and matching: first DIMACS implementation challenge, vol. 12. Amer. Mathematical Society (1993)
16. Goldberg, A.V., Tarjan, R.E.: Finding minimum-cost circulations by successive approximation. Mathematics of Operations Research 15(3), 430–466 (1990)
17. Klein, M.: A primal method for minimal cost flows with applications to the assignment and transportation problems. Management Science 14(3), 205–220 (1967)
18. Goldberg, A.V., Tarjan, R.E.: Finding minimum-cost circulations by canceling negative cycles. Journal of the ACM (JACM) 36(4), 873–886 (1989)
19. Okabe, A., Boots, B., Sugihara, K., Chiu, S.N.: Spatial tessellations: concepts and applications of Voronoi diagrams, vol. 501. Wiley (2009)
20. Okabe, A., Sugihara, K.: Spatial Analysis Along Networks: Statistical and Computational Methods. Statistics in Practice. Wiley (2012)
21. Erwig, M.: The graph voronoi diagram with applications. Networks 36(3), 156–163 (2000)
22. Okabe, A., Satoh, T., Furuta, T., Suzuki, A., Okano, K.: Generalized network voronoi diagrams: Concepts, computational methods, and applications. International Journal of Geographical Information Science 22(9), 965–994 (2008)
23. Győri, E.: On division of graphs to connected subgraphs. In: Combinatorics (Proc. Fifth Hungarian Colloq., Keszthely, 1976), vol. 1, pp. 485–494 (1976)
24. Lovász, L.: A homology theory for spanning trees of a graph. Acta Mathematica Academiae Scientiarum Hungarica 30(3-4), 241–251 (1993)
25. Győri, E.: Partition conditions and vertex-connectivity of graphs. Combinatorica 1(3), 263–273 (1981)
26. Dyer, M., Frieze, A.: On the complexity of partitioning graphs into connected subgraphs. Discrete Applied Mathematics 10(2), 139–153 (1985)
27. Diwan, A.A.: Partitioning into connected parts (slide 7) in graph partitioning problems. In: Research Promotion Workshop on Introduction to Graph and Geometric Algorithms (January 2011), http://goo.gl/b8fTN
28. Barth, D., Fournier, H.: A degree bound on decomposable trees. Discrete Mathematics 306(5), 469–477 (2006)
29. Fredman, M.L., Tarjan, R.E.: Fibonacci heaps and their uses in improved network optimization algorithms. Journal of the ACM (JACM) 34(3), 596–615 (1987)
30. OpenStreetMap, http://goo.gl/Hso0 (retrieved January 2013)

Mining Driving Preferences
in Multi-cost Networks '

Adrian Balteanu, Gregor Jossé, and Matthias Schubert

Institute for Informatics, Ludwig-Maximilians-Universität München, Oettingenstr. 67,
D-80538 Munich, Germany
{balteanu,josse,schubert}@dbs.ifi.lmu.de

Abstract. When analyzing the trajectories of cars, it often occurs that
the selected route differs from the route a navigation system would propose. Thus, to predict routes actually being selected by real drivers, trajectory mining techniques predict routes based on observations rather
than calculated paths. Most approaches to this task build statistical
models for the likelihood that a user travels along certain segments of a
road network. However, these models neglect the motivation of a user to
prefer one route over its alternatives. Another shortcoming is that these
models are only applicable if there is sufficient data for the given area
and driver. In this paper, we propose a novel approach which models the
motivation of a driver as a preference distribution in a multi-dimensional
space of traversal costs, such as distance, traffic lights, left turns, congestion probability etc.. Given this preference distribution, it is possible to
compute a shortest path which better reflects actual driving decisions.
We propose an efficient algorithm for deriving a distribution function of
the preference weightings of a user by comparing observed routes to a
set of pareto-optimal paths. In our experiments, we show the efficiency
of our new algorithm compared to a naive solution of the problem and
derive example weighting distributions for real world trajectories.

1 Introduction

When proposing a route to a driver, navigation systems rely on computing the
optimal path between goal and destination. However, when comparing the computed routes to real-world trajectories, it can be observed that drivers often
select paths which are significantly different from the proposed route. One reason for this effect is that the travel time between two places depends on a variety
of time-dependent and uncertain parameters like the behaviour of other vehicles or the synchronization between the travel progress and traffic lights. Thus,
instead of trying to predict the travel time directly, experienced drivers rather
try to select a path based on the trade-off between the risk of being delayed
and the opportunity to reach the destination in minimal time. Consider a risk
averse driver who wants to make certain that he reaches a goal in time but is
not necessarily in a hurry. In this case, our driver would usually prefer to avoid
areas with a large likelihood of congestion and a high density of traffic lights,

M.A. Nascimento et al. (Eds.): SSTD 2013, LNCS 8098, pp. 74–91, 2013.

even if this route is considerably longer than an alternative route without such risk factors. On the other hand, a more optimistic driver would surely choose the shortest path while assuming that delays at traffic lights or due to dense traffic will not be severe. To conclude, the path a driver actually chooses depends on personal preferences as well as on the optimality. To capture this effect, the research community has started to build models on the driving behavior of people. The idea of these approaches is that most of the routes a driver is selecting are travelled multiple times if they fit the personal preferences of a driver [1], [2],[3]. Thus, it is possible to build statistical models describing the preferred routes of a driver in a certain area. However, this method cannot be applied if there isn't any observed data for a certain area.

In this paper, we propose a new approach modeling driving preferences as a set of trade-off factors between two cost factors. For example, a route could be described by the travelled distance, the encountered traffic lights or the likelihood of a congestion. Our model is based on the assumption that the selection of a route depends on a combination of at least two cost types instead of a single one. For example, consider a path having 5 kilometers distance but 10 traffic lights. If there exists an alternative route having 6 kilometers distance but only 2 traffic lights, a user might prefer this path over the other because the 8 additional traffic lights outweigh the additional kilometer.

Formally, the preferences of a driver are captured by the ratio of one cost value relative to another one. Thus, a driver would select a certain route because it is optimal w.r.t. the cost combination which is implied by these preferences. However, since people do not necessarily display consistent preferences, our method estimates a preference distribution over all available observations. To estimate this distribution, it is important to distinguish between the cost ratios of the observed routes and the preferences of the driver. Since the observed route might be the only one connecting the start to the destination, the observed cost ratios might not even be close to the preferences. To still exploit the observation and gain knowledge about the preferences, we have to distinguish it from the set of alternative routes which have not been selected. To derive a reasonable set of alternative routes, we rely on the route skyline [4]. The route skyline contains the set of paths being optimal under any possible driving preference. Thus, a rational driver would always select the element of the route skyline being optimal for his preferences. To build a model of the preferences of a driver, our new method requires a set of trajectories and a corresponding multi-cost road network. Each sample trajectory is now compared to the route skyline for the same start and destination. Thus, we receive a cost vector for the observed trajectory and a set of cost vectors for the unselected skyline routes. The cost vectors are mapped into a preference space, making them independent from the absolute cost values. We now derive a preference area corresponding to all possible preference values making the selected route better than the alternative skyline routes. Now, by adding up these areas in the preference space, we receive a distribution over the preferences of a driver. If the preferences are consistent the distribution will display a rather dense area describing the tolerated trade-off between two types of costs.

This basic approach has an important drawback. The number of skyline routes might strongly vary with the distance between start and destination. This leads to a potentially large computational overhead when computing the route skyline. However, for determining the preference area of a trajectory it is not necessary to compute the complete route skyline. Instead we define the set of neighbors limiting the preference area and provide an algorithm only searching for the set of neighbors. Thus, adding a new observation to the preference distribution yields only a small overhead. To conclude, the contribution of this paper are:

- A novel model for describing driving preferences as the set of pairwise trade-offs between cost factors.
- A method for estimating the preference distribution for a user comparing observed trajectories with the set of pareto-optimal alternative routes.
- A more efficient algorithm that strongly decreases the computational cost for the required pareto-optimal routes.

The rest of the paper is organized as follows: In the section 2, we survey related work on multi-cost routing, route prediction and modelling driving preferences. Section 3 formalizes our model and describes the problem. The algorithms and methods for determining the preference distribution are described in section 4 and section 5. The experimental evaluation in section 6 examines the quality of the derived preference distributions and shows the driving preferences on real world data. The paper concludes with a summary and an outlook to ongoing research in section 7.

2   Related Work

In this section, we will shortly review existing work in modeling driving preferences and route prediction. Furthermore, we will give a short introduction on routing in multi-cost networks.

In [5] the authors describe one of the first approaches to route prediction for optimizing a shared fleet of cars. The method assumes that start and destination are given. Therefore, the setting shares some similarity to the work in this paper. However, the method does not personalize the prediction. [1] describes a system that uses GPS tracks to build a hidden Markov model for predicting routes and destinations. The system strongly relies upon the fact that users tend to prefer the same set of routes and often travel to the same locations. In [2], the authors focus on improving the input data by proposing a pipeline from raw GPS data to map-matched trajectories on a road network. An important aspect of the approach is that frequently driven routes have to be discovered and outlier trajectories should not be considered for learning models about preferred routes. In [3], the authors propose a model for route prediction on patterns of visited locations. The system is built upon a client server architecture considering the privacy aspects of the collected trajectory data. The T-Drive system [6,7] is based on the trajectories of 33,000 taxis in the city of Beijing which were analysed over a time-period of 3 months. The idea behind this system is to research the routes of professional drivers and build a model which imitates their behaviour when proposing a route to a driver. Technically, the system exploits the information by deriving time-dependent traversal times for central road segments. Furthermore, the system generates a landmark graph of preferred road segments called landmark road segments. The landmarks form the backbone for a specialized network containing streets being used by the taxi drivers. In combination with the improved time-dependent travel time estimation, the authors demonstrated that the system is capable to propose routes which required significantly less time than the routes being proposed by conventional routing. Though this method adapts the proposed routes to the preferences of taxi drivers, the system does not distinguish different preferences. Instead the general assumption in T-Drive is that all drivers prefer the fastest route and taxi drivers share the joint experience to achieve this goal. A system which actually models personal driving preferences is the TRIP system [8]. Similar to T-Drive TRIP is based on observed trajectories but does not join these observations to build a global model of professional driving behavior. Instead TRIP works with a much smaller set of personal trajectories and models each driver separately. Similar to T-Drive, the system employs all observed trajectories to generate time-dependent travel times. Furthermore, the system computes an inefficiency ratio for each user, modeling the importance of travel time in his routing decisions. The inefficiency ratio is then employed to lower the cost of road segments a driver has actually travelled before. Thus, the proposed route will contain more already known road segments if the driver has a low efficiency ratio.

Though all of these systems share a common motivation with the work being proposed in this paper, there is one very important difference. The preferences in previous approaches of learning driving preferences are modeled w.r.t. preferred localities in a given area. Thus, all of these approaches cannot be applied in areas where no trajectories are available. However, this is the case where a navigation system is most needed.

A major aspect of our method is to compare the observed trajectory to a set of reasonable alternative route which were rejected by the driver. Using more than one cost attribute causes the number of route alternatives to increase rapidly. Recently, several researches started to examine multi-cost or multi-attribute transportation networks [9,4,10]. In this type of network, the cost of a path is not only measured w.r.t. a particular cost function, like travel time. Instead, the cost of each path is measured by a complete vector of cost criteria like distance, fuel consumption, toll fees, criminal risks etc.. Considering these additional metrics enables the computer to find alternatives which are usually not optimal w.r.t. all criteria, but offer reasonable trade-offs. For example, the system might propose an alternative which requires 5 minutes of additional driving time, but might not pass through an area with a high criminal activity. A method that calculates this type of trade-off routes is route skyline processing proposed in [4]. A route skyline query processes all pareto-optimal routes between a start and a destination. In other words, the cost vectors of the result form a skyline in the multi-dimensional space of cost criteria. Thus, the optimal path for each

weighting of costs is contained in the skyline. We will formally introduce route
skylines and some of their implications in the next section. In [9], the authors
introduce preference queries in multi-cost networks. In particular, the paper proposes ranking and skyline queries to compute and sort a result set of possible
target destinations in a multi-cost transport network. Though this work is also
based on multi-cost networks, the proposed query returns a skyline of locations.
Thus, the method cannot be employed to determine a set of alternative routes
between two specified locations.

3   Preliminaries

We assume a street network to be representable as a directed, weighted graph
$G = (V, E, c, d)$ which is uniquely defined by its set of nodes V, its set of directed
edges E connecting pairs of nodes in the given order and a cost function $c :$
$E \to (\mathbb{R}_0^+)^d$. We restrict ourselves to the case $d = 2$, i.e. every non-trivial edge
$e = (v_1, v_2), v_1 \neq v_2 \in V$ is assigned a 2-dimensional cost vector consisting of
non-negative values $c_i(e) := (c(e))_i$ both of which stand for the cost regarding
an occurrence of the pre-selected attributes A_1, A_2 taken from the set of all
attributes. These attributes may for example be travel distance, travel time,
number of traffic lights, difference in altitude et cetera. A driver might, for
instance, trade off travel distance against number of lights since it is common to
prefer a greater distance over the chance of several red lights.

A set of non-trivial consecutives edges $e_1, \ldots, e_k$ is referred to as a path p. Any
such path can be assigned a unique cost vector, namely $c(p) = (c_1(p), c_2(p)) :=$
$\sum_{j=1}^{k} c(e_j)$. For two nodes $s \neq t, s, t \in V$, we denote the set of all paths from s
to t by $\mathcal{P}(s, t)$.

Let us note that unless otherwise stated, a set of paths or a single path always
refers to a graph $G = (V, E, c, 2)$. Furthermore, we assume start and end nodes
of a path to be distinct, and we presume lower cost values to be more desirable
than higher ones.

Our approach is based upon the plausible assumption that a driver navi-
gates consciously, by himself and does so using knowledge of the various possible
routes. Consider a driver who has selected the path p, then we think of p to
be optimal w.r.t. the personal cost function of the driver which is a trade-off
between the different attributes (number of traffic lights and travel distance, for
instance).

When choosing another route, it is likely that our driver will evaluate the
costs of this new path according to his previously used preference. We model the
driver's spectrum of preference independent of start and end nodes as well as
from absolute values. Any such preference gradient (as introduced below) can be
used as a weight vector when computing routes. This concept is the mathematical
formalization of weighing different paths' attributes before selecting a path.

Definition 1. *Given a path p and a gradient γ_2/γ_1, we define the **weight vec-
tor** as $\gamma := (\gamma_1, \gamma_2)$ and the **cost function** of p w.r.t. γ as $c_\gamma(p) := \|\gamma * c(p)\|_\infty$,*

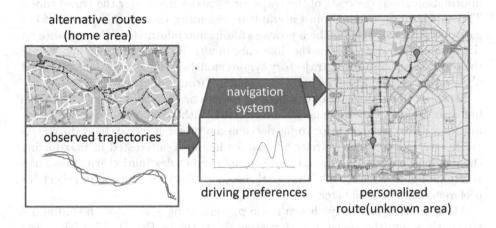

alternative routes
(home area)

observed trajectories

navigation
system

driving preferences

personalized
route(unknown area)

Fig. 1. Illustration of the concept of our solution

where $\|(v_1,...,v_n)\|_\infty := \max_{1 \leq i \leq n} |v_i|$ denotes the L_∞-norm and $(\cdot * \cdot)$ denotes pointwise multiplication.

It is important to note that there need not be any correlation as to the locality of the route the gradient was derived from and the route to be computed from it. Using Dijkstra's or the A^* algorithm where the queue of paths is sorted ascendingly by the values of the above defined cost function yields a computation of paths in the order specified by the preference vector.

Although the above stated assumptions are plausible, it is improbable for a driver to always navigate according to the same preference. This might be due to different moods (casual driving on weekends as opposed to efficiency-optimized driving during the week) or simply due to the lack of routes exactly corresponding to the preference. Thus, we assume a preference distribution which the driver implicitly constitutes over time with every route he chooses. The preference distribution models his characteristic behavior in traffic, hence any routing decision may be perceived as drawing a sample from the driver's distribution and selecting the shortest route w.r.t. that value. Our concept of deriving driving preferences and personalized routing suggestions are visualized in figure 1. The goal of our method is to mine the distributions of pairwise cost trade-offs. Given these distributions, we can derive personalized cost function for route computation. To learn the distribution functions, we compare a set of observed paths and compare them to the corresponding route skyline to find out for which trade-off the observed path was a reasonable selection.

4 Learning Preference Distributions

If a driver chooses a particular path, we refer to it as a trajectory and assume it is tracked by GPS. The raw GPS data itself, however, does not contain any information about the costs of the trajectory (except maybe for the travel time). This is why it is matched onto a graph representing the street network of the corresponding region [11], which provides additional information (for instance on the number of traffic lights or the difference in altitude). In the following, we use the terms chosen route and trajectory synonymously and mean the map-matched observation enriched by the information gained from the graph.

Essential to our concept of driving preferences are not only the chosen routes but also the ones that have not been selected. Without knowledge of the possibilities a driver weighed prior to his decision, no preferences can be deduced. As an illustration, consider a driver who is, for instance, interested in maximizing the driving flow but there is no route to his current destination with less than 10 traffic lights. Hence, whichever path he chooses will not properly reflect his preference due to of the lack of suitable options.

Thus, we transform any chosen route p, connecting $s \neq t \in V$, including its alternatives into the space of cost vectors for paths in $\mathcal{P}(s,t)$. Then the costs of p and those of the non-selected paths are compared, yielding what we refer to as a preference window of p w.r.t. $\mathcal{P}(s,t)$, i.e. a rectangle which reflects the motives for this particular route selection.

Definition 2. *Given a path p (connecting nodes $s \neq t \in V$), we introduce the notion of a **preference window** w.r.t. p, $W(p)$:*

$$W(p) := \left\{ \begin{pmatrix} \gamma_1 \\ \gamma_2 \end{pmatrix} \in (\mathbb{R}_0^+)^2 \mid c_\gamma(p) \leq \min_{q \in \mathcal{P}(s,t)} c_\gamma(q) \text{ where } \gamma_2 := 1 - \gamma_1 \right\}$$

For given start and end nodes there exist countless possible paths (arbitrarily long detours), of which definition 2 takes every single one into consideration. However, it is a plausible assumption that if a driver knew an optimal route which approximately suits his preference, he would choose it. On the contrary, he would most likely avoid any path which corresponds to his preference but has high values in both dimensions. If there is no better alternative within his preference window, the driver would probably decide in favor of a better route although it might not comply with his desired preference. Consequently, we assume a rational driver to always choose an optimal route and thereby reduce computational costs drastically. Our notion of optimality is defined below.

Definition 3. *For a set of paths $\mathcal{P}$ (connecting the same nodes $s \neq t$ within a graph $G = (V, E, c, d)$), we call $p \in \mathcal{P}$ a **skyline path** (of $\mathcal{P}$) (or a **pareto-optimum**), iff there exists no $q \in \mathcal{P} \setminus \{p\}$ and no $1 \leq i \leq d$, such that:*

$$\forall j \neq i : c_j(q) \leq c_j(p) \land c_i(q) < c_i(p).$$

*Given a set of paths (all connecting the nodes $s \neq t \in V$), the **route skyline**, $\mathcal{S}(s,d)$ consists of all the "undominated" paths in the above defined sense.*

Figure 2 visualizes the 2-dimensional cost vectors of a set of paths as well as the skyline paths therein.

In accordance with the aforementioned assumption we restrict ourselves to comparing the chosen path to the route skyline. Every suboptimal path might

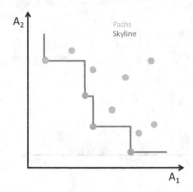

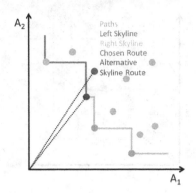

Fig. 2. Set of 2-dimensional cost vectors and skyline paths

Fig. 3. Chosen route and its skyline correspondent

indeed reflect a preference ratio but is dominated by at least one skyline path which might not yield the exact same preference area. However, it is likely that the skyline path yields a similar ratio as the suboptimal path (cf. figure 3). If a given trajectory is not pareto-optimal, we map it onto its skyline correspondent.

Definition 4. *Let p be a trajectory, then $s \in S$ is its **skyline correspondent** (where S denotes the set of route skyline connecting the same start and end nodes as p), iff $\delta(s) = \min_{t \in S} \delta(t)$, where*

$$\delta(t) := \left\| c(t) - \frac{\langle c(t), c(p) \rangle}{\|c(p)\|} \right\|.$$

is the length of the normal vector from p onto the straight line $\lambda \cdot c(s), \lambda \in \mathbb{R}$.

In order to simplify the presentation of a preference window, we introduce the so-called preference space which is 1-dimensional, in contrast to the cost space. Beforehand, we need the concept of lower and upper preference gradients.

Definition 5. *Given route skyline $S = S(s, d)$ and a trajectory $p \in S$ (connecting s and t) therein, we define the **lower and upper preference gradients** of p, α and β respectively, as follows:*

$$\beta := \min \left\{ \frac{\gamma_1}{\gamma_2} \mid \begin{pmatrix} \gamma_1 \\ \gamma_2 \end{pmatrix} \in W(p) \wedge \frac{\gamma_1}{\gamma_2} > \frac{c_2(p)}{c_1(p)} \right\}$$

$$\alpha := \max \left\{ \frac{\gamma_1}{\gamma_2} \mid \begin{pmatrix} \gamma_1 \\ \gamma_2 \end{pmatrix} \in W(p) \wedge \frac{\gamma_1}{\gamma_2} < \frac{c_2(p)}{c_1(p)} \right\}$$

*Based on this definition, we refer to $\mathcal{D}(p) := [\tan^{-1}(\alpha), \tan^{-1}(\beta)]$ as the **driving preference** w.r.t. p which is a subset of (or in rare special cases equal to) the **preference space** $[0, \pi/2]$.*

The reason for the preference space being 1-dimensional, is that the components of every vector within the preference window add up to 1, i.e. one of the components redundant. Figure 5 shows the preference gradients of a chosen route (cf. figure 4) transferred into the preference space.

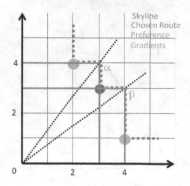
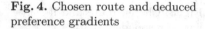

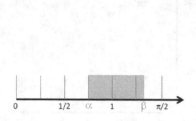

Fig. 4. Chosen route and deduced preference gradients

Fig. 5. Preference gradients converted into preference space

Our goal is not only to derive a driver's preference by evaluating one of his chosen routes but to generate a preference distribution from all recorded trajectories in order to model overall driving behavior. To avoid the computational costs of a continuous probability distribution, we approximate it with a histogram which has a pre-defined number of bins specified by the user. Hence, our preference distribution is defined as follows. Let us note that the histogram is normalized such that the sum over all bins equals 1.

Definition 6. *Given a set of trajectories* $\mathcal{P}$*, the joint* **preference distribution** *is an array of reals and of length* $k \in \mathbb{N}$ *(input parameter). Let*

$$I_j := \left[\frac{j\pi}{2k}, \frac{(j+1)\pi}{2k} \right], 0 \leq j \leq k-1$$

be a partition of the preference space into k *equal-sized intervals. Then the entries of* $\mathcal{D}(\mathcal{P})$ *are defined as follows:*

$$\mathcal{D}(\mathcal{P})[j] := \frac{\sum_{p \in \mathcal{P}} \mu(I_j, D(p))}{\sum_{0 \leq j \leq k-1} \sum_{p \in \mathcal{P}} \mu(I_j, D(p))}$$

where $\mu(I_j, D(p)) := \frac{\lambda(I_j \cap D(p))}{\lambda(I_j)}$

with the 1-dimensional lebesgue measure $\lambda(\cdot)$*.*

Before we present our basic algorithm 1 for computing a driving preference distribution, we need to introduce the concept of left and right skylines which is illustrated in figure 3.

Definition 7. *For a given route skyline* $\mathcal{S} = \mathcal{S}(s, d)$ *and a trajectory* $p \in \mathcal{S}$ *(connecting* s *and* t*) therein, we define the* **left and right (route) skyline** *w.r.t.* p*,* $\mathcal{S}_L(p) = \mathcal{S}_L$ *and* $\mathcal{S}_R(p) = \mathcal{S}_R$ *respectively:*

$$\mathcal{S}_L(p) := \{q \in \mathcal{S} \mid c_1(q) < c_1(p)\}$$
$$\mathcal{S}_R(p) := \{q \in \mathcal{S} \mid c_1(q) > c_1(p)\}$$

Algorithm 1. Basic Preference Distribution

Require: set of chosen routes $\mathcal{P}$
  for all trajectory $p \in \mathcal{P}$ connecting some s and t **do**
    compute skyline $\mathcal{S} = \mathcal{S}(s, t)$ and sort ascendingly by c_1
    if $p \notin \mathcal{S}$ **then**
      $c(p) \leftarrow$ skyline correspondent of $c(p)$
    end if
    for all $q \in \mathcal{S}_L(p)$ **do**
      determine trade-off between $c(p)$ and $c(q)$
      remember smallest trade-off
    end for
    for all $q \in \mathcal{S}_R(p)$ **do**
      determine trade-off between $c(p)$ and $c(q)$
      remember greatest trade-off
    end for
    add interval between stored trade-offs to distribution
  end for
  return normalized distribution

The skyline computation can be done efficiently as presented in [4]. Instead of using the rather abstract definitions 2 and 5 of a preference area, we use the above mentioned trade-offs. These are defined below and easy to compute. The following lemma proves that both concepts coincide.

Definition 8. *Let* $p \neq q$ *be skyline paths (connecting the same start and end nodes), then the **trade-off (gradient)** between* p *and* q *is defined as:*

$$
t(p, q) := \begin{cases} \pi/2, & \text{if } c_1(q) = 0 \\ \tan^{-1}(c_2(q)/c_1(p)), & \text{if } c_1(p) < c_1(q) \\ \tan^{-1}(c_2(p)/c_1(q)), & \text{if } c_1(q) < c_1(p) \\ 0, & \text{if } c_1(p) = 0 \end{cases}
$$

Lemma 1. *Let* $\mathcal{S}$ *be a route skyline and* p *a path therein, and let*

$$n_L \in \mathcal{S}_L(p) \quad \text{such that} \quad t(p, n_L) < t(p, q) \; \forall \; \mathcal{S}_L(p) \ni q \neq n_L$$
$$n_R \in \mathcal{S}_R(p) \quad \text{such that} \quad t(p, n_R) > t(p, q) \; \forall \; \mathcal{S}_R(p) \ni q \neq n_R$$

$$[\tan^{-1} t(p, n_R), \tan^{-1} t(p, n_L)] = \mathcal{D}(p)$$

Proof. Let q be any path within the same skyline as p, $p \neq q$. Hence either $c_1(q) < c_1(p)$ or $c_1(p) < c_1(q)$, equality would imply a contradiction to their distinctiveness. Let us distinguish these two cases:

(i) $c_1(p) < c_1(q)$ which is equivalent to $c_2(q) < c_2(p)$. Let $\gamma := (\gamma_1, \gamma_2)$ such that $\gamma_1/\gamma_2 < c_2(p)/c_1(p)$, i.e. $\gamma \in W(p)$ and if maximized, it equals α. $\|\gamma * p\|_\infty = \|\gamma * q\|_\infty$ holds, if $\gamma_2 c_2(p) = \gamma_1 c_1(q)$, since $\|\gamma * p\|_\infty = \gamma_2 c_2(p)$ and $\|\gamma * q\|_\infty = \gamma_1 c_1(p)$ by presumption. This equality implies $c_2(p)/c_1(q) = \gamma_1/\gamma_2$, which is maximal, iff $q = n_R$ *(cf. figure 7)*.

(ii) $c_1(p) > c_1(q)$ which is equivalent to $c_2(q) > c_2(p)$. As before, let $\gamma := (\gamma_1, \gamma_2)$ such that $\gamma_1/\gamma_2 < c_2(p)/c_1(p)$, i.e. $\gamma \in W(p)$ but now if minimized, it equals β. $\|\gamma * p\|_\infty = \|\gamma * q\|_\infty$ holds, if $\gamma_1 c_1(p) = \gamma_2 c_2(q)$, for analogous reasons as above. This equality implies $c_2(q)/c_1(p) = \gamma_1/\gamma_2$, which is minimal, iff $q = n_L$ *(cf. figure 7)*.

Applying $\tan^{-1}(\cdot)$ *proves the lemma.*

5   An Efficient Algorithm for Computing Preference Distributions

Computing the entire route skyline may even be costly in 2 dimensions since its cardinality is merely bounded by the total number of paths. Therefore, we present an algorithm which only requires a set of at most 2 skyline elements (per trajectory) to derive a preference area. The novel approach is based upon the observation made at the end of the section 4. Given a trajectory p, it suffices to know $n_L(p) \in \mathcal{S}_L(p)$ and $n_R(p) \in \mathcal{S}_R(p)$ since they already define the preference area, as proven in lemma 1. From now on, we refer to $n_L(p)$ and $n_R(p)$ (defined above) as the **left and right neighbors** of p, respectively.

In general the set of neighbors will contain 2 elements, i.e. the skyline paths whose cost vector projections onto one of the axes are closest to and better than the reference path (cf. Figure 6). Note that the set of neighbors, however, may also be empty ($|\mathcal{S}(p)| = 1$) or contain only one element ($\exists\, i \in \{1, 2\} : \forall q \neq p \in \mathcal{S}(p) : c_i(p) < c_i(q)$). In the latter case, we extend the preference area to the according axis, i.e. if $i = 1$ then $\mathcal{D}(p) = [\alpha, \pi/2]$, conversely if $i = 2$ $\mathcal{D}(p) = [0, \beta]$. This reflects the lack of an alternative route w.r.t. to dimension i, therefore the preference area is not bounded by a trade-off in this direction.

Instead of computing the complete route skyline as in algorithm 1, we now want to compute only potential neighbors of a given trajectory p. Our algorithm works similar to the ARSC algorithm introduced in [4] (for more information on the ARSC see Appendix 7). However, unlike this algorithm, we can stop traversing the graph before the priority queue storing all paths which still have to be extended is empty. To achieve this performance gain, we first order the queue of nodes containing unprocessed paths ascendingly by the maximum distance to $c(p)$. Now, we can guarantee that the skyline results are retrieved w.r.t. to their closeness to p. Then, we process the skyline and find p or its skyline correspondent.

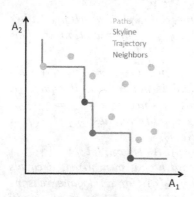

Fig. 6. Illustration of a set of neighbors w.r.t. a trajectory

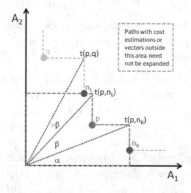

Fig. 7. Visualization of a trajectory, its neighbors and trade-off gradients

The next time the destination node t is on top of the queue, we find a neighbor n being responsible for the top-ranking within the queue. n is either part of $\mathcal{S}_L(p)$ or $\mathcal{S}_R(p)$. If it is part of $\mathcal{S}_L(p)$, we found an upper limit l_2 for c_2. This allows us to prune all further paths where the forward approximation of c_2 is larger than l_2. Correspondingly, we can derive l_1 and prune any path whose forward approximation of c_1 is bigger than l_1. This pruning area is depicted in figure 7. The algorithm can terminate if the value of the top element of the queue is larger than $\max(l_1, l_2)$.

In conclusion, both algorithms compute the same driving preference distributions when given a set of trajectories. However, the improved algorithm 2 severely reduces the number of computed paths in comparison to algorithm 1 (cf. 6).

6 Experimental Evaluation

In this section, we describe the results of our experimental evaluation. We examined three different aspects of the proposed method. The first aspect is whether the proposed algorithm is capable of reproducing a given preference distribution based on the preference intervals being observed by our algorithm. The second aspect is a study of real driving behavior being observed on real trajectories. The last aspect is a runtime analysis of our proposed fast algorithm compared

---

Algorithm 2. Fast Preference Distribution

Require: set of chosen routes $\mathcal{P}$
  for all trajectory $p \in \mathcal{P}$ connecting some s and t **do**
    initialize queue Q of nodes sorted by ascending max distance to p
    set $l_1, l_2 := \infty$ and $n_1, n_2 :=$ null
    while ¬Q.isEmpty ∧ Q.topValue < $\max(l_1, l_2)$ **do**
      currentNode = Q.topElement
      if currentNode = t **then**
        for all path $q \in$ currentNode **do**
          if $q = p \vee q$ outside of pruning area **then**
            continue
          else if $c_1(q) < c_1(p) \wedge c_2(q) < l_2$ **then**
            $n_1 \leftarrow q$ and $l_2 \leftarrow c_2(n_1)$
          else if $c_1(q) > c_1(p) \wedge c_1(q) < l_1$ **then**
            $n_2 \leftarrow q$ and $l_1 \leftarrow c_1(n_2)$
          end if
        end for
      else
        for all path $q \in$ currentNode **do**
          if forward approximation of q outside of pruning area **then**
            extensions ← buildExtensions q
            remove locally dominated extensions
            for all ext ∈ extensions **do**
              update Q with ext.lastNode if better w.r.t. max distance to p
            end for
          end if
        end for
      end if
    end while
    add interval between trade-offs $t(p, n_1)$ and $t(p, n_2)$ to distribution
  end for

---

  return normalized distribution

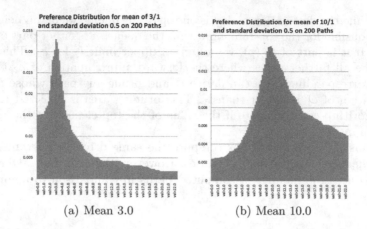

(a) Mean 3.0   (b) Mean 10.0

Fig. 8. Retrieved preference distributions for a mean ratio of 3/1 (left) and 10/1 (right.)

to the basic approach which relies upon computing the complete skyline. We implemented our approach in the MARiO framework [10] which gave us access to real-world maps from Open Street Map (OSM) [1]. We mainly tested our method on a version of the OSM map of Beijing containing 29747 nodes and 37632 links. For the real world study, we employed a map of the German city of Bad Toelz having 8282 nodes and 13620 links. As cost criteria, we employed the traveled distance and the number of encountered traffic lights which are already available in MARiO. All experiments were conducted on a laptop having an Intel Centrino 2 processor and 2 Gb of Ram.

In our first experimental setting, we show that our algorithm is capable of reconstructing driving preferences from observed trajectories. In order to have a ground truth of known distributions, we generated preference distributions based on Gaussians with given mean values and standard deviations. In order to derive a trajectory, a preference vector is drawn from this distribution. Then, the preference vector used it to compute an optimal path for a randomly select pair of nodes. Let us note that this setting may allow start and end nodes where only path exists. For the following tests, we generated 200 random trajectories and employed our method to reconstruct the distribution function. In a first experiment, we tested the method for different mean values, 3.0 and 10.0 having the same standard deviation 0.5. The result can be seen in figure 8. In both cases the derived distribution has its peak on the mean value of generating distribution. In the next experiment, we tested the tolerance w.r.t. different standard deviations. Therefore, we set the standard deviation to 0.5 as in the example above and additionally tested a much larger standard deviation of 3.0. The results are depicted in figure 9. It can be seen that even though the standard deviation is rather large, the observed preference distribution is still rather similar to the generating distribution.

---

[1] www.openstreetmap.org

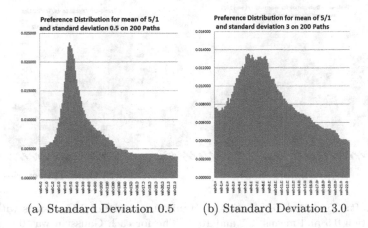

(a) Standard Deviation 0.5   (b) Standard Deviation 3.0

Fig. 9. Retrieved preference distributions for a mean ratio of 5/1 and standard deviations 0.5 (left) and 3.0 (right.)

In a final setting, we tested our algorithm in order to reconstruct a mixture of two Gaussians having a common standard deviation of 0.5 and mean values of 3.0 and 10.0. The weights of both Gaussians are set to 0.5. The results are shown in figure 10(a). The resulting histogram shows two clear peaks corresponding to the means of the contributing Gaussians. To conclude, the technique of building histograms from preference areas is capable of reconstructing preference distributions. Let us note, that we do not provide experiments on the amount of necessary sample paths. Such experiments would be extremely dependent on the quality of the employed trajectories and not of their quantity. Since a trajectory having multiple alternative routes yields a much smaller preference area, even a few well-selected observations might be sufficient to compute a very good approximation of the preference distribution.

In the next experiment, we wanted to examine the preference distributions on a real set of trajectories. In order to generate results being as realistic as possible, we need trajectories and network data of high quality. Though there is trajectory data publicly available, e.g. from [6], the data is often located in Beijing. However, the available OSM data for Beijing was rather incomplete w.r.t. to road segments and traffic lights. Thus, we tested our method on the OSM network of the German city of Bad Toelz which is more thorough. The used trajectories are the raw trajectories from OSM which were used to create the OSM map. We employed a simple topological map-matching approach to map the trajectories onto the graph. After sorting out cyclic paths (since they where obviously not driven with the intention to find an efficient path), we gathered 156 trajectories. For those trajectories, we derived a preference distribution which is depicted in figure 10(b). As cost criteria, we examined the path length in contrast to the passed traffic lights. The strongest peak of the derived distribution reflects the decision of driving the shortest path regardless of the number of encountered traffic lights. However, there are two further peaks in the distribution for values

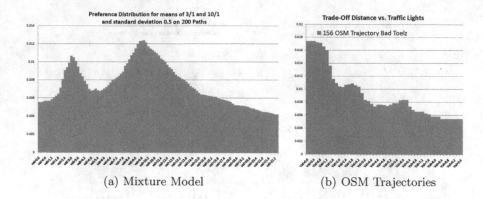

(a) Mixture Model   (b) OSM Trajectories

Fig. 10. Left: Computed distribution for a mixture model of 2 Gaussians with standard deviation 0.5 and means 3/1 and 10/1. The for each Gaussian was 0.5. **Right:** Preference distribution of 156 raw trajectories the OSM map of Bad Toelz is built upon. The distribution shows 3 different peaks which is natural since the trajectories are generated by various drivers.

2.8 and 6.2, showing that some of the paths where longer but had less traffic lights. Let us note that deriving several peaks is not surprising for this data set because a city in OSM is usually recorded by multiple persons. Unfortunately, the downloaded GPX file did not contain the owner of the trajectory. Thus, running tests for a particular person was not possible. To conclude, even for a group of drivers the preference distribution displays clear peaks which can be employed for proposing route alternatives which better mirror the trade-offs of the observed persons.

In a final experiment, we examine the performance increase when employing the fast algorithm directly determining neighbors in comparison to the basic approach of employing a given algorithm for route skyline computation. The route skyline in the basic approach was computed with the ARSC algorithm [4]. Let us note that though our method consists of more steps than determining the preference area for one trajectory, this step is the computationally most expensive. We measured the average runtime for computing one preference area and the amount of visited nodes on the Beijing graph for 1000 queries. Figure 11 shows the average results for one query. The amount of visited nodes was only 70 % percent of the basic approach. However, the ARSC algorithm uses an embedding for making an A^*-like search, when computing the route skyline. Thus, the amount of visited nodes is already rather small. However, our fast algorithm is about 6 times faster. This is due to the fact that the overhead for processing unnecessary paths and their extensions is significantly smaller. For determining preference areas in the Beijing network our algorithm needed only an average of 1 second per trajectory. Thus, it would be possible to maintain preference distributions even in embedded system with computational restrictions.

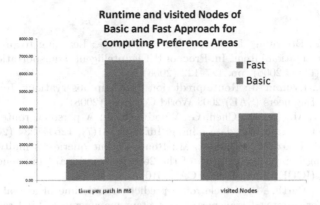

Fig. 11. Performance comparison between the basic approach and the fast algorithm directly generating the neighboring paths on the skyline. The fast algorithm performs between 5 to 6 times faster when computing the preference area of a trajectory.

7 Conclusions

In this work, we introduce a novel concept for modeling driving behavior. In contrast to existing approaches, our work is independent of location and absolute values of the actual driven routes. Furthermore, it allows and identifies different driving moods by generating a histogram approximating a driver's personal preference distribution. This histogram could serve to predict future driving behavior as well as to help navigation systems to adapt to the personal bias of certain cost attributes. Technically, our method relies upon the notion of route skylines and augments it by the new idea to compute only the required skyline paths. We propose a fast algorithm which incorporates a much more restrictive abort criterion significantly reducing the computational cost. Our experiments show the effectiveness and the efficiency of our algorithmic model. Given a random distribution, we are able to reproduce the input. The analysis of real-world trajectories implied varying drivers' trade-offs between travel distance and number of traffic lights for a group of drivers contributing to open street map.

For future work, we plan to incorporate more use cases from the e-traffic sector in order to gain insight into the trade-off between energy efficiency and travel time. This is essential for optimizing route planning for electric cars – a field which is rapidly gaining significance with rising ecological awareness. Moreover, we would like to expand our 2-dimensional model to an arbitrary number of attributes, ideally generating multi-variate preference distributions to mirror possible multi-attribute preference trade-offs.

Acknowledgements. This research has been supported by the IKT II program in the Shared-E-Fleet project. They are funded by the German Federal Ministry of Economics and Technology under the grant number 01ME12107. The responsibility for this publication lies with the authors.

References

1. Simmons, R., Browning, B., Zhang, Y., Sadekar, V.: Learning to predict driver route and destination intent. In: Proc. of IEEE Intelligent Transportation Systems Conference (ITSC 2006), pp. 127–132 (2006)
2. Froehlich, J., Krumm, J.: Route prediction from trip observations. In: Society of Automotive Engineers (SAE) 2008 World Congress (2008)
3. Chen, L., Lv, M., Ye, Q., Chen, G., Woodward, J.: A personal route prediction system based on trajectory data mining. Inf. Sci. 181(7), 1264–1284 (2011)
4. Kriegel, H.P., Renz, M., Schubert, M.: Route skyline queries: a multi-preference path planning approach. In: Proc. of the 26th International Conference on Data Engineering (ICDE), Long Beach, CA (2010)
5. Karbassi, A., Barth, M.: Vehicle route prediction and time of arrival estimation techniques for improved transportation system management. In: Proceedings of the Intelligent Vehicles Symposium 2003, pp. 511–516. IEEE (2003)
6. Yuan, J., Zheng, Y., Zhang, C., Xie, W., Xie, X., Sun, G., Huang, Y.: T-drive: driving directions based on taxi trajectories. In: Proc. of the 18th ACM SIGSPATIAL International Conference on Advances in Geographic Information Systems (ACM GIS), San Jose, CA, pp. 99–108 (2010)
7. Yuan, J., Zheng, Y., Xie, X., Sun, G.: Driving with knowledge from the physical world. In: Proc. of the 17th ACM International Conference on Knowledge Discovery and Data Mining (SIGKDD), San Diego, CA, pp. 316–324 (2011)
8. Letchner, J., Krumm, J., Horvitz, E.: Trip router with individualized preferences (trip): Incorporating personalization into route planning. In: Proc. of 8th Conference on Innovative Applications of Artificial Intelligence, AAAI 2006. The AAAI Press (2006)
9. Mouratidis, K., Lin, Y., Yiu, M.: Preference queries in large multi-cost transportation networks. In: Proc. of the 26th International Conference on Data Engineering (ICDE), Long Beach, CA, pp. 533–544 (2010)
10. Graf, F., Kriegel, H.P., Renz, M., Schubert, M.: Mario: Multi attribute routing in open street map. In: Proc. of the 12th International Symposium on Spatial and Temporal Databases (SSTD), Minneapolis, MN (2011)
11. Zheng, Y., Zhou, X. (eds.): Computing with Spatial Trajectories. Springer (2011)

APPENDIX: Advanced Route Skyline Computation

Algorithm 3. ARSC

Require: start node s, target node t, network Graph G
    initialize queue Q_{node} of nodes composing skyline candidates, sorted in ascending order w.r.t. a
    given score function and list S_{route} of fully expanded skyline routes
    while $\neg Q_{node}$.isEmpty **do**
      $v = Q_{node}$.top
      for all paths $p \in$ v.SRS compute vector p.LB consisting of lower bound estimations for each
      path attribute w.r.t. t **do**
        if p.LB is dominated by some route in S_{route} // PC 1 **then**
          remove p from v.SRS
        else
          if $\neg$p.isProcessed **then**
            p.isProcessed = true
            expand p by one hop in each direction and store all new paths in the list pNext
            for all p' in pNext **do**
              if p' ends at t **then**
                if p' is not dominated by any route in S_{route} **then**
                  S_{route}.insert(p') and remove all routes from S_{route} that are dominated by p'
                end if
              else
                vNext = p'.lastNode
                if p' is not dominated by any route in vNext.SRS // PC 2 **then**
                  vNext.SRS.insert(p') and remove all sub-routes from vNext.SRS that are
                  dominated by p'
                end if
              end if
              update Q_{node}
            end for
          end if
        end if
      end for
    end while
    return S_{route}

In accordance with [4], we present the Advanced Route Skyline Computation algorithm (ARSC) 3. It is based on two pruning criteria:

PC1 Lower bounding forward cost estimation for partially expanded routes: For arbitrary start and target nodes s, t respectively and sub-routes $\langle s, \ldots, v \rangle$ the function c_{LB} underestimates the cost for any complement $\langle v, \ldots, t \rangle$ using the Reference Node Embedding, i.e. $c(\langle s, \ldots, v \rangle) + c_{LB}(\langle v, \ldots, t \rangle) \leq c(\langle s, \ldots, v, \ldots, t \rangle)$

PC2 Sub-Route Skyline Criterion: Any sub-route p ending at a node v can be excluded from further expansion if it is dominated by another route p' ending at the same node v.

Thus, the ARSC computes route skylines efficiently, being able to prune routes avoiding costly expansion up to the target node. It requires an updatable priority queue Q_{node} of nodes. Every node n maintained in Q_{node} stores its own sub-route skyline in a list $n.SRS$.

Mining Sub-trajectory Cliques
to Find Frequent Routes*

Htoo Htet Aung, Long Guo, and Kian-Lee Tan

School of Computing, National University of Singapore

Abstract. Knowledge of the routes frequently used by the tracked objects is embedded in the massive trajectory databases. Such knowledge has various applications in optimizing ports' operations and route-recommendation systems but is difficult to extract especially when the underlying road network information is unavailable. We propose a novel approach, which discovers frequent routes without any prior knowledge of the underlying road network, by mining sub-trajectory cliques. Since mining all sub-trajectory cliques is NP-Complete, we proposed two approximate algorithms based on the *Apriori* algorithm. Empirical results showed that our algorithms can run fast and their results are intuitive.

1   Introduction

Advances in location-tracking technologies, such as the GPS, enable access to spatial-temporal movement data (trajectory data) of the tracked objects in question. Such movement data are usually archived in **T**rajectory **D**atabases (TJDBs) for further analysis to discover actionable knowledge and support decision making. For instance, the **A**utomatic **I**dentification **S**ystem (AIS) transmits the trajectory of a ship to maritime authorities, who use it to track and monitor the movement of the vessels in their territories. The authorities often archive the ship trajectories for further studies to obtain actionable knowledge, which is, in turn, used to optimize their ports' operations. Similarly, businesses in the public transportation industry (taxi and bus operators) and those in the logistics industry record and archive the movement data of their fleets in TJDBs for analysis aiming to improve the quality of their services.

Since spatial-temporal movement data (trajectories) of the tracked objects are archived in TJDBs, the tracks taken by the entities in question are also recorded in the TJDBs. Therefore, the knowledge of "frequent routes" — a frequent route can be loosely defined as a path, which many of the tracked objects take frequently — is embedded in the massive TJDBs and such knowledge is useful in many applications. For example, maritime authorities can use the frequent routes of vessels to optimize their port operations. Another interesting application of the knowledge of frequent routes is in route suggestions. Current traffic navigation systems (marketed as GPS devices) use the shortest-paths in

---

* An extended version of this paper is available as a technical report at
  http://www.comp.nus.edu.sg/~tankl/sstd13.pdf

M.A. Nascimento et al. (Eds.): SSTD 2013, LNCS 8098, pp. 92–109, 2013.

the road network to navigate their users to reach their destinations. This approach has several limitations since the shortest route is not necessarily the best route in terms of time taken to travel. Moreover, the shortest path may not be suitable for the tourists, when the recommended path passes undesirable areas such as areas having high crime rates. Knowledge of how to select the best route is often embedded in locals' trajectories as frequent routes since the locals know and often take the best route.

Mining frequent routes from a TJDB is not trivial for the following reasons:

- Firstly, in many applications, the underlying road network (or semantic and properties of spatial-regions) is not available. For instance, pedestrians are not confined to road networks and will walk arbitrarily (through buildings and open-spaces). Therefore, without the information of all the underlying routes, it is not possible to count the number of time each route is used.
- Secondly, two vehicles travelling along the same road rarely have two identical sequences of locations reported in the Trajectory Databases because the spatial space is continuous. Even if the movement is made on the exact same path, it is still not possible to directly match the sequence of locations (sub-trajectories) as the movements made may be at different speeds and the two vehicles may have different GPS sampling rates. Therefore, matching two sub-trajectories if they are taking the same route is not trivial and needs a complicate similarity metric.

Contributions. Since a road network or semantic of the regions of the spatial space the moving objects are traversing is often not available, we explored the option of grouping similar sub-trajectories together and extracting a frequent route from each group as this two-step method does not require to have the underlying road networks that the moving entities in question take. Our approach is different from the approaches proposed in [1–4] because:

- The methods appeared in [1, 2] convert the tracks logged in the TJDB into a finite sequences of regions (the set of regions is also finite) and perform sequence mining on the converted tracks using either a prior knowledge of the spatial-space or a pre-processing step, while our proposed solutions require neither a prior knowledge or pre-processing. In addition, the end results obtained from the techniques in [1, 2] contain sequences of regions, which have lost many subtle but important details of the frequent routes contrary to the detailed routes our algorithms report.
- The methods appeared in [3, 4] partition the tracks into line-segments and cluster the line-segments resulting in long frequent routes being reported as multiple line-segment clusters and, hence, in order to infer these long frequent routes, a post-processing step on the line-segment clusters is required, while our proposed solutions directly report long frequent routes.

In order to group similar sub-trajectories, i.e. sub-trajectories taking the same route, together in the same group, we used Fréchet distance as the similarity measure in grouping sub-trajectories. Since Fréchet distance is indifferent to speed, independent to the spatial distance between sampling points, and independent to sampling frequency, sub-trajectories having taken the same route belong to the same sub-trajectory group regardless of the speed they travelled, differences in sequence of locations the corresponding tracked objects reported, and differences in their GPS sampling rates.

Finding groups of sub-trajectories — or sub-trajectory clusters/cliques — using Fréchet distance is a known NP-Complete problem and there is no polynomial approximation for approximation factor less than 2 [5]. We approach the problem from data-driven perspective by proposing an output-sensitive approximation algorithm based on the *Apriori* algorithm. In the experiments using real-life data, our proposed algorithm performs faster and is more accurate than the known polynomial-time approximation algorithm proposed in [5]. In addition, we propose a divide and conquer algorithm based on our proposed algorithm both to maintain the memory requirements under a manageable amount and to achieve parallelism.

2   Related Works

In order to discover frequent routes, the paths that the tracked objects frequently use to travel possibly at different points in time, the earlier works [1,2] suggested to divide the spatial region into regions, transform the trajectories into sequences of regionsm and perform sequence mining on the resulting sequences. This approach has two obvious drawbacks. First, they need a pre-processing step and a prior knowledge of the underlying spatial region in order to divide the entire map into regions. Secondly, the granularity of resulting frequent route is reduced depending on the size of the regions. Therefore, they are not suitable in situations, where a prior knowledge of the spatial region is not available and/or a high granularity output frequent routes are required.

In the absence of underlying spatial information such as road network data, however, researchers suggested a two-step methods — (a) group similar sub-trajectories together and (b) extract a representative route from each group — to find frequent routes [3–5]. Lee *et al* [3] suggested to divide the trajectories into simplified trajectory-segments (line-segments), cluster the trajectory-segments, and calculate representative trajectories from each cluster. Their method, however, is not applicable to trajectories that cross each other often [6]. Zhu *et al* [4] proposed a similar method, in which partition is performed by a grid of uniformly sized cells and combining trajectory clusters found across cells.

Buchin *et al* [5] suggested that one can simply choose an arbitrary sub-trajectory in a group as its representative route (reference trajectory) if the group contains only sub-trajectories, which are **strictly similar** to each other. They proposed to use Fréchet distance (see below) as the similarity measure to ensure each sub-trajectory group contains strictly similar sub-trajectories. They define a sub-trajectory cluster as a set of sub-trajectories such that (a) it contains m distinct sub-trajectories, (b) the longest of them is not shorter than l, and (c) all its sub-trajectories are within a distance r from each other, where m, l, and r are user-defined parameters. They proved that, given a TJDB of size n and parameters, r, m, and l, finding longest sub-trajectory clusters is NP-Complete. They also proposed approximation algorithms having an approximation factor of 2 for several variants of the problem.

Fréchet Distance. Fréchet distance is regarded as a natural measure to quantify similarity of curves [7]. An early treatment on computing Fréchet distance for two polygonal curves is given in [8]. It reports that for two curves defined by p and q points, deciding whether the Fréchet distance between them is less than a given threshold r needs $O(pq)$ time. They defined a two-dimensional data structure called Free-space as a set of points that visualize whether a pair of two points in the given two polygonal curves are within a Euclidean distance of r and proved that the Fréchet distance between two polygonal curves is not more than r if and only if there is a monotone curve in the corresponding Free-space. Their results are extended for a set of polygonal curves in [9].

3   Sub-trajectory Cliques and the Frequent Routes

Definition 1. *Trajectory of an Object* — *The Trajectory of a given object o is a ordered sequence: $traj_o = \langle t_{\text{start}}(o), loc_o(t_{\text{start}}(o)) \rangle, \cdots, \langle t_{\text{end}}(o), loc_o(t_{\text{end}}(o)) \rangle$, where $t_{\text{start}}(o)$ and $t_{\text{end}}(o)$ denote the earliest and latest time-stamps that o reported its location respectively while $loc_o(t)$ denotes the location $loc \in \mathbb{R}^2$ of object o at time-stamp t.*

Definition 2. *Sub-trajectory* — *Given the trajectory of an object $traj_o$ and two time-stamps s and e such that $t_{\text{start}}(o) \leq s \leq e \leq t_{\text{end}}(o)$, the sub-trajectory of o between s and e is the ordered sequence, i.e. $\langle s, loc_o(s) \rangle, \cdots, \langle e, loc_o(e) \rangle$.*

Definition 1 defines a trajectory of an object while Def. 2 defines a sub-trajectory as a continuous portion of a trajectory. Figure 1(a) shows two example trajectories of two objects, a and b. Trajectory of a is $traj_a = \langle t_{a1}, (x_{a1}, y_{a1}) \rangle, \cdots, \langle t_{a5}, (x_{a5}, y_{a5}) \rangle$. Without lost of generality, we will assume the time is continuous and there is a method to derive the location of objects in any real time-stamp $t \in \{ts \in \mathbb{R} | t_{\text{start}}(o) \leq ts \leq t_{\text{end}}(o)\}$. We will also use the same term "sub-trajectory" to refer to the corresponding route of a sub-trajectory (without time-information). For instance, in Fig. 1(a), sub-trajectory of a between time-stamp t_{a2} and t_{a4} is $sub_a(t_{a2}, t_{a4}) = \langle t_{a2}, (x_{a2}, y_{a2}) \rangle, \cdots, \langle t_{a4}, (x_{a4}, y_{a4}) \rangle$ and its corresponding route is the poly-line: $(x_{a2}, y_{a2}), \cdots, (x_{a4}, y_{a4})$.

Definition 3. *Fréchet Distance between Two Sub-trajectories* — *Let $st_a = sub_a(s_a, e_a)$ and $st_b = sub_b(s_b, e_b)$ be two sub-trajectories. Also let $\mathcal{A} = \{\alpha | \alpha : [0, 1] \to [s_a, e_a]$ and α is monotone.$\}$ and $\mathcal{B} = \{\beta : \beta : [0, 1] \to [s_b, e_b]$ and β is monotone.$\}$ be two sets of re-parametrizations.*

The Fréchet distance $dist_{\text{Fr}}$ between the two sub-trajectories, sub_a and sub_b, is defined as the minimum $dist_{\max}$ over $\mathcal{A}$ and $\mathcal{B}$, where $dist_{\max}(\alpha, \beta)$ is the maximum distance between $loc_a(\alpha(x))$ and $loc_b(\beta(x))$ for all $x \in [0, 1]$.

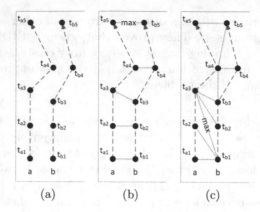

Fig. 1. (a) Two Routes and (b) a Possible Re-parametrization Pair of the Routes and (c) Another Re-parametrization Pair of the Routes

Definition 3 defines a distance measure between two sub-trajectories, known as Fréchet distance in the literature. Intuitively, the Fréchet distance between two given routes is the minimum length of leash required for a man and his dog, each walking on a route. Each is (intelligently) trying to minimize the required leash length under a constraint that they cannot walk backward (but they may stop walking to wait for the other).

We are going to briefly discuss how the Fréchet distance of the two example routes, a and b, in Fig. 1(a)[1] is calculated. Let us first assume route a (b) is taken by the man (his dog). Each re-parametrization $\alpha \in \mathcal{A}$ ($\beta \in \mathcal{B}$) corresponds to how the man (his dog) chooses to walk. The Fréchet distance between the two given routes is the minimum $dist_{\max}(\alpha, \beta)$ of all pairs of $(\alpha, \beta) \in \mathcal{A} \times \mathcal{B}$. Two such pairs, (α_1, β_1) and (α_2, β_2), are given in Table 1 and some of the location pairs given by the two re-parametrization pairs are visualized (shown as thin dotted lines) in Fig. 1(b) and Fig. 1(c).

For each pair (α, β), the **continuous** function $f_{(\alpha,\beta)} : [0,1] \rightarrow \mathbb{R}$ is defined as $f_{(\alpha,\beta)}(x)$ is the Euclidean distance between $loc_a(\alpha(x))$ and $loc_b(\beta(x))$. The leash length required for each pair of movement is the maximum value of the image of f. For instance, $f_{(\alpha_1,\beta_1)}(0.25) = dist(loc_a(t_{a2}), loc_b(t_{b2}))$ and $f_{(\alpha_2,\beta_2)}(0.25) = dist(loc_a(t_{a3}), loc_b(t_{b1}))$. Thus, the required leash length for the first pair, $dist_{\max}(\alpha_1, \beta_1)$, is given by $f_{(\alpha_1,\beta_1)}(1.00) = dist(loc_a(t_{a5}), loc_b(t_{b5}))$ — denoted as "max" in Fig. 1(b) — while that of the second pair, $dist_{\max}(\alpha_2, \beta_2)$, is by $f_{(\alpha_2,\beta_2)}(0.2) = dist(loc_a(t_{a3}), loc_b(t_{b1}))$ — denoted as "max" in Fig. 1(c).

Hence, we can say that, in the first pair of re-parametrizations (see Fig. 1(b)), the man and his dog walk in a more synchronized manner requiring shorter (maximum) leash length than the second pair (see Fig. 1(c)), in which the man

---

[1] The time-stamps in this example are arbitrary, i.e. two time-stamps with the same numerals do not necessarily equal and the time-interval between two time-stamps with the same alphabet and consecutive numerals are not necessarily the same.

Table 1. Two Possible Pairs of Re-parametrizations for the Two Sub-trajectories Shown in Fig. 1(a)

Re-parametrization Pair 1						Re-parametrization Pair 2					
x	$\alpha_1(x)$	$\beta_1(x)$	x	$\alpha_1(x)$	$\beta_1(x)$	x	$\alpha_2(x)$	$\beta_2(x)$	x	$\alpha_2(x)$	$\beta_2(x)$
0.00	t_{a1}	t_{b1}	0.50	t_{a3}	t_{b3}	0.00	t_{a1}	t_{b1}	0.50	t_{a3}	t_{b3}
⋮	⋮	⋮	⋮	⋮	⋮	⋮	⋮	⋮	⋮	⋮	⋮
0.01	t_{a1}	t_{b1}	0.51	t_{a3}	t_{b3}	0.01	t_{a2}	t_{b1}	0.51	t_{a4}	t_{b3}
⋮	⋮	⋮	⋮	⋮	⋮	⋮	⋮	⋮	⋮	⋮	⋮
0.02	t_{a1}	t_{b1}	0.75	t_{a4}	t_{b4}	0.02	t_{a3}	t_{b1}	0.75	t_{a4}	t_{b4}
⋮	⋮	⋮	⋮	⋮	⋮	⋮	⋮	⋮	⋮	⋮	⋮
0.25	t_{a2}	t_{b2}	0.76	t_{a4}	t_{b4}	0.25	t_{a3}	t_{b1}	0.76	t_{a4}	t_{b5}
⋮	⋮	⋮	⋮	⋮	⋮	⋮	⋮	⋮	⋮	⋮	⋮
0.26	t_{a2}	t_{b2}	1.00	t_{a5}	t_{b5}	0.26	t_{a3}	t_{b2}	1.00	t_{a5}	t_{b5}
⋮	⋮	⋮	⋮	⋮	⋮	⋮	⋮	⋮	⋮	⋮	⋮

has walked (on route a) far before his dog begins to move (on route b). The Fréchet distance is the shortest among the (maximum) leash lengths required for all manners of progressing the man and his dog can make on their routes (including those depicted in this example).

In essence, Fréchet distance considers only the direction (the man and his dog are not allowed to walk back) and the shape of the curves (it chooses the minimum among the leashes for all ways they can walk). Yet, it neither considers the speed (each parametrization represent a different speed) nor the number of sampled points (the re-parametrizations are defined as functions to a continuous range), i.e. the Fréchet distance between two curves in the above example is the same even if we remove the sample point at t_{a_2} and put multiple sample points between t_{b1} and t_{b_2} by linear interpolation. Therefore, we choose the Fréchet distance to measure the similarity between sub-trajectories.

Definition 4. *Sub-trajectory Clique* — *Consider a Trajectory Database $\mathcal{R}$, which contains the trajectories of a set of objects O. Given parameters m, l, and r, a set of sub-trajectories J forms a sub-trajectory clique (*TRAJCLIQ*) if (i) each $j \in J$ is within r Fréchet distance away from all sub-trajectories $j' \in J$, (ii) the lengths of all sub-trajectories $j \in J$ are longer than l and (iii) J contains at least m non-identical sub-trajectories.*

Definition 4 defines a sub-trajectory clique (TRAJCLIQ) which includes m non-identical sub-trajectories, which are at least l units in spatial length (not time duration). Two sub-trajectories are non-identical if either they are taken by different tracked objects or, in cases when they are taken by the same object, their starting points are more than r distance away (measured along the trajectory, not Euclidean distance). We choose the Fréchet distance, defined in Def. 3, to measure the similarity between routes because Fréchet distance ensures sub-trajectories in the same clique are spatially close, similar in shape, and similar in direction. Therefore, all nearby sub-trajectories of similar shape and direction to a sub-trajectory will belong to the same TRAJCLIQ even though the corresponding movements were taken at different speeds during different time-spans. In other words, a TRAJCLIQ groups the sub-trajectories, which are on the same route and, hence, a track-clique containing m sub-trajectories corresponds to a frequent route taken at least m times.

Figure 2(a) illustrates a TJDB containing trajectories of four objects, a, b, c, and d. For $m = 3$, the sub-trajectory of a from t_{a1} to t_{a4}, another sub-trajectory of a from t_{a7} to t_{a10} and the sub-trajectory of b from t_{b1} to t_{b4} form a Sub-trajectory clique (i.e. $sub_a(t_{a1}, t_{a4})$, $sub_a(t_{a7}, t_{a10})$, and $sub_b(t_{b1}, t_{b4})$ form a TRAJCLIQ) as the three sub-trajectories are within a Fréchet distance of r. Note that two sub-trajectories of a single object, i.e. object a, can involve in the same TRAJCLIQ as they are non-identical (object a has travelled more than r distance between $loc_a(t_{a1})$ and $loc_a(t_{a7})$). However, the sub-trajectory of a from t_{a5} to t_{a6}, that of c from t_{c1} to t_{c2}, and that of d from t_{d1} to t_{d2} do not form a TRAJCLIQ because the Fréchet distance between the sub-trajectories of c and d is more than r.

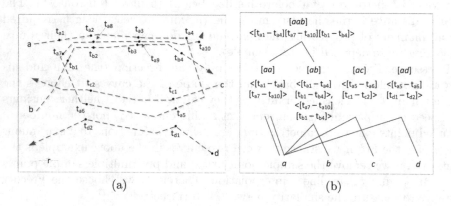

(a)   (b)

Fig. 2. (a) A Trajectory Database Containing Four Trajectories and (b) its Corresponding Search-space

Definition 5. *Frequent Route* — *A sub-trajectory clique is closed if and only if there is no other sub-trajectory clique covering it. The longest sub-trajectory in a closed sub-trajectory clique is defined as the corresponding **Frequent Route** (FreqRo) of the closed sub-trajectory clique.*

Definition 5 defines a Frequent Route as the longest sub-trajectory (spatial length, not time duration) of a closed TRAJCLIQ. A frequently-used route derived from a non-closed sub-trajectory is useless to the users since its information is conveyed in the frequent route derived from a closed TRAJCLIQ.

4 Methods to Mine Sub-trajectory Cliques to Extract Frequent Routes

Since mining all closed sub-trajectory cliques (TRAJCLIQ) is NP-Complete [5], we will describe a data-driven exact algorithm to discover all closed TRAJCLIQs and extract a corresponding frequent route from each TRAJCLIQ. The exact algorithm is based on the *apriori*-properties of the TRAJCLIQs. We will proceed to discuss why such an algorithm is not feasible and present two new algorithms to approximate TRAJCLIQs. The first approximation algorithm has an approximation factor of 2. The second algorithm divides the TJDB into distinct subsets so that an instance of the first algorithm can be applied to each subset, reducing memory requirement and/or enabling parallel processing.

4.1 Apriori-Based Frequent Route Miner

Since mining all closed TRAJCLIQ is an NP-Complete problem, we are going to analyze the feasibility of using an *Apriori*-based data-driven algorithm to discover closed TRAJCLIQs as intermediate results and extract Frequent Routes from them. First, we will present the *apriori*-properties of TRAJCLIQs.

Lemma 1. *Suppose a set of sub-trajectories $J = \{sub_1, \cdots, sub_p\}$ and its subset $J' = \{sub'_1, \cdots, sub'_q\}$ ($p \geq q$). If there is a re-parametrization α_i for each sub-trajectory $sub_i \in J$ such that for any $i, j \in \{1, \cdots, p\}$, $dist_{\max}(\alpha_i, \alpha_j) \leq r$, then there is a re-parametrization α'_k for each sub-trajectory $sub'_k \in J'$ such that for any $i, j \in \{1, \cdots, q\}$, $dist_{\max}(\alpha'_i, \alpha'_j) \leq r$.*

Proof. For a pair of sub-trajectories $sub'_i, sub'_j \in J'$, suppose there is no pair of re-parametrizations α'_i and α'_j, which gives the $dist_{\max}(\alpha'_i, \alpha'_j) \leq r$. By definition, (although not necessarily unique) there is a pair of re-parametrizations that gives minimum $d = dist_{\max}(\alpha'_i, \alpha'_j)$ — i.e. $r < d$.

However, since $J' \subseteq J$, $sub'_i \in J$ and $sub'_j \in J$. Therefore, α_i and α_j for sub-trajectories sub'_i and sub'_j gives $dist_{\max}(\alpha_i, \alpha_j) \leq r < d$ (contradiction). $\square$

Using Lemma 1, we can derive the *apriori*-properties of TRAJCLIQs as follow: if a set of sub-trajectories $J = \{sub_1, \cdots, sub_p\}$ containing p sub-trajectories forms a TRAJCLIQ, there is a re-parametrization α_i for each $sub_i \in J$, which gives $dist_{\max(\alpha_i, \alpha_j)} \leq r$ for any $sub_i, sub_j \in J$. Following Lemma 1, there is also (at least) a re-parametrization α'_i for each $sub_i \in J' \subseteq J$, which gives $dist_{\max(\alpha'_i, \alpha'_j)} \leq r$ for any $i, j \in J'$. Therefore, the subset J' of a set J of sub-trajectories forming a TRAJCLIQ also forms a TRAJCLIQ. In other words, J does not form a TRAJCLIQ if any of its subset does not.

The *Apriori*-based Frequent Route Miner (A-0), adapted from the *Apriori*-algorithm in [10], exploits the *apriori*-properties of the TRAJCLIQs to systematically discover the TRAJCLIQs formed by $(k+1)$ sub-trajectories only when those formed by its subsets exist. An outline of the A-0 is given in Algorithm 1. The function CLIQ($\mathcal{R}, l, r, O_k$) returns list of TRAJCLIQs, which may or may not be closed, formed the k sub-trajectories of the objects in O_k.

Algorithm 1. Apriori-based Frequent Route Miner (A-0)

Input: $\mathcal{R}$, r, m and l.
Output: A set of frequent routes $\mathcal{F}$.
 1: Set the set of object-lists $\mathcal{C}_1 \leftarrow \emptyset$, $\mathcal{U} \leftarrow \emptyset$, $\mathcal{F} \leftarrow \emptyset$ and $k \leftarrow 2$
 2: **for all** $(o, o') \in O \times O$ **do**
 3:     Set Object-list $O_2 \leftarrow [o, o']$ and Clique-list $L(O_2) \leftarrow \text{CLIQ}(\mathcal{R}, l, r, O_2)$
 4:     **if** $L(O_2)$ is not empty **then**
 5:         Set $\mathcal{C}_2 \leftarrow \mathcal{C}_2 \cup \{O_2\}$
 6: **while** $\mathcal{C}_k \neq \emptyset$ **do**
 7:     **for all** $O_k \in \mathcal{C}_k$ **do**
 8:         **if** $k \geq m$ **then**
 9:             Set $\mathcal{U} \leftarrow \mathcal{U} \cup L(O_k)$
10:     Set the set of Object-lists $\mathcal{C}_{k+1} \leftarrow \emptyset$
11:     **for all** $O_k, O'_k \in \mathcal{C}_k$ such that $k\text{-prefix}(O_k) = k\text{-prefix}(O'_k)$ **do**
12:         Set $O_{k+1} \leftarrow append(O_k, last(O'_k))$ and $L(O_{k+1}) \leftarrow \text{CLIQ}(\mathcal{R}, l, r, O_{k+1})$
13:         **if** $L(O_{k+1})$ is not empty **then**
14:             Set $\mathcal{C}_{k+1} \leftarrow \mathcal{C}_{k+1} \cup \{O_{k+1}\}$
15:     Set $k \leftarrow k + 1$
16: Set $\mathcal{U} \leftarrow \mathcal{U} - \{U | U$ is not a closed-TRAJCLIQ$\}$.
17: **for all** $Q \in \mathcal{U}$ **do**
18:     Set $\mathcal{F} \leftarrow \mathcal{F} \cup \{\text{Get-Frequent-Route}(U)\}$

The algorithm A-0 first initializes the Clique-list containing the TRAJCLIQs formed by sub-trajectories of all possible pairs of objects (lines 2 - 5). Starting with $k = 2$, the Clique-list of TRAJCLIQs formed by $k+1$ sub-trajectories of a list of objects (O_{k+1}) are built only when those of its sub-lists (O_k and O'_k) sharing the same k-prefix form TRAJCLIQs (lines 6 - 15). In doing so, if $k \geq m$, then the TRAJCLIQs in the Clique-list, which are formed by the k sub-trajectories of the current Object-list (O_k), are potential closed TRAJCLIQ. Thus, they are put into the set of TRAJCLIQs $\mathcal{U}$ (lines 7 - 9), which is later filtered to remove non-closed TRAJCLIQs (line 16). As the last step, the algorithm A-0 extracts a frequent route from each closed TRAJCLIQin $\mathcal{U}$ (lines 17 - 18).

In order to allow a TRAJCLIQto contain two (different) sub-trajectories of the same object, unlike other *Apriori*-based algorithms, A-0 uses a list no-tation for objects, which may contain an object multiple times. Figure 2(b) shows a portion of the search space the algorithm A-0 traverses to find the TRAJCLIQs in the data in Fig. 2(a) for given parameters r, l and $m = 3$. A-0 starts with finding TRAJCLIQ formed by two sub-trajectories of objects, starting from the TRAJCLIQs formed by sub-trajectories of the Object-list $[aa]$. Since sub-trajectories $sub_a(t_{a1}, t_{a4})$ and $sub_a(t_{a7}, t_{a10})$ forms a TRAJCLIQ, it is stored in the Clique-list $L([aa])$. It continues to find TRAJCLIQs formed by all the Object-lists containing two objects with prefix $[a]$, i.e. $[ab]$, $[ac]$ and $[ad]$. Then A-0 tries to find TRAJCLIQs formed by all Object-lists containing two objects with prefixes $[b]$, $[c]$ and $[d]$, which do not exist in the TJDB depicted in Fig. 2(a).

The first row ($k = 2$) of the Table 2 lists the state of the variables after this initialization step. In the next step ($k = 3$), A-0 tries to find the TRAJCLIQs formed by sub-trajectories of $[aaa]$, $[aab]$, $[aac]$ and $[aad]$ since their subsets are in $\mathcal{C}_2$. It only finds a TRAJCLIQ formed by sub-trajectories of Object-list $[aab]$ and put $[aab]$ into $\mathcal{C}_3$ (see the second row in Tab. 2). Since $k \geq m$, $L([aab])$ is put into the result set $\mathcal{U}$. The algorithm exits the loop at $k = 4$ as $\mathcal{C}_4$ becomes empty and output the frequent route extracted from $L([aab])$.

Table 2. A Trace of A-0 Running on the Trajectory Database in Fig. 2(a)

k	$\mathcal{C}_k$	$L(O_k)$	$\mathcal{U}$
2	$[aa]$	$\langle sub_a(t_{a1}, t_{a4}), sub_a(t_{a7}, t_{a10}) \rangle$	—
	$[ab]$	$\langle sub_a(t_{a1}, t_{a4}), sub_b(t_{b1}, t_{b4}) \rangle,$	
		$\langle sub_a(t_{a7}, t_{a10}), sub_b(t_{b1}, t_{b4}) \rangle$	—
	$[ac]$	$\langle sub_a(t_{a5}, t_{a6}), sub_c(t_{c1}, t_{c2}) \rangle$	—
	$[ad]$	$\langle sub_a(t_{a5}, t_{a6}), sub_d(t_{d1}, t_{d2}) \rangle$	—
3	$[aab]$	$\langle sub_a(t_{a1}, t_{a4}), sub_a(t_{a7}, t_{a10}), sub_b(t_{b1}, t_{b4}) \rangle$	$L([aab])$
4	ϕ	ϕ	$L([aab])$

Finding Sub-trajectory Cliques of k Sub-trajectories. In algorithm A-0, lines 3 and 12 call a sub-routine (CLIQ) in order to find all sub-trajectory cliques composed of k sub-trajectories. CLIQ uses Free-space (defined below) to extract the sub-trajectory cliques.

Definition 6. *Free-space — Given a distance threshold r and two sub-trajectories sub_i and sub_j containing p and q segments respectively, their corresponding Free-space is the set:* $F(i, j) = \{(x, y) \in [0, p] \times [0, q] : dist(loc_i(x), loc_j(y)) \leq r\}$.

Definition 6 defines a free-space as a two-dimensional map that maintains, which parts of the two sub-trajectories in questions are within distance r. Figure 3 illustrates a Free-space (cell) of two sub-trajectories i and j containing one segment each for a distance r as the white region. The two points marked by black rectangles in the Fig. 3 is closer than the distance r. Therefore the point in the Free-space that represents these two points of the trajectories falls in the white region while that of two points marked by black circles, in the grey region. For sub-trajectories containing p and q segments, their Free-space for a given r is a two-dimensional array of such Free-space cells.

For two-dimensional GPS trajectories, the white region of a Free-space cell is always an intersect of an eclipse and a rectangle [8]. The white region, therefore, is defined by eight points called critical points (illustrated as black stars in Fig. 3). Adjacent Free-space cells share the critical points between them. For example, the critical points on the right boundary of the Free-space cell at $(x - 1, y)$ is the same as those on the left of Free-space cell at (x, y).

Alt and Godau [8] proved that the Fréchet distance between two sub-trajectories is less than r if and only if there is a monotone path in their Free-space. The sub-routine CLIQ uses a brute-force process to find all monotone paths

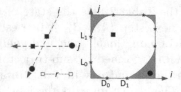

Fig. 3. Two Sub-trajectories and Their Corresponding Free-space

(monotone in all coordinates) in a k dimensional-space in the k-dimensional Free-space of the trajectories of k given objects in O_k. However, for $k \geq 3$, despite being polynomial, this process is not feasible for real-life datasets containing hundreds of trajectories with thousands of segments. In addition, the generalized process needs $k(k-1)/2$ free-space to be kept in the memory for all $k \geq 2$. Therefore, we turn our focus to approximation algorithms.

4.2 Approximation of Sub-trajectory Cliques

We observe that if a set of sub-trajectories $J = \{sub_1, sub_2, \cdots, sub_p\}$ forms a TRAJCLIQ, the Fréchet distance between the first sub-trajectory, sub_1, and other sub-trajectories, $sub_2, \cdots, sub_p$ is at most r. We will use this observation to approximate TRAJCLIQs formed by $k \geq 3$ sub-trajectories. We will denote the first sub-trajectory in a set of sub-trajectories as the "reference sub-trajectory" (or simply "reference trajectory") of the TRAJCLIQ it forms. The approximation factor of our proposed solution is 2 since Fréchet distance follows triangular inequality [7].

The approximation algorithm we develop, namely *Apriori*-based Approximate Frequent Route Miner (A-1), is essentially following the same steps of A-0. The only difference between the exact algorithm (A-0) and the approximate one (A-1) is that A-1 uses an approximation sub-routine instead of the exact CLIQ in line 12. For the trajectories in Fig. 2(a), A-1 finds the TRAJCLIQs of Object-list containing two objects (the same as A-0). Then, for $k = 3$, it tries to approximate TRAJCLIQs of sub-trajectories in Object-lists, whose subsets are found in O_2. A-1 finds two (approximate) TRAJCLIQs, $J_1 = \{sub_a(t_{a1}, t_{a4}), sub_a(t_{a7}, t_{a10}), sub_b(t_{b1}, t_{b4})\}$ and $J_2 = \{sub_a(t_{a5}, t_{a6}), sub_c(t_{c1}, t_{c2}), sub_d(t_{d1}, t_{d2})\}$ using $sub_a(t_{a1}, t_{a4})$ and $sub_a(t_{a5}, t_{a6})$ as their reference trajectories respectively. It does not find any TRAJCLIQs for $k = 4$.

Reducing the Number of False-Positives. Since the Fréchet distance between the sub-trajectory $sub_c(t_{c1}, t_{c2})$ and the sub-trajectory $sub_d(t_{d1}, t_{d2})$ is larger than r, the set of sub-trajectories J_2 that our algorithm A-1 approximates as a TRAJCLIQ is a false positive the approximation introduces. We further exploit the *apriori*-properties of TRAJCLIQs to reduce the number of false positives in the approximation results. When the approximation sub-routine builds the TRAJCLIQs formed by sub-trajectories $\{s_1, s_2, \cdots, s_{k+1}\}$ of $O_{k+1} = [o_1, o_2, o_3, \cdots o_{k+1}]$, it also checks whether $\{o_2, o_3, \cdots, o_{k+1}\}$ also forms

TRAJCLIQs. This simple check prunes false positives as illustrated in the following example. Consider the sub-trajectories $s_a = sub_a(t_{a5}, t_{a6})$, $s_b = sub_c(t_{c1}, t_{c2})$, and $s_d = sub_d(t_{d1}, t_{d2})$ in Fig. 2(a). The Fréchet distance between s_c and s_d is larger than r and s_c and s_d do not form a TRAJCLIQ. Therefore, before the approximation sub-routine builds the sub-trajectory cluster $\langle s_a, s_c, s_d \rangle$, it checks whether s_c and s_d form a TRAJCLIQand, since they do not, the approximation sub-routine simply prunes the TRAJCLIQs $\langle s_a, s_c, s_d \rangle$.

This pruning mechanism may not be able to remove all false-positives since A-1 is still approximating a TRAJCLIQ by checking only the Fréchet distances between the reference trajectory and the other sub-trajectories in a candidate. The approximation factor also remains at 2. However, we expect A-1 to have fewer false-positives in its results compared to other approximation algorithms.

4.3   A Divide and Conquer Scheme

Although the *Apriori*-based Approximate Frequent Route Miner (A-1) can run fast, it still needs a large amount of memory both for the *Apriori* process and the Free-space it needs to calculate and store. For larger datasets and real-life computing settings, in which the available main memory is limited, this memory requirement may become a big challenge. Therefore, we devise a Divide and Conquer Scheme to mitigate the memory requirement issue of A-1.

We observe that, given an arbitrary bounding box B, all sub-trajectories it confines (sub-trajectories are completely within it) cannot have a Fréchet distance less than or equal to r to those sub-trajectories not confined by the bounded box B', which is B extended by r on all sides. For instance, in Fig. 4, the whole trajectory of a is in the shaded bounding box. Therefore, it cannot have its Fréchet distance with the sub-trajectories of x (such as $sub_x(t_{x2}, t_{x4})$) that goes beyond the extended bounding box shown by the thick borders. In other words, any sub-trajectory of a cannot form a TRAJCLIQ with $sub_x(t_{x2}, t_{x4})$.

From this observation, we deduce that if we divide the spatial-space into multiple zones with stripes of width r between them, the (class of) sub-trajectories, which pass the stripes, cannot form a TRAJCLIQ with the sub-trajectories confined in the zone. In the above example depicted in Fig. 4, there are four zones and two stripes — one horizontal and one vertical. The sub-trajectories passing either the horizontal and vertical stripes cannot form TRAJCLIQs with any sub-trajectories confined in the four zones, i.e. sub-trajectory $sub_x(t_{x1}, t_{x4})$ passes the vertical strip and, hence, it cannot form a TRAJCLIQ with any sub-trajectories confined in one of the zones like $sub_x(t_{a1}, t_{a2})$, $sub_y(t_{y4}, t_{y5})$, and $sub_p(t_{p1}, t_{p2})$.

Algorithm A-2 utilizes the Divide and Conquer Scheme and works as follow. The spatial-space is divided into zones with stripes of width r between the zones of size $\lambda \times \lambda$, starting with λ at the user-defined zone-size λ_0. For each zone z, which has at least a trajectory wholly confined in, the sub-trajectories in its extended zones z^+ are processed using algorithm A-1. After each extended zone z^+ is processed, trajectories wholly contained in z is removed from $\mathcal{R}$ as it cannot form any more TRAJCLIQs with other sub-trajectories (passing the strips) in the next pass. After all zones are processed, the zone-size is enlarged by multiplying

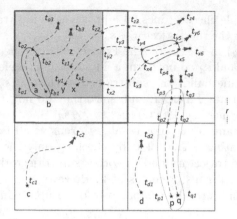

Fig. 4. An Illustration of the Divide and Conquer Scheme

λ with user-defined zone-size multiplier θ. When a single zone span the entire spatial space, A-2 outputs a frequent route for each closed-TRAJCLIQ.

We will illustrate how the algorithm A-2 works using the example trajectories in Fig. 4. Initially, it divides the spatial-space into four zones. For the top-left zone, trajectories a and b, and the maximal sub-trajectories of x, y, and z confined in the extended bounded box (shown by thick borders), i.e. $sub_a(t_{a1}, t_{a3})$, $sub_b(t_{b1}, t_{b3})$, $sub_x(t_{x1}, t_{x3})$, $sub_y(t_{y1}, t_{y3})$ and $sub_z(t_{z1}, t_{z3})$, are processed using A-1 and finds the TRAJCLIQ formed by sub-trajectories $sub_a(t_{a1}, t_{a2})$ and $sub_b(t_{b1}, t_{b2})$. Before moving on to the next zone, A-2 removes trajectories of a and b from $\mathcal{R}$. Then, A-2 skips the top-right zone as no trajectory is confined in it. A-2 moves on to bottom-left zone (finds no TRAJCLIQ and removes trajectory c) and bottom-right zone (finds the TRAJCLIQ $U'_{p,q}\langle sub_p(t_{p1}, t_{p3})$ and $sub_q(t_{q1}, t_{q3})\rangle$ and removes trajectory d). After all four zones are processed, the algorithm A-2 tries to find TRAJCLIQs formed by sub-trajectories that remain. In this stage, two TRAJCLIQs, TRAJCLIQ $U_{x,y} = \langle sub_x(t_{x4}, t_{x5}),\ sub_y(t_{y4}, t_{y5})\rangle$ and $U_{p,q} = \langle sub_p(t_{p1}, t_{p4}),\ sub_q(t_{q1}, t_{q4})\rangle$ are found. Since $U_{p,q}$ covers $U'_{p,q}$, $U'_{p,q}$ is removed from the results. By design, algorithm A-1 and A-2 give the same output and, hence, have the same accuracy.

For large Trajectory Databases, A-2 would provide a reasonable trade-off because it divides both the search space of *Apriori* processing and the portions of Free-space needed in the memory to approximate TRAJCLIQs. Moreover, the divisions of the TJDBs A-2 provides can be processed independently using A-1 and, hence, A-2 can enable parallel mining of Frequent Routes.

5   Experimental Evaluations

We implemented algorithms, A-1, A-2, Sweep [5], and Traclus [3] in Java. Sweep is a 2-distant approximation algorithm for mining sub-trajectory cliques, i.e. its approximation factor is 2. We made minor changes in Sweep to ensure all sub-trajectories in each reported approximated TRAJCLIQs have length l. We ran our experiments on a Linux server equipped with 2.27GHz Intel Xeon CPU E5607 and 32GB of RAM. We had an R-Tree index for each trajectory-segment.

We used five datasets — Statefair, Orlando, New York, NCSU and KAIST — of human movement [11] datasets. In addition, we also used datasets, Trucks [12] consisting of 50 trajectories of trucks moving in Athens and Ships consisting of 458 trajectories of ships moving in Singapore in September 5, 2011. All distance units are in metre. Table 3 shows the details of the datasets we used.

Table 3. A Summary of the Experiment Settings and Performance of Frequent Route Mining Algorithms

Dataset	Number of Objects	Time-span	m	r	l	λ_0	Sweep	A-1	A-2	Traclus
			Parameters				Run-time (seconds)			
Statefair	19	3hr	3	30	500	5,000	22	9	11	870
NCSU	35	21hr	3	30	500	5,000	129	63	65	6,722
New York	39	22hr	3	30	500	5,000	178	103	113	3,427
Orlando	41	14hr	3	30	500	5,000	196	96	101	18,088
KAIST	92	23hr	3	30	500	5,000	2,617	1,237	1,369	71,443
Trucks	50	33 days	3	20	2,000	5,000	5,709	3,323	1,416	88,123
Ships	458	4hr	3	100	1,500	10,000	500	262	610	—

Table 3 shows the default parameters and run-time performance of each algorithm for different datasets. Sweep uses the same set of parameters m, l, and r as algorithms A-1 and A-2, while, for Traclus, we use all the parameter values as suggested in [3]. We stopped Traclus for Ships dataset after it took several days running. In all datasets, we observe that our proposed algorithms, A-1 and A-2, finished faster than Sweep and Traclus. Traclus took the most time because the distance measure it uses cannot make use of any spatial index. A-1 and A-2 performed faster than Sweep because of their pruning mechanisms. A-1 took less time compared to A-2 for all datasets except Trucks because the Free-space data structure fits in the main memory (favouring A-1) and performance of A-2 depends on the whether the dataset is divided into zones, which can confine as many (whole) trajectories as possible — hence, on λ_0 and θ.

Figure 5(a) shows the frequent routes we extracted from Ships dataset using algorithm A-1. A-1 reported frequent routes used at least five times in four hour period, which we consider as significant giving that ships do tend to follow the exact same routes. Figure 5(b) shows the frequent routes discovered by A-1 in Trucks dataset. It clearly shows the major road sections in Athens, which the trucks in the dataset used multiple times. For both datasets, Sweep also reported a very similar results. However, its results contain more false-positives than A-1 and A-2 as A-1 and A-2 exploits the *apriori*-properties of TRAJCLIQs to prune some false-positives while Sweep does not have any pruning mechanism.

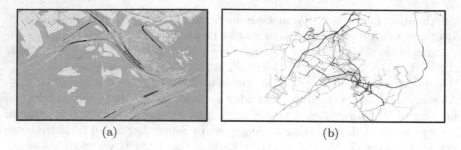

(a)   (b)

Fig. 5. (a) Frequent (Used at least three times and five times) Routes Discovered by Algorithm A-1 (in colors and in black) Superimposed on all Trajectories (in grey) in the Ships Dataset and (b) Frequent (Used at least five times) Routes Discovered by Algorithm A-1 (in black) Superimposed on All Trajectories (in light pink) in the Trucks Dataset.

Figure 6 compares the significant portions of the results produced by Traclus with the frequent routes, which were used more than five times, produced by A-1 for the same area. The results produced by Traclus, using its default parameters, contain a few trajectory clusters, from which we need to extract representative trajectories. The representative routes are often short and straight lines. Traclus failed to find longer and more complicated routes like A-1 did — Traclus did not find the route travelling east to west in the eastern side of the map. The results of A-1 include frequent routes across junctions (involving turns), which Traclus failed to find — notice the routes in the western side of the map, which involves turns and crosses junctions. Through closer inspection, we noticed that, when A-1 found a longer frequent route crossing multiple junctions, Traclus found only a portion of that frequent route or no frequent route at all.

To summarize the comparison of the algorithms' outputs, we learnt that our proposed algorithms produced results, which agrees to human intuitions. On the other hand, Traclus, using suggested parameters, failed to find some frequent

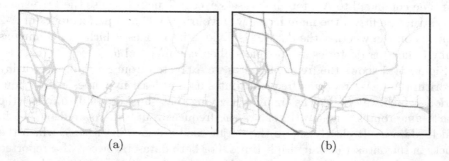

(a)   (b)

Fig. 6. (a) Trajectory Clusters of the Trucks Discovered by Traclus and (b) Frequent Routes of the Trucks Discovered by Algorithm A-1 in the Same Area

routes, which our proposed algorithms found. Since Traclus failed to deliver intuitive output, we omitted Traclus in further experiments.

Figure 7(a) and Figure 7(b) shows the total running time of algorithms for datasets, New York and KAIST, using default l and r while varying m. For both New York and KAIST, A-1 and A-2 finished faster than Sweep, with A-1 being the fastest. The changes in the value of m did not significantly affect the run-time of all algorithms although the run-times were slightly reduced when $m = 4$ as larger m values result in fewer frequent routes required to be outputted.

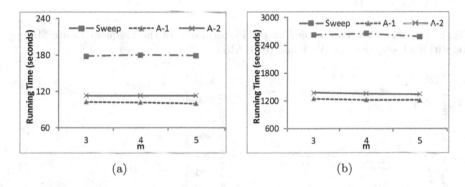

(a)   (b)

Fig. 7. Impact of the Parameter m on the Performance of the Frequent Route Mining Algorithms Using (a) New York and (b) KAIST

Figure 8(a) and Fig. 8(b) show the total running time for datasets, New York and KAIST, with default m and r while varying l. For both datasets, as l increases, the total running time for both A-1 and A-2 decreases, which is more pronounced for A-2 on KAIST. It is because as l increases, the number of TRAJCLIQs and frequent routes decreases and A-1 and A-2 are output sensitive algorithms. We observed A-1 always performed faster than the others and A-2 performed faster than Sweep except in KAIST when $l = 300m$. A-2 took longer time in KAIST for $l = 300m$ because KAIST contains longer trajectories, which A-2 cannot prune early, forming using a large set of shorter frequent routes A-2 has to wastefully process multiple time.

Figure 9(a) and 9(b) compare the performance of the algorithms for New York and KAIST using default m and l while using different values for r. We see that the running times for all algorithms increase when r increases because increasing r also increases the number of TRAJCLIQs and, hence, the number of frequent routes in the results. A-1 and A-2 still outperformed Sweep and the run-time of Sweep increases faster than A-1 and A-2 when r increases. Regardless of the value of r given, A-1 finished faster than A-2.

To summarize our experiment results, for real-life settings, our proposed algorithms (A-1 and A-2) provide more intuitive results and perform faster than Traclus. They also perform faster than the polynomial time approximation algorithm

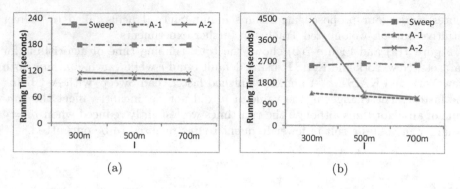

Fig. 8. Impact of the Parameter l on the Performance of the Frequent Route Mining Algorithms Using (a) New York and (b) KAIST

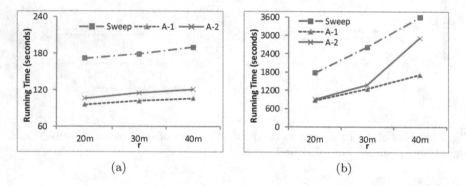

Fig. 9. Impact of the Parameter r on the Performance of the Frequent Route Mining Algorithms Using (a) New York and (b) KAIST

Sweep due to their pruning power based on *apriori*-properties of TRAJCLIQs. Although A-1 performed faster than A-2 in our experiments, with larger Trajectory Databases and limited amount of memory, A-2 would give a reasonable trade-off between run-time and memory requirements in finding frequent routes.

6   Conclusion

In order to find frequent routes when detailed information of underlying road network is not available, we had proposed a two-step method, (a) to find cliques of sub-trajectories using Fréchet distance as the similarity measure and (b) to extract a frequent routes from each sub-trajectory cliques. We designed two pragmatic approximation algorithms — the first one based on *Apriori* algorithm while the second sub-divides the input (large) TJDB into (smaller) subsets so that an instance of the first algorithm can process each subset independently. In the experiments we conducted, both algorithms performed reasonably fast and provided intuitive results.

References

1. Giannotti, F., Nanni, M., Pinelli, F., Pedreschi, D.: Trajectory pattern mining. In: Proceedings of the 13th ACM SIGKDD International Conference on Knowledge Discovery and Data Mining, KDD 2007, pp. 330–339. ACM, New York (2007)
2. Morzy, M.: Mining frequent trajectories of moving objects for location prediction. In: Perner, P. (ed.) MLDM 2007. LNCS (LNAI), vol. 4571, pp. 667–680. Springer, Heidelberg (2007)
3. Lee, J.G., Han, J., Whang, K.Y.: Trajectory clustering: a partition-and-group framework. In: Proceedings of the 2007 ACM SIGMOD International Conference on Management of Data, SIGMOD 2007, pp. 593–604. ACM, New York (2007)
4. Zhu, H., Luo, J., Yin, H., Zhou, X., Huang, J.Z., Zhan, F.B.: Mining trajectory corridors using frèchet distance and meshing grids. In: Zaki, M.J., Yu, J.X., Ravindran, B., Pudi, V. (eds.) PAKDD 2010, Part I. LNCS, vol. 6118, pp. 228–237. Springer, Heidelberg (2010)
5. Buchin, K., Buchin, M., Gudmundsson, J., Löffler, M., Luo, J.: Detecting commuting patterns by clustering subtrajectories. In: Hong, S.-H., Nagamochi, H., Fukunaga, T. (eds.) ISAAC 2008. LNCS, vol. 5369, pp. 644–655. Springer, Heidelberg (2008)
6. Kashyap, S., Roy, S., Lee, M.L., Hsu, W.: Farm: Feature-assisted aggregate route mining in trajectory data. In: Proceedings of the 2009 IEEE International Conference on Data Mining Workshops, ICDMW 2009, pp. 604–609. IEEE Computer Society, Washington, DC (2009)
7. Driemel, A., Har-Peled, S., Wenk, C.: Approximating the frèchet distance for realistic curves in near linear time. In: Proceedings of the 2010 Annual Symposium on Computational Geometry, SoCG 2010, pp. 365–374. ACM, New York (2010)
8. Alt, H., Godau, M.: Computing the frèchet distance between two polygonal curves. Int. J. Comput. Geometry Appl. 5, 75–91 (1995)
9. Dumitrescu, A., Rote, G.: On the frèchet distance of a set of curves. In: Proceedings of the 16th Canadian Conference on Computational Geometry, CCCG 2004, pp. 162–165. Concordia University, Montréal (2004)
10. Agrawal, R., Srikant, R.: Fast algorithms for mining association rules in large databases. In: Bocca, J.B., Jarke, M., Zaniolo, C. (eds.) Proceedings of 20th International Conference on Very Large Data Bases, VLDB 1994, pp. 487–499. Morgan Kaufmann, Santiago de Chile (1994)
11. Jetcheva, J.G., Chun Hu, Y., Palchaudhuri, S., Kumar, A., David, S., Johnson, B.: Design and evaluation of a metropolitan area multitier wireless ad hoc network architecture, pp. 32–43 (2003)
12. http://www.rtreeportal.org

Discovering Influential Data Objects over Time

Orestis Gkorgkas[1], Akrivi Vlachou[1,2,*], Christos Doulkeridis[3,**], and Kjetil Nørvåg[1]

[1] Norwegian University of Science and Technology (NTNU), Trondheim, Norway
[2] Institute for the Management of Information Systems, R.C. "Athena", Athens, Greece
[3] Department of Digital Systems, University of Piraeus, Greece
{orestis,vlachou,cdoulk,noervaag}@idi.ntnu.no

Abstract. In applications such as market analysis, it is of great interest to product manufacturers to have their products ranked as highly as possible for a significant number of customers. However, customer preferences change over time, and product manufacturers are interested in monitoring the evolution of the popularity of their products, in order to discover those products that are consistently highly ranked. To take into account the temporal dimension, we define the *continuous influential query* and present algorithms for efficient processing and retrieval of continuous influential data objects. Furthermore, our algorithms support incremental retrieval of the next continuous influential data object in a natural way. To evaluate the performance of our algorithms, we conduct a detailed experimental study for various setups.

1 Introduction

In online marketplaces, top-k queries are typically used to present a limited number of products ranked according to the user's preferences. This is extremely helpful for the user as it enables decision-making, without the need to inspect large amounts of possibly uninteresting results. In addition, the user is not overwhelmed by the available information and can retrieve results that satisfy her information need. As a result, an increasing amount of research has focused on efficient techniques for top-k query processing lately [6].

From the perspective of the product manufacturers top-k queries are of great interest as well, since the visibility of a product clearly depends on the number of different top-k queries for which it belongs to the result set. The reason for this is twofold: 1) users usually consider only a few highly ranked products and ignore the remaining ones, and 2) products that appear in the top-k result sets are far more likely to be chosen by a potential customer, because these products satisfy the customers' preferences. Recently, *reverse top-k queries* [14] were proposed to study the visibility of a given product. A reverse top-k query returns the set of user preferences (i.e., customers) for which a

* The work of Akrivi Vlachou was supported by the Action "Supporting Postdoctoral Researchers" of the Operational Program "Education and Lifelong Learning" (Action's Beneficiary: General Secretariat for Research and Technology), and is co-financed by the European Social Fund (ESF) and the Greek State.
** The research of Christos Doulkeridis was supported under the Marie-Curie IEF grant number 274063 with partial support from the Norwegian Research Council.

M.A. Nascimento et al. (Eds.): SSTD 2013, LNCS 8098, pp. 110–127, 2013.
© Springer-Verlag Berlin Heidelberg 2013

given product is in the result set of the respective top-k queries. Intuitively, a product that appears in as many as possible top-k result sets, has a higher visibility and therefore also a higher impact on the market. This has naturally lead to the definition of the *most influential products* based on the cardinality of their reverse top-k result sets [17]. Identifying the most influential products from a given set of products is important for market analysis, since the product manufacturer can estimate the impact of her products in the market.

However, an important aspect of a product's influence that has not been taken into account yet is its variance over time as the user preferences change. The customers' criteria can differ significantly over time for various reasons. For example, in online marketplaces, new customers pose queries and new preferences are collected. In addition, customers that have already posed queries will disconnect after some time. As user preferences change over time, a product which appears consistently in the top-k results of as many customers as possible, thus satisfying many customers' criteria at any time, has a higher impact on the market than a product that is absent from those results. Therefore, these products are the best candidate products to advertise to potential customers, and it is important to identify such products efficiently.

In this paper, we study for the first time the problem of finding the product that belongs consistently to the most influential products over time, the *continuous influential products*. This is an important problem for many real-life applications. For example, the products advertised on the first page of an online marketplace should be the products that have the greatest impact on the market, i.e., the products that are the most popular among the customers. Since customers change all the time, the products that consistently belong to the most influential products over time are more probable to attract many potential customers at any time. It is therefore essential to identify the objects (products) that have high impact over a period of time and despite the fluctuation of preferences these objects remain among the most influential objects. From now on we will use the terms *product* and *object* interchangeably.

In the following, we first define formally the problem of continuous influential products and provide a baseline algorithm that sequentially scans all time intervals in order to retrieve the most continuous influential product. Then, we provide a bounding scheme in order to facilitate early termination of our algorithms and avoid processing time intervals that do not alter the result set. Summarizing, the main contributions of this paper are:

- We study, for the first time, the problem of identifying the data object that has the highest impact over time.
- An appropriate score of influence (called *continuity score*) based on the reverse top-k query is defined to capture the product impact over a period of time.
- We derive upper and lower bounds for the continuity score of a given object that lead to efficient algorithms for retrieving the most continuous influential product. Two different algorithms are presented that provide early termination based on the bounds, but follow different strategies in order to terminate as soon as possible.
- We conduct a detailed experimental study for various setups and demonstrate the efficiency of our algorithms.

The rest of this paper is organized as follows: In Section 2 we provide the necessary preliminaries, while in Section 3 we formulate the problem statement. Section 4 presents a baseline algorithm for finding the data object that belongs consistently to the most influential products. Section 5 provides the foundation for our bounding scheme and describes the two threshold-based algorithms. Our experimental results are presented in Section 6. Section 7 provides an overview of related work. Finally, in Section 8 we conclude the paper.

2   Preliminaries

Let $\mathcal{D}$ be a dataspace with n dimensions $\{d_1, \ldots, d_n\}$ and S be a set of data objects on $\mathcal{D}$. A data object is represented as a point $o = \{o[1], \ldots, o[n]\}$ where $o[i]$ is the value of the attribute d_i.

2.1   Time-Invariant Case

Given a monotonic scoring function $f : S \to \mathbb{R}$, a top-k query returns the k best objects $o \in S$ ranked based on their scores $f(o)$. The most important and commonly used case of scoring functions is the weighted sum function, also called linear. For a given data object o and a weighting vector $\mathbf{w}$, its score $f_{\mathbf{w}}(o)$ is equal to the weighted sum of the individual values of o: $f_{\mathbf{w}}(o) = \sum_{i=1}^{n} w[i]o[i]$, where $w[i] \geq 0$ $(1 \leq i \leq n)$. The value of each dimension $w[i]$ of the vector $\mathbf{w}$ is a weighting (preference) score on dimension d_i. Without loss of generality we assume that (a) minimum values are preferable, and (b) for each vector $\mathbf{w}$ it holds that $\sum_{i=1}^{n} w[i] = 1$. We denote the result set of top-k query defined by a weighting vector $\mathbf{w}$ as $TOP_k(\mathbf{w})$.

Definition 1. (Top-k query): *Given a positive integer k and a user-defined weighting vector $\mathbf{w}$, the result set $TOP_k(\mathbf{w})$ of the top-k query is a ranked set of objects such that $TOP_k(\mathbf{w}) \subseteq S$, $|TOP_k(\mathbf{w})| = k$ and $\forall o, o' : o \in TOP_k(\mathbf{w})$, $o' \in S - TOP_k(\mathbf{w})$ it holds that $f_{\mathbf{w}}(o) \leq f_{\mathbf{w}}(o')$.*

Given a data set S of objects, a set W of weighting vectors, an object q and an integer k, a reverse top-k query returns all weighting vectors $\{\mathbf{w}\} \in W$ for which $q \in TOP_k(\mathbf{w})$. We denote the result set of weighting vectors $\{\mathbf{w}\}$, as $RTOP_k(q) = \{\mathbf{w}\}$.

Definition 2. (Reverse top-k query [14]): *Given an object q, a positive number k and two data sets S and W, where S represents data objects and W is a data set of weighting vectors, a weighting vector $\mathbf{w} \in W$ belongs to the reverse top-k result set $RTOP_k(q)$ of q, if and only if $\exists o \in TOP_k(\mathbf{w})$ such that $f_{\mathbf{w}}(q) \leq f_{\mathbf{w}}(o)$.*

We can also define the *influence score* of a data object by simply setting a single value k that determines the scope of the reverse top-k queries that are taken into account for identifying influential data objects.

Definition 3. (Influence score [17]): *Given a positive integer k, a data set S, and a set of preferences (weighting vectors) W, the* influence score *of a data object o is defined as the cardinality $|RTOP_k(o)|$ of the reverse top-k query result set of object o.*

Based on the definition of influence score, we define the ranked set of m most influential data objects.

Definition 4. (Top-m most influential data objects [17]): *Given a positive integer k, a data set S, and a set of preferences (weighting vectors) W, the result set $ITOP_k^m$ of the top-m influential query is a ranked set of objects such that $ITOP_k^m \subseteq S$, $|ITOP_k^m| = m$ and $\forall o, o' : o \in ITOP_k^m, o' \in S - ITOP_k^m$ it holds that $|RTOP_k(o)| \geq |RTOP_k(o')|$.*

2.2 Temporal Model

We model the time domain $\mathcal{T}$ as an ordered set of V disjoint time intervals that cover the complete domain, i.e., $\mathcal{T} = \{\mathcal{T}_1, \mathcal{T}_2, \ldots, \mathcal{T}_V\}$ and $\mathcal{T}_i \cap \mathcal{T}_j = \emptyset$ for $i \neq j$. We denote the start and end of time interval $\mathcal{T}_i$ with $t_s(\mathcal{T}_i)$ and $t_e(\mathcal{T}_i)$ respectively. Then, it also holds that $t_e(\mathcal{T}_i) = t_s(\mathcal{T}_{i+1})$, and that $t_s(\mathcal{T}_1)$ and $t_e(\mathcal{T}_V)$ denote the start and end of $\mathcal{T}$ respectively. Obviously, the number of time intervals V is user-specified and application-dependent, and its exact value depends on the desired level of detail for monitoring temporal changes.

In order to model the interval that a user is online, we associate the weighting vector representing the user preferences with a time interval. Thus, given a weighting vector $\mathbf{w}$ and by abusing notation slightly, we denote the start of this interval as $t_s(\mathbf{w})$ and its end as $t_e(\mathbf{w})$. We are now ready to define the validity of a weighting vector with respect to a time domain $\mathcal{T}$ that consists of time intervals.

Definition 5. (Validity of weighting vector): *Given a time domain $\mathcal{T} = \{\mathcal{T}_1, \mathcal{T}_2, \ldots, \mathcal{T}_V\}$ and a weighting vector $\mathbf{w}$, the validity of $\mathbf{w}$ with respect to $\mathcal{T}$ is the interval $[t_s(\mathcal{T}_i), t_e(\mathcal{T}_j))$, where $t_s(\mathbf{w}) \in \mathcal{T}_i$ and $t_e(\mathbf{w}) \in \mathcal{T}_j$.*

Based on Definition 5, we consider as the validity period of a weighting vector $\mathbf{w}$ the interval defined by the start and end of the time intervals ($\mathcal{T}_i$ and $\mathcal{T}_j$) that enclose $t_s(\mathbf{w})$ and $t_e(\mathbf{w})$ respectively. Henceforth, we will use $t_s(\mathbf{w})$ to refer to $t_s(\mathcal{T}_i)$ and $t_e(\mathbf{w})$ to refer to $t_e(\mathcal{T}_j)$.

3 Problem Formulation

Given a time domain $\mathcal{T} = \{\mathcal{T}_1, \mathcal{T}_2, \ldots, \mathcal{T}_V\}$, we define a total order $\prec$ such that $\mathcal{T}_i \prec \mathcal{T}_j$ if $t_e(\mathcal{T}_i) \leq t_s(\mathcal{T}_j)$ for any $\mathcal{T}_i, \mathcal{T}_j \in \mathcal{T}$. Furthermore, we use $ITOP_k^m(\mathcal{T}_i)$ to refer to the result set of the top-m most influential objects by taking into account only the weighting vectors that are valid in the interval $\mathcal{T}_i$.

In order to identify products that are consistently highly ranked for multiple users as time passes, we define the *continuity score* of an object $o \in S$.

Definition 6. (Continuity score): *Given a data set S, a set of weighting vectors W, and a time domain $\mathcal{T} = \{\mathcal{T}_1, \mathcal{T}_2, \ldots, \mathcal{T}_V\}$, the continuity score $cis(o)$ of an object $o \in S$ is the maximum number of consequent intervals $\mathcal{T}_i$ for which o belongs to the top-m most influential data objects, i.e., $o \in ITOP_k^m(\mathcal{T}_i)$.*

The continuity score of an object is practically a measure of the object's aggregated influence over time. As we aim to discover the object with highest continuity score, we define the most continuous influential data object in a straightforward way.

Definition 7. (Most continuous influential data object): *Given a data set S, a set of weighting vectors W, and a time domain $\mathcal{T} = \{\mathcal{T}_1, \mathcal{T}_2, \ldots, \mathcal{T}_V\}$, the most continuous influential data object $o \in S$ is the object for which it holds that $\nexists o' \in S$ such that $cis(o') > cis(o)$.*

We are now ready to formally define the problem of discovering the most influential object over time. Another closely related problem is the one of discovering a ranked set of the most influential object over time.

Problem 1. *(Most continuous influential object)*: Given a data set S, a set of weighting vectors W, and a time domain $\mathcal{T} = \{\mathcal{T}_1, \mathcal{T}_2, \ldots, \mathcal{T}_V\}$, find the most continuous influential object $o \in S$.

Problem 2. *(Top-N continuous influential objects)*: Given a data set S, a set of weighting vectors W, a time domain $\mathcal{T} = \{\mathcal{T}_1, \mathcal{T}_2, \ldots, \mathcal{T}_V\}$, and an integer N, find the ranked set of the N most continuous influential object $\{o_1, o_2, \ldots, o_N\} \in S$.

In this paper, we focus our attention to Problem 1 and present our algorithms for solving this problem. However, our algorithms can be extended in a straightforward way to solve also Problem 2. For the sake of simplicity we omit the details here.

4   Sequential Interval Scan

A baseline algorithm for solving Problem 1 is to compute the $ITOP_k^m(\mathcal{T}_i)$ sets for all time intervals $\mathcal{T}_i$ of $\mathcal{T}$ and simply follow a counting approach of the appearance of any data object o in consequent intervals. Then, the most continuous influential object is the one that appears in the $ITOP_k^m(\mathcal{T}_i)$ sets for the maximum number of consequent intervals. In the following, we refer to this algorithm as *Sequential Interval Scan (SIS)*.

Intuitively, in each iteration (lines 2–10 of Algorithm 1), *SIS* examines the next consequent interval $\mathcal{T}_i \in \mathcal{T}$ and computes the set of most influential objects $ITOP_k^m(\mathcal{T}_i)$ within $\mathcal{T}_i$. For each retrieved object $o \in ITOP_k^m(\mathcal{T}_i)$, we maintain its current continuity score, which is derived based on the processed intervals so far. We use the concept of *alive object* to refer to any object retrieved in a previous interval $\mathcal{T}_j (j \leq i)$ that is influential in all intervals between $\mathcal{T}_j$ and $\mathcal{T}_i$ and also belongs to the most recently processed $ITOP_k^m(\mathcal{T}_i)$ set; we also refer to objects that stopped being influential at some intermediate interval between $\mathcal{T}_1$ and $\mathcal{T}_i$ as *dead objects*. To ensure correctness, *SIS* needs to maintain the alive objects in a list A and only a single dead object d, which is the one with the maximum continuity score among all other dead objects (lines 4–6). Then, the retrieved influential objects in $\mathcal{T}_i$ are examined, and if an object belongs to A (i.e., was and remains alive) then its score is increased by 1 (line 8), otherwise we add it to A (line 9). After having examined all intervals, the algorithm terminates and reports the object with maximum score among the alive objects and the dead object (line 10).

The main shortcoming of *SIS* is that it needs to evaluate the $ITOP_k^m$ query for all $|V|$ time intervals. In the following, we study how to derive appropriate score bounds, in order to find the most continuous influential object without processing all queries.

Algorithm 1. *Sequential Interval Scan (SIS)*

Input: S:data set; k, m: the parameters of the $ITOP_k^m$ queries; $\mathcal{T} = \{\mathcal{T}_1, \ldots, \mathcal{T}_V\}$.
Output: o: the most continuous influential object.

1 $A \leftarrow \emptyset$, $d \leftarrow$ null   (A: alive objects, d: dead object)
2 **for** $i = 1 \ldots V$ **do**
3 | $\mathcal{I} \leftarrow ITOP_k^m(\mathcal{T}_i)$
4 | **forall the** $o \in A$ *and* $o \notin \mathcal{I}$ **do**
5 | | $A \leftarrow A - \{o\}$ (remove dead objects)
6 | | $d \leftarrow$ objMaxScore($\{d\} \bigcup \{o\}$)
7 | **forall the** $o \in \mathcal{I}$ **do**
8 | | **if** $o \in A$ **then** o.incScore()   (increase score)
9 | | **else** $A \leftarrow A \bigcup \{o\}$ (add new objects)
10 $o \leftarrow$ objMaxScore($A \bigcup \{d\}$)
11 **return** o

---

5 Algorithms with Early Termination

SIS relies on processing multiple consequent intervals of $\mathcal{T}$ to produce the most continuous influential object. In fact, all our algorithms rely on the evaluation of multiple $ITOP_k^m$ queries in different intervals $\mathcal{T}_i$, in order to find the most continuous influential object, however these intervals are not necessarily consequent. In this sense, our algorithms treat the $ITOP_k^m$ computation as black-box, hence any existing techniques that solve efficiently the problem of indentifying influential objects can be directly exploited by our algorithms.

Let us assume that at some point during query processing, a subset of (not necessarily consequent) intervals of $\mathcal{T}$ have been processed. We define the following sets for any retrieved data object o.

Definition 8. *Given a data object o, a set of processed intervals $\{\mathcal{T}_i\}$ and a set of corresponding results sets $\{ITOP_k^m(\mathcal{T}_i)\}$, we define:*

- $\mathcal{T}^+(o)$ *is the set of intervals* $\{\mathcal{T}_i\}$, *such that* $\mathcal{T}_i \in \mathcal{T}^+(o)$ *if* $o \in ITOP_k^m(\mathcal{T}_i)$
- $\mathcal{T}^-(o)$ *is the set of intervals* $\{\mathcal{T}_i\}$, *such that* $\mathcal{T}_i \in \mathcal{T}^-(o)$ *if* $o \notin ITOP_k^m(\mathcal{T}_i)$
- $\mathcal{LB}(o)$ *is a maximal sequence of intervals* $\{\mathcal{T}_i, \mathcal{T}_{i+1}, \ldots, \mathcal{T}_j\}$, *such that* $\forall \mathcal{T}_z \in \mathcal{LB}(o)$: $\mathcal{T}_z \in \mathcal{T}^+(o)$
- $\mathcal{UB}(o)$ *is a maximal sequence of intervals* $\{\mathcal{T}_i, \mathcal{T}_{i+1}, \ldots, \mathcal{T}_j\}$, *such that* $\forall \mathcal{T}_z \in \mathcal{UB}(o)$: $\mathcal{T}_z \in \mathcal{T} - \mathcal{T}^-(o)$

We emphasize that according to Definition 8, $\mathcal{T}^+(o)$ and $\mathcal{T}^-(o)$ are sets of intervals, i.e., they may contain non-consequent intervals. Instead, the sequences $\mathcal{LB}(o)$ and $\mathcal{UB}(o)$ contain consequent intervals, and moreover they are of maximal size, i.e., there exists no other longer sequence of intervals with the same properties respectively.

By exploiting the above sets and sequences, we derive an upper and a lower bound on the score of any candidate most continuous influential object.

Lemma 1. (Score bounds): *The continuity score of object o is bounded by the lower bound L(o) and the upper bound U(o), i.e., $L(o) \leq cis(o) \leq U(o)$, where $L(o) = |\mathcal{LB}(o)|$ and $U(o) = |\mathcal{UB}(o)|$ are the lengths of the sequences $\mathcal{LB}(o)$ and $\mathcal{UB}(o)$ respectively.*

Proof. By contradiction. Let us assume that $cis(o) < L(o)$. Then it holds that there exists a sequence of processed intervals of length $|\mathcal{LB}(o)|$ such that for each time interval $\mathcal{T}_i$ of $\mathcal{LB}(o)$ it holds that $\mathcal{T}_i \in \mathcal{T}$ and $o \in ITOP_k^m(\mathcal{T}_i)$, which leads to a contradiction since $cis(o)$ is defined by the sequence of maximum length (according to Definition 6). Similarly, the assumption $cis(o) > U(o)$ leads to a contradiction, because for each time interval $\mathcal{T}_i$ of the sequence that defines $cis(o)$, it holds that $\mathcal{T}_i \notin \mathcal{T}^-(o)$ for any set of processed intervals $\{\mathcal{T}_i\}$. In other words, the sequence of intervals whose length defines $cis(o)$ is always smaller or equal to the sequence $\mathcal{UB}(o)$ whose length defines $U(o)$, hence $cis(o) \leq U(o)$ which is a contradiction.

The lower bound $L(o)$ of o is equal to the continuity score of the object o, if we take into account only the time intervals that have been processed so far. The upper bound $U(o)$ of o is the continuity score of the object o, if we assume that for any time interval $\mathcal{T}_i$ that does not belong to $\mathcal{T}^-$ the object o belongs to $ITOP_k^m(\mathcal{T}_i)$ (because optimistically for all unprocessed time intervals, o may belong to the most influential objects).

Theorem 1. (Early termination condition) *The data object o is the most continuous influential object, if for any other data object o' it holds that $L(o) \geq U(o')$.*

Proof. By contradiction. Let us assume that o is not the most continuous influential object, even though it holds that $L(o) \geq U(o')$. Thus, there must exist another object o' which is the most continuous influential object (i.e., $cis(o) < cis(o')$). Then, it holds that $L(o) \leq cis(o) \leq U(o)$ and $L(o') \leq cis(o') \leq U(o')$. From these inequalities, we derive that $L(o) \leq cis(o) < cis(o') \leq U(o')$ and finally that $L(o) < U(o')$, which is a contradiction.

The intuition of the above condition for early termination is that if an object has a continuity score based on some processed time intervals that is definitely higher than the score of any other object, then it can be safely reported as the most continuous influential object, because the score of any other object cannot increase sufficiently in the remaining time intervals.

Algorithm *SIS* is oblivious of the derived bounds and examines all time intervals following a brute-force approach. Hence, we propose two algorithms, termed *Early Termination Interval Scan (TIS)* and *Early Termination Best-First Interval (TBI)*, that exploit the bounds to provide early termination. However, despite using the same concept of bounding, *TIS* and *TBI* follow different strategies in order to terminate as soon as possible. *TIS* aims to maximize as quickly as possible the lower bound of the current most continuous influential object o and therefore examines time intervals sequentially. Instead, *TBI* aims to reduce the upper bound of any object o by breaking the longest unprocessed sequence of time intervals.

Algorithm 2. *Early Termination Interval Scan (TIS)*

Input: S:data set; k, m: the parameters of the $ITOP_k^m$ queries; $\mathcal{T} = \{\mathcal{T}_1, \ldots, \mathcal{T}_V\}$.
Output: o: the most continuous influential object.

1   $A \leftarrow \emptyset$, $d \leftarrow$ null   (A: alive objects, d: dead object)
2   $i = 1$, upperBound $= 0$, lowerBound $= -1$
3   **while** *lowerBound* $<$ *upperBound* **do**
4     $\mathcal{I} \leftarrow ITOP_k^m(\mathcal{T}_i)$
5     $i = i + 1$
6     $o \leftarrow$ objMaxScore($A \bigcup \{d\}$)
7   lowerBound $= o$.score()
8   **forall the** $o \in A$ *and* $o \notin \mathcal{I}$ **do**
9        $A \leftarrow A - \{o\}$ (remove dead objects)
10        $d \leftarrow$ objMaxScore($\{d\} \bigcup \{o\}$)
11   **forall the** $o \in \mathcal{I}$ **do**
12   **if** $o \in A$ **then** o.incScore()   (increase score)
13   **else** $A \leftarrow A \bigcup \{o\}$ (add new objects)
14     $o' \leftarrow$ objMaxScore($A - \{o\}$)
15   upperBound $= max(o'$.score()$+(V - i), d$.score())
16 **return** o

5.1 Early Termination Interval Scan

In this section, we describe the Early Termination Interval Scan (*TIS*) algorithm. Similar to the *SIS* algorithm, *TIS* processes sequentially the time intervals of time domain $\mathcal{T}$. However, the significant advantage of *TIS* lies in the fact that it can terminate early and report the most continuous influential object o without processing the $ITOP_k^m$ query for all V time intervals $\mathcal{T}_i$.

Intuitively, the main objective of *TIS* is to increase the lower bound of any retrieved object, by scanning the time intervals sequentially. Notice that only consequent time intervals may lead to a higher lower bound. *TIS* takes advantage of the fact that the time intervals are processed sequentially and computes the $L(o)$ and $U(o)$ without maintaining the sets $\mathcal{T}^-(o)$ and $\mathcal{T}^+(o)$. The lower bound is defined as the continuity score of the current most continuous influential object, which can be computed by maintaining only the alive and dead objects similar to *SIS*. For *TIS*, the upper bound is defined as the maximum value of the score of the dead object or the second highest score of the alive objects plus the number of remaining time intervals.

Although these bound definitions of *TIS* are simpler than the ones of lower and upper bound in Lemma 1, it can be shown that they are equivalent. The reason for their simplicity is that *TIS* examines intervals sequentially, which is a special case of interval selection and the computation of the bounds can be simplified. Instead, the bound definitions of Lemma 1 and Definition 8 apply in the general case of selecting any interval for processing next (not necessarily in a sequential manner).

Algorithm 2 contains the pseudocode of *TIS*. In each iteration, the next interval of the time domain $\mathcal{T}$ is examined and the result set $ITOP_k^m(\mathcal{T}_i)$ is computed. For each retrieved object a score is maintained which is the maximum number of consequent intervals for which this object belongs to the respective $ITOP_k^m$ sets. The retrieved data objects that belong to the most recent $ITOP_k^m$ set are considered to be alive, while we also keep track of the dead object with the highest score.

In more detail, as long as the termination condition does not hold (lines 3-15), the $ITOP_k^m$ set for the next time interval is computed and the alive and dead objects are updated (lines 8-10, 12, 13), similarly to the case of the *SIS* algorithm. Furthermore, in each iteration, the current most continuous influential object o is found (line 6). The current score of o defines the lower bound (line 7), as any other point must have a higher score to become the most continuous influential. Also, the alive object o' with the second highest score is found (line 14)[1]. The maximum possible score of any object (regardless of whether it has been retrieved or not) is equal to maximum value between the score of the dead object and the score of o' plus the number of remaining unprocessed intervals. This is because any object that is still alive in the best case scenario may be in the $ITOP_k^m$ set for all remaining time intervals. Also, the score of the dead object cannot be increased further. Notice that if the same object appears in the $ITOP_k^m$ set, it is considered to be a new alive object. Any new alive object can appear only in the $V - i$ remaining time intervals. Thus, if the termination condition holds, no object can exceed the score of the currently most continuous influential object and the algorithm safely reports this object as the result.

It should be noted that *TIS* reports the most continuous influential object over a time domain, however it does not report its score accurately. One can draw parallels with Fagin's NRA algorithm [4], which produces the top-k objects from ranked lists but without guaranteeing accuracy of scores. In order to calculate the exact continuity score of the most continuous influential object, we need to proceed until we find an interval where the object does not belong to the $ITOP_k^m$ set.

5.2   Early Termination Best-First Interval

In the following, we describe the Early Termination Best-first Interval (*TBI*) algorithm. The most important difference to *TIS* is that *TBI* follows a different strategy with respect to interval selection, namely *TBI* does not process intervals sequentially.

For each retrieved object o, *TBI* maintains the two sets $\mathcal{T}^+(o)$ and $\mathcal{T}^-(o)$ that correspond to the processed time intervals for which o belongs to or not to the most influential data objects respectively. This information is sufficient to derive the lower bound $L(o)$ and upper bound $U(o)$ of o. The algorithm first computes the influential objects $ITOP_k^m(\mathcal{T}_1)$ and $ITOP_k^m(\mathcal{T}_V)$. The following example demonstrates the information maintained by *TBI* at this point.

Example 1. *Let us assume that* $V = 6$, $m = 2$, *and that* $ITOP_k^m(\mathcal{T}_1) = \{o_1, o_2\}$ *and* $ITOP_k^m(\mathcal{T}_6) = \{o_2, o_3\}$. *Then,* TBI *maintains the following sets:* $\mathcal{T}^+(o_1) = \{\mathcal{T}_1\}$, $\mathcal{T}^-(o_1) = \{\mathcal{T}_6\}$, $\mathcal{T}^+(o_2) = \{\mathcal{T}_1, \mathcal{T}_6\}$, $\mathcal{T}^-(o_2) = \emptyset$, $\mathcal{T}^+(o_3) = \{\mathcal{T}_6\}$, $\mathcal{T}^-(o_3) = \{\mathcal{T}_1\}$. *In addition, the derived bounds are:* $L(o_1) = 1$, $U(o_1) = 5$, $L(o_2) = 1$, $U(o_2) = 6$, $L(o_3) = 1$, $U(o_3) = 5$.

---

[1] In the extreme case where $A - \{o\} = \emptyset$ we assume that $o'.score = 0$.

TBI iteratively selects a time interval that has not been processed yet and computes the influential objects in the selected time interval. Then, the bounds of retrieved objects can be updated as indicated in the following.

Example 2. *Continuing the previous example, assume that the next interval that is processed is $\mathcal{T}_3$ and $ITOP_k^m(\mathcal{T}_3) = \{o_2, o_4\}$. Then, the following sets are maintained: $\mathcal{T}^+(o_1) = \{\mathcal{T}_1\}$, $\mathcal{T}^-(o_1) = \{\mathcal{T}_3, \mathcal{T}_6\}$, $\mathcal{T}^+(o_2) = \{\mathcal{T}_1, \mathcal{T}_3, \mathcal{T}_6\}$, $\mathcal{T}^-(o_2) = \emptyset$, $\mathcal{T}^+(o_3) = \{\mathcal{T}_6\}$, $\mathcal{T}^-(o_3) = \{\mathcal{T}_1, \mathcal{T}_3\}$, $\mathcal{T}^+(o_4) = \{\mathcal{T}_3\}$, $\mathcal{T}^-(o_4) = \{\mathcal{T}_1, \mathcal{T}_6\}$. In addition, the bounds are updated as follows: $L(o_1) = 1$, $U(o_1) = 2$, $L(o_2) = 1$, $U(o_2) = 6$, $L(o_3) = 1$, $U(o_3) = 3$, $L(o_4) = 1$, $U(o_4) = 4$.*

The remaining challenge is how to select the most beneficial time interval for the next influential query to be processed, i.e., the time interval that will lead the algorithm to terminate as quickly as possible. *TBI* follows a best-first approach by selecting the time interval that will split the longest $\mathcal{UB}(o)$ sequence for any o in the queue. Intuitively, this "breaks" long sequences of unknown time intervals, in an attempt to reduce the upper bound of any data object.

In more detail, the next interval to be processed is selected in the following way. Given a candidate data object o and the corresponding $\mathcal{UB}(o) = \{\mathcal{T}_i, ..., \mathcal{T}_j\}$, the middle time interval $\mathcal{T}_z$ is computed such that $z = i + \lceil \frac{j-i}{2} \rceil$. If $\mathcal{T}_z \notin \mathcal{T}^+(o)$ then $\mathcal{T}_z$ is the next interval. Otherwise it means that $\mathcal{T}_z$ has been already processed and in this case the sequence $\{\mathcal{T}_i, ..., \mathcal{T}_z\}$ is tried to be split by finding the middle interval $\mathcal{T}_{z'}$ of it. If also $\mathcal{T}_{z'} \in \mathcal{T}^+(o)$, then the middle interval of $\{\mathcal{T}_z, ..., \mathcal{T}_j\}$ is examined if it qualifies for being the next interval. This is done recursively by examining always the longest sequence until an interval is found that does not belong to $\mathcal{T}^+(o)$. Note that it is guaranteed that such an interval exists, because otherwise $L(o) = U(o)$ and the algorithm terminates. Intuitively, computing $ITOP_k^m(\mathcal{T}_z)$ may break the longest sequence $\mathcal{UB}(o)$ in two smaller sequences if $o \notin ITOP_k^m(\mathcal{T}_z)$, thus reducing the upper bound, which will allow the algorithm to terminate faster.

During query processing, *TBI* keeps the retrieved data objects in a priority queue. The queue is sorted in descending order based on the upper bound $U(o)$ of each object o, so that immediate access to the object with the highest upper bound is provided. Algorithm 3 presents the pseudocode of *TBI*. First, the intervals $\mathcal{T}_1$ and $\mathcal{T}_V$ are processed and the retrieved objects are inserted in the queue (lines 1–4). The lower and upper bounds are initiated based on the object located at the head of the queue (lines 5, 6). In each iteration, we remove from the queue the object o (candidate object) with maximum upper bound $U(o)$ (line 8). Note that the candidate object is not necessary the object with the highest continuity score based on the processed partitions (which is the lower bound), and there may exist another object o' that has a higher score (lower bound) currently. But it is guaranteed that the algorithm cannot terminate at this iteration even if o' was processed next, because it holds that $L(o') \leq U(o')$ and $U(o') \leq U(o)$ so that the termination condition cannot hold. Thus, *TBI* does not process unnecessary time intervals.

After selecting the candidate o with the highest upper bound, *TBI* recursively selects the middle interval to be processed (line 9) and processes the query (line 10). Afterwards, the queue is updated (line 11), which means that every object in $ITOP_k^m(\mathcal{T}_i)$ is either added to the queue (if it is the first time that it was retrieved) or the existing

---

Algorithm 3. *Early Termination Best-first Interval (TBI)*

Input: S:data set; k, m: the parameters of the $ITOP_k^m$ queries; $\mathcal{T} = \{\mathcal{T}_1, \ldots, \mathcal{T}_V\}$.
Output: o: the most continuous influential object.

1 $\mathcal{I} \leftarrow ITOP_k^m(\mathcal{T}_1)$
2 queue.update($\mathcal{I}$)
3 $\mathcal{I} \leftarrow ITOP_k^m(\mathcal{T}_V)$
4 queue.update($\mathcal{I}$)
5 upperBound $= U(queue.peek())$
6 lowerBound $= L(queue.peek())$
7 **while** *lowerBound<upperBound* **do**
8     $o \leftarrow queue.dequeue()$
9     $i = \text{nextInterval}(\mathcal{UB}(o))$　　　　　(find next interval)
10    $\mathcal{I} \leftarrow ITOP_k^m(\mathcal{T}_i)$
11   queue.update($\mathcal{I}$)
12   upperBound $= U(queue.peek())$
13   lowerBound $= L(o)$
14   queue.enqueue(o)　　　　　(add o back to queue)
15 **return** o

---

object is updated by changing the corresponding $\mathcal{T}^+$ set. Moreover, for every object in the queue that does not belong in $ITOP_k^m(\mathcal{T}_i)$, the set $\mathcal{T}^-$ is updated.

The algorithm terminates when it holds that the candidate object o has $L(o) \geq U(o'), \forall o' \in$ queue. This is the *termination condition* (line 7), which means that o has a higher lower bound than the upper bound of the current head object o' in the queue.

In principle, we can also free part of the memory during the processing of the algorithm, by evicting candidate points that will never become the most continuous influential object. The condition for eviction is if a candidate object o has $U(o) \leq L(o')$, where o' is another candidate object.

6 Experimental Evaluation

In this section, we present the results of the experimental evaluation. All algorithms are disk-based and implemented in Java, and the experiments run on 2x Intel Xeon X5650 Processors (2.66GHz), 128GB. The index structure used was an R-tree with a buffer size of 100 blocks and the block size is 4KB.

6.1 Experimental Setup

Data sets. For the data set S we employ both real and synthetic data collections, namely uniform (UN), correlated (CO) and anticorrelated (AC). For the uniform data set, the data object values for all n dimensions are generated independently using a uniform distribution. The correlated and anticorrelated data sets are generated as described in [3].

In addition, we use two real data sets. NBA consists of 17265 5-dimensional tuples, representing a player's performance per year. The attributes are average values of: number of points scored, rebounds, assists, steals and blocks. HOUSE (Household) consists of 127930 6-dimensional tuples, representing the percentage of an American family's annual income spent on 6 types of expenditure: gas, electricity, water, heating, insurance, and property tax.

For the data set W of the weighting vectors, two different data distributions are examined, namely uniform (UN) and clustered (CL). The clustered data set W is generated as described in [14] and models the case that many users share similar preferences. In more detail, first C_W cluster centroids that belong to the $(n$-1)-dimensional hyperplane defined by $\sum w[i] = 1$ are selected randomly. Then, each coordinate is generated on the $(n$-1)-dimensional hyperplane by following a normal distribution on each axis with variance σ_W^2, and a mean equal to the corresponding coordinate of the centroid. We consider $V = 100$ time intervals and assign a vector $\mathbf{w}$ to a time interval $\mathcal{T}_i$ ($i \in [1, 100]$) uniformly at random.

We conduct a thorough sensitivity analysis varying the dimensionality (2-5d), the cardinality (10K-100K) of S, the cardinality (100K-500K) of W the value of k (5-15), the value of m (5-15), and the number of intervals V (50-150). Unless explicitly mentioned, we use the default setup of: $|S| = 50K$, $|W| = 300K$, d=3, k=10, m=10, V=100, and uniform distribution for S and W. For the clustered data set W we use $C_W = 5$ and $\sigma_W = 0.1$, and try different values of σ_W.

Algorithms. We evaluate: a) sequential interval scan (*SIS*), b) early termination interval scan (*TIS*), and c) early termination best-first interval (*TBI*). All algorithms use the computation of the top-m most influential data objects as a black-box. In particular, the branch-and-bound algorithm proposed in [17] is employed for the underlying computation of influential objects.

Metrics. Our metrics include: a) the total execution time, b) the number of I/Os, and c) the number of processed time intervals by each algorithm. Notice that we do not measure the I/Os that occur by reading W, since this is the same for every algorithm and does not affect their comparative performance. For our experiments on synthetic data, we report the average of each metric over 10 different instances of the data set. We generate the different instances by keeping the parameters fixed and changing the seeds of the random number generator. We adopt this approach in order to factor out the effects of randomization.

6.2 Performance of Query Processing

Effect of Data Set Size $|S|$. Fig. 1 illustrates the performance of all algorithms when we vary the data set cardinality. For all metrics, *TBI* outperforms both *TIS* and *SIS*. In terms of time (Fig. 1(a)), *TBI* is significantly faster than the other algorithms, and more importantly its gain increases as the data set size increases. This is strong evidence that *TBI* scales gracefully with $|S|$. Similar observations can be made for the I/O metric depicted in Fig. 1(b). Fig. 1(c) depicts the number of processed intervals by each algorithm, which is a factor that affects all other metrics. *SIS* always processes the complete set of V intervals. *TIS* improves the performance of *SIS*, by exploiting the bounds and allowing for early termination. It should be clarified that *TIS* cannot process fewer than $V/2$ intervals to produce the correct result. Thus, in this setup ($V = 100$), *TIS* would in

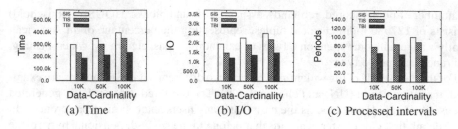

Fig. 1. Effect of varying data cardinality $|S|$

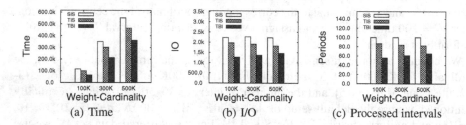

Fig. 2. Effect of varying cardinality of weighting vectors $|W|$

best case process 50 intervals. Still, *TBI* outperforms all other algorithms, which indicates that its best-first strategy for selecting the next interval performs more efficiently.

The advantage in the performance of *TIS* against *SIS* lies on the fact that *TIS* terminates when it is certain that the object with the highest continuity score cannot be surpassed. The advantage of *TBI* over *TIS* lies on the fact that TIS alters the upper and lower bounds each time by 1 interval while TBI splits the largest unseen interval in half. In the best case, every $2^{\lambda+1} - 1$ steps the upper bound will have been reduced to $|V|/(2^{\lambda} + 1)$, while for *TIS* the upper bound in the best case will have been reduced to $|V| - (2^{\lambda+1} - 1)$. Obviously in the early steps of *TBI* the upper bound and lower bound converge faster than in *TIS*.

Effect of Varying Cardinality of Weighting Vectors $|W|$. In Fig. 2, we study the effect of increasing the size of $|W|$. First, with respect to time (depicted in Fig. 2(a)), we observe that time increases linearly with $|W|$ for all algorithms. This is expected, since the size of W determines the number of user preferences, which is the number of potential top-k queries that may be evaluated. When the induced I/Os are considered, we see in Fig. 2(b) that all algorithms show a stable performance irrespective of $|W|$. Recall that we only measure the I/O induced on data set S, and this metric does not depend on W. Hence, this explains the stability of the measured I/O values. Fig. 2(c) shows the processed intervals by each algorithm. Also in this setup, *TBI* performs better than its competitors. It can be also observed that the size of W does not affect the number of processed intervals. The observations made for varying the data cardinality hold also here. The increased computation cost with respect to time is due to the fact that the complexity of the $ITOP_k^m$ queries increases when the weight cardinality rises.

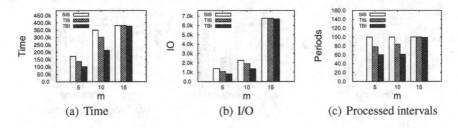

Fig. 3. Effect of varying m

(a) Time   (b) I/O   (c) Processed intervals

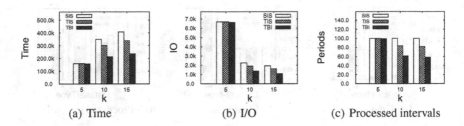

(a) Time   (b) I/O   (c) Processed intervals

Fig. 4. Effect of varying k

Effect of Varying m. Fig. 3 shows the effect of increasing the number of retrieved influential objects. *TBI* has a significant performance advantage over *SIS* and *TIS* when the value of m is relatively small. When m increases, we observe that all algorithms demonstrate similar performance. The reason for this behavior is that for larger values of m we observe that there exist data objects that have maximum continuity score equal to V. In other words, some data objects are influential in all V intervals. In this degenerate case, no algorithm can perform better than *SIS*, since all intervals must be processed in order to safely report the most continuous influential object.

Effect of Varying k. As k increases, all algorithms need more time to produce the results set as depicted in Fig. 4(a). For smaller values of k, all algorithms perform similarly because again there exist objects with maximum continuity score, which can only be reported when all intervals have been processed. For higher values of k, *TBI* performs better than all other algorithms.

Effect of Varying V. Based on Fig. 5(a), we observe that *TIS* has a bigger advantage over *SIS* for small number of intervals, while *TBI* benefits more from large number of intervals. The reason is that the more the time intervals the smaller the possibility for an object to be influential in all of them. This fact is exploited by *TBI* which manages to reduce the upper bound fast in the first loops of its execution, and thus the lower bound and the upper bound converge fast and allow *TBI* to finish earlier that *SIS* and *TIS*. Contrary to the upper bound, the lower bound is expected to increase slowly when the time domain is partitioned with high granularity since many objects (including the one with the highest continuity score) are likely to disappear and re-appear from the $ITOP_k^m$ influential sets, and consequently the convergence between the bounds is delayed.

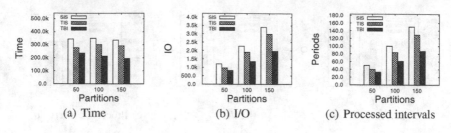

| (a) Time | (b) I/O | (c) Processed intervals |

Fig. 5. Effect of varying V

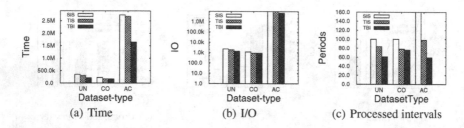

| (a) Time | (b) I/O | (c) Processed intervals |

Fig. 6. Effect of varying the data distribution of S

Effect of Different Data Distributions of S. Fig. 6 compares the performance of the three algorithms when the set of data objects S follow different distributions, namely uniform (UN), correlated (CO) and anti-correlated (AC). Notice that we use log-scale in Fig. 6(b). Clearly, the cost of all algorithms (in terms of time and I/O) increases for AC. This is due to the more expensive processing of the underlying computation for influential data objects in the case of AC. However, as depicted in Fig. 6(c), the difference between the algorithms is significant in terms of processed intervals. Also, notice that *TBI* is not significantly affected by the challenging AC data distribution and processes comparable number of intervals, irrespective of the data distribution of S.

Effect of Clustered Data Set W. Fig. 7 shows the results of using a clustered data set W for different values of σ_W. Smaller values of σ_W correspond to more clustered data sets, or in other words the weighting vectors are more compact with respect to the cluster centroids. For smaller values of σ_W, *TBI* performs better than the other algorithms. However, an interesting observation is that when σ_W increases, the performance of *TIS* tends to be similar to *TBI*.

Table 1. Experimental results of real data sets NBA and HOUSE

Algorithm	NBA data set			HOUSE data set		
	Time(sec)	I/O	Proc. Intervals	Time(sec)	I/O	Proc. Intervals
SIS	822.77	8119	100.0	903.00	9476	100.0
TIS	712.74	6988	85.9	867.33	9189	95.2
TBI	454.60	4508	55.6	865.58	9235	97.1

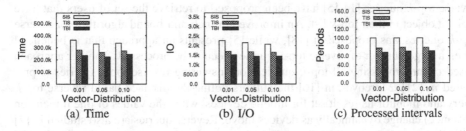

Fig. 7. Effect of varying the standard deviation for clustered data set W

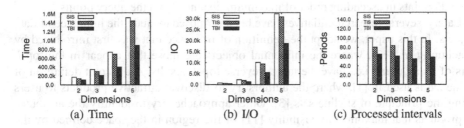

Fig. 8. Effect of varying dimensionality n

Effect of Increasing Dimensionality. Fig. 8 illustrates the results for varying the number of dimensions. With respect to time (Fig. 8(a)) and I/O (Fig. 8(b)), the performance of all algorithms degrades with increased dimensionality. However, notice that *TBI* is less affected by the increased dimensionality, compared to the other algorithms. With respect to the number of processed intervals (Fig. 8(c)), we observe that this metric increases with dimensionality in the case of *TIS*. When *TBI* is considered, the metric drops for increased values of n. This means that *TBI* manages to process fewer intervals as n grows, however each top-k processing costs on average more for increased n, which explains why both time and I/O increase for *TBI* too.

Experiments with Real Data Sets. Table 1 shows the results obtained for the two real data sets employed in our study (NBA and HOUSE). In both cases, the observed values follow the results and conclusions drawn from synthetic data. *TBI* outperforms the other two algorithms for the NBA data set. *TBI* needs almost half the time of *SIS* to identify the most influential object. In the case of HOUSE data set the difference between *TIS* and *TBI* is marginal but both outperform *SIS*. The higher the dimensionality of the problem the smaller is the probability that the most influential object will be influential for a long time interval. This fact reduces the advantage of *TBI* over *TIS* and the two algorithms have similar performance.

7 Related Work

Top-k queries have been well-studied in the last years to enable ranked retrieval of objects based on user preferences (for a thorough overview we refer to [6]). Recently, reverse top-k queries [14,15] have been proposed to retrieve the set of users that have a given object in their top-k list. An improved branch-and-bound algorithm for reverse top-k queries was proposed in [18], while [5] presents an approach that is beneficial when a large number of reverse top-k queries need to be processed. Another approach based on preprocessing all top-k queries for answering reverse top-k queries is presented in [21]. Moreover, in [16] the authors define the distance-based reverse top-k query and monitor its result set for mobile devices, when the values of one dimension (distance) change dynamically as devices move. Reverse queries are also studied in [2] following a unified approach. The authors examine the Inverse ϵ-Range, Inverse k-NN and Inverse Dynamic Skyline queries using a three-filter approach. The first two filters use only the query points whose number is usually small and the third query accesses the data points in ascending order of maximum distance from the query points.

Lately, several research initiatives have been proposed to study the influence of data objects. In this paper, we adopt the definition of influence that was first introduced by Vlachou et al. [17], where the influential objects are those that appear in the top-k lists of many users, i.e., have the larger reverse top-k results. A different definition of influence is used in [1], where the authors try to discover attractive products to users using the principle of skyline sets [3]. Other approaches try to identify the attributes of products that maximize its visibility [11] or the region in the space defined by the products' attributes where a product can be promoted [19,20].

Jestes et al. [7] study the problem of performing top-k queries on a time window. They assume that the values of the objects change over time and instead of performing instant top-k queries they retrieve the top-k objects by ranking them after aggregating their scores in a query interval. Lee et al. [9] discuss the idea of objects that appear continuously in top-k queries over data streams. They focus on discovering objects that appear continuously on a moving window of time. In [13] the authors study techniques for *durable top-k search* in document archives, where the aim is to identify documents that are consistently in the top-k results of a given query. Kontaki et al. [8] study the problem of discovering the objects that remain the most dominant over a data stream. Our main difference towards these approaches is that they consider the ranking functions to be static while the values of the objects are changing while we consider the exact opposite. Other work related to top-k and time includes processing of top-k queries on temporal data where the aim is finding the top-k objects at a particular time [10], as well as monitoring top-k queries over sliding windows [12].

8   Conclusions

In this paper, we studied for the first time the problem of finding the *most continuous influential products* that belong consistently to the most influential products over time. To this end, we defined the continuous influential query, where the influence score is defined based on reverse top-k queries and it changes as user preferences change over a long time period. In order to be able to efficiently discover the continuous influential products, we studied the properties of the proposed continuity score and derived appropriate upper and lower bounds. In turn, this lead to the design of efficient algorithms with the salient property of early termination. To evaluate our approach, we conducted a thorough experimental study that demonstrates the efficiency of our algorithms.

References

1. Arvanitis, A., Deligiannakis, A., Vassiliou, Y.: Efficient influence-based processing of market research queries. In: Proc. of CIKM, pp. 1193–1202 (2012)
2. Bernecker, T., Emrich, T., Kriegel, H.-P., Mamoulis, N., Renz, M., Zhang, S., Züfle, A.: Inverse queries for multidimensional spaces. In: Pfoser, D., Tao, Y., Mouratidis, K., Nascimento, M.A., Mokbel, M., Shekhar, S., Huang, Y. (eds.) SSTD 2011. LNCS, vol. 6849, pp. 330–347. Springer, Heidelberg (2011)
3. Börzsönyi, S., Kossmann, D., Stocker, K.: The skyline operator. In: Proc. of ICDE, pp. 421–430 (2001)
4. Fagin, R., Lotem, A., Naor, M.: Optimal aggregation algorithms for middleware. In: Proc. of PODS, pp. 102–113 (2001)
5. Ge, S., Hou, U.L., Mamoulis, N., Cheung, D.W.: Efficient all top-k computation: A unified solution for all top-k, reverse top-k and top-m influential queries. TKDE 25(5), 1015–1027 (2013)
6. Ilyas, I.F., Beskales, G., Soliman, M.A.: A survey of top-*k* query processing techniques in relational database systems. ACM Comp. Surv. 40(4) (2008)
7. Jestes, J., Phillips, J.M., Li, F., Tang, M.: Ranking large temporal data. PVLDB 5(11), 1412–1423 (2012)
8. Kontaki, M., Papadopoulos, A.N., Manolopoulos, Y.: Continuous top-k dominating queries. TKDE 24(5), 840–853 (2012)
9. Lee, M.L., Hsu, W., Li, L., Tok, W.H.: Consistent top-k queries over time. In: Zhou, X., Yokota, H., Deng, K., Liu, Q. (eds.) DASFAA 2009. LNCS, vol. 5463, pp. 51–65. Springer, Heidelberg (2009)
10. Li, F., Yi, K., Le, W.: Top-k queries on temporal data. VLDB Journal 19(5), 715–733 (2010)
11. Miah, M., Das, G., Hristidis, V., Mannila, H.: Standing out in a crowd: Selecting attributes for maximum visibility. In: Proc. of ICDE, pp. 356–365 (2008)
12. Mouratidis, K., Bakiras, S., Papadias, D.: Continuous monitoring of top-k queries over sliding windows. In: Proc. of SIGMOD, pp. 635–646 (2006)
13. Hou U, L., Mamoulis, N., Berberich, K., Bedathur, S.J.: Durable top-k search in document archives. In: Proc. of SIGMOD, pp. 555–566 (2010)
14. Vlachou, A., Doulkeridis, C., Kotidis, Y., Nørvåg, K.: Reverse top-k queries. In: Proc. ICDE, pp. 365–376 (2010)
15. Vlachou, A., Doulkeridis, C., Kotidis, Y., Nørvåg, K.: Monochromatic and bichromatic reverse top-k queries. TKDE 23(8), 1215–1229 (2011)
16. Vlachou, A., Doulkeridis, C., Nørvåg, K.: Monitoring reverse top-k queries over mobile devices. In: Proc. of MobiDE, pp. 17–24 (2011)
17. Vlachou, A., Doulkeridis, C., Nørvåg, K., Kotidis, Y.: Identifying the most influential data objects with reverse top-k queries. PVLDB 3(1-2), 364–372 (2010)
18. Vlachou, A., Doulkeridis, C., Nørvåg, K., Kotidis, Y.: Branch-and-bound algorithm for reverse top-k queries. In: Proc. of SIGMOD (2013)
19. Wu, T., Sun, Y., Li, C., Han, J.: Region-based online promotion analysis. In: Proc. of EDBT, pp. 63–74 (2010)
20. Wu, T., Xin, D., Mei, Q., Han, J.: Promotion analysis in multi-dimensional space. PVLDB 2(1), 109–120 (2009)
21. Yu, A., Agarwal, P.K., Yang, J.: Processing a large number of continuous preference top-*k* queries. In: Proc. of SIGMOD, pp. 397–408 (2012)

Finding Traffic-Aware Fastest Paths
in Spatial Networks

Shuo Shang, Hua Lu, Torben Bach Pedersen, and Xike Xie

Department of Computer Science, Aalborg University, Denmark
{sshang,luhua,tbp,xkxie}@cs.aau.dk

Abstract. Route planning and recommendation have received significant attention in recent years. In this light, we propose and investigate the novel problem of finding traffic-aware fastest paths (TAFP query) in spatial networks by considering the related traffic conditions. Given a sequence of user specified intended places O_q and a departure time t, TAFP finds the fastest path connecting O_q in order to guarantee that moving objects (e.g., travelers and bags) can arrive at the destination in time. This type of query is mainly motivated by indoor space applications, but is also applicable in outdoor space, and we believe that it may bring important benefits to users in many popular applications, such as tracking VIP bags in airports and recommending convenient routes to travelers. TAFP is challenged by two difficulties: (i) how to model the traffic awareness practically, and (ii) how to evaluate TAFP efficiently under different query settings. To overcome these challenges, we construct a traffic-aware spatial network $G_{ta}(V, E)$ by analysing uncertain trajectory data of moving objects. Based on $G_{ta}(V, E)$, two efficient algorithms are developed based on best-first and heuristic search strategies to evaluate TAFP query. The performance of TAFP has been verified by extensive experiments on real and synthetic spatial datasets.

1 Introduction

The continuous proliferation of GPS-enabled mobile devices (e.g., car navigation systems, smart phones and PDAs) and online map services (e.g., Google-maps[1], Bing-maps[2] and MapQuest[3]) enable people to acquire their current geographic positions in real time and to interact with servers to query spatial information regarding their trips [23]. In the meantime, with the rapid development of indoor positioning systems (e.g., Wi-Fi, RFID, and Bluetooth), the movements of objects in indoor spaces are increasingly tracked and recorded [10] [11], which makes it possible to plan and optimize the travel routes of moving objects in indoor spaces. The potential market of these location based services in the near future enables many novel applications. An emerging application is **traffic-aware fastest path queries** (TAFP for short) in spatial networks, which is designed

---

[1] http://maps.google.com/
[2] http://www.bing.com/maps/
[3] http://www.mapquest.com

M.A. Nascimento et al. (Eds.): SSTD 2013, LNCS 8098, pp. 128–145, 2013.

to find the fastest path that connects a sequence of user intended places in order. This type of query is mainly motivated by indoor space applications, but is also applicable in outdoor space, and we believe that it may bring important benefits to users in many popular mobile applications such as tracking VIP bags in airports and recommending convenient routes to travelers. To give examples, we describe the following two application scenarios:

Indoor Scenario: At an international airport, bags are collected from travelers and delivered to the corresponding aircrafts. The movements of bags are constrained in a large transfer network, which is made up by hundreds of conveyer belts, trolleys and buggies. The travel routes of bags are roughly planned when they are collected, such as "Check-in 8:55am ⟩ Screen machine 9:30am → Tilt-tray sorter 9:50am → Flight SK1217 10:20am". The travel path between any two adjacent intended places is uncertain. Large amounts of bags may create many traffic jams in the transfer network and the delivery of some bags may be delayed. Suppose the travel routes of all bags in delivery are available, our task is to find the fastest paths for VIP bags to guarantee their arrival times. This work is motivated by the BagTrack project[4], which is dedicated to improving bag handling in aviation industry globally.

Outdoor Scenario: Trajectory sharing and search are pervasive nowadays. Travellers can easily upload their trajectories to some specialized web sites such as Bikely[5] and Share-My-Routes[6]. In practice, most of the existing trajectory data are stored in compressed format (lossy compression leads to the uncertainty). Compared to original trajectories, compressed trajectories have clear advantages in data processing, transmitting, and storing [13]. By analysing historical travel trajectories of commuters, it is possible to construct a comprehensive traffic model to describe the traffic conditions in road networks, and to help users plan a fastest travel route.

The TAFP query is applied in spatial networks, since in a large number of practical scenarios, objects move in a constraint environment (e.g., transfer networks in the indoor scenario, and road networks in the outdoor scenario) rather than an Euclidean space. TAFP is challenging due to three reasons. First, it is necessary to construct a practical traffic model to detect the potential traffic jams (time-delays) and describe the traffic conditions in spatial networks, by analysing uncertain trajectories (i.e., roughly planned routes in the indoor scenario, and compressed trajectories in the outdoor scenario) of moving objects. Second, the uncertainty should be taken into account during the uncertain trajectory reconstruction, thus a probabilistic model and an efficient trajectory reconstruction algorithm are required. Third, we need to develop efficient algorithms to evaluate the TAFP under different query settings. It worth to note that TAFP is mainly motivated by indoor space applications and actively uses the trajectories of currently moving objects. In contrast, T-drive [23] is applicable to

---

[4] http://daisy.aau.dk/bagtrack
[5] http://www.bikely.com/
[6] http://www.sharemyroutes.com/

outdoor space and finds fastest paths based on traffic patterns discovered from the *historical* taxi trajectories. Thus, these two problems are not comparable.

To overcome these challenges, we construct a traffic-aware spatial network $G_{ta}(V, E)$ to detect potential time-delays and model traffic conditions in spatial networks. First, we propose an efficiency algorithm to reconstruct uncertain trajectories. An uncertain trajectory is reconstructed into several possible paths, and all possibilities are considered. Then, we map these possible paths onto spatial networks. For each vertex $v \in G_{ta}(V, E)$, we maintain a set of traffic records to describe its traffic conditions[7]. Based on $G_{ta}(V, E)$, two efficient algorithms based on best-first and heuristic search strategies are developed to evaluate the TAFP query. To sum up, the main contributions of this paper are as follows:

- We define a novel type of query: traffic-aware fastest path query according to the proposed spatio-temporal metrics. It provides new features for advanced spatio-temporal information systems, and may benefit users in many popular mobile applications such as path planning and recommendation.
- We propose a comprehensive probabilistic model to evaluate the uncertainty when reconstructing uncertain trajectories, and a traffic-aware spatial network based on uncertain trajectory data to model the traffic conditions practically (Section 2 and 3).
- We develop two algorithms based on best-first and heuristic search strategies to process the TAFP query efficiently (Section 4).
- We conduct extensive experiments on real and synthetic spatial data sets to investigate the performance of the proposed algorithms (Section 5).

The rest of the paper is organized as follows. Section 2 introduces spatial networks and uncertain trajectories used in this paper, as well as uncertain trajectory reconstruction algorithm. The construction of traffic-aware spatial network is detailed in Section 3. The TAFP query processing is addressed in Section 4, which is followed by the experimental results in Section 5. This paper is concluded in Section 7 after some discussions of related work in Section 6.

2   Uncertain Trajectory Reconstruction

2.1   Preliminaries

Spatial Networks. (i.e., transfer networks in the indoor space, and road networks in the outdoor space) are modeled by a connected and undirected graph $G(V, E)$, where V is the set of vertices and E is the set of edges. A weight is assigned to each edge to represent application specific factors such as traveling

---

[7] The concept of traffic-aware spatial network was briefly introduced in our previous work [17], including the definitions and examples of time-delay and the establishment of traffic-aware spatial network by using possible paths, which covers the majority of the contents in Sections 3.1 and 3.2. We improve and perfect the definitions, and detail the examples in this paper.

time. Given two vertices a and b in a spatial network, the network distance between them is the length of their shortest network path (i.e., a sequence of edges linking a and b where the accumulated weight is minimal). Data points (e.g., trajectory sample points) are embedded in networks and they may be located in edges. If the network distances to the two end vertices of an edge are known, it is straightforward to derive the network distance to any point in this edge. Thus, we assume that all data points are in the vertices for the simplification of description.

Uncertain Trajectory. There are two categories of uncertain trajectory data: roughly planned travel routes in the indoor space, and compressed trajectories in the outdoor spaces. In the indoor space, travel routes of bags are roughly planned such as "Check-in 8:55am $\rightarrow$ Screen machine 9:30am $\rightarrow$ Tilt-tray sorter 9:50am $\rightarrow$ Flight SK1217 10:20am", where sample points (intended places) are the vertices on a spatial network. In the outdoor space, raw trajectory samples obtained from GPS devices are typically of the form of $\langle longitude, latitude, timestamp \rangle$. How to select historical trajectory data of travelers and how to map the tuple $\langle longitude, latitude, timestamp \rangle$ onto a spatial network are interesting research problems themselves, but they are outside the scope of this paper. We assume that trajectory data are selected and all trajectory sample points have already been aligned to the vertices on the spatial network by some map-matching algorithms [2] [3] [7] [20]. Between any two adjacent sample points a and b, the exact travel path of moving objects is uncertain in both indoor and outdoor spaces. The spatio-temporal attributes of an uncertain trajectory are defined as follows.

Definition 1: Uncertain Trajectory
An uncertain trajectory τ of a moving object in a spatial network G is a finite sequence of timestamped positions: $\tau = \langle p_1, p_2, ..., p_n \rangle$, where p_i is a vertex in G, and $p_i.t$ is its timestamp, for $i = 1, 2, .., n$.

2.2 Uncertain Trajectory Reconstruction Algorithm

Given a spatial network $G(V, E)$, each vertex $v \in G.V$ is allocated a threshold $v.k$ to describe its traffic processing capability. That means at most $v.k$ moving objects can be processed at vertex v in one minute, and each individual moving object will take $\frac{1}{v.k}$ minutes processing time.

Given an uncertain trajectory segment $\tau_{seg}(p_i, p_j)$ connecting two adjacent sample points p_i and p_j, it is difficult to find the exact path $P(p_i, p_j)$ between them due to the rough route planning (indoor) or trajectory compression loss (outdoor). Here, we propose a random walk based probabilistic model to evaluate the uncertainty of trajectories, and all possibilities are considered. Conceptually, we assume that the movement of object o between p_i and p_j is according to a random walk, and its moving space is constrained by two thresholds. First, the maximum moving time between p_i and p_j is constrained by $(p_j.t - p_i.t)$, where $p_i.t$ and $p_j.t$ are the timestamps of p_i and p_j, respectively. Second, there should be no loop in $P(p_i, p_j)$, which means that one vertex cannot appear twice in one path. The length and probability of path $P = \langle p_1, ..., p_k \rangle$ are defined as follows:

$$P.length = \sum_{i=1}^{k-1}((p_i, p_{i+1}).weight + \frac{1}{p_{i+1}.k}) \tag{1}$$

$$P.prob = \prod_{i=1}^{k-1}(p_i, p_{i+1}).prob \tag{2}$$

Here, $(p_i, p_{i+1}).weight$ and $(p_i, p_{i+1}).prob$ are the weight and probability of edge (p_i, p_{i+1}), respectively. Suppose a moving object o is at vertex p_i and it may select edge (p_i, p_{i+1}) as its following moving direction. This probability is defined as the probability of edge (p_i, p_{i+1}).

Algorithm 1. Uncertain Trajectory Reconstruction

Data: two adjacent sample points p_i, p_j
Result: $Pathlist(p_i, p_j)$

1 $path \leftarrow null$;
2 **begin**
3   **Procedure**(p_i)
4   **for** *each vertex* $n \in p_i.adjacentList$ **do**
5   **if** $n \in path$ **then**
6   Continue;
7   **if** $path.dist + sd(p_i, n) > (p_j.t - p_i.t)$ **then**
8   Continue;
9          $path.vertex \leftarrow path.vertex + n$;
10         $path.lenth \leftarrow path.length + sd(p_i, n) + 1/n.k$;
11         $path.prob \leftarrow path.prob/p_i.validNeighbors.size$;
12   **if** $n = p_j$ **then**
13             $path.add(n)$;
14             $Pathlist.add(path)$;
15             $path.remove(n)$;
16   Continue;
17   **Procedure** (n);
18     $path.remove(p_i)$;
19 **end**
20 **return** $Pathlist$;

All valid sub-paths of the trajectory segment $\tau_{seg}(p_i, p_j)$ can be retrieved according to Algorithm 1. Here, p_i and p_j are two adjacent sample points and Procedure(p_i) is a recursive function. A depth-first traversal is conducted to find all possible sub-paths connecting p_i and p_j. A network expansion [4] starts from p_i, and all adjacent vertices of p_i are scanned iteratively (line 4). For each vertex n adjacent to p_i, we check whether it conflicts the two thresholds: n cannot appear twice in the same path (lines 5-6), and the path length cannot

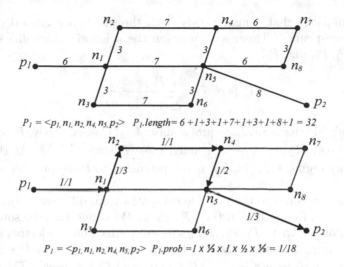

$P_1 = <p_1, n_1, n_2, n_4, n_5, p_2>$ $P_1.length= 6 +1+3+1+7+1+3+1+8+1 = 32$

$P_1 = <p_1, n_1, n_2, n_4, n_5, p_2>$ $P_1.prob =1 \times \frac{1}{3} \times 1 \times \frac{1}{2} \times \frac{1}{3} = 1/18$

Fig. 1. Probabilistic model

be greater than $(p_j.t - p_i.t)$ (lines 7-8). If n satisfies both thresholds, the related information of $path(n)$ are recorded, including its vertices, length and probability (lines 9-11). Once the destination p_j has been scanned by the expansion, a valid sub-path is found and stored in $Pathlist(p_i, p_j)$ (lines 12-16). Otherwise, a new recursive function based on n will be conducted (line 17). Finally, all the possible sub-paths connecting p_i and p_j are returned to users (line 20).

An example of trajectory segment reconstruction is shown in Figure 1, where p_1 and p_2 are two adjacent trajectory sample points, and $n_1, n_2, ..., n_8$ are vertices in a spatial network G. Each edge is assigned a weight to represent the travel time along this edge, and each vertex is assigned a threshold to describe its traffic processing capability. In Figure 1, the traffic processing capability for each vertex is 1, thus the processing time for each individual object is also 1. The maximum moving time between p_1 and p_2 is $p_2.t - p_1.t = 33$.

There exist three sub-paths P_1, P_2 and P_3 that satisfy both thresholds mentioned above. For $P_1 = \langle p_1, n_1, n_2, n_4, n_5, p_2 \rangle$, according to Equation 1, its length is computed as $P_1.length = 6 + 1 + 3 + 1 + 7 + 1 + 3 + 1 + 8 + 1 = 32$. The probability computation is more complex than the length computation. (p_1, n_1) is the first edge of P_1 and its probability is $\frac{1}{1}$. We assume that the movement of objects is according to a random walk. At n_1, a moving object m has four choices: $\{p_1, n_2, n_3, n_5\}$ for the next step. Vertex p_1 is invalid in this case, since a loop $\langle p_1, n_1, p_1 \rangle$ is formed. As a result, the rest three vertices share the same probability $\frac{1}{3}$. Then, at n_2, n_4 holds $\frac{1}{1}$ probability to be the next vertex, since n_1 will form a loop. Following this procedure, the probability of P_1 is computed based on Equation 2: $P_1.prob = \frac{1}{1} \times \frac{1}{3} \times \frac{1}{1} \times \frac{1}{2} \times \frac{1}{3} = \frac{1}{18}$. Similarly, we compute the length and probability of $P_2 = \langle s_1, n_1, n_5, s_2 \rangle$ and $P_3 = \langle s_1, n_1, n_3, n_6, n_5, s_2 \rangle$. P_2 has the length 24 and the probability $\frac{1}{1} \times \frac{1}{3} \times \frac{1}{4} = \frac{1}{12}$. P_3 has the length 32 and the probability $\frac{1}{1} \times \frac{1}{3} \times \frac{1}{1} \times \frac{1}{1} \times \frac{1}{3} = \frac{1}{9}$.

Other sub-paths that cannot satisfy both thresholds are labeled as invalid paths and are pruned. Then, we normalize the original probabilities of valid sub-paths $P_1, P_2,$ and P_3.

$$P_i.probN = \frac{P_i.prob}{\sum_{i=1}^{k} P_i.prob} \qquad (3)$$

Here, $P_i.probN$ is the normalized probability of P_i. Consequently, $Prob_N(P_1) = \frac{1/18}{1/4} = \frac{2}{9}$, $Prob_N(P_2) = \frac{1/12}{1/4} = \frac{1}{3}$ and $Prob_N(P_3) = \frac{1/9}{1/4} = \frac{4}{9}$. At this stage, the trajectory segment $T_{seg}(s_i, s_j)$ is reconstructed to three possible sub-paths P_1, P_2, P_3.

After that, we combine the possible sub-paths from different trajectory segments to create a full path $P(s, d) = \langle P_1, P_2, ..., P_k \rangle$ connecting the source s and the destination d, where $P_1, P_2, ..., P_k$ are sub-paths from trajectory segments $T_{seg1}, T_{seg2}, ..., T_{segk}$, respectively. The length and probability of $P(s, d)$ can be computed as $P(s, d).length = \sum_{i=1}^{k}(P_i.length)$ and $P(s, d).prob = \prod_{i=1}^{k} P_i.prob$, respectively.

3   Traffic-Aware Spatial Network

3.1   Time-Delay

For any vertex v in a spatial network $G(V, E)$, if the number of moving objects to be processed by v is over its processing capability $v.k$, a time-delay (or a traffic jam, depending on the number of objects) will occur. Suppose the processing capability of v is $v.k$ per minute, and each individual moving object will take $\frac{1}{v.k}$ minutes processing time. If the gap between any two moving objects is less than $\frac{1}{v.k}$ minutes, a time-delay will be triggered. The time-delay for object o at vertex v is estimated as follows.

$$T_d(o, v, T_a(o, v)) = \begin{cases} \sum_{o_i \in O_p} o_i.prob \cdot (\frac{1}{v.k} - T_a(o, v) \\ + T_a(o_i, v)) + \sum_{o_j \in O_w} \frac{o_j.prob}{v.k} & \text{if } C_1 \\ \\ 0 & \text{if } C_2 \end{cases} \qquad (4)$$

C_1: v is occupied at time point $T_a(o, v)$, and O_p is the set of moving objects being processed by v at this time point. In addition, there are also $|O_w|$ moving objects waiting to be processed.
C_2: v is available at time point $T_a(o, v)$, thus the corresponding time-delay is 0. Here $T_a(o, v)$ is the arrival time of object o at vertex v.

The total time cost of object o at vertex v is the sum of waiting time (time-delay) and processing time. Thus the leaving time of moving object o from vertex v can be computed as follows.

$$T_l(o, v) = T_a(o, v) + T_d(o, v, T_a(o, v)) + \frac{1}{v.k} \qquad (5)$$

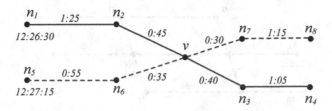

Fig. 2. Time-delay

An example is shown in Figure 2, where $P_1 = \langle n_1, n_2, v, n_3, n_4 \rangle$ and $P_2 = \langle n_5, n_6, v, n_7, n_8 \rangle$ are two reconstructed probabilistic trajectories for moving objects o_1 and o_2 respectively. P_1 has 50% probability while P_2 has 75% probability. Vertex v is an intersection and its traffic processing capability is $v.k = 4$ per minute, which means processing an individual moving object will take $\frac{1}{4}$ minute = 15 seconds. The expected arrival time of o_1 at v is 12:28:40, while o_2's expected arrival time at v is 12:28:45. When o_2 arrives at v, the processing of o_1 is not finished until 10 seconds later. Referring to Equations 4 and 5, the time-delay of moving object o_2 at vertex v can be estimated as $T_d(v, o_2) = 50\% \times 0{:}10 = 0{:}05$. The estimated leaving time of o_2 from v is 12:28:45 + 0:05 + 0:15 = 12:29:05, and the expected arrival time of o_2 at vertices n_7 and n_8 will be delayed correspondingly.

Note that at some time points, there may exist more than one moving object with the status of "being processed by vertex v" (referring to Equation 4). For example, in Figure 2, moving object o_1 is being processed within the time range [12:28:40, 12:28:55], and the estimated processing time of o_2 is within the time range [12:28:50, 12:29:05]. In the overlapped time range [12:28:50, 12:28:55], o_1 and o_2 are both in the status of "being processed by v". Suppose there is a new moving object o_3, and the arrival time of o_3 at v is within the overlapped time range [12:28:50, 12:28:55], to estimate the time-delay of o_3, the influences from both o_1 and o_2 should be taken into account. In addition, if more than one moving object arrive at one vertex at the same time point, we will arrange their schedules randomly.

3.2 Traffic-Aware Spatial Network

To detect potential time-delays and model traffic conditions practically, we construct a traffic-aware spatial network $G_{ta}(V, E)$ by analysing the uncertain trajectories over spatial networks. In our definition, a traffic-aware spatial network $G_{ta}(V, E)$ is a spatial network, where each vertex $v \in G_{ta}.V$ has been assigned a threshold $v.k$ to describe its traffic processing capability. There are also a set of traffic information records attached to v to describe its traffic conditions. Each record is in the form of (object, probability, expected arrival time, time-delay, processing time).

Given a spatial network $G(V, E)$ and a set of uncertain trajectories T, the construction of a traffic-aware spatial network takes two steps. First, we reconstruct

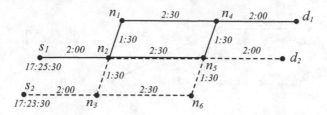

Fig. 3. Traffic-aware spatial network construction

the uncertain trajectories in T according to a random walk probabilistic model, and map the reconstructed trajectories (with probability) onto $G(V, E)$. Second, we sort the vertices in reconstructed trajectories according to their timestamps, and then refine their time-delays.

An example is shown in Figure 3, where $n_1, n_2, ..., n_6$ are vertices and their traffic processing capabilities are all 1 moving object per minute. There are 2 uncertain trajectories $\tau_1 = \langle s_1, d_1 \rangle$ and $\tau_2 = \langle s_2, d_2 \rangle$ of moving objects o_1 and o_2, respectively. Trajectory τ_1 is reconstructed into two possible paths $P_1 = \langle s_1, n_2, n_1, n_4, d_1 \rangle$ and $P_2 = \langle s_1, n_2, n_5, n_4, d_1 \rangle$, and each of them has 50% probability; while τ_2 is reconstructed into two possible paths $P_3 = \langle s_2, n_3, n_2, n_5, d_2 \rangle$ and $P_4 = \langle s_2, n_3, n_6, n_5, d_2 \rangle$, and each of them also has 50% probability. Then, we map all possible paths onto the spatial network, and compute the expected arrival times (ignore the time-delay) as the timestamps for vertices in all possible paths. For instance, the departure time of P_1 is 17:25:30, and the expected arrival time of n_2, n_1, n_4 and d_1 are 17:27:30, 17:30:00, 17:33:30, and 17:36:30, respectively.

We maintain a dynamic priority heap containing these timestamps. At each step, we only refine the timestamp on the top of the heap. In Figure 3, $n_3 \in P_3, P_4$ will be refined firstly, followed by $n_2 \in P_1, P_2$. Then, we detect the first time-delay at vertex $n_2 \in P_3$. The expected arrival time of moving object o_1 at n_2 is 17:27:30 with 100% probability (P_1 and P_2), while the expected arrival time of o_2 at n_2 is 17:28:00 with 50% probability (P_3). Referring to Equations 4 and 5, the time-delay of $o_2 \in P_3$ at n_2 is 0:30, and the estimated leaving time is 17:29:30. Due to the time-delay at n_2, the expected arrival times for the rest vertices in P_3 (i.e., n_5 and d_2) have to be adjusted correspondingly.

Table 1. Traffic Information Records

vertex	traffic records
n_1	(o_1, 50%, 17:30:00, 0, 1:00)
n_2	(o_1, 100%, 17:27:30, 0, 1:00), (o_2, 50%, 17:28:00, 0:30, 1:00)
n_3	(o_2, 100%, 17:25:30, 0, 1:00)
n_4	(o_1, 100%, 17:33:30, 0, 1:00)
n_5	(o_1, 50%, 17:31:00, 0, 1:00), (o_2, 50%, 17:31:30, 0:15, 1:00), (o_2, 50%, 17:32:00, $0:22\frac{1}{2}$, 1:00)
n_6	(o_1, 50%, 17:29:00, 0, 1:00)

The refinement follows this procedure, step by step, until the priority heap is empty and all *time-stamp*s have been refined. For each vertex $v \in V$, we maintain a set of traffic records to describe its traffic conditions. The refinement results of Figure 3 are demonstrated in Table 1.

The complete procedure of establishing a traffic-aware spatial network is detailed in Algorithm 2. Given a spatial network $G(V, E)$ and a set of uncertain trajectories T, for each uncertain trajectory τ, we reconstruct its possible paths and map them onto $G(V, E)$ (lines 1-6). In the next, we select the vertex v with the minimum timestamp and refine it (lines 8-9). If a time-delay is detected, the timestamps of the rest vertices following v in the same path are updated correspondingly (lines 10-11).

Algorithm 2. Traffic-Aware Spatial Network Construction

Data: spatial network $G(V, E)$, uncertain trajectory set T
Result: $G_{ta}(V, E)$
1 **while** $T \neq \emptyset$ **do**
2 | $\tau \leftarrow T.pop()$;
3 | reconstruct τ;
4 | **while** $Pathlist(\tau) \neq \emptyset$ **do**
5 | | $P \leftarrow Pathlist(\tau).pop()$;
6 | |_ map P onto $G(V, E)$;

7 **while** $H \neq \emptyset$ **do**
8 | $v \leftarrow Heap.pop()$;
9 | refine v;
10 | **if** $v.delay$ *is triggered* **then**
11 | |_ update the timestamps of $v.rest$;

3.3   Traffic Records Indexing and Search

Each vertex $v \in G_{ta}(V, E)$ maintains a set of traffic records to describe its traffic conditions (refer to Table I). We notice that there may exist a large number of records (e.g, hundreds of or thousands of records, depending on the number of uncertain trajectories) at one vertex. Such massive traffic data may prevent the computation from being addressed in real time. To accelerate the computing process, we establish a B-tree for the column "expected arrival time". Note that the utilization of the well known B-tree is only for improving the search efficiency. Other indexing approaches can also be easily adapted. Given a moving object o and its expected arrival time $T_a(o, v)$ at vertex v, if a moving object o_i in the traffic records of v satisfies Equation 6, it will be labeled as "a waiting object" (i.e., when moving object o arrives at v, o_i is waiting for being processed) and put into data set O_w. Otherwise, if o_i satisfies Equation 7, it will be labeled as "a processing object" (i.e., when o arrives at v, o_i is being processed) and put into data set O_p. According to Equation 4, the time-delay of moving object o at vertex v can be obtained.

$$T_a(o_i, v) < T_a(o, v) < T_a(o_i, v) + T_d(o_i, v, T_a(o_i, v)) \tag{6}$$

$$\begin{cases} T_a(o_i, v) + T_d(o_i, v, T_a(o_i, v)) < T_a(o, v) \\ T_a(o, v) < T_a(o_i, v) + T_d(o_i, v, T_a(o_i, v)) + \frac{1}{v.k} \end{cases} \tag{7}$$

4   TAFP Query Processing

In this section, we define a novel type of query: traffic-aware fastest path (TAFP) query, and develop two algorithms to process it efficiently.

Definition 2: Traffic-Aware Fastest Path Query
Given a traffic-aware spatial network $G_{ta}(V, E)$, a sequence of user intended places O_q and a departure time t, TAFP query finds the fastest path P that connects O_q in order, such that $\forall P' \in Pathlist(O_q)(T_t(P, t) \leq T_t(P', t))$, where $Pathlist(O_q)$ is a data set that contain all paths connecting O_q in order.

4.1   Best-First Search Strategy

Dijkstra's algorithm [4] is a conventional method to address the shortest/fastest path problem in spatial networks, but it fails to solve TAFP due to the non-awareness of moving object processing time and time-delay at each vertex in $G_{ta}(V, E)$. We develop a novel search algorithm named D_p to answer TAFP in real time, which inherits the *best-first* search strategy from Dijkstra's algorithm (i.e., Dijkstra's algorithm always selects the vertex with the minimum distance label for expansion). In D_p, the distance label of each vertex v in a network expansion tree is defined as

$$v.dist = c.dist + w(c, v) + T_d(v, (T_l(c) + w(c, v))) + \frac{1}{v.k} \tag{8}$$

where v and c are vertices in $G_{ta}(V, E)$, and c is the parent vertex of v (i.e., $c = v.pre$) in the expansion tree. In contrast to Dijkstra's algorithm, the distance label in D_p contains three parts: the weight (travel cost) of edge (c, v), the time-delay of v at the time point $(T_l(c) + w(c, v))$, and the individual moving object processing time at v.

The D_p algorithm is detailed in Algorithm 3. Initially, the distance label of each vertex $v \in V$ is set to $+\infty$ (line 2). Then, we compute the fastest path $P(o_i, o_{i+1})$ that connects $o_i, o_{i+1} \in O_q$ by using D_p. By combining the computation results, we find the fastest path P that connects O_q in order, and return the fastest path P (lines 3-7).

The D_p algorithm is designed to find the fastest path between any two vertices $o_i, o_j \in G_{ta}(V, E)$. The query input of D_p includes a source point o_i, a destination point o_j and a departure time t (line 9). In each iteration, we select the vertex

---

Algorithm 3. Finding TAFP using D_p Algorithm

Data: $G_{ta}(V, E)$, O_q and t
Result: P and $T_t(P, t)$

1   $t_0 \leftarrow t$; $P \leftarrow \emptyset$; $O_s \leftarrow \emptyset$; $P(o_i, o_{i+1}) \leftarrow \emptyset, i \in [0, |O_q| - 2]$;
2   $v.dist \leftarrow +\infty, \forall v \in V$;
3   **for** $i = 0$; $i < |O_q| - 1$; $i++$ **do**
4      $D_p(o_i, o_{i+1}, t_i)$;
5      $t_{i+1} \leftarrow t_i + T_t(P(o_i, o_{i+1}), t_i)$;
6      $P \leftarrow P \cup P(o_i, o_{i+1})$;

7   **return** P;
8   **begin**
9   **Function** $D_P(o_i, o_j, t)$
10     $O_s.push(o_i)$;
11   **while** $Q_s \neq \emptyset$ **do**
12   select $v \in O_s$ with the minimum $v.dist$;
13       $O_s.remove(v)$;
14   **if** $v = o_j$ **then**
15   **while** $v.pre \neq null$ **do**
16           $P(o_i, o_j).push(v)$;
17           $v \leftarrow v.pre$;
18         $T_t(P(o_i, o_j), t) \leftarrow o_j.dist - t$;
19   **return** $P(o_i, o_j)$ & $T_t(P(o_i, o_j), t)$;
20   **for** *each vertex* $n \in c.adjacentList$ **do**
21         $T_a(n) \leftarrow v.dist + w(c, v)$;
22   **if** $n.dist > v.dist + w(v, n) + T_d(n, T_a(n)) + 1/n.k$ **then**
23           $n.dist \leftarrow v.dist + w(v, n) + T_d(n, T_a(n)) + 1/n.k$;
24           $n.pre \leftarrow v$;
25           $O_s.push(n)$;

26 **end**

---

v with the minimum distance label $v.dist$ from the heap O_s. Once $v = o_j$, the destination o_j has been reached and the fastest path $P(o_i, o_j)$ is found (lines 14-17). For each adjacent vertex of v, if its current distance label is less than the new distance label $v.dist + w(v, n) + T_d(n, T_a(n)) + 1/n.k$ (i.e., the distance of the path from o_i to n via v), it will be replaced by the new distance label, and n will be put into the heap O_s (lines 20-25).

4.2 Heuristic Search Strategy

Similar to Djkstra's algorithm, D_p lacks an effective heuristic search strategy that helps it focus on the paths more likely to be the optimal choice, and constrain the search space and improve the query performance. We develop a novel A_p algorithm to further enhance the efficiency of TAFP, which is an extension of A* algorithm [8]. A_p shares the same structure as D_p, but the distance label of each vertex v in a network expansion tree is defined as follows.

$$v.dist = c.dist + w(c,v) + T_d(v, T_a(v)) + \frac{1}{v.k} + sd(v,d) \tag{9}$$

where c is the parent vertex of v in the expansion tree, and $sd(v,d)$ is the shortest path distance between vertex v and the query destination d (i.e., a sequence of edges linking v and d where the accumulated weight is minimal).

The distance label of A_p has two parts: $(c.dist + w(c,v) + T_d(v, T_a(v)) + 1/v.k)$ is to describe the travel cost from the query source to vertex v via c, while $sd(v,d)$ is the heuristic part to estimate the exact travel cost from vertex v to the destination d. Intuitively, the value of $sd(v,d)$ is always less than the real travel cost from v to d, since the moving object processing time and time-delay are not considered in the value of $sd(v,d)$. Thus the correctness of A_p algorithm can be guaranteed. The search process of finding TAFP using A_p algorithm is conducted by substituting Equation 9 into Algorithm 3 (lines 22-23).

To acquire the shortest path distance $sd(a,b)$ between any two vertices $a, b \in V$ efficiently, we assume that the network paths between all pairs have been pre-computed and encoded to reduce storage cost from $O(|V|^3)$ to $O(|V|^{1.5})$ [16]. This encoding takes advantage of the fact that the shortest paths from a vertex u to all of the remaining vertices can be decomposed into subsets based on the first edges on the shortest paths to them from u. The simplest way of representing the shortest path information is to maintain an array A of size $|V| \times |V|$, where $A[u,v]$ contains the first vertex on the shortest path from u to v. Using A, the construction of the shortest path between u and v is performed by repeatedly visiting $A[u',v]$, where $u' = u$ for the first time and $u' = A[u',v]$ in subsequence, and the time is proportional to the length to the path.

5   Experimental Results

Next, we conducted extensive experiments on real and synthetic spatial data sets to demonstrate the performance of TAFP query. The two data sets used in our experiments were the Beijing Road Network (BRN)[8] and synthetic Indoor Transfer Network (ITN), which contain 28,342 vertices and 6,105 vertices, respectively, stored as adjacency lists. In BRN, we adopted the real trajectory data collected by MOIR project [12]. In ITN, synthetic trajectory data were used. All algorithms were implemented in Java and tested on a Linux platform with Intel Core i7-3520M Processor (2.90GHz) and 8GB memory. The experimental results were averaged over 20 independent trials with different query inputs. The main performance metrics were CPU time and the number of visited vertices. The number of visited vertices was selected as a metric for two reasons: (i) it can describe the exact amount of data access; (ii) it can reflect the real disk I/O cost to a certain degree. The parameter settings are listed in Table 2. By default, the number of uncertain trajectories were 4,000 and 2,000 in BRN and ITN respectively to construct the traffic-aware spatial network.

---

[8] http://www.iscas.ac.cn/

Table 2. Parameter settings

	BRN	ITN
Uncertain Trajectory Length	2 – 10	2– 10
Number of Reconstructed Paths	10,000 – 40,000 (default 10,000)	10,000 – 40,000 (default 10,000)
Number of Intended Places $|O_q|$	2 – 10	2 – 10

5.1 Performance of Uncertain Trajectory Reconstruction

First of all, we studied the performance of the uncertain trajectory reconstruction (UTR) in Algorithm 1 when uncertain trajectory length (the number of sample points) varies (Figure 4). The length of uncertain trajectory was set from 2 to 10 in both BRN and ITN. Longer trajectories lead to more computation efforts, thus the CPU time and the number of visited vertices are expected to be higher in both BRN and ITN. Generally, the maximum CPU time is under 700 ms in BRN and under 400 ms in ITN, while the maximum number of visited vertices is less than 2000 in BRN and is around 1500 in ITN.

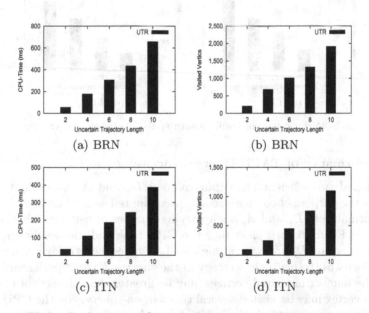

(a) BRN   (b) BRN

(c) ITN   (d) ITN

Fig. 4. Performance of uncertain trajectory reconstruction

5.2 Traffic-Aware Spatial Network $G_{ta}(V, E)$ Construction

Figure 5 demonstrates the performance of Traffic-Aware spatial network $G_{ta}(V, E)$ construction, including the construction times (Figures 5(a) and 5(c)) and network sizes (Figures 5(b) and 5(d)) in BRN and ITN, respectively. The number of reconstructed possible paths (reconstructed trajectories) is from

10,000 to 50,000 in both BRN and ITN. Obviously, more trajectory data leads to higher construction time, and larger size of the constructed traffic-aware spatial network. The traffic-aware spatial network construction is an **off-line** process due to the huge amount of possible paths, and the traffic records are stored in disk. For each vertex $v \in G_{ta}(V, E)$, we maintain a pointer pointing to the positions of the related traffic records in disk. Once the vertex v is scanned by a network expansion, the related traffic records can be accessed efficiently. .

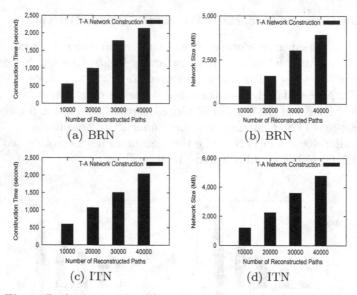

Fig. 5. Performance of traffic-aware spatial network construction

5.3   Performance of TAFP Query Processing

We developed two efficient algorithms named D_p and A_p to evaluate TAFP query. For the purpose of comparison, D_p was adopted as the baseline algorithm. The performance of D_p and A_p with varying number of user intended places is presented in Figure 6. It is clear that the CPU time and the number of visited vertices required by D_p are twice as high as A_p. These results clearly demonstrate the superiority of the heuristic strategy in the shortest/fastest path search. Note that (i) the number of visited vertices may be greater than the size of vertex set V, since a vertex may be visited several times in one query; (ii) the CPU time is not fully corresponding to the Disk I/O times. In some cases, the increment of computation cost may offset the benefits from the reduction of Disk I/O times. .

6   Related Work

6.1   Shortest Path Queries in Spatial Networks

In static spatial networks, Dijkstra's algorithm [4] and the A* algorithm [8] are two classical methods to address the shortest path problem. In Dijkstra's

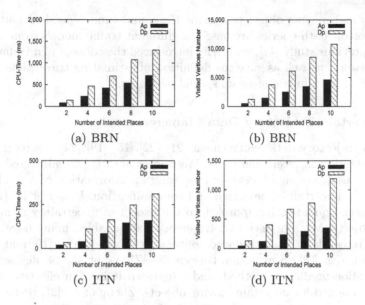

Fig. 6. Performance of TAFP query processing

algorithm, an expansion wavefront is expanded from the source s. A priority heap H is adopted to maintain the unscanned vertexes. At each expansion, vertex v with the minimum distance label (usually the network distance from s to v) will be selected, and removed from the priority heap, and labeled as scanned vertex. In the next, all the unscanned neighbor vertex to v will be put into H, until the destination d has been reached and the shortest path from s to d is found. In A* algorithm, the value of $sd(v_s, v) + d_E(v, v_d)$, where $d_E(v, v_d)$ is the Euclidean distance between v and v_d, is used as the distance label of vertex v, to estimate the network distance from v_s to v_d via v. A* algorithm is an heuristic searching approach, and its performance is greater than Dijkstra's algorithm in general.

Time-dependent road network [5] and probabilistic road network [9] are two representative dynamic spatial networks. A time-dependent road network can be established according to the speed pattern, which motivates the-time dependent shortest path query [5]. Time-dependent shortest path query is a variant of dynamic shortest path problem, which is designed to find the best departure time for users, to minimize the global traveling time from a source to a destination over a large road network, where the traffic conditions are dynamically changing from time to time. The challenge of this problem lies in the dynamic edge delay. In probabilistic road networks, each edge is assigned a set of probabilistic data to describe the traveling cost along this edge, and probabilistic shortest path queries [9] ask for (i) the fastest path constrained by a probability threshold, and (ii) the path with the highest probability constrained by a travel time threshold.

Despite the bulk of literature on shortest path queries [4] [8] [14] [1] [6] [5] [9], none of the existing work can address the proposed TAFP query due to two reasons: Dijkstra's algorithm and A* algorithm are not aware of the

time-delay and moving object processing time, and time-dependent and probabilistic shortest path queries are based on different traffic models from ours.

In our previous study [17], we briefly introduced the concept of a traffic-aware
spatial network, as well as gave the definition of estimating travel time along a
user specified path in traffic-aware spatial networks.

6.2   Uncertain Trajectory Data Management

Uncertain trajectory data management [21] [22] [15] [19] [24] has received significant attention in recent ten years. Wolfson et al. [21] [22] addressed the update problem of moving objects by proposing an information cost model that
captured the uncertainty, deviation and communication. Pfoser et al. [15] proposed a formal quantitative approach to the aspect of uncertainty in modeling
moving objects. Massive uncertain trajectory data enabled many novel spatial
queries. Trajcevski et al. introduced continuous range queries [19] and nearest
neighbor queries [18] on uncertain trajectories. Zhang et al. [24] devised an efficient location-prediction method, and integrated it into an effective indexing
structure designed for uncertain moving objects. Zheng et al. [25] presented the
probabilistic range queries on uncertain trajectories in road networks and the
corresponding efficient solutions.

7   Conclusions and Future Directions

In this work, we proposed and studied a novel problem called traffic-aware fastest
path (TAFP) query. Given a sequence of user intended places O_q and a departure
time, the TAFP query finds the fastest path that connects O_q in order. We
believe that this type of query can bring significant benefits to users in many
popular applications, such as path planning and recommendation. To detect
potential time-delays and model traffic conditions practically, we constructed a
traffic-aware spatial network $G_{ta}(V, E)$ by analysing uncertain trajectory data
of moving objects. Based on $G_{ta}(V, E)$, we developed two algorithms using best-
first and heuristic search strategies respectively to address the TAFP query in
real time. Finally, we conducted extensive experiments on real and synthetic
datasets to demonstrate the performance of the TAFP query. In the future,
it is of interest to consider objects moving patterns in the uncertain trajectory
reconstruction, and to consider parallel computing techniques in the construction
of traffic-aware spatial networks.

References

1. Ahuja, R.K., Orlin, J.B., Pallottino, S., Scutellà, M.G.: Dynamic shortest paths
   minimizing travel times and costs. Networks 41(4), 197–205 (2003)
2. Alt, H., Efrat, A., Rote, G., Wenk, C.: Matching planar maps. In: SODA, pp.
   589–598 (2003)
3. Brakatsoulas, S., Pfoser, D., Salas, R., Wenk, C.: On map-matching vehicle tracking
   data. In: VLDB, pp. 853–864 (2005)

4. Dijkstra, E.W.: A note on two problems in connection with graphs. Numerische Math. 1, 269–271 (1959)
5. Ding, B., Yu, J.X., Qin, L.: Finding time-dependent shortest paths over large graphs. In: EDBT, pp. 205–216 (2008)
6. George, B., Kim, S., Shekhar, S.: Spatio-temporal network databases and routing algorithms: A summary of results. In: Papadias, D., Zhang, D., Kollios, G. (eds.) SSTD 2007. LNCS, vol. 4605, pp. 460–477. Springer, Heidelberg (2007)
7. Greenfeld, J.: Matching gps observations to locations on a digital map. In: 81th Annual Meeting of the Transportation Research Board (2002)
8. Hart, P.E., Nilsson, N.J., Raphael, B.: A formal basis for the heuristic determination of minimum cost paths. IEEE Transactions on Systems Science and Cybernetics 4(2), 100–107 (1968)
9. Hua, M., Pei, J.: Probabilistic path queries in road networks: traffic uncertainty aware path selection. In: EDBT, pp. 347–358 (2010)
10. Jensen, C.S., Lu, H., Yang, B.: Graph model based indoor tracking. In: Mobile Data Management, pp. 122–131 (2009)
11. Jensen, C.S., Lu, H., Yang, B.: Indexing the trajectories of moving objects in symbolic indoor space. In: Mamoulis, N., Seidl, T., Pedersen, T.B., Torp, K., Assent, I. (eds.) SSTD 2009. LNCS, vol. 5644, pp. 208–227. Springer, Heidelberg (2009)
12. Liu, K., Deng, K., Ding, Z., Li, M., Zhou, X.: Moir/mt: Monitoring large-scale road network traffic in real-time. In: VLDB, pp. 1538–1541 (2009)
13. Muckell, J., Hwang, J.-H., Lawson, C., Ravi, S.: Algorithms for compressing gps trajectory data: An empirical evaluation. ACM GIS (2010)
14. Orda, A., Rom, R.: Shortest-path and minimum-delay algorithms in networks with time-dependent edge-length. J. ACM 37(3), 607–625 (1990)
15. Pfoser, D., Jensen, C.S.: Capturing the uncertainty of moving-object representations. In: Güting, R.H., Papadias, D., Lochovsky, F.H. (eds.) SSD 1999. LNCS, vol. 1651, p. 111. Springer, Heidelberg (1999)
16. Samet, H., Sankaranarayanan, J., Alborzi, H.: Scalable network distance browsing in spatial databases. In: Proceedings of SIGMOD, pp. 43–54 (2008)
17. Shang, S., Lu, H., Pedersen, T.B., Xie, X.: Modeling of traffic-aware travel time in spatial networks. In: MDM, 4 pages (2013)
18. Trajcevski, G., Tamassia, R., Ding, H., Scheuermann, P., Cruz, I.F.: Continuous probabilistic nearest-neighbor queries for uncertain trajectories. In: EDBT, pp. 874–885 (2009)
19. Trajcevski, G., Wolfson, O., Hinrichs, K., Chamberlain, S.: Managing uncertainty in moving objects databases. ACM Trans. Database Syst. 29(3), 463–507 (2004)
20. Wenk, C., Salas, R., Pfoser, D.: Addressing the need for map-matching speed: Localizing globalb curve-matching algorithms. In: SSDBM (2006)
21. Wolfson, O., Chamberlain, S., Dao, S., Jiang, L., Mendez, G.: Cost and imprecision in modeling the position of moving objects. In: ICDE, pp. 588–596 (1998)
22. Wolfson, O., Sistla, A.P., Chamberlain, S., Yesha, Y.: Updating and querying databases that track mobile units. Distributed and Parallel Databases 7(3), 257–387 (1999)
23. Yuan, J., Zheng, Y., Zhang, C., Xie, W., Xie, X., Sun, G., Huang, Y.: T-drive: driving directions based on taxi trajectories. In: GIS, pp. 99–108 (2010)
24. Zhang, M., Chen, S., Jensen, C.S., Ooi, B.C., Zhang, Z.: Effectively indexing uncertain moving objects for predictive queries. PVLDB 2(1), 1198–1209 (2009)
25. Zheng, K., Trajcevski, G., Zhou, X., Scheuermann, P.: Probabilistic range queries for uncertain trajectories on road networks. In: EDBT, pp. 283–294 (2011)

Geodetic Distance Queries on R-Trees
for Indexing Geographic Data

Erich Schubert, Arthur Zimek, and Hans-Peter Kriegel

Ludwig-Maximilians-Universitt Mnchen
Oettingenstr. 67, 80538 Mnchen, Germany
{schube,zimek,kriegel}@dbs.ifi.lmu.de
http://www.dbs.ifi.lmu.de

Abstract. Geographic data have become abundantly available in the recent years due to the widespread deployment of GPS devices for example in mobile phones. At the same time, the data covered are no longer restricted to the local area of a single application, but often span the whole world. However, we do still use very rough approximations when indexing these data, which are usually stored and indexed using an equirectangular projection. When distances are measured using Euclidean distance in this projection, a non-neglibile error may occur. Databases are lacking good support for accelerated nearest neighbor queries and range queries in such datasets for the much more appropriate geodetic (great-circle) distance. In this article, we will show two approaches how a widely known spatial index structure – the R-tree – can be easily used for nearest neighbor queries with the geodetic distance, with no changes to the actual index structure. This allows existing database indexes immediately to be used with low distortion and highly efficient nearest neighbor queries and radius queries as well as window queries.

1 Introduction

Nowadays, we are much more used to seeing maps than using a globe. But once we look at maps of the world, all map projections have some error: they cannot preserve the three components of geographical information, distance, area, and angles, equally well. The projections that we are most used to are the Mercator projection and the even simpler equi-rectangular projection. Google maps, for example, is based upon the Mercator projection. Figure 1 shows three different projections used for maps. Figure 1a is a variation of the usual Mercator projection, which has a huge influence on our understanding of the world. Figure 1b is an alternate projection developed 1923 by John Paul Goode, which is an equal-area projection. The difference in the ability to preserve area can in particular be seen for the Antarctica, but also Greenland, Canada, Alaska and Russia show massive area distortions in the Mercator projection. The Mercator projection is obtained by wrapping a cylinder around the earth and projecting the earth surface onto this cylinder. This yields a good map where the cylinder touches or intersects the earth, but along the axis of the cylinder the distortion is infinite – the north and south poles do not project to the cylinder at

M.A. Nascimento et al. (Eds.): SSTD 2013, LNCS 8098, pp. 146–164, 2013.

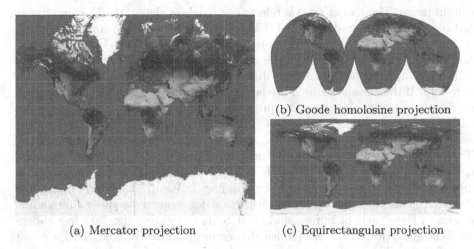

(b) Goode homolosine projection

(a) Mercator projection   (c) Equirectangular projection

Fig. 1. Three different projections of the earth surface. Images from Wikimedia Commons, by user "Strebe"[1]

all. Therefore, Mercator maps are commonly truncated at the poles. Figure 1a for example is truncated to a latitude of $\pm 82°$. While the Mercator projection does not preserve area and distances, it preserves the shapes quite well – but even more importantly, it preserves angles, which makes it useful for navigation and this probably is the main reason for the popularity of this projection. The equirectangular projection (Figure 1c) is probably the simplest projection, where latitude and longitude translate directly into y and x. In contrast to Mercator, it does not preserve angles. However, being able to trivially translate latitude and longitude to pixels on this map projection makes it a popular choice in GIS raster applications, and this is the default projection for placing custom texture overlays on the earth surface, e.g. in Google Earth and NASA WorldWind.

There are hundreds of geodetic reference datums, some of which are historic, but many are still in use for good reasons, e.g. for land survey. We tend to assume that any geographic position is well-defined, but for example due to tectonic plate shift we have to accept the fact that even the largest mountains move with respect to each other over time. The most popular geographic coordinate system is that of geographic latitude and longitude, with respect to the World Geodetic System 1984 (WGS84) reference ellipsoid. Geographic position is then measured with three components, known as latitude, longitude, and elevation. Elevation is commonly given with respect to the reference ellipsoid's surface, so that values of 0 are approximately at sea level. In the following, we will use λ to denote latitude and ϕ to denote longitude. We will not be using elevation, since it often is not available for a data set. Furthermore, in order to fully specify position, one would also need to take the time of the measurement into account. For all these reasons, we have to live with some error in geo positioning and, thus, distance

---

[1] For detailed copyright information on these images (CC-BY-SA 3.0), see
   https://commons.wikimedia.org/wiki/Category:Images_of_map_projections

computations. There exist several reference models of the earth. The simplest is that of a sphere with radius $r = 6371$ km, but there exist more complex models such as the GRS80 (Geodetic Reference System 1980) and the WGS84 ellipsoids; the latter is commonly used with the global positioning system GPS.

In many applications, Euclidean distance is chosen to measure distances on such data. If the variables stored are latitude and longitude, this equals measuring distances in the plate carrée (equirectangular) projection – which preserves neither distances nor area nor angles. The errors resulting from this are considerably larger than one might assume, even at short distances: For example at $45°N$, the latitude of Minneapolis, Turin and Bordeaux, $1°$ of longitude equals, in Euclidean distance, approximately $0.707°$ of latitude (traveling $1°$ north is approximately 111 km, traveling $1°$ east is just 78.7 km). Therefore, this naïve approach leads to a non-negligible distortion for nearest neighbor and radius queries in large parts of Europe and the US. Transforming the data to a different *local* geodetic system can reduce this error significantly if we need a small part of the world only, but this approach does not work for global data sets.

A much more adequate choice for computing distances is the great-circle distance, also known as orthodromic distance and called geodetic distance when using the Earth's radius. This is the shortest distance on the *surface* of a sphere (or ellipsoid). In the simpler case of a sphere of radius r – not an ellipsoid – it can be computed using $\Delta\lambda := |\lambda_1 - \lambda_2|$:

$$d_{\mathrm{arccos}}(\phi_1, \lambda_1, \phi_2, \lambda_2) := r \arccos\left(\sin\phi_1 \sin\phi_2 + \cos\phi_1 \cos\phi_2 \cos(\Delta\lambda)\right)$$

For increased precision and support for spheroid models such as WGS84, it may be desirable to use the haversine formula [1] or Vincenty's formula [2].

Additionally to computing distances between points, one could also want to compute distances between points and a desired track: the so called "cross-track distance" (XTD, also "cross track error") is the sideways deviation from a desired path of travel, for example due to wind affecting an airplane. In the simpler, spherical, model, it can be computed as

$$d_{\mathrm{xtd}} := r \arcsin\left(\sin(d_{sp}/r) \cdot \sin(\theta_{sp} - \theta_{sd})\right), \tag{1}$$

where d_{sp} is the distance from the starting point s to the current position p, θ_{sp} is the initial bearing from the starting point s to the current position p, and θ_{sd} is the initial bearing from the starting point s to the desired destination d.

The bearing (or forward azimuth) is the initial direction one needs to travel to the destination on the great-circle path (i.e., on the shortest path). Note that, when traveling along this path, the measured direction will actually change, so it does matter at which point we compute the bearing. The spherical formula for the bearing from a to b is:

$$\theta_{ab} = \arctan \frac{\sin(\Delta\lambda)\cos\phi_b}{\cos\phi_a \sin\phi_b - \sin\phi_a \cos\phi_b \cos(\Delta\lambda)} \tag{2}$$

The Cross-Track-Distance is visualized in Figure 2 for a track from Munich to New York City. The left image, Figure 2a is the equirectangular projection used

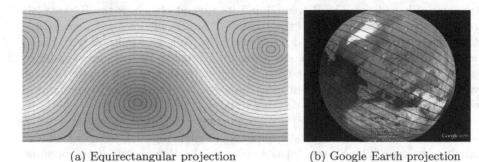

(a) Equirectangular projection   (b) Google Earth projection

Fig. 2. Cross-Track-Distance from the course Munich to New York

by the visualization and may appear distorted. But when the map is projected onto the sphere, as done in Figure 2b using Google Earth, it appears regular as expected. Red colors indicate a deviation to the left or a position along the track before the starting point, which are often represented by negative numbers. Black lines indicate contours of 1/36 the earth circumference.

In the following, we discuss aspects of the problem w.r.t. existing work and state our contributions (Section 2). We introduce our method in Section 3 and present an experimental evaluation in Section 4. We conclude in Section 5.

2   Related Work

Considering the background, state-of-the-art, and open problems, we discuss three aspects: first, data indexing methods proposed in scientific literature; second, the functionality commonly available in actual database engines; third, more specifically the problems of geodetic data when combined with these indexes.

2.1   Data Indexing in Scientific Work

Among the simplest index structures for spatial data are quadtrees [3], that recursively split an overflowing cell into four parts (by splitting in the middle of the x and y axes) until the cells contain at most the desired number of objects. A similar idea is the base of k-d trees [4], which split into two parts similar to a B-tree, but rotate through the d axes at each level. By splitting at the median (on an optimal tree) instead of the middle, a k-d tree can be a balanced tree. However, it does not allow easy dynamical rebalancing. Repeated insertion may cause the tree to become unbalanced and require the index to be rebuilt to retain good performance. Hierarchical Triangular Mesh [5] can be seen as an adoption of quadtrees to indexing the surface of a sphere. The initial approximation is an octahedron, and each triangle is then repeatedly decomposed into the four smaller triangles obtained by splitting each edge in half. While quadtrees can be linearized and then stored efficiently in a B+-tree (which effectively turns the quadtree into a Z-curve [6]), these indexes do not make native use of the block structure of harddisks. The R-tree [7] and its variants (such as the R*-tree [8]) are very popular spatial index structures for the use in databases, since they offer three main benefits over k-d trees: primarily, they are designed with paged memory in mind, and thus can easily be implemented as disk-based indexes. Secondly, they also consider non-singular objects with a spatial extent of their own, such as polygons. But even more importantly, they allow for dynamical rebalancing when inserting and deleting data.

In this respect, the R*-tree [8] is an important extension of the R-tree when handling dynamic data, since its focus is on rebalancing and, this way, on optimizing the tree. The SS-tree [9] is similar to the R-tree, but uses bounding spheres instead of rectangles for its bounding volume hierarchy. The M-tree [10] includes the distance to each child in the parent page and uses the triangle inequality to prune search candidates. Both do not rely on projections or coordinates, but index the data solely by their distances. However, this requires the tree being built for the specific distance function to be used for querying; It can not be used to answer queries with arbitrary distance functions. While the query process is quite straightforward, the construction and incremental update of these trees is much harder than in the case of the coordinate-oriented R-trees. The SS+-tree [11] uses k-means clustering to find a good split; however k-means only optimizes for (squared) Euclidean distance, and does not search for a minimum cover, but minimizes in-cluster variances. A bulk-loading strategy [12] for the M-tree is based on sampling and k-means style clustering to avoid having to compute all pairwise distances. The Slim-Tree [13,14] suggests the use of the minimum spanning tree to split nodes and uses R*-tree-inspired reorganization techniques to "slim down" the tree. These difficulties of efficiently computing a good split constitute the largest drawback of the M-tree. There are also hybrid techniques, such as an R-tree which also stores the center and radius of the page (SR-Tree [15]). Here, the split strategies of the R-tree can be used, and the covering radius serves as an additional pruning heuristic. Again, this radius can be only used for a specific distance function, chosen at index construction time.

2.2   Data Indexing in Practical Use

Not all of the techniques discussed above are used in practise. The most used techniques probably are Quadtrees (because of their simplicity and the ability to map them to existing B+-tree indices for harddisk storage) and R*-trees. However, while many database engines (including but not limited to Oracle, IBM Informix, Microsoft SQL Server, PostgreSQL, SQLite) list these indices on their feature sheets, the actual support for using them in query evaluation varies. The main functionality that seems to be widely supported is that of multidimensional range queries, which is for example useful when displaying parts of a map. Few seem to allow index-accelerated distance-based queries, and it is even more unclear, if they do, which distance functions are supported. Most database engines support only Euclidean distance queries.

Microsoft SQL Server uses a multi-level grid-based approach closely related to Quadtrees that requires filter refinement that can (since SQL Server 2012) also accelerate nearest neighbor queries [16,17] (at least for Euclidean distance). PostGIS/PostgreSQL have support for R-trees and M-trees implemented on top of their GiST architecture. However, these indexes can currently only be used with bounding box-based operators and not with queries that use Euclidean ST_Distance [18], yet the spherical ST_Distance_Sphere. Built-in operators of the database such as <#> and <-> also appear to support Euclidean distances only. IBM Informix [19] uses a data partitioning scheme with a predefined set of Voronoi cells (see also [20]), based on the population density of the areas. Furthermore, it supports the R-tree with custom strategy functions via an API. It is not clear from the documentation whether the nearest neighbor search functions are provided for any of the predefined geodetic data types. However, the algorithms presented in this paper can be implemented straightforwardly in this API. Oracle Spatial supports quadtrees and R-trees. Only the latter can be used for geodetic distance queries. An empirical evaluation at Oracle found R-trees to "consistently outperform quadtrees by a factor of 3" [21] for distance queries on the GIS data used in the study and to offer equivalent or better performance in almost all cases without parameter tuning. According to [22], the approach used by Oracle is to project the data into a 3 dimensional, "earth-centered-earth-fixed" (ECEF) coordinate system space and index the data there, but details on the exact method used do not seem to be available.

2.3   Handling Geodetic Data with Non-geodetic Indexes

As we have seen, many popular index structures are either designed or implemented only with Euclidean distance in mind. When geodetic data are naïvely stored in such an index, it will result in errors in distance computations. In Section 1 we already noted that the error at $45°N$, the latitude of Minneapolis, Turin and Bordeaux, the same Euclidean distance going east is just 70.7% of that when going south instead. Therefore, we should not use Euclidean distance with data in the non-Cartesian coordinate system of latitude and longitude. Many widely available index structures will only support such distance functions.

In order to get a more reasonable precision using Euclidean distance, the data must be transformed into a locally equidistant projection such as European Datum 1987 (ED87) or the European Terrestrial Reference System (ETRS) for data in Europe. Close to the fundamental point of these coordinate systems, the Euclidean distance can be reasonably used without large distortion. However, on a global data set there exists no optimal fundamental point, and distortions will occur when using the Euclidean distance.

2.4   Summary and Contributions

Indexing of spatial data is not a new domain. Many methods have been around for a long time. A method that clearly has proved itself in the "test of time" is the R*-tree, which seems to be used by every major database engine. Yet, when processing geodetic data, the index support in common database engines is often surprisingly limited. The R*-tree is – in contrast to the M-tree – actually distance agnostic and by no means limited to the Euclidean distance or L_p norms. The R*-tree implementations in ELKI [23] for example also support Canberra distance, Histogram intersection distance, and Cosine similarity as long as the data does not contain the origin. In the following we will introduce two approaches to index data with respect to the geodetic distance based on R-trees. Both approaches have strengths and weaknesses, some of which we will look at in detail.

For the simpler approach (which likely is similar to what Oracle Spatial uses [22,24]), we can project the data into a 3-dimensional ECEF coordinate system and use an R-tree or M-tree with Euclidean distance. The Euclidean distance then is a lower bound for the geodetic distance, and we can therefore get a good approximation of the query set, which then can be refined with geodetic distance. The main drawback of this approach is that the resulting tree is less useful when querying map sections based on a latitude/longitude rectangle; it also needs additional memory to index the data this way.

The second approach we introduce will work on the unmodified data (i.e. latitude and longitude coordinates) and an unmodified R-tree. What is needed to enable this kind of queries is the minimum distance from a query point to an index rectangle, which is a key contribution of this paper. The main benefits of this approach are that the same index can be used to answer typical map window queries that consist of latitude and longitude ranges, allowing for a dual-use index. Furthermore, since the index is using the original, 2D data, it requires less memory (and thus less I/O) than the other approach. The drawback is that, since it involves trigonometric computations, it is more CPU intensive.

Both approaches can easily be implemented in any existing database that already supports the R-tree: there are no changes needed to the index or index construction. In the first approach the index only needs to be able to accelerate 3D Euclidean distance queries. In the second approach it needs to allow for custom distance functions.

3   Indexing Geodetic Data

Some index structures such as the M-tree can be built for any metric distance function. Being the shortest path on the surface, the great-circle distance is a proper metric. We will discuss two alternate approaches here.

3.1   Indexing Geodetic Data Using 3D Euclidean Coordinates

While the geographic latitude and longitude are probably the most popular datum for geographic positions, there exist other coordinate systems. One of these sticks out because it actually uses Cartesian coordinates – which is also the main drawback, because it is not at all map oriented. Instead of giving coordinates on the earth surface, it uses three axes originating from the Earth's centre of mass. The x and y axes span the equatorial plane, with the x-axis pointing to the prime meridian, the y-axis pointing to $90°E$, the z-axis pointing straight north, i.e. it coincides with the average rotational axis of the Earth. Figure 3a

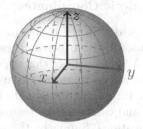
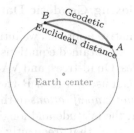

(a) ECEF coordinate system: x is to $0°N, 0°E$, y is to $0°N, 90°E$, z to $90°N$

(b) Euclidean distance is a lower bound for geodetic (great-circle) distance

Fig. 3. ECEF Cartesian index space

visualizes the axes with respect to the earth sphere. This coordinate system can be referred to as earth-centered earth-fixed (ECEF) or simply "XYZ" coordinate system. The name originates from the coordinate system being centered on the earth mass point and fixed to be invariant to the earths rotation.

While Euclidean distances in this coordinate system do not follow the Earth's surface, they have an interesting property. As sketched in Figure 3b, the Euclidean distance is the chord length in the great circle used by the geodetic distance. This yields two important properties of the ECEF coordinate system:

– Euclidean distance in ECEF is a *lower bound* for the geodetic distance:

$$L_{2,\text{ECEF}}(a, b) \leq d_{\text{geodetic}}(a, b) \tag{3}$$

– Euclidean distance in ECEF is *strictly monotone* to geodetic distances, i.e.

$$L_{2,\text{ECEF}}(a, b) < L_{2,\text{ECEF}}(x, y) \Leftrightarrow d_{\text{geodetic}}(a, b) < d_{\text{geodetic}}(x, y) \tag{4}$$

The first property guarantees that for a query radius r, all objects (although also some more) are found that are in the desired range the geodetic distance. The second property gives an even stronger guarantee for retrieving the k nearest neighbors: here, no additional objects should be included (notwithstanding numerical issues).

Therefore any index that can support Euclidean distance in 3 dimensions can be used to index geodetic data, after transformation into ECEF Cartesian coordinates. This includes (but is not limited to) gridfiles, octrees, the M-tree, and the R-tree family. For our experiments, we will focus on the M-tree and R-trees. This approach is probably not novel. In [22,24] the authors mention that Oracle Spatial uses a 3D R-tree to index geodetic data, but without giving further details or properties – it might as well be a 3D R-tree on latitude, longitude and elevation. IBM Informix documentation also mentions 3D bounding boxes for geodetic data, but only mentions intersection queries. Therefore it is not clear if above properties have been realized and are used yet. The PostgreSQL pgSphere project seems to include this transformation, but does not appear to make use of it for indexing yet.

3.2   Indexing Geodetic Data Using 2D Geodetic Coordinates

The alternative approach we introduce here is designed with the R-tree in mind, although the obtained equations could also prove useful with other indexes such as grid-files, Quad-trees, and VA-File [25] indexes. A key benefit of this approach is that it can use a regular R-tree [7], R*-tree [8], or any of its many variants, as index *without modifications* to the actual index structure. In particular, the index is built on the latitude and longitude coordinates, and can therefore be used for window queries that frequently arise in map applications. Similarly, the search can trivially be bounded with such a window. In contrast to the M-tree [10], which needs to be built for a specific distance function, the R-tree family of indexes are unspecific, but index the coordinates using bounding boxes. Because the same tree can be used with very different distance functions, R-trees can be considered general purpose spatial indexes, whereas M-trees are highly specific.

Each object as well as each index page in an R-tree is represented by a minimum bounding rectangle (MBR), which in this case means it is represented by a quadruple $(\lambda_{\min}, \phi_{\min}, \lambda_{\max}, \phi_{\max})$ (plus other attributes, if present; in the following we will assume to only index latitude and longitude). In order to query the R-tree, we need to compute a lower bound for the distance of the query point to an arbitrary – unknown – object within the given rectangle. For L_p norms, this distance computation is very efficient, which makes the R-tree attractive to use. But also for other – even some non-metric – distances, such a lower bound can be specified. For L_p norms, the minimum distance can be computed using simple case distinctions in each dimension:

$$
\mathrm{mindist}_{L_p}(o, \mathrm{MBR}) := \left(\sum_i \begin{cases} (\min(\mathrm{MBR}, i) - o_i)^p & \text{if } o_i < \min(\mathrm{MBR}, i) \\ (o_i - \max(\mathrm{MBR}, i))^p & \text{if } o_i > \max(\mathrm{MBR}, i) \\ 0 & \text{otherwise} \end{cases} \right)^{1/p}
$$

Unfortunately, in geodetic data, the formula will become more complicated. This is largely due to the fact that an MBR in the equirectangular projection – which does not preserve distances – when projected to the surface of the earth yields a much more complex shape. However, to compute the minimum distance, we will still need to distinguish the same 3×3 cases, just as for the L_p norms. The 3×3 case distinction in 2-dimensional Euclidean space is shown in Figure 4a: if the query point is inside the rectangle – area 0 – the minimum distance will be 0. In the areas N, E, S and W, the shortest path is along the normal vector to the closest edge, while in the corner areas $1 \ldots 4$, the shortest path is to the nearest corner of the MBR. The transfer of this model to geodetic data is shown in Figure 4b for an example on the northern hemisphere. The north and south edges of the rectangle are parallels of the equator, while the east and west edges are meridians. Note that the north and south poles in the equirectangular projection are not a single point, but actually the complete northern and southern edges of the projected map. At first, the situation appears to be highly asymmetric. However this largely is an artifact of the geographic coordinate system and the equirectangular projection, in which great-circle paths are only straight lines if

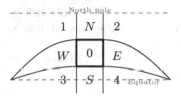

(a) Case distinction in Euclidean data

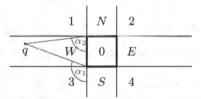

(b) Case distinction in geodetic data

(c) Spherical interpretation

Fig. 4. Case distinctions for point-to-MBR distance in geodetic data

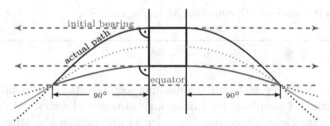

Fig. 5. Angle-based test in Euclidean space

Fig. 6. Case distinctions for point-to-MBR distance in geodetic data, detail

they are meridians or the equator, and curves of the type $y = \arctan_{360}(a \cdot \sin(x - x_0))$ otherwise. While it is not evident from the 2D projection of the case distinction (Figure 4b), the N and S areas are actually also triangular, since the north and south edge of the projection represent a single point each. When mapped onto a sphere, the regions look approximately similar, as visualized in Figure 4c. Note that the four triangles (N, S vs. W, E) nevertheless do *not* have the same mathematical properties: only the east and west edges of the MBR are on great-circles, while the north and south edges are lines of constant latitude. So from the point of view of spherical geometry, the north and south triangles have one bent edge each.

Algorithm 1. Min. Dist. Point to MBR by Azimuth (non-optimized)

Data: c circumference of earth (spherical model)
Data: (ϕ_q, λ_q) query point
Data: $(\phi_l, \lambda_l, \phi_h, \lambda_h)$ index MBR
if $\phi_l \leq \phi_q \leq \phi_l$ then
    if $\lambda_q \leq \lambda_l$ then return $c \cdot (\lambda_l - \lambda_q)/360°$;   /* South of MBR */
    if $\lambda_q \geq \lambda_t$ then return $c \cdot (\lambda_q - \lambda_t)/360°$;   /* North of MBR */
    return 0;   /* Inside MBR */
else if $mod_{360}(\phi_l - \phi_q) \leq mod_{360}(\phi_q - \phi_h)$ then   /* West of MBR */
    $\theta_h \leftarrow$ Azimuth$(\phi_l, \lambda_h, \phi_q, \lambda_q)$;
    if $\theta_h \geq 270°$ then   /* North-West */
    | return $Great\text{-}Circle\text{-}Distance(\phi_l, \lambda_h, \phi_q, \lambda_q)$
    end
    $\theta_l \leftarrow$ Azimuth$(\phi_l, \lambda_l, \phi_q, \lambda_q)$;
    if $\theta_l \leq 270°$ then   /* South-West */
    | return $Great\text{-}Circle\text{-}Distance(\phi_l, \lambda_l, \phi_q, \lambda_q)$
    end
    return $|Cross\text{-}Track\text{-}Distance(\phi_l, \lambda_l, \phi_l, \lambda_h, \phi_q, \lambda_q)|$;   /* West */
else   /* East of MBR */
    $\theta_h \leftarrow$ Azimuth$(\phi_h, \lambda_h, \phi_q, \lambda_q)$;
    if $\theta_h \leq 90°$ then   /* North-East */
    | return $Great\text{-}Circle\text{-}Distance(\phi_h, \lambda_h, \phi_q, \lambda_q)$
    end
    $\theta_l \leftarrow$ Azimuth$(\phi_h, \lambda_l, \phi_q, \lambda_q)$;
    if $\theta_l \geq 90°$ then   /* South-East */
    | return $Great\text{-}Circle\text{-}Distance(\phi_h, \lambda_l, \phi_q, \lambda_q)$
    end
    return $|Cross\text{-}Track\text{-}Distance(\phi_h, \lambda_h, \phi_h, \lambda_l, \phi_q, \lambda_q)|$;   /* East */
end

Fortunately, both the test and the distance computation for points in areas N and S remains as simple as for Euclidean distance – the shortest great-circle path to the rectangle is a meridian. In order to distinguish the other cases, we first need to test whether we are on the left or on the right side by rotating the mean longitude of the rectangle by 180° – the meridian opposite of the rectangle. The key to distinguishing the cases 1, W and 3 (and, identically, 2, E, 4) then is the *azimuth* (north-based, also referred to as bearing) from the two corners towards the query point. The azimuth plays roughly the role of the angle in Euclidean geometry. In order to test whether a point is in area W in Euclidean space, we can compute the angle at the north-west and south-west corners of the MBR. This idea is sketched in Figure 5: only if the angles at the south-west α_1 are larger than 90° and the angle at the north-west corner α_2 is less than 90°, then the query point is in area W. By substituting the azimuth for the angle, we can perform the same test in the spherical domain:

A shortest path to the west edge of the MBR must be a great circle path that arrives at an azimuth of 90° to the edge. If and only if the azimuth at the south corner is larger than 90° and the azimuth at the north corner is smaller than 90°, then there exists a point on the meridian in between where the course is exactly 90°. The difference between Euclidean space and spherical geometry shows when we travel a path that started at an initial bearing of 90°: it will not be a straight line in the equirectangular projection. This is visualized in Figure 6: with an initial bearing of 270° (to north, 90° with respect to the south pole!) – indicated by the red lines – the blue curves are obtained. Conversely, for points on the blue lines, an initial bearing of 270° is obtained for one of the corners.

Algorithm 1 uses this idea to compute the minimum distance from the query point to an MBR by using the azimuth for case distinction. Note that for practical use, this pseudocode should be optimized by inlining the great-circle and cross-track distances in order to share all redundant trigonometric computations. When implementing this in a database, the number of trigonometric computations must be kept low as they are rather expensive. However, we can further improve this algorithm: we do not need the exact values of the azimuth, but we only need to know whether it will be smaller or larger than 90°. If we can compute the blue lines in Figure 6 directly, we can easily test whether a point is between the two blue lines. As noted before, each great-circle (that is not a meridian) can be expressed as $\lambda = \arctan_{360}(a \cdot \sin_{360}(\phi - \phi_0))$. If we know the parameters ϕ_0 and a we can easily test whether a point is north or south of these lines. ϕ_0 is the longitude where the great-circle crosses the equator, and the maximum longitude is achieved when $\phi - \phi_0 = 90°$, with $\lambda = \arctan_{360}(a)$. Since we want the curves to be orthogonal to the meridian, this is where they must have a maximum or minimum. Therefore, we can choose $\phi_0 = \phi_r - 90$ and $a = \tan_{360}(\lambda_r)$ for a given reference point r. A point q is south of the great-circle path that goes orthogonally to the meridian through (ϕ_r, λ_r) if

$$\lambda_q < \arctan_{360}(\tan_{360}(\lambda_r) \cdot \sin_{360}(\phi_q - \phi_r + 90))$$

or, equivalently,

$$\tan_{360}(\lambda_q) < \tan_{360}(\lambda_r) \cdot \cos_{360}(\phi_r - \phi_q),$$

in which we can reuse $\tan_{360}(\lambda_r)$ for the second test and preserve numerical precision slightly better. Since this test is faster to compute (since it involves fewer trigonometric functions), it allows for a further optimized version of the algorithm, which is given in Algorithm 2. For points close to the rectangle, we can just test whether they are above the great-circle through the upper corner of the MBR, below the great-circle through the lower corner, or in between. However, these two lines will intersect when crossing the equator at $\Delta\phi = 90°$ from the MBRs edge. Starting at this distance, we will instead look at the great-circle through the middle of the MBR, and use this for distinguishing the remaining two cases: at this distance we know that one of the two corners must be closest.

Figure 7 visualizes the minimum distance from an example bounding box around Bavaria. Again, in the equirectangular projection in Figure 7a, it appears to be irregular, but projected onto the sphere in Figure 7b the shape of rectangles with increasingly rounded corners becomes visible.

(a) Equirectangular projection   (b) Google Earth projection

Fig. 7. Minimum distance from a bounding box around Bavaria

3.3 Non-spherical Earth Models

The equations we presented were so far all for a spherical earth model for simplicity. For geographic data it is however a best practise to use for example a spheroid model such as WGS84. For some of the equations, formulas for the spheroid model are readily available, for others they are not. For the new algorithm, the minimum distance to the MBR, it may be simpler and more efficient to stick to a spherical model, and just use the *minimum* radius of the spheroid model to obtain a slightly less tight minimum distance. This will underestimate the minimum bound by at most 0.3% and thus should have negligible impact on the pruning power of the tree; while saving many trigonometric computations.

4 Experiments

4.1 Test Environment

We implemented the distance functions in ELKI [23], which can be easily extended with custom distance functions for R*-tree queries and includes ready-to-use query benchmarking classes. The test machines have AMD Phenom II X4 3.0 GHz CPUs and 8GB of RAM.

4.2 Data Sets

For our benchmarking experiments we use data from DBpedia 3.7, a parsed version of Wikipedia. From this data set we obtained 442775 points of interest around the world, and for 109577 of these we also obtained a region of interest which we use as query radius. The second data set we use is road accidents data set from the UK government, spanning the years 2005 to 2011 containing data on 1.2 million road accidents in the UK. The third data set consists of 6.3 million radiation measurements taken in Japan after the Fukushima nuclear disaster.[2]

---

[2] The data sets are publicly available at http://dbpedia.org/, http://data.gov.uk/ and http://safecast.org/

Algorithm 2. Optimized Minimum Distance Point to MBR

Data: c circumference of earth (spherical model)
Data: (ϕ_q, λ_q) query point
Data: $(\phi_l, \lambda_l, \phi_h, \lambda_h)$ index MBR
if $\phi_l \leq \phi_q \leq \phi_l$ **then**
 if $\lambda_q \leq \lambda_l$ **then return** $c \cdot (\lambda_l - \lambda_q)/360°$;   /* South of MBR */
 if $\lambda_q \geq \lambda_t$ **then return** $c \cdot (\lambda_q - \lambda_t)/360°$;   /* North of MBR */
 return 0;   /* Inside MBR */
else if $mod_{360}(\phi_l - \phi_q) \leq mod_{360}(\phi_q - \phi_h)$ **then** /* West of MBR */
 $\tau \leftarrow \tan_{360}(\lambda_q)$;
 if $mod_{360}(\phi_l - \phi_q) \geq 90°$ **then**   /* Large $\Delta\phi$ */
  if $\tau \leq \tan_{360}((\lambda_l + \lambda_h)/2) \cos_{360}(\phi_l - \phi_q)$ **then**
   return $Great\text{-}Circle\text{-}Distance(\phi_l, \lambda_h, \phi_q, \lambda_q)$; /* North-West */
  else
   return $Great\text{-}Circle\text{-}Distance(\phi_l, \lambda_l, \phi_q, \lambda_q)$; /* South-West */
  end
 end
 if $\tau \geq \tan_{360}(\lambda_h) \cos_{360}(\phi_l - \phi_q)$ **then**
  return $Great\text{-}Circle\text{-}Distance(\phi_l, \lambda_h, \phi_q, \lambda_q)$; /* North-West */
 if $\tau \leq \tan_{360}(\lambda_l) \cos_{360}(\phi_l - \phi_q)$ **then**
  return $Great\text{-}Circle\text{-}Distance(\phi_l, \lambda_l, \phi_q, \lambda_q)$; /* South-West */
 return $|Cross\text{-}Track\text{-}Distance(\phi_l, \lambda_l, \phi_l, \lambda_h, \phi_q, \lambda_q)|$; /* West */
else   /* East of MBR */
 $\tau \leftarrow \tan_{360}(\lambda_q)$;
 if $mod_{360}(\phi_q - \phi_h) \geq 90°$ **then**   /* Large $\Delta\phi$ */
  if $\tau \leq \tan_{360}((\lambda_l + \lambda_h)/2) \cos_{360}(\phi_h - \phi_q)$ **then**
   return $Great\text{-}Circle\text{-}Distance(\phi_h, \lambda_h, \phi_q, \lambda_q)$; /* North-East */
  else
   return $Great\text{-}Circle\text{-}Distance(\phi_h, \lambda_l, \phi_q, \lambda_q)$; /* South-East */
  end
 end
 if $\tau \geq \tan_{360}(\lambda_h) \cos_{360}(\phi_h - \phi_q)$ **then**
  return $Great\text{-}Circle\text{-}Distance(\phi_h, \lambda_h, \phi_q, \lambda_q)$; /* North-East */
 if $\tau \leq \tan_{360}(\lambda_l) \cos_{360}(\phi_h - \phi_q)$ **then**
  return $Great\text{-}Circle\text{-}Distance(\phi_h, \lambda_l, \phi_q, \lambda_q)$; /* South-East */
 return $|Cross\text{-}Track\text{-}Distance(\phi_h, \lambda_h, \phi_h, \lambda_l, \phi_q, \lambda_q)|$; /* East */
end

4.3 Efficiency

To study the behavior of the 2D- and the 3D-model in existing index structures, we use the R*-tree (both incrementally built and bulk loaded) and the M-tree (incrementally built only). Since the M-tree supports any metric distance function – and the geodetic distance is metric – it can be used with the geodetic distance. Figure 8 shows the results for 100 nearest neighbor and range queries on the DBpedia data set. From a mere CPU perspective (Figure 8a and Figure 8b), the 3-dimensional ECEF approach appears to be best, due to the rather costly trigonometric functions needed for the direct indexing approach. However, this may be misleading in an actual database context, because for larger data sets the input and output cost must be taken into account. And with this, the reduced memory requirements of the direct indexing approach pay off, which allow storing about 40% more objects per page. Figure 8c and Figure 8d show the I/O cost (in number of page accesses times page size, to make values across different page sizes more comparable) for querying the trees. The stronger pruning power of the 2D rectangles manifests itself in requiring fewer distance computations (Figure 8e and Figure 8f). Figure 8g shows the time needed to build the index, while Figure 8h visualizes the resulting index sizes. Note that the bulk loading actually has less work to do (fewer sorting passes) with larger page sizes, while for the M-tree construction, which requires the computation of all pairwise distances, construction time grows quadratically.

Except for the expensive build time of the M-tree implementation we used, all indexes offered a significant performance improvement over a linear scan which took 176.6 ms CPU time per 100NN query, 442775 distance computations and needs to read about 9 MB of data. The bulk-loaded geodetic R*-tree with 512 b pages (i.e. storing up to 9 entries on directory pages, 12 objects on leaf pages, 13 MB total size) took on average just 0.222 ms CPU per query, 344 distance computations and read (uncached) 12 kB (23 pages) of data.

4.4 Accuracy

The UK traffic accidents and the Safecast data sets are good examples for regionally constrained data sets that can reasonably be handled with an appropriate local projection: for the traffic accidents we can use for example the UK Ordnance Survey 1936 (OSGB36) datum or UTM Zone 30N, which is a transversal Mercator projection that is expected to have low error in the UK. For the Safecast data set, UTM Zone 54S covers this part of Japan well. The projection library PROJ.4[3] we used for transforming the data refused to project coordinates outside of their design range (i.e. would not project the traffic accidents data to UTM 54S or SafeCast data to UTM 30N). For DBpedia, neither could be used. We compare the nearest neighbors obtained by a linear scan over the data using geodetic distance to the nearest neighbors found using different settings: Euclidean distance in (non-Cartesian) latitude and longitude coordinates, but also after the transformation to the local coordinate system. Furthermore, we also used our index with geodetic distance. For a sample of 100000 objects, we computed the 100 nearest neighbors each and compute precision with respect to the ground truth from a linear scan. Table 1 gives the precision as well as average query CPU times. It can clearly be seen that the naïve Euclidean approach suffers from significant distortion. Where applicable, the approaches based on a localized Cartesian coordinate system work well – at least within their design range of 6° of longitude (in fact, they probably are more accurate than the simple

---

[3] Available at http://trac.osgeo.org/proj/

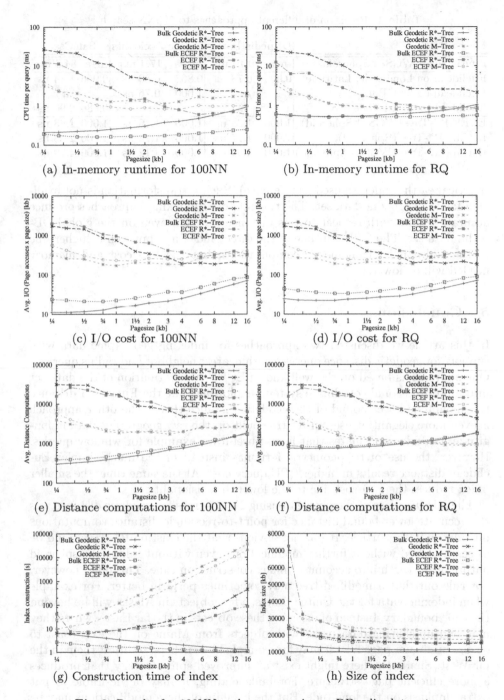

(a) In-memory runtime for 100NN

(b) In-memory runtime for RQ

(c) I/O cost for 100NN

(d) I/O cost for RQ

(e) Distance computations for 100NN

(f) Distance computations for RQ

(g) Construction time of index

(h) Size of index

Fig. 8. Results for 100NN and range queries on DBpedia data set

Table 1. Precision of different approaches to 100NN search

Method	DBpedia Places		Traffic Accidents		SafeCast	
Linear Scan (WGS84 Spheroid)	n/a	9.10 ms	n/a	17.4 ms	n/a	84.3 ms
Euclidean on Longitude, Latitude	0.903	0.78 ns	0.868	0.80 ns	0.956	1.14 ns
Euclidean on OSGB36		n/a	1.000	0.76 ns		n/a
Euclidean on UTM Zone 54S		n/a		n/a	0.986	0.90 ns
ECEF R*-Tree (WGS84 Spheroid)	1.000	1.28 ns	1.000	1.16 ns	1.000	1.63 ns
LatLng R*-Tree (Spherical)	1.000	1.77 ns	1.000	1.74 ns	1.000	2.08 ns
LatLng R*-Tree (WGS84 Spheroid)	1.000	4.59 ns	1.000	4.12 ns	1.000	4.79 ns

spherical earth model we used as reference). Such a transformation is not available for the DBpedia data set. The two proposed indexing approaches produce accurate results with respect to this distance function even on the global DBpedia data set. The CPU runtime of the direct indexing approach is higher due to the additional trigonometric computations, but the I/O cost and memory needed will be lower.

5   Conclusions

In this article we discussed two approaches for indexing geographic data with support for geodetic distance queries (both nearest neighbor and radius queries). One approach is based on the well-known ECEF transformation of the data set into a 3-dimensional coordinate system, exploiting that the Euclidean distance is a monotone lower bound of the geodetic distance there. The other approach is even more elegant: it uses an R*-tree built on the raw geographic data in longitude, latitude coordinates that is also useful for example for window queries. However, the use of trigonometric formulas instead of the much simpler Euclidean distance results in higher CPU query cost. At the same time, the smaller index reduces the I/O cost due to the lower dimensionality.

The missing piece of the puzzle to using 2D R*-trees for geodetic distance was the accurate lower-bound distance for point-to-rectangle distance computations introduced in this article. It was not obvious that the existing R*-trees could be as easily reused without further modifications even without taking the spheroid nature of the earth into account at index construction time. This does, however, not rule out that a modified tree may sometimes perform better. For example when indexing data for the United States, a few objects in Alaska will be beyond the 180° boundary. Instead of inserting these objects in the far East – where they may end up sharing data pages with objects from Maine on the east coast to ensure minimum page fill – it may for example be beneficial to insert them in the far West. Similarly, there might exist an improved split strategy that produces a more efficient page structure. For bulk loading, it may be desirable to put extra emphasis on the longitude. But the benefits of these modifications may be highly application dependent and not be of general use, whereas the introduced distance computation lays the foundation for querying the resulting indexes, no matter how they were constructed.

References

1. Sinnott, R.: Virtues of the haversine. Sky and Telescope 68, 158–159 (1984)
2. Vincenty, T.: Direct and inverse solutions of geodesics on the ellipsoid with application of nested equations. Survey Review 23(176), 88–93 (1975)
3. Finkel, R.A., Bentley, J.L.: Quad trees. A data structure for retrieval on composite keys. Acta Informatica 4(1), 1–9 (1974)
4. Bentley, J.L.: Multidimensional binary search trees used for associative searching. Commun. ACM 18(9), 509–517 (1975)
5. Kunszt, P., Szalay, A., Thakar, A.: The hierarchical triangular mesh. Mining the Sky, 631–637 (2001)
6. Morton, G.M.: A computer oriented geodetic data base and a new technique in file sequencing. Technical report, International Business Machines Co. (1966)
7. Guttman, A.: R-Trees: A dynamic index structure for spatial searching. In: Proc. SIGMOD, pp. 47–57 (1984)
8. Beckmann, N., Kriegel, H.P., Schneider, R., Seeger, B.: The R*-Tree: An efficient and robust access method for points and rectangles. In: Proc. SIGMOD, pp. 322–331 (1990)
9. White, D.A., Jain, R.: Similarity indexing with the SS-tree. In: Proc. ICDE, pp. 516–523 (1996)
10. Ciaccia, P., Patella, M., Zezula, P.: M-Tree: an efficient access method for similarity search in metric spaces. In: Proc. VLDB, pp. 426–435 (1997)
11. Kurniawati, R., Jin, J.S., Shepherd, J.A.: The SS+-tree: An improved index structure for similarity searches in a high-dimensional feature space. In: Proc. SPIE, vol. 3022, pp. 110–120 (1997)
12. Ciaccia, P., Patella, M.: Bulk loading the M-tree. In: Proc. ADC (1998)
13. Traina Jr., C., Traina, A., Seeger, B., Faloutsos, C.: Slim-trees: High performance metric trees minimizing overlap between nodes. In: Zaniolo, C., Grust, T., Scholl, M.H., Lockemann, P.C. (eds.) EDBT 2000. LNCS, vol. 1777, pp. 51–65. Springer, Heidelberg (2000)
14. Traina Jr, C., Traina, A., Faloutsos, C., Seeger, B.: Fast indexing and visualization of metric data sets using slim-trees. IEEE TKDE 14(2), 244–260 (2002)
15. Katayama, N., Satoh, S.: The SR-tree: An index structure for high-dimensional nearest neighbor queries. In: Proc. SIGMOD, pp. 369–380 (1997)
16. Microsoft Corporation: Whitepaper New Spatial Features in SQL Server 2012 (2012)
17. Fang, Y., Friedman, M., Nair, G., Rys, M., Schmid, A.E.: Spatial indexing in microsoft sql server 2008. In: Proc. SIGMOD, pp. 1207–1216 (2008)
18. PostGIS project: Postgis 2.0 manual, http://postgis.net/docs/manual-2.0/
19. IBM Informix: IBM Informix Geodetic DataBlade Module User's Guide
20. Lukatela, H.: Hipparchus geopositioning model: An overview. In: Proc. Auto. Cartography, vol. 8, pp. 87–96 (1987)

21. Kothuri, R.K.V., Ravada, S., Abugov, D.: Quadtree and R-tree indexes in oracle spatial: a comparison using GIS data. In: Proc. SIGMOD, pp. 546–557 (2002)
22. Hu, Y., Ravada, S., Anderson, R.: Geodetic point-in-polygon query processing in oracle spatial. In: Pfoser, D., Tao, Y., Mouratidis, K., Nascimento, M.A., Mokbel, M., Shekhar, S., Huang, Y. (eds.) SSTD 2011. LNCS, vol. 6849, pp. 297–312. Springer, Heidelberg (2011)
23. Achtert, E., Kriegel, H.P., Schubert, E., Zimek, A.: Interactive data mining with 3d-parallel-coordinate-trees. In: Proc. SIGMOD (2013)
24. Hu, Y., Ravada, S., Anderson, R., Bamba, B.: Topological relationship query processing for complex regions in oracle spatial. In: Proc. ACM GIS, pp. 3–12 (2012)
25. Weber, R., Schek, H.J., Blott, S.: A quantitative analysis and performance study for similarity-search methods in high-dimensional spaces. In: Proc. VLDB, pp. 194–205 (1998)

Energy Efficient In-Network Data Indexing
for Mobile Wireless Sensor Networks

Mohamed M. Ali Mohamed[1], Ashfaq Khokhar[1], and Goce Trajcevski[2]

[1] University of Illinois at Chicago, ECE Department, USA
{mali25,ashfaq}@uic.edu
[2] Northwestern University, EECS Department, USA
goce@eecs.northwestern.edu

Abstract. In-network indexing is a challenging problem in wireless sensor networks (WSNs), particularly when sensor nodes are mobile. In the past, several indexing structures have been proposed for WSNs for answering in-network queries, however, their maintenance efficiency in the presence of mobile nodes is relatively less understood. Assuming that mobility of the nodes is driven by an underlying mobility control algorithm or application, we present a novel distributed protocol for efficient maintenance of distributed hierarchical indexing structures. The proposed protocol is generic, in the sense that it is applicable to any hierarchical indexing structure that uses binary space partitioning (BSP), such as k-d trees, Quadtrees and Octrees. It is based on locally expanding and shrinking convex regions such that update costs are minimized. Based on SIDnet-SWANS simulator, our experimental results demonstrate the effectiveness of the proposed protocol under different mobility models, mobility speeds, and query streams.

Keywords: Distributed Algorithms, Mobility, Wireless Sensor Networks, Data Indexing, Query Processing.

1 Introduction

Wireless Sensor Networks (WSNs) have been proposed as effective and efficient distributed systems for monitoring varieties of phenomena in different application domains [1]. In particular, the ability of sensor nodes in WSNs to self organize and provide coverage for monitoring a given region or activity makes them highly useful for scenarios involving harsh conditions or remote surveillance. Typically, individual sensor nodes cooperate in real-time monitoring of phenomena over a given geographic region in two end-of-spectrum modalities: (1) either periodically reporting the sensed values to a given sink (possibly coupled with in-network aggregation); or (2) reporting detections of pre-defined events, i.e., exceeding of a certain temperature-threshold – possibly over spatial extents. Broadly speaking, the purpose of indexing structures in WSN is to facilitate the process of collaboration for monitoring the sensed field, the detection/reporting events of interest, as well as providing in-network storage for answering queries about the sensed phenomena.

M.A. Nascimento et al. (Eds.): SSTD 2013, LNCS 8098, pp. 165–182, 2013.

Mobile sensor nodes [2, 21] greatly increase the adaptability of the WSNs from different perspectives: (1) ensuring a level of Quality of Service (QoS) in response to phenomena fluctuation, in the sense of providing better spatial resolution of sampling in desired/targeted areas; (2) enabling a control over (balancing) the levels of connectivity and coverage. We note that the motion of the nodes may vary in different applications but, from a general perspective, it can be predictable [3], random [4], or controlled [5]. For example, in the data coverage problem in WSN [22], controlled mobility of the sensor nodes is utilized in different applications to achieve more efficacious coverage.

An illustrating example of the motivation for this work is shown in Fig. 1. In Fig. 1(a), a sensed field with randomly deployed sensor nodes is shown. Part (b) of the same figure shows the nodes location distribution, after the occurrence of an event of interest in the southeast corner of the field, where the application or mobility control algorithm (as [29]) has steered more sensor nodes towards that corner, in order to collect more precise information, while still maintaining coverage and network connectivity across the region. Due to this mobility of the nodes required by the application, the underlying distributed indexing structure may become highly skewed, unless it is adjusted to reflect the new distribution of the nodes in a balanced way. The main question addressed in this work is how to efficiently adapt the indexing structures that manage in-network query processing and aggregation in such mobility scenarios, in response to the change of nodes' distribution, such that the overall maintenance cost is minimized. We emphasize that the actual mobility information as to which nodes should move in what direction is given by the application. Also, it is the application responsibility to guarantee minimum number of nodes needed to provide connectivity and coverage. In order to show our work, we use [29] as the dictating application for mobility.

Existing data indexing approaches in WSNs, centralized [6] (i.e., all the data are gathered to one centralized sink node), or distributed [7, 8] – presume that the sensor nodes are *static*, i.e., their locations do not change. Being centralized, they have two-fold disadvantage: (1) increasing traffic towards the sink node, which creates a communication bottleneck; and (2) decreasing the network lifetime, especially in the

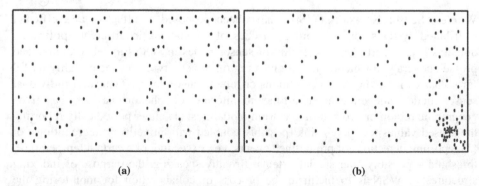

(a)   (b)

Fig. 1. Part (a) - a set of sensor nodes randomly deployed. Part (b) – nodes distribution after occurrence of an event of interest in the southeast corner.

vicinity of the sink node. Some data gathering algorithms employ mobile sink node(s) that traverses the network to gather the data [9-11]. However, a potential drawback of such approaches is the latency/delay.

Organizing the network information across a distributed indexing structure, while knowing that the reporting nodes are not necessarily in the location they reported information from is highly challenging, particularly when the nodes' resources, such as storage, computing power, communication range/bandwidth, and battery capacity are limited. In order to accommodate mobility of the nodes in a general indexing framework, the indexing structure should adapt to the vicissitude of nodes distribution across the field. Once the mobility information for each node is available, the indexing structure should adapt to the change in a distributed fashion to avoid skewed, unbalanced structures. Furthermore, such adaptation should induce minimal overhead.

In this paper we present a novel protocol that enables several existing in-network data indexing structures to incorporate mobile nodes with high transparency. Our approach is applicable to data structures that use Binary Space Partitioning (BSP) [12, 13], where the field is divided into contiguous, non-overlapping, convex regions (e.g., k-d trees, Quadtrees, Octrees [23]). The proposed protocol runs in a distributed fashion, resolving the consequences of the nodes mobility (i.e., relocation) within their regions by locally shrinking or expanding the convex regions, reducing the need to transfer information about this motion across the network or the indexing structure. After a small "transient regime", in the worst case scenario the message cost to re-stabilize the index over the geographic field of interest is linear in the number of the indexing structure nodes. Our simulation results on SIDnet-SWANS simulator [24] show that the cost of maintaining the indexing structure under different mobility scenarios remains sub-linear. In our experiments, over 83% of the mobility in the field is resolved locally, without the need of informing the rest of the network with this mobility. The results also show an overall improvement in the latency of data queries. Note that we do not compare our work to the solutions that use mobile sink nodes, because they use different energy optimization functions which include the consumed energy of mobility. Besides that these solutions focus on data gathering rather than indexing.

The rest of the paper is organized as follows. In Section 2, we start with a preliminary discussion on BSP based hierarchical structure, and outline one such structure that we have used for in network indexing of static WSNs [27, 28]. We use this indexing structure to explain our proposed mobility management protocol. The details of the proposed protocol and its performance analysis are presented in Section 3, followed by a discussion of the applied experiments and simulation results in Section 4, followed by a discussion of related work in section 5. Section 6 concludes the paper and discusses future work.

2   Preliminaries

To better understand the proposed mobility management protocol, we now briefly overview the features of BSP based hierarchical indexing structures. In such structures, recursive splitting is applied to a given space into convex sets via hyper-planes. As a hierarchical data structure, each node in the BSP tree represents a space that is subdivided among its child nodes. At each level, the number of children of each node represents the fan out, denoted as k. The root node of the tree represents first split of the whole space, and the deeper a node in the tree is, the more local (smaller) space split it represents. Each level in the tree contains nodes that embody the whole space partitioned at a certain level of detail. A higher level in the tree represents coarser partitioning (i.e, smaller number of larger subspaces), whereas a lower level in the tree represents more detailed finer scale partitioning (i.e, larger number of smaller subspaces).

In our previous work [27, 28], we have developed efficient abstractions of data and spatial fields in a hierarchical BSP framework. These abstractions are performed for representing the sensed values and positions of the sensor nodes, which we call physical-space abstraction and data-space abstraction, respectively, both rooted at the corresponding sink. Figures 2 and 3 depict numerical examples of physical-space indexing at a leaf node of the indexing tree, as well as assembling/compressing the data in intermediate non-leaf nodes at a given level in a fixed-size array (see [28] for details). We used Wavelet Transform (WT) [30] to combine and compress the data from the children-nodes.

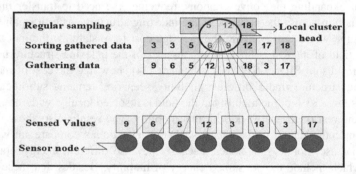

Fig. 2. Processing of sensed values at indexing tree leaf node (Local Cluster Head)

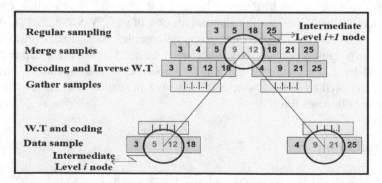

Fig. 3. Processing of sensed values at indexing tree intermediate node(s)

Similarly, the data-space in each region is distributed among a group of nodes managing that region. At the lowest level, the sibling leaf nodes in each locality categorize the reported sensor nodes locations according to data range. Each inner (non-leaf) node receives maps from the nodes in the lower level covering the same data range, having multiple maps received for the same data range across a group of contiguous regions. Fig. 4 shows an example of the process of creating hierarchical maps for the data range (26 – 50) in a given region, and zooming it out to upper level of the hierarchy with a 1:4 factor.

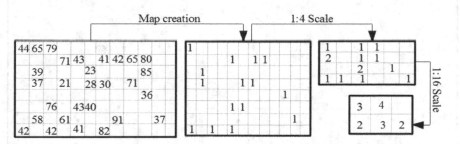

Fig. 4. Data space abstraction process. A set of sensed values in a region (to the left) have a map created for the data range [26-50], then zoomed out twice with 1:4 factor.

Queries to the WSN originate at the sink node, which has a coarse representation for the physical-space and data-space of the whole field, with a specific level of accuracy. If the accuracy requirement for the given query cannot be satisfied by the sink node, it forwards it to its child node(s) according to the query constraints of the physical-space and/or data-space, and the process is recursively repeated until a node(s) is reached capable of providing answer with a required level of accuracy. At this stage the query response is backtracked across the same route in the indexing tree [27, 28].

3   Managing Index Structures with Mobile Nodes

We now proceed with the details of the protocol for adapting the hierarchical indexing structure to capture the mobility of the nodes. The protocol has three distinct stages for which we present the corresponding algorithms and discuss the respective complexities.

3.1   Initial Configuration

Assume that logically there are two types of nodes, senor nodes that sense the field and indexing structure nodes that contain the keys to help maintain the indexing structure. Physically, a node can be a sensor node as well as a node in the indexing structure. Further assume that the number of nodes in the indexing structure is n, and the fan-out of each inner node is k, such that the height of the indexing structure is $O(log_k\ n)$. The initial setup of the protocol assigns an integer rank for each border/hyperplane corresponding to node in the indexing BSP tree , equal to the depth of the node in the tree (i.e., its level-distance from the root). Fig. 5 illustrates a field with randomly deployed sensor with the corresponding borders rank (color-coded with the same colors according to the splitting order). Each leaf node is responsible for (the sensed values of) a group of m sensor nodes within its vicinity. Sensor nodes periodically (with fixed cycle length) report their sensed values and locations to their respective cluster head.

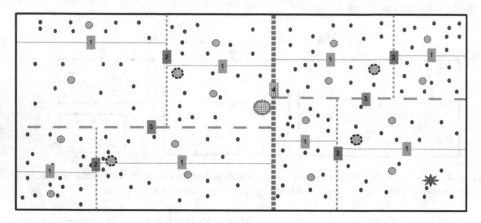

Fig. 5. The field contains ($n = 103$) randomly deployed (small size/red color) sensor nodes. Example indexing structure in this figure is based on orthogonal bisections, performed recursively, such that 16 (thin solid line/green color) local cluster heads are at the first level. Second level of the indexing structure consists of four (thicker dashed line/blue color) intermediate level cluster heads. Last is the (thickest dotted line/yellow color) sink node. Border line shapes follow same nodes drawing/color. In this (initial) configuration, an event of interest is observed in the South-East corner.

We reiterate that the motion/displacement of the nodes occurs due to a specific objective (e.g., better coverage due to an observed event in a given geographic region) and, as a result, some leaf node(s) in the indexing structure finds more sensor nodes entering to its vicinity and requesting to join. For example, an event of interest may require more sensor nodes to be moved towards, in order to monitor and report more precise data, as depicted in Fig. 5. Also, note that sibling or child/parent node may not be within single hop of each other. In such case, multihop routing of message will be assumed.

3.2   Processing a Request to Incorporate New Mobile Node

Each leaf node has a specified capacity $m' > m$. A leaf node will accept the joining of new sensor nodes coming into its vicinity until reaching the threshold m'. Congestion happens when a new join request is received at leaf node that has reached its maximum capacity m'. The leaf node then initiates a request to reduce the size of its space of responsibility by changing the position of one of its surrounding borders/hyperplanes.

The process of border change starts with a communication aiming at changing the spatial splitting locally. The leaf node in the indexing structure experiencing congestion starts by locating the border of its surrounding sides corresponding to the lowest rank convex region. It sends to its sibling node(s) on the other side of the lowest rank border, a *change_border_request*. When sibling leaf node receives the *change_border_request* message, it starts assessing if it can change the specified border in order to accommodate some of the sensor nodes currently managed by the requesting sibling. The calculation in this case is based on the capacity of the leaf node that received the request. A response is sent back to the requesting node after the calculation. If all the involved leaf nodes have large populations, then they cannot accommodate more incoming sensor nodes, causing them to reject the request. In such case, since the change cannot be handled locally, a new request for changing borders is propagated in the hierarchy to the node corresponding to the next higher rank - i.e., the requesting leaf node sends the request message to its parent node. Upon receiving the request, the parent node checks if the total number of sensor nodes covered by its children is at the capacity limits. If not, it initiates a request to its sibling on the other side of the smallest rank border of its region. The same assessment algorithm runs at the sibling node, which consequently sends the response back. In case of rejection, the same process is recursively applied - in the worst case, reaching the root of the hierarchy (the sink). The algorithm executed locally by the participating node is formalized below:

Algorithm 1: Forward Mobility Request
Input: *Rank of the border required to change, The count of sensor nodes associated to the requesting indexing node (or its subtree for non-leaf nodes)*
Output: *A border_change_response OR, in case the whole region is congested, it issues a new border_change_request (if request is received from a child node).*

> **Receive** *border_change_request (Receiver, Sender.Rank, Sender.nodesCount)*
> > *If (Sender.depth == Receiver.depth)*   // *If the request is received from a sibling node*
> > > *extraNodesCount = Sender.nodesCount - Receiver.optimalNodesCountForCluster*
> > > *If (Receiver.nodesCount + extraNodesCount < Receiver.maximumNodesCountForCluster)*
> > > > *newBorderLocation = calculateNewBorderLocation(Sender.Rank, extraNodexCount)*
> > > > **send** *border_change_response(Sender, accepted, Rank, newBorderLocation)*
> > > > **apply** *border_change_inform(This, Rank, newBorderLocation)*
> > > *Else*
> > > > **send** *border_change_response(Sender, rejected, Rank, Receiver.nodeCount)*
> > > *EndIf*
> > *Else*   // *If the request is received from a child node*
> > > *Receiver.UpdateNodesCount(Sender, Sender.nodesCount)*
> > > *If(Receiver.nodeCount < Receiver.maximumNodesCountForCluster)*
> > > > *newBorderLocation = calculateNewBorderLocation(Sender.Rank)*
> > > > **send** *border_change_response(Sender, accepted, Rank, newBorderLocation)*
> > > > *Foreach childNode other than Sender*

> **send** *border_change_inform(childNode, Rank, newBorderLocation)*

    Else

> > *Rank = Sender.Rank + 1*
> >
> > **send** *border_change_request (Sibling, Rank, Receiver.nodesCount)*
> >
> > *requestingBorderChange = True*
>
>   *EndIf*

    EndIf

  End_Receive border_change_request

e,ii

The local behavior of the nodes participating in the border-adjustment is formalized in Algorithm 2 below.

Algorithm 2: Receive Mobility Response

Input: *Rank of the border to be changed, The response (accept or reject), The new border location (in case of acceptance)*

Output: *Applies the border change for the node, in case of acceptance, Or initiate new request in case of rejection.*

> **Receive** *border_change_response (Receiver, response, Rank, newBorderLocation)*
>
>   *if (response == accepted)*
>
> > **apply** *border_change_inform(This, Rank, newBorderLocation)*
> >
> > *requestingBorderChange = False*
>
>   *Else*
>
> > *Rank = Sender.Rank + 1*
> >
> > *nodeCount = Sender.nodeCount + Receiver.nodesCount*
> >
> > **send** *border_change_request (Parent, Rank, nodesCount)*
>
>   *EndIf*
>
> *EndReceive border_change_response*

Complexity: In the worst-case scenario, the request needs to be propagated all the way to the sink node. For a BSP indexing tree consisting of n nodes, with a fan-out factor k, at each level, at most $k - 1$ request message(s) will be transmitted to change the lowest rank border, and $k - 1$ rejection message(s) will be received. In the 2D planar case, $k = 2$ for k-d trees and $k = 4$ if quadtrees are used.

Since, by construction, the height of the BSP with n nodes and fan-out factor k is $log_k n$, the number of messages required $2 * (k - 1) * (log_k n - 1)$, bounding the message complexity of the forwarding stage to $O(log_k n)$. We note that the overall network-wide running time complexity is the same, since each participating node is executing constant operations to check its current capacity.

3.3   Response Propagation

When a border change decision is taken in non-leaf nodes, all their affected child-nodes are notified, recursively propagating the changes until the affected leaf nodes. Leaf nodes, in turn, inform the affected sensor nodes to change their reporting destination. While this border change information message is flowing through the structure, each recipient node recalculates its population according to the new change to ensure that it is within its capacity. If not, the node finding congestion in its region initiates a new *change_border_request* message and sends it to its sibling node. The important observation is that this particular message is guaranteed to affect borders that are in the sub-tree of the originally changed border, which caused this new congestion, because the capacity has already been checked/verified at the parent or ancestor node.

The determining of the new border location is based on the population size of the requesting (congested) and responding nodes. For that, we rely on the structural properties of the tree's boundary between the nodes at the same level. Namely, we move the border of the node that has a capacity to incorporate new sensors in a direction perpendicular to the current border's position towards the requesting node position, resulting in shrinking the requesting node's area, and accordingly getting more sensor nodes out of its region towards the accepting node's region. The new border location in the low level requests (i.e, requests between leaf nodes) is determined by the requesting node, which knows exactly the location of all its sensor nodes. In higher level requests, the border location change is proportional to the desired new population size of the congested region. After the change takes place, the node that asked for the border change recalculates its new population to ensure it is within its capacity limits. If not, the node reissues a new border_change_request, accordingly. Fig. 6 shows the reconfiguration of the borders after sensor nodes have moved towards an event of interest in the southeast corner of the field.

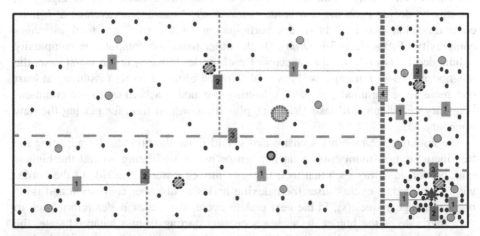

Fig. 6. Borders reconfiguration after sensor nodes are moved towards an event of interest in the southeast corner of the field

The last step of the protocol involves notifying the mobile motes about the new borders of the tree, so that they know which node-ID to use when reporting the sensed values. This if formalized below:

Algorithm 3: Apply and Propagate Mobility Response
Input: *Rank of the border to be changed, The response (accept or reject), The new border location (in case of acceptance)*
Output: *Applies the border change for the node, in case of acceptance, Or initiate new request in case of rejection.*

> *Receive border_change_inform (Receiver, Rank, newBorderLocation)*
>   *Receiver.border[Rank] = newBorderLocation*
>   *If (Receiver.depth == MaximumDepth)   // Leaf node*
>   *Foreach sensorNode*
>   *If sensorNode.Location is out of leaf node new region*
>   *send detach_sensor(sensorNode)*
>   *EndIf*
>   *Else   // Non-leaf node*
>   *Foreach childNode other than Sender*
>   *send border_change_inform(childNode, Rank, newBorderLocation)*
>   *EndIf*
> *EndReceive border_change_inform*

Complexity: Algorithm 3 executes when Algorithms 1 and 2 have terminated, and is applied to all the children of the subtree rooted at the node at which Algorithm 2 has terminated. In the worst-case scenario, the execution of Algorithms 1 and 2, will cause the request to be forwarded all the way to the sink node. This, in turn, means that each of the n nodes in the tree will have to be notified about borders change (and, eventually, decide upon the new border's location). Assuming an average of h hops communication between the nodes participating in the tree, the total message-complexity of Algorithm 3 is $O(hn)$. On the other hand, the computation complexity is bounded by $O(log\ m)$ – the capacity of each node. Namely, in the worst case, the neighboring nodes (siblings) will have a difference of m – 1 motes (assuming at least one mote for a minimal occupancy). Sorting the nodes according to the common-boundary coordinate will take $O(log\ m)$, plus the constant time for placing the new boundary.

We note that the mobility scenario that would make the protocol for adjusting the tree incur its maximum cost, is having sensor nodes oscillating around the highest rank border, in a way such that their majority moves towards one side of the border within one update cycle causes the indexing nodes to discover congestion and issue border_change_request(s). In the next update cycle, the sensor nodes return back to the other side of the border. In such a scenario, starting from a balanced state, the algorithm behavior would start by a first request at the node(s) adjacent to the highest rank border to change their lowest rank border, which gets accepted at the same level. After the accepting node(s) reach their capacity, while sensor nodes are still crossing the highest rank border towards the adjacent cluster(s), the next request will need to be elevated on level in the indexing tree. On the higher level, the same operation will take place until the managed region is congested.

3.4   Data Indexing under Mobility

The aim of a in-network data indexing system is to arrange and store the sensed data in a distributed fashion. Indexing tree manages the sensor nodes where each group of sensors report their sensed values and positions to a node of the indexing tree. The recipient indexing nodes store the received information, process them, and elevate approximate constructs across the indexing hierarchy. Mobility causes some of the sensor nodes to move apart from their reporting node(s) of the indexing structure, and hence, get into other node(s) vicinity. This causes unbalance in number of senor nodes reporting to the nodes of the indexing structure. Such unbalance results in the reported data across the indexing structure.

In physical-space abstraction, two approaches can be followed. The first approach is to increase the size of the update message according to the count of the sensor nodes population attached to each node of the indexing structure, in order to keep same sampling distance between the update message values. This would not increase the overall size of physical-space update messages traversed, because the total number of sensor nodes in the field is the same. However, it will create a skew in the size flowing in each branch of the indexing tree, where the larger population branches will have larger size update messages than the other branches. The second approach is keeping the update messages size unchanged, at the expense of increase in the accuracy loss across the indexing hierarchy. In other words, upon receiving a physical-space query, there might be a bigger chance of not being able to satisfy its accuracy requirements from the higher level nodes of the indexing tree, and having to forward the query to next level(s) for achieving the required accuracy. The advantage for physical-space abstraction because of the mobility handling algorithm is that the change in number of nodes is bounded by the capacity of each leaf node in the indexing tree m'.

In data-space abstraction, the change occurring is not because of the motion of sensor nodes, but rather because of the modification of borders location to balance the indexing tree. Due to this change, the bitmap constructs used to represent each data-space are increased/decreased in size, in order to represent the new cluster space. Contrary to the physical-space abstraction, which has its skew factor bounded by the capacity of the indexing structure leaf nodes m', the area of a single cluster can increase to approach the size of the whole field. This can only be bounded with the logic of the mobility algorithm, physical constraints of the sensors (i.e., robots moving them), and the field physical barriers. In such extreme case, the large regions can be represented with lower granularity, so the cell size would be coarser than the same level other nodes. This would require high accuracy queries for this region to be forwarded all the way to the leaf nodes. The other solution is to forward the update of such lower level large size cluster(s) as an array of positions rather than a bitmap, and insert them into the bitmap in the higher level node(s) of the indexing tree.

4   Experimental Results

The proposed mobility management protocol was implemented on SIDnet-SWANS simulator for WSN [24]. IEEE 802.15.4 protocol is used for the MAC layer, and

Shortest Geographical Path Routing for the routing layer. The power consumption characteristics are based on Mica2 Motes specifications, MPR500CA. Each sensor node is assumed to have a GPS to obtain the location information. In this section, we present the simulation results and discuss the performance.

The simulations were run for a 300-nodes network, where nodes were randomly deployed in a 500x500 square meters geographic area. The nodes' mobility was assumed under two different mobility models: random and controlled. The controlled mobility refers to a scenario where sensor nodes are moved based on an underlying application requirement. For our simulations we used the algorithm presented in [29] to compute the coordinates of mobile nodes at each step. In the case of random mobility the new location of each node is computed using a random direction. In addition, we also tested the mobility management protocol under different speeds, ranging from 0.5 m/s to 2 m/s, which is practically used in several WSN systems [25, 26]. In our future work we plan to simulate higher speed nodes as well.

For our experiments, we have constructed a K-D tree based hierarchical indexing structure over the sensed field. The index nodes are considered to be static, but would rather be moved according to the borders change, to maintain connectivity with the other nodes in their region. The cycle time in our simulations is 5 seconds, i.e., every 5 seconds, nodes inform their value as well positions to their immediate cluster heads (indexing tree leaf nodes).

We measure the performance of the protocol in terms of following parameters: mobility request latency, mobility resolution factor, and query latency. Mobility request latency refers to the time it takes for the protocol to adjust the structure to reflect the nodes new positions. Mobility resolution factor (MRF) reflects the percentage of requests that required changes beyond the first level of the indexing hierarchy.

Fig. 7 plots the average mobility request latency under different mobility speeds. The performance of both mobility cases is quite stable, where the latency is almost consistent with the change of sensor nodes velocity. The mobility request latency for the controlled mobility scenario (i.e. nodes move towards an events of interest while maintaining coverage [29]) is around 15% higher than the random mobility request latency. This is because the number of mobility request received by the cluster heads in the case of controlled mobility is higher, compared to the random mobility. Note that in the case of random mobility, overall more sensor nodes maybe moving. However, a significant number of consecutive mobility steps may cancel each other, thus keeping the sensor nodes within the same local region. On the other side, in the controlled mobility scenario each sensor node is moving on a specific path towards the target point. Accordingly, with each time step, a node progresses towards moving into or outside of a specific local region, thus requiring mobility adjustment in the indexing structure.

In Fig. 8, MRF is shown for different mobility scenarios. The general trend of the MRF is larger for the controlled mobility algorithm, as the nodes following a specific path are able to cause more disturbance in all the regions they pass by, which creates unbalance in multiple local regions. Because of this unbalance, adjustment to mobility may require adjustment at more than one of the hierarchy. The maximum MRF shown

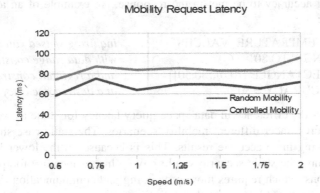

Fig. 7. Average latency of incorporating mobile node in the indexing structure Vs. sensor node speed

for all cases is less than 17%. Which means that the mobility management protocol is able to resolve successfully over 83% of the mobility requests at the lowest level of the indexing tree, without the need of having this mobility information traverse the whole indexing structure.

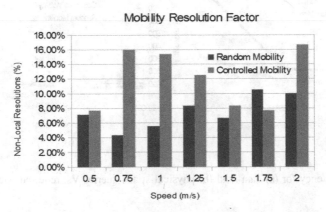

Fig. 8. Mobility Resolution Factor (MRF): The percentage of mobility requests that the mobility protocol is unable to resolve at the lowest level of the indexing structure

Figures 9 and 10 compare the latency of different data queries to the mobility managed structure (under random and controlled mobility) and the static structure where the indexing structure does not change itself to accommodate mobility and thus becomes relatively unbalanced. We present results for three different types of queries.

Physical-space queries inquire values sensed in a specific region. Data-space queries inquire locations of sensor nodes sensing data in a specific data range. A hybrid query inquires either sensed values, or sensor nodes locations, giving constraints of both region and data range. In approximate querying, the user defines a desired level of accuracy to be met in the response. An example of an approximate hybrid query is:

SELECT TEMPRATURE_VALUES	*inquiring sensed values*
BETWEEN 70° TO 80°	*with data range constraint*
INSIDE RECTANGLE {[0,0],[30,50]}	*and a regional constraint*
WITH ACCURACY = 80%	*at a desired accuracy level*

Fig. 9a shows the difference in data-space query latency for static as well mobility manages structures under different mobility scenarios. The static case shows higher costs for achieving more accurate results. This is because on the lower level of the indexing structure, the static scenario would have a higher memory footprint for the congested regions, which requires more processing and communication time. In Fig. 9b, physical space query latency of the static indexing structure almost matches the mobility managed structure under the random mobility scenario for lower accuracy levels, which is slightly higher than the controlled mobility scenario. However for exact queries (i.e., 100% accuracy), which require the indexing structure to get the data from its leaf nodes, static scenario incurs higher query latency costs.

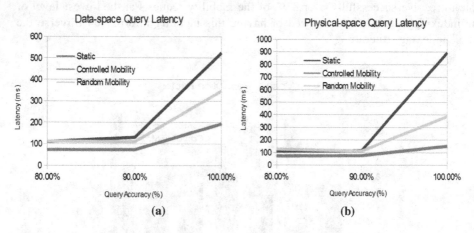

Fig. 9. Query latency for (a) data- and (b) physical-space queries Vs. required query response accuracy

In Fig. 10, the hybrid query latency can be viewed as a combination of latencies of both physical-space and data-space queries, where it is clear that the incurred latency is higher for the static case when requiring higher accuracy level. These results show the efficiency of appropriately handling mobility, and its effect on query latency for most cases of mobility scenario, where the static indexing would not be able to provide same latency for queries inquiring higher accuracy, especially for the queries inquiring exact responses.

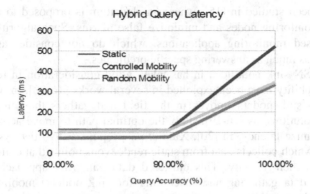

Fig. 10. Query latency for hybrid queries Vs. required accuracy

5   Related Work

Data indexing in WSN has been studied over the past decade, and several algorithms with different perspectives were proposed to solve it. The vast majority of these algorithms did not consider the mobility of sensor nodes. Centralized solutions, as in [6], proposed transmitting data across paths in the network using lifting technique and wavelet based compression. In such methods the network usually suffers from congestion around the sink node, which creates a communication bottleneck, and decreases the lifetime of the nodes in the area around the sink node. Several distributed data indexing algorithms were proposed [7, 8, 14]. In [7], a hierarchical data structure is constructed and data is mapped to the indexing structure using geographic hash tables (GHT). This algorithm creates redundancy in data transmission, where the same raw data is reported to multiple nodes in the indexing structure. Meliou et al.[9] proposed an algorithm with a novel idea for data indexing of sensed values in a hierarchical data structure using approximate modeling. Gaussian models were used in this system to abstract large amount of sensed values and elevate them across the hierarchy, leading to more efficient reporting at the cost of accuracy loss across the hierarchy. Such system lacks the representation of sensor nodes positions, and assumes that Gaussian models are suitable for all types of sensed phenomena, which is not generic enough for a wide range of sensed phenomena not of Gaussian distribution nature. Also, Gaussian models are successful in representing the average behavior of a region, but they lose the information about the extreme (maximum and minimum) sensed values, which are of high interest for many WSN applications. Another distributed algorithm proposed by Xiao et al.[14] which indexes the WSN data across a spanning tree according to a key for each node of the spanning tree. However this algorithm supports mobility of sensor nodes, it falls short in the maintenance cost of the data updates, as a sensor node may have to update its information at an indexing node that is far from its location. On the other side, if the key is arranged in a way that favors position of sensor node for local region reporting, the system doesn't support data-space indexing efficiently. Monitoring the WSN for events have been studied in [15], where an algorithm is proposed to use an optimal number of monitoring nodes and minimize false alarms. Such algorithms are useful for event based monitoring applications, which do not consider aggregating the network data as much as answering specific predicates.

Mobile WSN sink node idea in has taken good consideration in recent research. Controlled mobility have been exploited in several works [16-20], in which the – one or multiple – sink node(s) moves in the field and gathers the sensed data. Non-hierarchical solutions, as [16-19], study the optimal path to move across the field, in order to minimize latency. In [20], Xing *et al.* propose at two tier system of mobile sink node(s) which collects data from static rendezvous points that collect sensed data locally within their vicinity. This clustered data gathering approach increases the efficiency of data gathering and scheduling for sink node(s) mobility, however it doesn't provide a full hierarchical solution. It does not present a distributed data indexing solution, but rather an optimized data gathering algorithm based on clustering. Moreover, the energy minimization criteria is significantly different in such solutions, because the amount of energy spent on mobility is orders of magnitude higher than the energy spent on communication and computation.

6   Conclusion and Future Work

In this paper we presented a protocol to manage and maintain in-network indexing structures in WSN under the constraint of mobile nodes. The protocol is applicable to BSP tree structures, where it is based on assigning incrementing values for space splitting borders of the BSP tree. The protocol is based on shrinking and expanding the indexed regions according to the residing number of nodes, in order to keep a balanced load for the indexing structure. The complexity of the proposed solution does not exceed a linear order in the size of the indexing structure. Our results show the capability of handling over 83% of mobility within their local regions of occurrence, without the need of communicating this information across the network. The average latency of balancing the structure in the presence of mobility is in reasonable range. The results also show improvement for query latency results, especially for the higher accuracy queries. In our future work, we plan to incorporate mobility models that involve higher mobility speed and uniform direction. In addition, we also plan to study mobility management under higher dimensional indexing structures that do no involve orthogonal bisections. An extension of our work is to consider the mobility of the nodes participating in the indexing structure itself. Another extension is to incorporate the aspect of optimizing the coverage for multiple-events monitoring.

Acknowledgments. This research has been supported in part by the NSF grants CNS 0910988, 0910952 and III 1213038.

References

1. Zhao, F., Guibas, L.: Wireless Sensor Networks: An Information Processing Approach. Morgan Kaufmann (2004)
2. Ekici, E., Gu, Y., Bozdag, D.: Mobility-Based Communication in Wireless Sensor Networks. IEEE Comm. Magazine 44(7), 56–62 (2006)
3. Shah, R.C., Roy, S., Jain, S., Brunette, W.: Data MULEs: Modeling a Three-Tier Architecture for Sparse Sensor Networks. In: 2003 IEEE Workshop Sensor Network Protocols and Applications, SNPA 2003 (May 2003)
4. Chakrabarti, A., Sabharwal, A., Aazhang, B.: Using Predictable Observer Mobility for Power Efficient Design of Sensor Networks. In: Proc. 2nd International Workshop on Information Processing in Sensor Networks (2003)
5. Somasundara, A., Ramamoorthy, A., Srivastava, M.: Mobile Element Scheduling for Efficient Data Collection in Wireless Sensor Networks with Dynamic Deadlines. In: Proc. 25th IEEE International Real-Time System Symposium (2004)
6. Ciancio, A., Pattem, S., Ortega, A., Krishnamachari, B.: Energy-Efficient Data Representation and Routing for Wireless Sensor Networks Based on a Distributed Wavelet Compression Algorithm. In: Proceedings of the 5th International Conference on Information Processing in Sensor Networks, IPSN 2006, pp. 309–316 (2006)
7. Greenstein, B., Estrin, D., Govindan, R., Ratnasamy, S., Shenker, S.: DIFS: A Distributed Index for Features in Sensor Networks. Ad Hoc Networks 1, 333–349 (2003)
8. Meliou, A., Guestrin, C., Hellerstein, J.: Approximating Sensor Network Queries Using In-Network Summaries. In: Proceedings of the International Conference on Information Processing in Sensor Networks, IPSN 2009, pp. 229–240 (2009)
9. Goldenberg, D., Lin, J., Morse, A., Rosen, B., Yang, Y.: Towards Mobility as a Network Control Primitive. In: Proc. ACM MobiHoc (2004)
10. Wang, Z., Basagni, S., Melachrinoudis, E., Petrioli, C.: Exploiting Sink Mobility for Maximizing Sensor Networks Lifetime. In: Proc. 38th Ann. Hawaii Int'l Conf. System Sciences, HICSS 2005 (2005)
11. Gandham, S., Dawande, M., Prakash, R., Venkatesan, S.: Energy Efficient Schemes for Wireless Sensor Networks with Multiple Mobile Base Stations. In: Proc. IEEE Global Telecomm. Conf., GlobeCom 2003 (2003)
12. Fuchs, H., Kedem, Z., Naylor, B.: On Visible Surface Generation by A Priori Tree Structures. SIGGRAPH 1980 Proceedings of the 7th Annual Conference on Computer Graphics and Interactive Techniques, pp. 124–133. ACM, New York (1980)
13. Thibault, C., Naylor, F.: Set operations on polyhedra using binary space partitioning trees. In: SIGGRAPH 1987 Proceedings of the 14th Annual Conference on Computer Graphics and Interactive Techniques, pp. 153–162. ACM, New York (1987)
14. Xiao, L., Ouksel, A.: Scalable Self-Configuring Integration of Localization and Indexing in Wireless Ad-hoc Sensor Networks. In: IEEE International Conference on Mobile Data Management, MDM, vol. 151 (2006)
15. Liu, C., Cao, G.: Distributed Monitoring and Aggregation in Wireless Sensor Networks. In: Proc. of Infocom, pp. 1–9 (2010)
16. Gandham, S., Dawande, M., Prakash, R., Venkatesan, S.: Energy Efficient Schemes for Wireless Sensor Networks with Multiple Mobile Base Stations. In: Proc. IEEE Global Telecomm. Conf., GlobeCom 2003 (2003)
17. Luo, J., Hubaux, J.: Joint Mobility and Routing for Lifetime Elongation in Wireless Sensor Networks. In: Proc. IEEE INFOCOM (2005)

18. Wang, Z., Basagni, S., Melachrinoudis, E., Petrioli, C.: Exploiting Sink Mobility for Maximizing Sensor Networks Lifetime. In: Proc. 38th Ann. Hawaii Int'l Conf. System Sciences, HICSS 2005 (2005)
19. Hanoun, S., Creighton, D., Nahavandi, S.: Decentralized mobility models for data collection in wireless sensor networks. In: IEEE International Conference on Robotics and Automation, ICRA (2008)
20. Xing, G., Li, M., Wang, T., Jia, W., Huang, J.: Efficient rendezvous algorithms for mobility-enabled wireless sensor networks. IEEE Transactions on Mobile Computing (2012)
21. Pileggi, F., Fernandez-Llatas, C., Meneu, T.: Evaluating mobility impact on wireless sensor network. In: UkSim 13th International Conference on Modelling and Simulation, pp. 461–466. IEEE (2011)
22. Mulligan, R., Ammari, H.: Coverage in Wireless Sensor Networks: A Survey. Network Protocols and Algorithms 2(2) (2010)
23. Samet, H.: The Design and Analysis of Spatial Data Structures. Addison-Wesley (1990)
24. Ghica, O., Trajcevski, G., Scheuermann, P., Bischoff, Z., Valtchanov, N.: Sidnet-swans: A simulator and integrated development platform for sensor networks applications. ACM SenSys (2008)
25. Pon, R., Batalin, M., Gordon, J., Kansal, A., Liu, D., Rahimi, M., Shirachi, L., Yu, Y., Hansen, M., Kaiser, W., Srivastava, M., Sukhatme, G., Estrin, D.: Networked Infomechanical Systems: A Mobile Embedded Networked Sensor Platform. In: Proc. Fourth Int'l Symp. Information Processing in Sensor Networks, IPSN 2005 (2005)
26. Dantu, K., Rahimi, M., Shah, H., Babel, S., Dhariwal, A., Sukhatme, G.: Robomote: Enabling Mobility in Sensor Networks. In: Proc. Fourth Int'l Symp. Information Processing in Sensor Networks, IPSN 2005 (2005)
27. Mohamed, M., Khokhar, A.: Dynamic indexing system for spatio-temporal queries in wireless sensor networks. In: 12th IEEE International Conference on Mobile Data Management MDM, vol. 2, pp. 35–37 (2011)
28. Mohamed, M., Khokhar, A., Trajcevski, G., Ansari, R., Ouksel, A.: Approximate hybrid query processing in wireless sensor networks. In: Proceedings of the 20th International Conference on Advances in Geographic Information Systems (SIGSPATIAL 2012), pp. 542–545. ACM, New York (2012)
29. Caicedo, C., Zefran, M.: A coverage algorithm for a class of non-convex regions. In: IEEE Conference on Decision and Control, pp. 4244–4249 (2008)
30. Chui, C.: An Introduction to wavelets. Academic Press Prof. Inc., San Diego (1992)

PL-Tree: An Efficient Indexing Method
for High-Dimensional Data

Jie Wang, Jian Lu, Zheng Fang, Tingjian Ge, and Cindy Chen

Department of Computer Science,
University of Massachusetts Lowell,
Olsen Hall, 198 Riverside St., Lowell, MA 01854, U.S.
http://cs.uml.edu

Abstract. The quest for processing data in high-dimensional space has resulted
in a number of innovative indexing mechanisms. Choosing an appropriate indexing method for a given set of data requires careful consideration of data properties, data construction methods, and query types. We present a new indexing
method to support efficient point queries, range queries, and k-nearest neighbor
queries. Our method indexes objects dynamically using algebraic techniques, and
it can substantially reduce the negative impacts of the "curse of dimensionality".
In particular, our method partitions the data space recursively into hypercubes of
certain capacity and labels each hypercube using the Cantor pairing function, so
that all objects in the same hypercube have the same label. The bijective property
and the computational efficiency of the Cantor pairing function make it possible
to efficiently map between high-dimensional vectors and scalar labels. The partitioning and labeling process splits a subspace if the data items contained in it
exceed its capacity. From the data structure point of view, our method constructs
a tree where each parent node contains a number of labels and child pointers, and
we call it a **PL-tree**. We compare our method with popular indexing algorithms
including R*-tree, X-tree, quad-tree, and iDistance. Our numerical results show
that the dynamic PL-tree indexing significantly outperforms the existing indexing
mechanisms.

1 Introduction

Large-scale applications on high-dimensional data need efficient querying mechanisms
to quickly retrieve information. These data objects may contain tens and even hundreds of dimensions. Reducing dimensionality is a standard approach, which has met
with certain success in applications where the most significant information resides on a
small number of dimensions. However, dimensionality cannot always be reduced without losing critical information; even if it can be reduced, the remaining data objects may
still contain tens of dimensions. A number of indexing mechanisms have been developed for indexing multidimensional data. Among them, the R*-trees [2] and X-trees [4]
which are evolved from R-trees [11], have become the dominating indexing methods
because they are geometrically suited for spatial data. R-trees suffer from exponential
blowups of MBRs as the dimensionality increases. Attempts at reducing overlapping
bounded boxes have resulted in R*-trees; attempts at avoiding splits have resulted in

M.A. Nascimento et al. (Eds.): SSTD 2013, LNCS 8098, pp. 183–200, 2013.

X-trees. However, there are often significant overlaps among the sibling nodes in these indices, especially when the dimensionality is high. The Quadtree [22] is another commonly used indexing mechanism that is based on the hierarchical decomposition of data space. The Quadtree structure is overlap free and partitions the d-dimensional data space recursively into 2^d hypercubes.

We note that breaking the spell of the "curse of dimensionality" for indexing high-dimensional data is not impossible with a new line of thinking and our attempt is an algebraic approach. In particular, we present a new indexing method to support efficient point queries, range queries, and k-nearest neighbor (KNN) queries. Our method, called **PL-tree** (Partition and Label tree), is designed to substantially reduce the negative impacts of the "curse of dimensionality", with the following additional properties:

1. *Overlap free.* It partitions the data space into hypercubes and maps each hypercube to a unique label.
2. *Strong scalability.* It scales up well in terms of both dimensionality and data size, measured by the number of pages accessed and the total elapsed time for queries.
3. *Distribution insensitivity.* It works well regardless of the distribution of the data being indexed.

The main idea of the PL-tree is to partition the space recursively and map multiple dimensional vectors into scalar labels. It works as follows: (1) partition the original space into hypercubes, also called subspaces; (2) map objects in a subspace to a unique fixed point; (3) label each object in a subspace using the Cantor pairing function [6] on the fixed point of the subspace so that all objects in the same subspace have the same label and objects in different subspaces have different labels; (4) if the number of data objects contained in a subspace is greater than a pre-determined bound (e.g. the size of a database page), continue this process recursively. The structure of a PL-tree is a tree of labels with a number of children at each node, where each label uniquely identifies the set of objects contained in the same subspace of the node. PL-tree indexing cuts down searching redundancy substantially for range queries, and is especially suited for indexing large volumes of high-dimensional data. The uniqueness property of labeling plays a critical role in processing range queries efficiently (which cannot be achieved by hashing). Moreover, label generation using the Cantor pairing function is time efficient, and incurs little space overhead.

In this paper we present algorithms to construct a PL-tree and carry out point queries, range queries, and KNN queries. For dynamic data sets we also present algorithms for inserting and deleting data. We carry out detailed performance evaluations through a large number of experiments and show that PL-trees on high-dimensional synthetic and real-world data outperform the popular indexing methods.

The rest of the paper is organized as follows. We provide a brief overview of the background and related work in Section 2. In Section 3, we describe the PL-tree and present algorithms for constructing PL-trees, carrying out queries, and inserting (deleting) data to (from) an existing PL-tree. We show experiment results and performance analysis in Section 4. We conclude the paper with final remarks in Section 5.

2   Related Work

In the past several decades, many indexing methods for multidimensional data have been proposed.

R-Tree-Based Methods. The R-tree family is the most popular indexing structure including many multidimensional indexing methods [1, 2, 4, 11, 15, 20, 25]. An R-tree is a dynamic, balanced indexing structure which models data partition using Minimal Bounding Rectangles (MBRs). A node in an R-tree is split into two nodes if it contains too many MBRs. The splitting method with different heuristic optimizations varies among different R-tree variants. R-trees [11] split an MBR by minimizing the areas of the resulting MBRs, while R*-trees [2] also consider the overlaps. Hilbert R-trees [15] group similar MBRs using a "good" ordering based on the Hilbert curve. PR-trees [1] use priority rectangles on bulk loaded data where rectangles are represented as 4-dimensional points. However, these R-tree variants are mainly for indexing low-dimensional data. Several R-tree-based structures such as TV-trees and X-trees are designed to handle high-dimensional data. TV-trees [20] reduce dimensionality by ordering dimensions on their importance so that only important information among data objects is stored. X-trees [4] introduce the concept of supernodes to minimize overlaps in high dimensional space which keeps the directory as hierarchical as possible and at the same time avoids splits in the directory.

Space-Partition Methods. In addition to R-tree-based methods using MBRs to model the space partition, many methods employ the regular-partitioning of multidimensional data space. The Quadtree [24] is such a method which recursively divides the d-dimensional data space into 2^d sub-spaces. The Grid File [21] partitions space into buckets for indexing k-dimensional data, using a directory which contains a k-dimensional array and k one-dimensional arrays. The VA-File [27] (Vector Approximation File) is an array of b-bit strings which divides the data space into 2^b rectangular cells and uses a b-bit string for each cell. The Pyramid technique [3] is proposed to support efficient range queries which is based on a special partitioning strategy and is optimized for high-dimensional data. However, Fonseca and Jorge [9] point out that if the database is not uniformly distributed, the efficiency of range queries using Pyramid Technique cannot be guaranteed.

Feature-Based Methods. The feature-based similarity search is also an important search paradigm in database applications. SS-trees [28] are proposed for this purpose, which use Minimum Bounding Spheres (MBSs) rather than MBRs as bounding regions. SR-trees [16] are proposed to retain the advantages of what MBSs and MBRs can offer, but require more storage space for storing information of both MBRs and MBSs. A-trees [23] apply Relative Approximation to the hierarchical structure of SR-trees by introducing the concept of VBRs (Virtual Bounding Rectangles) which contain approximated MBRs and data objects.

Metric-Based Methods. There are also metric-based indexing structures [29] [7]. The metric-based methods differ from other indexing mechanisms in that they are based only on the relative distance between the data points. The VP-tree [29] is a static indexing method using binary tree based on the omni search strategy, where data points are indexed according to their distance to a set of vantage points. The M-tree [7] is the most efficient metric-based indexing structure known so far. It follows the idea of the Bisector Tree (BST) and the Geometric Near-Neighbor Access Tree (GNAT) to group data points around a set of representatives.

Dimension-Reducing Methods. In iMinMax(θ) [22], a data point in d-dimension is mapped to a $1D$ line using its maximum or minimum value of all dimensions. A query to the d-dimensional data is then mapped into d subqueries, with one query for each dimension. The NB-trees [9] calculate the Euclidean norm of a n-dimensional data point and inserts it into a B^+-tree. The iDistance [14] index is also based on B^+-Tree. The iDistance [14] index improves the efficiency of kNN queries by reducing dimensionality. It uses a clustering algorithm to choose reference points, and calculates the distance between each point and its closest reference point. This distance and an additional scaling value is called iDistance. However, iDistance only works well with point data.

We have discussed many indexing methods above, some methods are only suited to the point data and some can handle both point and spatial objects. Surveys of common indexing methods can be found in [5, 10]. The index can be constructed in different ways: If data are mostly static such as GIS databases, an efficient index of good structure can be constructed using bulkloading algorithms. For instance, Kim and Patel [17] showed that for kNN queries, STR bulkloaded [19] R*-trees outperform Quadtree, but the dynamic constructed R*-trees perform worse than Quadtree. However, some database applications are highly dynamic, such as stock or moving object databases. These databases need to be constructed using a dynamic algorithm. Moreover, many previous works evaluate different indices on particular queries. Kothuri et al. [18] compare R-trees and Quadtrees using a variety of range queries on 2D GIS spatial data and show that R-trees outperform Quadtrees in general. Hoel and Samet [13] evaluate the performance of traditional spatial overlap join on various R-tree variants and the Quadtree, and show that R+trees and Quadtrees outperform R*-trees using 2D GIS spatial data. Corral et al. [8] compare R*-trees, X-trees, and VA-File, and show that R*-trees outperform X-trees and VA-File on closet pair queries.

3   PL-Tree Indexing

In this section, we first introduce our dimensionality reduction method that maps multiple dimensional data into a scalar value. Secondly, we present PL-tree indexing algorithms on point data. Finally in Sect. 3.6, we explain how to construct PL-tree on spatial data and how to answer queries.

3.1   Multidimensional Space Mapping

Without loss of generality, we assume that data in a k-dimensional space are in the non-negative quadrant of a coordinate system; we refer to this coordinate system as the *home system*. Let $\mathbb{R}_0$ and $\mathbb{N}$ denote the sets of non-negative real numbers and integers, $D = (d_1, \ldots, d_k)$ be a point in the home system where $d_i \in \mathbb{R}_0$ for $i = 1, \ldots, k$. and $U = (u_1, \ldots, u_k)$ be a rescaling vector, where $u_i \in \mathbb{R}_0$ for $i = 1, \ldots, k$. A *scaling function* $S_U : \mathbb{R}_0^k \to \mathbb{N}^k$ is defined to map a real vector D to an integral vector:

$$S_U(D) = (\lfloor d_1/u_i \rfloor, \ldots, \lfloor d_k/u_i \rfloor), i = 1, \ldots, k.$$

A paring function is a bijection from integral k-dimensional points to integers and the Cantor paring function is a standard and commonly used paring function. We define the Cantor paring function $f_k : \mathbb{N}^k \to \mathbb{N}$ as follows:

$$f_k(i_1, \ldots, i_k) = \begin{cases} f_2(i_1, f_{k-1}(i_2, \ldots, i_k)), & \text{if } k > 2, \\ \frac{1}{2}(i_1 + i_2)(i_1 + i_2 + 1) + i_1, & \text{if } k = 2. \end{cases}$$

By using the *scaling function* and the *pairing function* we reduce a k-dimensional data point D to an integer by $L = f_k(S_U(D))$. Based on such reduction, we present the algorithms for PL-tree indexing in the following sections, and we also discuss the choice of the *re-scaling vector U*.

3.2   PL-Tree Index Structure

A leaf node of a PL-tree index contains a bounding box B and the multidimensional data identifiers included in B. Directory (non-leaf) nodes are in the form of (U, B, E), where U is a re-scaling vector; B is the bounding box of the hypercube represented by the node; and E is a list of entries of form (L, ptr), sorted by the value of L (L is an integer label and ptr points to node). In PL-tree nodes, the following processes may be operated in the home system of the corresponding hyperspace.

Partitioning. When the number of objects contained in a hypercube exceeds the limit (i.e., a page's capacity), we partition it into smaller sub-hypercubes according to the *re-scaling vector U*, so that the side length of a sub-hypercube is u_i units on the *i-th* dimension. Then we re-scale the sub-hypercubes into unit hypercubes consisting of 1 unit on each side in a new coordinate system. We refer to this new coordinate system as a U-system.

Mapping. Let $C_l = (l_1, \ldots, l_k)$ be the lowest corner point of a hypercube C in the U-system. In the home system, for any point $(p_1, \ldots, p_k)$ in C, we define *non-upper-boundary* points as points satisfying $u_i \cdot l_i \leq p_i < u_i \cdot (l_i + 1)$. Let $D = (d_1, \ldots, d_k)$ be a *non-upper-boundary* point, by the *scaling function* we obtain that $C_l = S_U(D)$. It follows that all *non-upper-boundary* points in C can be mapped to the same fixed point C_l. In Fig. 1, the 3 points in the middle cube are mapped to the point(3,3) which is $C_l = (1, 1)$ in the U-system.

Labeling. We define $C_L = \{D \mid S_U(D) = C_l\}$ as the set of *non-upper-boundary* points in C. According to the pairing function, we can label all points in C_L with a label $L = f_k(C_l)$, then the point set C_L becomes the child node with the label of L. Since f_k is a bijection, it is straightforward to show that L is unique with respect to C_L.

Re-coordination. When we move from a hypercube into its sub-hypercubes, we consider the sub-hypercube as a new home-system, by a re-coordinating process. For any D in a sub-home with a *re-scaling vector U*, we re-coordinate it to: $D_r = (d_1 - d_1' \cdot u_1, \ldots, d_k - d_k' \cdot u_k)$, where $d_i' = \lfloor d_i/u_i \rfloor = l_i$, for $i = 1, \ldots, k$. For example, in Fig. 1 the 3 points in the middle are re-coordinated to (1,2), (2.5,1.5), and (1.2,0.5) in the home system of the middle sub-square.

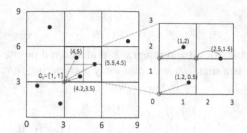

Fig. 1. PL-tree partition. The figure illustrates a PL-tree of 7 points in a 2D space, where each node can contain at most 2 points. In the home system of the root node $U = (3.0, 3.0)$ is used, and the home system of the middle square uses $U = (1.5, 1.5)$.

3.3   Point Data Indexing

Static Data Indexing

Given a dataset, we take the procedure of partitioning, mapping, labeling, and re-coordinating on the data space. The data with the same label become the children of the parent node. The procedure is repeated recursively on each hypercube until all the points of a sub-hypercube can fit in a page. Fig. 1 demonstrates the PL-tree partitioning and one of its sub-hypercube as a child node in a 2D space.

Dynamic Data Indexing

Searching   The search algorithm traverses the tree from the root using Alg.1. In the worst case, the number of node accesses for a single search is $O(h)$ where h is the expected height of the tree (see Sect. 3.5). Note that the labels at a node are compact and sorted. Thus, a particular label can be located very fast through a binary search.

Insertion.   The insertion algorithm (see Alg. 2) is similar to searching. It inserts a data entry D into a leaf node if D is found in a leaf, or it creates a new leaf including D if the label of D is not found in a directory node. There is also an overflow treatment when a leaf is full.

Deletion.   To remove a data record from a PL-tree, a similar searching algorithm is invoked; once the record is found in a leaf node, it is removed from the leaf node. We remove the leaf node and its label from its parent node if the left node becomes empty. A merge process may be executed after a deletion.

Query Processing

A point query on PL-tree is straightforward according to the search algorithm (see Alg. 1). For range queries which ask for a particular range (or hyper cuboid) of data points, let Q denote the query range. Hypercubes in the home system that intersect with Q can be classified into *innerblocks* and *outerblocks*. An *innerblock* is a hypercube that is fully contained in Q, while an *outerblock* partially intersects with Q. We devise a procedure RQP (Algorithm. 3) to carry out range queries. Given a query range Q, RQP identifies the innerblocks and outerblocks by checking the coordinates of each

---

Algorithm 1. Search(PLTNode N, Data D)

1: **if** N is a leaf node **then**
2:     Compare all entries $N.E$ with D
3: **else**
4:     $L \leftarrow N.\text{Label}(D)$
5:     $pos \leftarrow N. \text{Search\_in\_children}(L)$
6:     **if** L is found **then**
7:         re-coordinate D
8:         **return** Search($N.E[pos].ptr, D$)
9:     **else**
10:         **return** False
11:     **end if**
12: **end if**

---

label in the U-system. For each innerblock, RQP returns all the data points in it; For each outerblock, RQP casts the intersected portion as a range query on the corresponding sub-hypercube and processes the query recursively.

Given a query point p and a value k, a kNN query returns k data points which are closet to the query point p based on a distance function. We use a similar algorithm described in [12] to carry out kNN queries in PL-trees. The algorithm in [12] is implemented using R-trees; we adapted it to use PL-trees. To find the k-nearest neighbors for a query point p, PL-trees maintain a priority queue consisting of objects sorted based on their MINDIST from p. When the next object is retrieved from the priority queue, if it is the bounding box of a node, it is expanded by pushing its children into the priority queue; if it is a data entry, the data entry is reported as the next nearest neighbor to p. This process is repeated until k data entries are reported. According to the analysis in [12], the expected number of leaf node accesses is $O(k + \sqrt{k})$ and the expected number of objects in the priority queue is $O(\sqrt{k})$.

3.4 Compact PL-Tree Storage

In our insertion algorithm, while splitting a leaf node, if the data in the leaf are evenly distributed, the algorithm will create many new leaves with each leaf node containing only a few data points. These new "sparse" leaves seriously affect the efficiency of range queries, because too many "sparse" pages are accessed during the query. The problem becomes worse with increasing dimensionality. Hence we propose a compact storage which significantly reduces such "sparse" effect.

The basic idea is to store many leaf nodes in the same page. As a result, we need to store labels of the hypercubes corresponding to these nodes to identify each node. We refer to the labels that identify nodes as *home labels*, which indicates the location of the node with an offset value. We refer to the original labels with child pointers as *child labels*. Fig.2 illustrates the compact node structure, there are two arrays of entries, child label entries and home label entries. In a practical implementation, the two arrays of directory nodes grow respectively from the two ends of the page. Fig.3 is an example of a PL-tree and the corresponding compact pages, the value of offset$_i$ indicate the location

Algorithm 2. Insert(PLTNode N, Data D)	**Algorithm 3.** RQP(PLTNode N, Range Q)
1: **if** N is a leaf node **then**	1: $Res \leftarrow$ empty set
2: Insert $D.id$ into N	2: **if** N is a leaf node **then**
3: **if** N.full() **then**	3: check all data in N and add the inter-
4: $N.U$ = U-Calculator()	sected ones into Res
5: **for all** data d in N **do**	4: **else**
6: re-coordinate d and Insert(N, d)	5: **for all** E in $N.E$ **do**
7: **end for**	6: $C_l \leftarrow N.$ DeLabel($E.L$)
8: **end if**	7: let B be unit hypercube based on C_l
9: **else**	8: **if** B is in Q in U-system **then**
10: $L \leftarrow N$.Label(D)	9: $Res = Res \cup$ FetchAll($E.ptr$)
11: $pos \leftarrow N.$ Search_in_children(L)	10: **else**
12: **if** L is found **then**	11: **if** B intersects Q in U-system
13: re-coordinate D	**then**
14: Insert($N.E[pos].ptr,D$)	12: $Res = Res \cup$ RQP($E.ptr, Q$)
15: **else**	13: **end if**
16: create $NewChild$ with L in N	14: **end if**
17: Insert($NewChild, D$)	15: **end for**
18: **end if**	16: **end if**
19: **end if**	17: RETURN Res

of the first entry of the node with the home label L_i. In leaf nodes, we store the *home label* for each data entry.

The compact structure slightly complicates the PL-tree algorithms. For example, if we search for $D = O6$ in the PL-tree of Fig.3, we need one more parameter of home label for each search. Starting from the root node using Search(root, D, 0), the Search algorithm calculates the label 3 of D in root. Then Search(directory-node, D, 3) is called, and we calculate label 10 and find it from [5,10,97]. In the leaf node, we compare D with the objects with home label of 10 ($O6$-$O9$) and finally find $O6$ as the result. Nonetheless, the compact structure significantly improves the performance of queries (see Section 4.6).

3.5   Re-scaling Vector and Algorithm Performance

An interesting parameter of PL-trees is the *re-scaling vector U*. Our goal is to choose U such that the PL-tree is as height balanced as possible.

Height Estimation of PL-trees. Considering the points in a k-dimensional space, we assume that the root space is a hypercube with edges of the same length L and the smallest distance between any two points is l. If each time a PL-tree partitions hypercubes by splitting each dimension into d parts, then for a node of depth h, the diagonal of the node should be larger than l since a node must contain at least one point. Thus, we have $\sqrt{k(\frac{L}{d^{h-1}})^2} \geq l$ and it follows that $h \leq \log_d \frac{\sqrt{k}L}{l} + 1$, which is $O(\log L/l)$ if L/l is much larger than k. Hence in the worst case, the height of the PL-tree is logarithmic to the ratio of L to l. That is, the height of PL-trees is related to the density of the data.

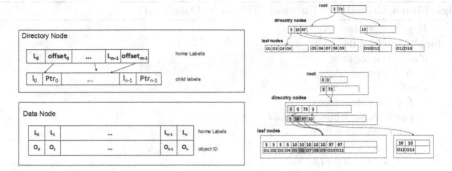

Fig. 2. Structure of PL-Tree Nodes   **Fig. 3.** Compact PL-tree with 7 nodes

Binwidth Optimization. If we partition the space randomly in PL-Trees, the hypercubes with dense data will be of deep height. Also, there are many blank areas in the hyperspace for non-uniformly distributed data. To make PL-trees as height balanced as possible, we choose U to partition the data of the hypercube as evenly as possible. Our partition strategy is, by choosing a proper U, the hypercube is partitioned into subhypercubes such that the sub-hypercubes have about the same number of data entries. Considering the data projections onto each dimension, choosing u_i for the i-th dimension is a *binwidth optimization* problem. Let $X = x_1, \ldots, x_n$ be a data set, the binwidth optimization is to find an optimal Δ such that the range of X is divided into N bins of width Δ. Let k_i be the number of elements in X that fall in the i-th bin. We first calculate the value of N that minimizes the MISE (Mean Integrated Squared Error) [26], which is a measure of the goodness-of-the-fit of the bin histogram to an unknown data rate. Then we choose the optimal N^* from 2 to N to minimize the modified variance-to-mean ratio V/M and the optimal u_i is $\frac{\max(X) - \min(X)}{N^*}$, where $M = \frac{1}{N'} \sum_{i=0}^{N} k_i$, $V = \frac{1}{N'} \sum_{i\text{-th bin is not empty}} (k_i - M)^2$, and N' is the number of non-empty bins.

3.6   Spatial Data Indexing

Spatial data can be considered as a k-dimensional spatial object represented by an MBR denoted by R. R can be uniquely determined by its lowest corner point R_l and highest corner point R_h, written as: $R_l = (a_1, \ldots, a_k)$, and $R_h = (b_1, \ldots, b_k)$.

Static Spatial Data Indexing

Alg. 4 shows the pseudocode for creating a PL-tree index from static spatial data. The process is recursive, similar to point data, but has the following changes:

Partitioning. We need to consider the size of the object while determining U to ensure each object covers at most 2^k hypercubes. Let S_I denote a static set of spatial objects. We calculate U as follows: calculate U' according to Section 3.5 and $U'' = \{u_1, \ldots, u_k\}$, where $u_i = \max_{(R_l, R_h) \in S_I}(b_i - a_i)$, then let $U = \max\{U', U''\}$.

Algorithm 4. CreateIndex_Cuboid(PLTNode N, DataSet A)

1:   $N.U \leftarrow$ U-Calculator(A)
2:   **for all** $data$ in A **do**
3:     **if** $data$ is huge **then**
4:       $N.HList$.add($data$)
5:     **else**
6:       $L \leftarrow N$.Label($data$)
7:       $pos \leftarrow$ N.Search_in_children(L)
8:       **if** L is not found **then**
9:         $temp \leftarrow$ new PLTNode
10:        insert $(L, temp)$ to $N.E$
11:        $temp$.add_data($data$)
12:      **else**
13:        $temp \leftarrow N.E[pos].ptr$
14:        $temp$.add_data($data$)
15:      **end if**
16:    **end if**
17: **end for**
18: **for all** $child$ of N **do**
19:   **if** $child$.size $> split\_threshold$ **then**
20:     CreateIndex_Cuboid($child$, $child$.dataset)
21:   **end if**
22: **end for**
23: **return** N

It ensures that every R must be confined in a region consisting of 2^k adjacent hyper-cubes that share a common point in the center of the region.

Mapping & Labeling. For any $R = (R_l, R_h) \in S_I$, we map R to a label L using $R' = (S_U((a_1, \ldots, a_k)), S_U((b_1, \ldots, b_k))) = ((a'_1, \ldots, a'_k), (b'_1, \ldots, b'_k)) = (R'_l, R'_h)$ and $L = f_2(R'_l, R'_h)$. The calculation still follows the bijective property of f_k such that the mapping from bounding boxes to labels is bijective.

Re-coordination. Instead of re-coordinating a single point, we re-coordinate both R_l and R_h for a bounding box R.

Huge Objects. Even if we choose a big U for indexing spatial data, it is still possible that there are relatively huge objects (e.g. one object is of half size of the hypercube). Thus, we use an extra link list (HugeObjectList) to store such huge objects.

Dynamic Spatial Data Indexing

When the insertion algorithm dynamically inserts a new spatial data entry R into the existing index structure, R may be too large for the size of U. We could re-calculate U and re-coordinate all existing data, but this is time consuming. Instead, we logically cut R, along the lines of the existing hypercubes, into several smaller hyper cuboids. We then logically replace R with the derived smaller hyper cuboids for indexing. A logical cut implies that the cut does not physically generate smaller objects to replace the original object. Each derived smaller hyper cuboid is simply a pointer to the original object. The coordinates of each derived hyper cuboid are calculated on the fly, which may be disposed after indexing. Thus, what we finally get is a set of pointers at different positions in the index pointing to the same original data. The original data will not be indexed. Alg. 5 and 6 show the pseudocodes of insertion and deletion for handling dynamic spatial data. We omit the search algorithm as it is similar to point data. Compared to the algorithms for indexing point data, inserting a spatial data may cost much more due to the extra insertions from the logical cut and the maintainance of HugeObjectList.

Query Processing

The query processing for spatial data is similar to that for point data; the only difference is that the objects in HugeObjectsList of each nodes are also considered. Due to space constraints, we do not provide the detailed algorithms here. Moreover, due to the logical cut, there may be some duplicate object pointers in the result for range queries and kNN queries. Therefore, we need to remove the duplicates before returning the result.

4   Performance Evaluation

In this section, we present the experimental results on point, range and kNN queries for multidimensional data. All indices are constructed using dynamic indexing methods. We compare PL-trees with R*-trees and X-trees. The R*-tree is the most common method for indexing multidimensional data, while the X-tree is an improved variant of the R*-tree for high-dimensional data. We also compare PL-trees with Quadtrees and some other indexing methods.

4.1   Data Sets and Configurations

We use both synthetic and real world data. For synthetic data, we generate a uniformly distributed data set containing 1,000,000 points in 15-dimensional space. Moreover, we use the following real world data for different evaluations:

1. USPP. A Point data set containing 15,206 populated places in the U.S.
2. TIGER (http://www.census.gov/geo/www/tiger/). A spatial dataset containing 556, 696 non-uniformly distributed polygons.
3. LLMPP. A High-dimensional data set from Lymphoma/Leukemia Molecular Profiling Project (http://llmpp.nih.gov/lymphoma/), which contains 1,843,200 items of 17 integer and 15 float attributes. We only took the 15 float attributes and randomly select 500,000 data points.
4. MAPS. The MAPS Catalog data containing photometric and astrometric data from the Palomar Observatory Sky Survey (http://aps.umn.edu/catalog/). It has 90 million items of 39 integer attributes, and we select one field P105 (40,398 points).

We implement the algorithms in C++ on an Intel Core i5 2.53G machine running Windows 7 with a 4GB memory. We apply LRU for buffer pool which is in the size of 1000 pages. The page size is 4KB for data sets of dimensionality no more than 8 and 8KB for higher dimensionalities. We focus on the following evaluation metrics: number of pages accessed, index size, and total query time.

Algorithm 5 . Insert_Cuboid(PLTNode N, DATA $data$)	**Algorithm 6 .** Delete_Cuboid(PLTNode N, DATA $data$)
1: **if** $data$ has larger size than $N.U$ **then**	1: **if** $data$ has larger size than $N.U$ **then**
2: $LogicalCuts \leftarrow$ partition($data, N.U$)	2: $LogicalCuts \leftarrow$ partition($data, N.U$)
3: **for all** lc in $LogicalCuts$ **do**	3: **for all** lc in $LogicalCuts$ **do**
4: Insert_Cuboid(N, lc)	4: Delete_Cuboid(N, lc)
5: **end for**	5: **end for**
6: **else**	6: **else**
7: **if** N is a leaf node **then**	7: **if** N is a leaf node **then**
8: N.add_data($data, L$)	8: **if** $data$ IsHuge($N.U$) **then**
9: **if** IsHuge($data, N.U$) **then**	9: remove $data$ from $N.HList$
10: $N.HList$.add($data$)	10: **else**
11: **else**	11: remove $data$ from $N.E$
12: insert $data$ into $N.E$	12: **end if**
13: **if** N is full **then**	13: **if** $N.E$ and $N.HList$ are empty **then**
14: IndexCreate_Cuboid($N,N.E$)	14: remove N from $N.parent$
15: **end if**	15: **end if**
16: **end if**	16: **else**
17: **else**	17: $L \leftarrow N.$ Label($data, N.U$)
18: $L \leftarrow N.$ Label($data$)	18: $pos \leftarrow N.$Search_in_children(L)
19: $pos \leftarrow N.$ Search_in_children(L)	19: **if** L is not found **then**
20: **if** L is not found **then**	20: **return** NOT_FOUND
21: $temp \leftarrow$ new PLTNode	21: **else**
22: insert ($L, temp$) into N	22: Delete_Cuboid($N.E[pos].ptr, data$)
23: Insert_Cuboid($temp, data$)	23: **end if**
24: **else**	24: **end if**
25: Insert_Cuboid($N.E[pos].ptr, data$)	25: **end if**
26: **end if**	
27: **end if**	
28: **end if**	

4.2 Index Size

One of the great features of a PL-Tree is that the size of index is dimensionality independent since PL-Trees store scalar labels instead of the multi-dimensional information. R-Tree-based methods store MBRs whose size increases linearly with dimensionality. By contrast, a label is a constant size. We create indices on synthetic points with different dimensionalities ($D = 2, 3, 4, 5, 6, 8, 10, 12$) and compare the sizes of index files. Fig. 4(a) shows that the index size of R-Tree-based indices increases about linearly with dimensionality. As expected, the size of a PL-Tree index is dimensionality independent. A PL-tree for the 2-dimensional dataset is 18872 KB, which is 0.62 times as large as an R*-Tree; while for the 12-dimensional dataset it is 14274 KB, which is only 0.08 times as large as an R*-Tree.

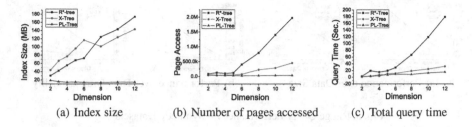

(a) Index size   (b) Number of pages accessed   (c) Total query time

Fig. 4. Performance Comparisons on Synthetic Datasets

4.3 Results for Point Queries

We evaluate point queries using synthetic data sets of various dimensionalities. We fix the data size at 500,000 records and create indices with different dimensionalities. We then randomly select 10,000 points to carry out point queries. For point queries on a hierarchical tree index, the query cost directly corresponds to the height of the tree. However, this is only true if there is no overlap between hypercubes of directory nodes. Due to the overlap of R-tree-based indices, Fig.4(b) shows that R*-Trees incur 50 times as many page accesses as PL-Trees for $D = 12$. X-Trees are designed as the hybrid of linear array-like and hierarchical R-Tree-like directory, and provide a much better performance by avoiding the splits to reduce overlaps. PL-Trees guarantee that there is only one path from root to leaf for a single point query, which implies that the number of page access only corresponds to the depth of the leaf nodes.

4.4 Results for Range Queries

Synthetic Datasets

In the first experiment, we compare the performance of range queries on synthetic data. We fix the selectivity to 0.1%, and carry out 10 random range queries. By varying the number of data points and the dimensionality respectively, we evaluate the impact of data size and data dimensionality on the range query performance. We measure both the number of page accesses and the total elapsed time.

We first fix the dimensionality to 12 and vary the data size from 100,000 to 1,000,000 records. Fig. 5(a) shows that PL-trees always perform much better than R*-trees and slightly better than X-trees on larger datasets. Then, we fix the data size to 1,000,000 and vary the dimensionality ($D = 3, 6, 9,$ and 12). Fig. 5(b) shows that X-trees perform slightly better when D is 6 or lower, but PL-tree performs better when D is higher. The reason is that, for relatively low dimensionalities, X-trees have significantly less overlap than R*-trees, and less intersection check than PL-trees. However, for high dimensionalities ($D = 9$ and 12) we observed that the performance of X-Trees decreases rapidly while PL-trees maintain a rather smooth curve despite the dimensionality growth. The results show that PL-trees scale well on both size and dimensionality.

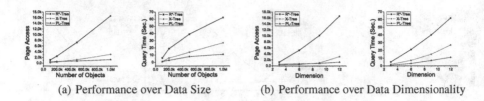

(a) Performance over Data Size   (b) Performance over Data Dimensionality

Fig. 5. Range Query Performance on Synthetic Datasets

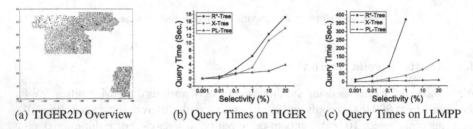

(a) TIGER2D Overview   (b) Query Times on TIGER   (c) Query Times on LLMPP

Fig. 6. Range Query Performance on Real-World Datasets

Real-World Datasets

We also evaluate indices on TIGER and LLMPP data sets to further examine the performance of PL-trees for various dimensionalities. We vary the selectivities from 0.001% to 20%. Fig.6 shows that as expected, PL-trees still outperform R*-Trees and X-Trees, especially for high selectivity. This is because a larger query range tends to incur more overlaps for R*-trees and X-trees, which lead to more I/O costs. The speedup factor of PL-trees over R*-trees and X-Trees reaches up to 7.5 and 3.6, respectively, when the selectivity is 20%.

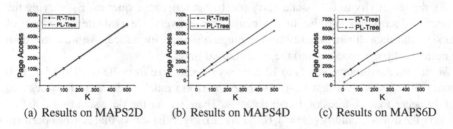

(a) Results on MAPS2D   (b) Results on MAPS4D   (c) Results on MAPS6D

Fig. 7. kNN Query Performance on Real-World Datasets

4.5   Results for kNN Queries

In this section, we compare the performance of PL-trees with R*-trees on kNN queries. We use the first 2, 4, and 8 attributes of MAPS data to get three real datasets: MAPS2D,

Table 1. Comparisons with Quadtrees

	Quadtree	PL-tree	Ratio
Creating Time (sec.)	0.368	4.700	11.2
Creating Disk I/O	40027	738871	54.15%
Avg. Read I/O of PQ	5.829	4.194	76.33%
Avg. Read I/O of kNN			99.07%
Avg. Time of RQ (sec.)	0.4234	0.2211	0.522
Avg. IntrCheck of RQ	77324	17265.8	22.32%
Avg. Read I/Oof RQ	71076.8	41375.8	58.21%

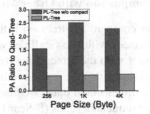

Fig. 8. Compact PL-trees

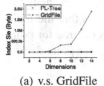 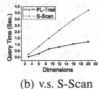

(a) v.s. GridFile   (b) v.s. S-Scan   (a) Build Time   (b) Query Time

Fig. 9. Comparison to GridFile and S-Scan   **Fig. 10.** Comparison to iDistance

MAPS4D, and MAPS6D. We randomly generate 1000 points for kNN queries and count the number of pages accessed. Fig .7, shows that the numbers of page accesses are proportional to the values of k, and PL-trees outperform R*-trees in that their numbers of accessed pages are only 83.6%-98.4%, 41.2%-81.3%, and 28.0%-50.5% of those of the R*-trees for 2D, 4D, and 6D datasets, respectively. We also observe that the performance of PL-trees over R*-trees increases with dimensionality.

4.6   Comparison with Quadtree

Quadtrees bear some similarity with PL-trees. However, PL-trees enjoy flexibilities in space partitioning and have a novel usage of Cantor pairing functions for labels. In this section, we evaluate these two indices on USPP data. We create indices on USPP using different page sizes (from 256B to 4KB). For point queries, we select 1% points of the data sets and search them in indices. To carry out kNN queries, we randomly generate 1000 points for processing kNN queries with different k. We use different selective factors from 0.1% to 30% for evaluating range queries. Table 1 shows that PL-trees need less disk I/O but much more CPU times due to the calculation of U, and perform better than Quadtrees on point queries but similarly on kNN queries. For range queries, PL-trees outperform Quadtrees in all metrics. Especially, PL-trees require significantly less intersection checking since inner-blocks are not checked (see Section 3.3). Moreover, to illustrate the benefit introduced by the compact structure (see Section 3.4), we also compare the performance on range queries for PL-trees with and without the compact structure. Fig.8 shows the ratio of page accessed number of PL-trees to that of the Quadtree, it shows that PL-trees outperform Quadtrees with the compact structure, but not without the compact structure. This verifies that the compact structure is very effective in improving performance.

4.7 Comparison with Other Methods

We also compare PL-trees with the GridFile, S-Scan, and iDistance. We first compare PL-trees with the GridFile, we create the index for 10,000 randomly generated data points with the dimensionality ranging from 2 to 14 and compare the index sizes. Fig. 9(a) shows that the space overhead of the GridFile increases dramatically especially for high dimensional data. We also evaluate the range query performance of PL-trees and pure sequential scans (S-Scan) on 100,000 synthetic data points in different dimensions with selectivity of 0.1%. Fig. 9(b) shows that PL-trees are far superior to S-Scan. Finally, we also compare PL-trees with iDistance. which is of high efficiency for point queries and kNN queries. We create the indices on 6-dimensional point data in different sizes from 50,000 to 300,000 points. We randomly select 1000 points to evaluate the performance of point queries on PL-trees and iDistance. Fig. 10 shows that the build time of an iDistance index is much longer but iDistance outperforms PL-tree for point queries. However, iDistance has a serious drawback that it only works for point data.

5 Conclusions

In this paper, we propose a new indexing method for high dimensional data. PL-trees label objects in high-dimensional space by using the Cantor pairing function which maps a high-dimensional vector into a scalar label bijectively. Due to the "curse of dimensionality", many existing indexing methods do not perform well for high dimensional data. Our indexing employs a novel usage of the Cantor paring functions to cope with high dimensionalities. Crucially, one can restore the corresponding multidimensional data from a scalar value since the Cantor pairing function is invertible. Our results show that our new indexing method scales up nicely with dimensionality and data size, and outperforms the state-of-the-art indexing methods in many ways.

Acknowledgments. Jie Wang was supported in part by the NSF, under the grants CCF-0830314, CNS-1018422, and CNS-1247875. Tingjian Ge was supported in part by the NSF, under the grants IIS-1149417 and IIS-1239176.

References

1. Arge, L., de Berg, M., Haverkort, H.J., Yi, K.: The priority R-tree: A practically efficient and worst-case optimal R-tree. In: Proceedings of ACM/SIGMOD Annual Conference on Management of Data (SIGMOD), pp. 347–358 (2004)
2. Beckmann, N., Kriegel, H.-P., Schneider, R., Seeger, B.: The R*-tree: An efficient and robust access method for points and rectangles. In: Proceedings of ACM/SIGMOD Annual Conference on Management of Data (SIGMOD), pp. 322–331 (1990)
3. Berchtold, S., Böhm, C., Kriegel, H.-P.: The pyramid-technique: Towards breaking the curse of dimensionality. In: Proceedings of ACM/SIGMOD Annual Conference on Management of Data (SIGMOD), pp. 142–153 (1998)
4. Berchtold, S., Keim, D.A., Kriegel, H.-P.: The X-tree: An index structure for high-dimensional data. In: Proceedings of International Conference on Very Large Data Bases (VLDB), pp. 28–39 (1996)

5. Böhm, C., Berchtold, S., Keim, D.A.: Searching in high-dimensional spaces: Index structures for improving the performance of multimedia databases. ACM Comput. Surv. 33(3), 322–373 (2001)
6. Cantor, G.: Contributions to the Founding of the Theory of Transfinite Numbers. Dover, New York (1955); Original year was 1915
7. Ciaccia, P., Patella, M., Zezula, P.: M-tree: An efficient access method for similarity search in metric spaces. In: Proceedings of International Conference on Very Large Data Bases (VLDB), pp. 426–435 (1997)
8. Corral, A., Cañadas, J., Vassilakopoulos, M.: Processing distance-based queries in multidimensional data spaces using r-trees. In: Manolopoulos, Y., Evripidou, S., Kakas, A.C. (eds.) PCI 2001. LNCS, vol. 2563, pp. 1–18. Springer, Heidelberg (2003)
9. Fonseca, M.J., Jorge, J.A.: Indexing high-dimensional data for content-based retrieval in large databases. In: Proceedings of International Conference on Database Systems for Advanced Applications (DASFAA), pp. 267–274 (2003)
10. Gaede, V., Günther, O.: Multidimensional access methods. ACM Comput. Surv. 30(2), 170–231 (1998)
11. Guttman, A.: R-trees: A dynamic index structure for spatial searching. In: Proceedings of ACM/SIGMOD Annual Conference on Management of Data (SIGMOD), pp. 47–57 (1984)
12. Hjaltason, G.R., Samet, H.: Distance browsing in spatial databases. ACM Trans. Database Syst. 24(2), 265–318 (1999)
13. Hoel, E.G., Samet, H., Tree, R.: Benchmarking spatial join operations with spatial output. In: Proceedings of the 21st International Conference on Very Large Data Bases, pp. 606–618 (1998)
14. Jagadish, H.V., Ooi, B.C., Tan, K.-L., Yu, C., Zhang, R.: idistance: An adaptive b+-tree based indexing method for nearest neighbor search. ACM Trans. Database Syst. 30, 364–397 (2005)
15. Kamel, I., Faloutsos, C.: Hilbert R-tree: An improved R-tree using fractals. In: VLDB, pp. 500–509 (1994)
16. Katayama, N., Satoh, S.: The SR-tree: An index structure for high-dimensional nearest neighbor queries. In: Proceedings of ACM/SIGMOD Annual Conference on Management of Data (SIGMOD), pp. 369–380 (1997)
17. Kim, Y.J., Patel, J.: Performance comparison of the r*-tree and the quadtree for knn and distance join queries. IEEE Transactions on Knowledge and Data Engineering 22(7), 1014–1027 July
18. Kothuri, R.K.V., Ravada, S., Abugov, D.: Quadtree and r-tree indexes in oracle spatial: a comparison using gis data. In: Proceedings of the 2002 ACM SIGMOD International Conference on Management of Data, SIGMOD 2002, pp. 546–557. ACM, New York (2002)
19. Leutenegger, S., Lopez, M., Edgington, J.: Str: a simple and efficient algorithm for r-tree packing. In: Proceedings of the13th International Conference on Data Engineering, pp. 497–506 (April 1997)
20. Lin, K.-I., Jagadish, H.V., Faloutsos, C.: The TV-tree: An index structure for high-dimensional data. VLDB Journal 3(4), 517–542 (1994)
21. Nievergelt, J., Hinterberger, H., Sevcik, K.C.: The grid file: An adaptable, symmetric multi-key file structure. ACM Trans. Database Syst. 9(1), 38–71 (1984)
22. Ooi, B.C., Tan, K.-L., Yu, C., Bressan, S.: Indexing the edges - a simple and yet efficient approach to high-dimensional indexing. In: Proceedings of the Nineteenth ACM SIGMOD-SIGACT-SIGART Symposium on Principles of Database Systems, PODS 2000, pp. 166–174. ACM, New York (2000)
23. Sakurai, Y., Yoshikawa, M., Uemura, S., Kojima, H.: The A-tree: An index structure for high-dimensional spaces using relative approximation. In: Proceedings of International Conference on Very Large Data Bases (VLDB), pp. 516–526 (2000)

24. Samet, H., Webber, R.E.: Storing a collection of polygons using quadtrees. ACM Trans. Graph. 4(3), 182–222 (1985)
25. Sellis, T.K., Roussopoulos, N., Faloutsos, C.: The R+-tree: A dynamic index for multi-dimensional objects. In: Proceedings of International Conference on Very Large Data Bases (VLDB), pp. 507–518 (1987)
26. Shimazaki, H., Shinomoto, S.: Kernel bandwidth optimization in spike rate estimation. Journal of Computational Neuroscience 29(1-2), 171–182 (2010)
27. Weber, R., Schek, H.-J., Blott, S.: A quantitative analysis and performance study for similarity-search methods in high-dimensional spaces. In: Proceedings of International Conference on Very Large Data Bases (VLDB), pp. 194–205 (1998)
28. White, D.A., Jain, R.: Similarity indexing with the SS-tree. In: Proceedings of International Conference on Data Engineering (ICDE), pp. 516–523 (1996)
29. Yianilos, P.N.: Data structures and algorithms for nearest neighbor search in general metric spaces. In: Proceedings of Annual ACM-SIAM Symposium on Discrete Algorithms (SODA), pp. 311–321 (1993)

Stream-Mode FPGA Acceleration
of Complex Pattern Trajectory Querying

Roger Moussalli[1], Marcos R. Vieira[2], Walid Najjar[1], and Vassilis J. Tsotras[1]

[1] University of California, Riverside, USA
{rmous,najjar,tsotras}@cs.ucr.edu
[2] IBM Research, Brazil
mvieira@br.ibm.com

Abstract. The wide and increasing availability of collected data in the form of trajectory has lead to research advances in behavioral aspects of the monitored subjects (e.g., wild animals, people, vehicles). Using trajectory data harvested by devices, such as GPS, RFID and mobile devices, complex pattern queries can be posed to select trajectories based on specific events of interest. In this paper, we present a study on FPGA-based architectures processing complex patterns on streams of spatio-temporal data. Complex patterns are described as regular expressions over a spatial alphabet that can be implicitly or explicitly anchored to the time domain. More importantly, variables can be used to substantially enhance the flexibility and expressive power of pattern queries. Here we explore the challenges in handling several constructs of the assumed pattern query language, with a study on the trade-offs between expressiveness, scalability and matching accuracy. We show an extensive performance evaluation where FPGA setups outperform the current state-of-the-art CPU-based approaches by over three orders of magnitude. Unlike software-based approaches, the performance of the proposed FPGA solution is only minimally affected by the increased pattern complexity.

1 Introduction

Due to their relative ease of use, general purpose processors are commonly favored at the heart of many computational platforms. These processors are deployed in environments with varying requirements, ranging from personal electronics to game consoles, and up to server-grade machines. General purpose CPUs follow the Von-Neumann model, which execute instructions sequentially. Nevertheless, in this model performance does not always linearly scale in multiprocessor environments, mostly due to the challenges of data sharing across cores. As it is non-trivial for these CPUs to satisfy the increasing time-critical demands of several applications, they are often coupled with application- or domain-specific parallel accelerators, such as Graphics Processing Units (GPUs) and Field Programmable Gate Arrays (FPGAs), which strive given a certain class of instructions and memory access patterns.

M.A. Nascimento et al. (Eds.): SSTD 2013, LNCS 8098, pp. 201–222, 2013.

FPGAs consist of a fully configurable hardware platform, providing the flexibility of software (e.g., programmability) and the performance benefits of hardware (e.g., parallelism). The advantages on performance of such platforms arise from the ability to execute thousands of parallel computations, relieving the application at hand from the sequential limitations of software execution on Von-Neumann based platforms. The processor "instructions" are now the logic functions processing the input data. Depending on the application, one big advantage of FPGAs is the ability to process streaming data at wire speed, thus resulting in a minimal memory footprint. The aforementioned advantages are shared with Application Specific Integrated Circuits (ASIC). FPGAs, however, can be reconfigured and are more adaptable to changes in applications and specifications, and hence exhibit a faster time to market. This comes at a slight cost in performance and in area, where one functional circuit would run faster on a tailored ASIC and require fewer gates.

As traditional platforms are increasingly hitting limitations when processing large volumes of streaming data, researchers are investigating FPGAs for database applications. Recent work has focused on the adoption of FPGAs for data stream processing in different scenarios. In [18] a stream filtering approach is presented for XML documents. [30] investigated the speedup of the frequent item problem using FPGAs. In [33], the FPGA is employed for complex event detection using regular expressions. [23] proposed a predicate-based filtering on FPGAs where user profiles are expressed as a conjunctive set of boolean filters. [16] describes an FPGA-based stream-mode decompression engine targeting Golomb-Rice encoded inverted indexes.

In this paper, we describe an FPGA-based setup allowing users to query spatio-temporal databases in a very powerful and intuitive way. Figure 1 depicts a generic overview of the various steps performed in spatio-temporal querying setups. Streams of trajectory data are harvested from devices, such as GPS and cellular devices. Coordinates are then translated into semantic regions that partition the spatial domain; these regions can be grid regions representing areas of interests (e.g., neighborhoods, school districts, cities). Our work is based on the FlexTrack framework [31,32], which allows users to query trajectory databases using flexible patterns. A flexible pattern query is specified as a combination of sequential spatio-temporal predicates, allowing the end user to search for specific parts of interests in trajectory databases. For example, the pattern query *"Find all taxi cabs (trajectories) that first were in downtown Munich in the morning, later passed by the Olympiapark around noon, and then were closest to the Munich airport"* provides a combination of temporal, range and Nearest-Neighbor (NN) predicates that have to be satisfied in the specific order. Essentially, flexible patterns cover that part of the query spectrum between the single spatio-temporal predicate queries, such as the range predicate covering certain time instances of the trajectory life (e.g., *"Find all trajectories that passed by the Deutsches Museum area at 11pm"*), and similarity/clustering based queries, such as extracting similar movement patterns from a trajectories that cover the

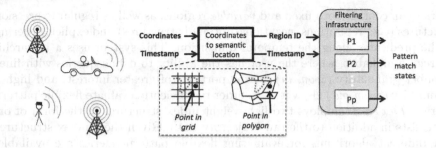

Fig. 1. Generic overview of various steps performed in spatio-temporal querying setups

entire life span of the trajectory (e.g., *"Find all trajectories that are similar to a given query trajectory according to some similarity measure"*).

Flexible pattern queries can also have "variable" spatial predicates, and thus substantially enhancing the flexibility and expressive power of the *FlexTrack* framework. An example of a variable-enhanced query is *"Find all trajectories that started in a region, then visited the downtown Munich, then at some later point returned to the first visited region"*.

This work serves as a proof-of-concept on the performance benefits of evaluating flexible pattern queries using FPGAs. Here we focus on the challenges of supporting hundreds (up to thousands) of variable-enhanced flexible patterns on FPGAs in streaming (fully-pipelined) fashion. Using FPGAs all pattern query predicates are evaluated in parallel over sequential streams of trajectories, hence resulting in over three orders of magnitude speedup over CPU-based approaches. This performance property also holds even when compared to CPU-based setups where the pre-processing of trajectories is performed beforehand using specialized indexes. To the best of our knowledge, this work is the first detailing FPGA support for variable-enhanced flexible pattern queries.

The remainder of this paper is organized as follows: related work is described in Section 2; the *FlexTrack* query language is detailed in Section 3; the proposed FPGA-based querying architecture is detailed in Section 4; the experimental evaluation is provided in Section 5; and the conclusions appear in Section 6.

2   Related Work

Single predicate queries (e.g., Range and *NN* queries) for trajectory data have been widely studied in the past (e.g., [2,20,28]). In order to make the query evaluation process more efficient [8], trajectories are first approximated using Minimum Bounding Regions (MBR) and then indexed using hierarchical spatiotemporal indexing structures, like the MVR-tree [27] and TPR-tree [29]. However, these solutions are only efficient to evaluate single predicate queries. For moving object data, patterns have been examined in the context of query language and modeling issues [5,14,24], as well as query evaluation algorithms [7,4,19].

The *FlexTrack* system [31,32], which our work is based on, provides a more general and powerful query framework than previous approaches. In *FlexTrack*, queries can contain both fixed and *variable* regions, as well as regular expression structures (e.g., repetitions, negations, optional structures) and explicit ordering of the predicates along the temporal dimension. This system uses a hierarchical region alphabet, where the user has the ability to define queries with finer alphabet granularity (*zoom in*) for the portions of greater interest, and higher granularity (*zoom out*) elsewhere. In order to efficiently evaluate flexible pattern queries, *FlexTrack* employs two lightweight index structures in the form of ordered lists in addition to the raw trajectory data. Given these index structures four different algorithms for evaluating flexible pattern queries are available, which are detailed in the next section.

The use of hardware platforms for pattern matching has been recently explored by many studies [26,13,12,33]. Most of these works focus on deep packet inspection and security as applications of interest. Using FPGAs, speedups of up to two orders of magnitude is achieved compared to CPU-based approaches, as every data element in stream can be processed in a single hardware cycle. The works in [17,15,18] present a novel dynamic programming, push down automata approach, using FPGAs and GPUs, for matching XML Path and Twig patterns in XML documents. Using the massively parallel solution running on parallel platforms, up to three orders of magnitude speedup was achieved versus state-of-the-art CPU bases approaches.

In [26] an NFA implementation of regular expressions on FPGAs is described. [13] proposes generating hardware code from Perl Compatible Regular Expressions. The work in [12] focuses on DFA implementations of regular expressions, while merging commonalities among multiple DFAs. [33] proposes the use of regular expressions for the representation of spatio-temporal queries. An FPGA implementation is detailed, allowing the sharing of query evaluation engines among several trajectories, with a minor impact on performance. In [3], it is investigated the use of GPUs for the fast computation of proximity area views over streams of spatio-temporal data. Our work mainly differs from all the above works from the perspective of the query language, described in Section 3. Specifically, we describe an investigation of the FPGA-based support of variable-enhanced patterns.

3   The *FlexTrack* System

We now provide a briefly description of the pattern query language syntax, as well as the key elements in the *FlexTrack* framework (for more details, see [31,32]).

3.1   Flexible Pattern Query Language

A trajectory T_{id} is defined as a list of locations collected for a specific moving object over an ordered sequence of timestamps, and is stored as a sequence of n pairs $\{(ls_1, ts_1), \ldots (ls_n, ts_n)\}$, where $ls_i \in \mathbb{R}^d$ is the object location recorded at timestamp ts_i $(ts_{i-1} < ts_i)$.

The *FlexTrack* uses a set of non-overlapping regions Σ_l that are derived from partitioning the spatial domain. Such regions correspond to areas of interest (e.g. *school districts, airports*) and form the alphabet language $\Sigma = \bigcup_l \Sigma_l = \{A, B, C, ...\}$. The *FlexTrack* query language defines a spatio-temporal predicate $\mathcal{P}$ by a triplet $\langle op, \mathcal{R}[,t]\rangle$, where $\mathcal{R}$ corresponds to a predefined spatial region in Σ or a *variable* in Γ ($\mathcal{R} \in \{\Sigma \cup \Gamma\}$), op describes the topological relationship (e.g. *meet, overlap, inside*) that the trajectory and the spatial region $\mathcal{R}$ must satisfy over the (optional) time interval $t := (t_{from} : t_{to}) \mid t_s \mid t_r$. A predefined spatial region is explicitly specified by the user in the query predicate (e.g. "the downtown area of Munich'). In contrast, a *variable* denotes an arbitrary region using the symbols in $\Gamma = \{@x, @y, @z, ...\}$. Conceptually, *variables* work as placeholders for explicit spatial regions and can be bound to a specific region during the query evaluation.

The *FlexTrack* language defines a pattern query $\mathcal{Q} = (\mathcal{S} \,[\cup\, \mathcal{D}])$ as a combination of a sequential pattern $\mathcal{S}$ and an optional set of constraints $\mathcal{D}$. A trajectory matches $\mathcal{Q}$ if it satisfies both $\mathcal{S}$ and $\mathcal{D}$ parts. The $\mathcal{D}$ part of $\mathcal{Q}$ allows us to describe general constraints. For instance, constrains can be distance-based constraints among the *variables* in $\mathcal{S}$ and the predefined regions in Σ. And $\mathcal{S} := \mathcal{S}.\mathcal{S} \mid \mathcal{P} \mid !\mathcal{P} \mid \mathcal{P}^{\#} \mid ?^{+} \mid ?^{*}$ corresponds to a sequence of spatio-temporal predicates, while $\mathcal{D}$ represents a collection of constraints that may contain regions defined in $\mathcal{S}$. The wild-card ? is also considered a variable, however it refers to any region in Σ, and not necessarily the same region if it occurs multiple times within a pattern $\mathcal{S}$.

The use of the same set of *variables* in describing both the topological predicates and the numerical conditions provides a very powerful language to query trajectories. To describe a query in *FlexTrack*, the user can use fixed regions for the parts of the trajectory where the behavior should satisfy known (strict) requirements, and *variables* for those sections where the exact behavior is not known but can be described by *variables* and the constraints between them.

In addition to the query language defined previously, we introduce the variable region set constraint defined in $\mathcal{D}$. A region set constraint (e.g., $\{@x : A, D, E\}$) is optional per variable, and can be only applied to variable predicates, having the purpose of limiting the region values that a given variable can take in Σ.

Consider the following query pattern and region set over $@x$, $\mathcal{Q} = (\mathcal{S} = \{A.B.@x.C.?^{+}.@x\}$, $\mathcal{D} = \{@x : A, D, E\})$. Here, $@x$ is constrained by the regions $\{A, D, E\}$. In practice, a variable can be limited to the neighboring regions of the fixed query predicates. Other constraints can be set by the user, hence, limiting the number of matches of interest. From a performance perspective, the use of variable region set constraints greatly simplifies hardware support for variable predicates separated by wildcards $?^{+}$ or $?^{*}$, as detailed in Section 4.

3.2   Flexible Pattern Query Evaluation

The *FlexTrack* system employs two lightweight index structures in the form of ordered lists that are stored in addition to the raw trajectory data. There is one *region-list* (*R-list*) per region in Σ, and one *trajectory-list* (*T-list*) per trajectory in the database. The *R-list* $\mathcal{L}_\mathcal{I}$ of a given region $\mathcal{I} \in \Sigma$ acts as an inverted index that contains all trajectories that passed by region $\mathcal{I}$. Each entry in $\mathcal{L}_\mathcal{I}$ contains a trajectory identifier T_{id}, the time interval (*ts-entry*:*ts-exit*] during which the trajectory was inside $\mathcal{I}$, and a pointer to the *T-list* of T_{id}. Entries in a *R-list* are ordered first by T_{id}, and then by *ts-entry*.

In order to fast prune trajectories that do not satisfy pattern $\mathcal{S}$ the *T-list* is used. For each trajectory T_{id} in the database, the *T-list* is its approximation represented by the regions it visited in the partitioning space Σ. Each entry in the *T-list* of T_{id} contains the region and the time interval (*ts-entry*:*ts-exit*] during which this region was visited by T_{id}, ordered by *ts-entry*. In addition, entries in *T-list* maintain pointers to the *ts-entry* part in the original trajectory data. With the above described index structures, there are four different strategies for evaluating flexible pattern queries:

1. *Index Join Pattern* (*IJP*): This method is based on a merge join operation performed over the *R-lists* for every fixed predicate in $\mathcal{S}$. The *IJP* uses the *R-lists* for pruning and the *T-lists* for the *variable* binding. This method is the one chosen as comparison to our proposed solution, since it usually achieves better performance for a wide range of different types of queries;

2. *Dynamic Programming Pattern* (*DPP*): This method performs a subsequence matching between every predicate in $\mathcal{S}$ (including *variables*) and the trajectory approximations stored as the *T-lists*. The *DPP* uses mainly the *T-lists* for the subsequence matching and performs an intersection-based filtering with the *R-lists* to find candidate trajectories based on the fixed predicates in $\mathcal{S}$;

3. *Extended-KMP* (*E-KMP*): This method is similar to *DPP*, but uses the Knuth-Morris-Pratt algorithm [11] to find subsequence matches between the trajectory representations and the query pattern;

4. *Extended-NFA* (*E-NFA*): This is an NFA-based approach to deal with all predicates of our proposed language. This method also performs an intersection-based pruning on the *R-lists* to fast prune trajectories that do not satisfy the fixed spatial predicates in $\mathcal{S}$.

4   Proposed Hardware Solution

4.1   Compiling Queries to Hardware

In this work, pattern queries are evaluated in hardware on an FPGA device. As trajectories are compared against hundreds, and potentially thousands, of pattern queries, manually developing custom hardware code becomes an extremely tedious (and error prone) task. Unlike software querying platforms, where a single (or set of) generic kernel can be used for the evaluation of any query pattern, hardware is at an advantage when each query pattern is mapped to a customized circuit. Customized circuitry has the benefits of only utilizing the needed resources out of all (limited) on-chip resources. Furthermore, the throughput of

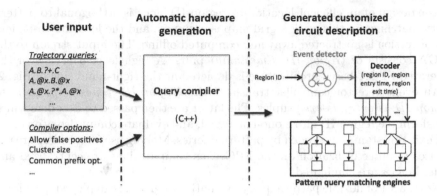

Fig. 2. Query-to-hardware tool flow

the query evaluation engines is limited by the operational frequency (hardware clock) which can in-turn be optimized to maximize performance.

For this purpose, a software tool written in C++ was developed from scratch (more than 6,500 lines of code), taking as input a set of user-specified pattern queries Q, and automatically generating a customized Hardware Description Language (HDL) circuit description (see Fig. 2). A set of compiler options can be specified, such as the degree of matching accuracy (reducing/eliminating false positives), and whether to make use of certain resource utilization (common prefix) and performance (clustering) optimizations.

Utilizing a query compiler provides the flexibility of software (ease of expression of queries from a user perspective), and the performance of hardware platforms (higher throughput), while no compromises are introduced.

4.2   High Level Architecture Overview

As depicted in Fig. 2, assuming an input stream of pairs $\langle location, timestamp\rangle$, the first step consists of translating the location onto semantic data; specifically, the region-IDs are of interest, using which the query patterns are expressed. The computational complexity of translating locations to regions depends on the nature of the map, and are discussed below:

1. **Regions defined by a grid map:** in this case, simple arithmetic operations are performed on the locations. These can be performed at wire speed (no stalling) on an FPGA;
2. **Polygon-shaped regions:** in this case, there are several well-defined point-in-polygon algorithms and their respective hardware implementations available (e.g., see [6,9,10,25]). However, none of these can operate at wire speed when the number of polygons is large. Here, the locations of vertices are stored off-chip in carefully designed data structures. The latter are traversed to locate the minimal set of polygons against which to test the presence of the locations.

As the design of an efficient location-to-region-ID block is orthogonal to pattern query matching, in this work a grid map is assumed, and the location-to-region-ID conversion is abstracted away and computed offline. The input stream to the FPGA consists of ⟨region-ID, timestamp⟩ pairs. A high level overview of the generated FPGA-based architecture is depicted at the right-hand side of Fig. 2.

An event detector controller translates the ⟨region-ID, timestamp⟩ pairs to ⟨region-ID, ts-entry, ts-exit⟩ tuples. The latter are then passed to decoders which transform the region-ID into a one-hot signal, and evaluate comparisons on entry and exit timestamps as needed by pattern queries. Making use of decoders greatly reduces resource utilization on the FPGA, as computations are centralized and redundancies are eliminated.

Next, a set of flexible pattern query evaluation engines are deployed, providing performance benefits through the following two parallelization opportunities:

1. **Inter-pattern parallelism:** where the evaluation of all pattern queries is achieved in parallel. This parallelism is available due to the embarrassingly parallel nature of the pattern matching problem;
2. **Intra-pattern parallelism:** where the match states of all nodes within a pattern are evaluated in parallel.

The throughput of pattern query matching engines is limited to one event per cycle. Given the current assumed streaming mechanism, events are less frequent than region-IDs.

Lastly, once a trajectory is done being streamed into the FPGA, the match state of each pattern query is stored in a separate buffer. This in turn allows the match states to be streamed out of the FPGA from the buffer as a new trajectory is queried (streamed in), hence, exploiting one more parallelism opportunity.

A description of the hardware query matching engines follows. While the discussion focuses on predicate evaluation, timing constraints are evaluated in a similar manner in the region-ID decoder, and are hence left-out of the discussion for brevity.

4.3   Evaluating Patterns with No Variables

We now describe the case of pattern queries with no variables. This approach is borrowed from the NFA-based regular expression evaluation as proposed in [13,26]. Figure 3(a) depicts the matching engine respective to the pattern query $A.B.?^*.A$, and Fig. 3(b) details the matching steps of that query given a stream of region-ID events. Each query node is implemented as:

1. A one-bit buffer (implemented using a flip-flop, depicted in grey in Fig. 3(b)), indicating whether the pattern has matched up to this node. All nodes are updated simultaneously, upon each region-ID event detected at the input stream;
2. Logic preceding this buffer, to update the match state (buffer contents).

As each buffer indicates whether the pattern has matched up to that predicate, a query node can be in a matched state if, and only if:

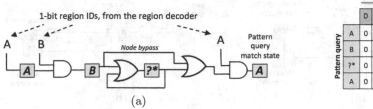
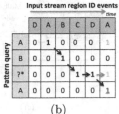

(a)   (b)

Fig. 3. (a) Query matching engines respective to the pattern query $A.B.?^*.A$, and (b) an event-by-event overview of the matching of the query.

1. All previous (non-wildstars ?*) predicates up to itself have matched. Wildstars are an exception since they can be skipped by definition (zero or more). To perform this check, it suffices to check the match state of the first previous non-wildstar node (see the node bypass in Fig. 3(a));
2. The current event (as noted by the region-ID decoder) relates to the region of that respective node. Wildcards are an exception, since by definition, they are not tied to a region-ID. Centralizing the comparisons and making use of a decoder helps considerably reducing the FPGA resource utilization respective to this inter-node logic (see the AND-gates in Fig. 3(a)). This is in contrast to reading the multi-bit encoded region-ID and performing a comparison locally;
3. It is a wildstar/wildplus ($?^*/?^+$), and it was in a match state at some point earlier. Wildstar and wildplus are **aggregation** nodes that, once matched, will hold that match state (see the OR-gate prior to the ?* node in Fig. 3(a)).

Looking closer at Fig. 3(b), each cell reflects the match state of a query node. All cells in a column are updated in parallel upon an event at the input stream. A '1' in a cell indicates that the query has matched up to that node; for a query to be marked as matched, a '1' should propagate from the first node (top row) to the last node (bottom row). As wildstar (and wildplus) nodes act as aggregators, they hold a *matched* state once activated; hence, a '1' can propagate "horizontally" only at wildstar (and wildplus) nodes. Grey cell contents indicate *matched* states that did not contribute to the detected matched query state in red color, but could contribute to later matches. The '1' depicted in red color in Fig. 3(b) indicates that the query was detected in the input stream.

4.4 Evaluating Patterns with Variables and without Wildstar/Wildplus Predicates

Supporting variables in pattern query matching requires an added level of memory saving. The basic rule of variables is that *all instances of a given variable need to match the same region-ID for a variable to be in a match state*. When no aggregator nodes $?^+/?^*$ are used, the distance between these two region-IDs occurring is the number of nodes between the variable instances in the query.

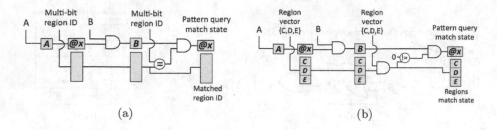

(a)   (b)

Fig. 4. Query matching engines respective to the pattern query $A.@x.B.@x$, (a) without and (b) with a region set constraint $\{C, D, E\}$ on $@x$. To handle variables in hardware, the first instance of a given variable in a query forwards, alongside the incoming match state, (a) the event detector's output encoded (multi-bit) region-ID, and (b) a one-hot signal consisting of bits respective to each region in the set of the variable. Every later instance of that variable in the query (here, the last query node) would match the event detector's ((a) encoded, and (b) multiple decoded) region-ID to the forwarded region-ID. If these match, then the region-ID is again forwarded, and the variable instance indicates a *matched* state.

One possible way for software systems to handle this would be to store, at each variable node (*in a matched state*), all the region-IDs encountered throughout the stream. A post-processing step would carefully intersect, for each variable, all stored region-IDs vectors. While that is a valid approach, storing region-IDs for each variable node of each pattern query is problematic as streams are longer. Furthermore, this is not needed unless aggregator nodes ?$^+$/?* occur in between variable occurrences; these cases are detailed in Sections 4.5 and 4.6. As FPGAs allow the deploying of custom matching engines for each pattern, matching pattern queries at streaming (no-stall) mode can be achieved here, with no post processing.

To handle variables in hardware, the first instance of a given variable in a pattern query forwards the event detector's output encoded (multi-bit) region-ID alongside the incoming match state (see the second node in Fig. 4(a)). Some cycles later (depending on the location of variable instances in the pattern), every instance of that variable in the query would match the event detector's region-ID to the forwarded region-ID. If these match, then the region-ID is again forwarded, and the variable instance indicates a *matched* state. Stated in other terms, at a variable node (instance) in a query, a match state is indicated if the current region was encountered earlier (given a fixed implied distance), and all match state propagation checks in between were valid (implying the distance).

Note that an encoded region-ID is used since it is smaller in bit size than a decoded ID, and any region can potentially satisfy the pattern query variable (i.e., variables are essentially a subset of wildcards). Also note that non-variable predicates buffer the forwarded region-ID, though no manipulation of the latter is required. Additionally, one set of region-ID buffers is required per variable, starting from the first occurrence of that variable.

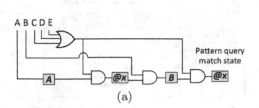

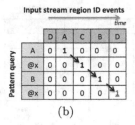

Input stream region ID events

	D	A	C	B	D
A	0	1	0	0	0
@x	0	0	1	0	0
B	0	0	0	1	0
@x	0	0	0	0	1

(a)   (b)

Fig. 5. (a) Query matching engine respective to the pattern query $A.@x.B.@x$ $\{@x : C, D, E\}$, such that the variable region set constraint is implemented as a "relaxed" OR. This relaxation helps save considerable hardware resources (compare to Fig. 4(b)). (b) An event-by-event overview of the matching of the query resulting in a false positive, due to the OR-based implementation of the variable region set constraint.

The same solution is applicable to pattern queries containing variables with region sets. Figure 4(b) shows the matching logic for the pattern $A.@x.B.@x$ where $@x$ is constrained by the regions $\{C, D, E\}$. Here, instead of storing the encoded region-ID in the variable buffers, the latter would hold, for each region in the set, a single bit. At the first occurrence of a variable, the buffer holds a one-hot vector, because input stream events are relative to one region only. Upon later instances of that variable, AND-ing the incoming region set buffer with specific bits of the region-ID decoder output will help indicating for which regions (if any) the pattern matches.

The above approach is similar to replicating the matching engine for each region in the variable region set constraint. For instance, the query in Fig. 4(b) can be seen as three queries, namely $A.C.B.C$, $A.D.B.D$ and $A.E.B.E$. However, the above approach offers much better scalability when multiple variables are used per pattern: replicating the pattern for each combination of variable regions would result in an exponential increase in resource utilization versus employing the aforementioned style of propagating buffers. Another advantage of the propagating region set variable buffers, when dealing with wildstar/wildplus pattern predicates, is described in the following.

We now describe an alternative "relaxed" implementation of the variable region set constraint, with the goal of saving considerable hardware resources, though at the expense of introducing false positives. Instead of keeping a propagating buffer holding information on each region in the set, the match state can be updated if *any* of the regions in the set are decoded using a simple OR-gate. Figure 5(a) depicts the gate-level implementation of the query $A.@x.B.@x$ $\{@x : C, D, E\}$, such that the variable region set constraint is implemented as an OR. Thus, history keeping is minimized, as no exact region information is kept per variable. While this mechanism introduces false positives (as described in Fig. 5(b)), the latter can be tolerable depending on the application. Otherwise, a post-processing software step can be performed only on the patterns marked as matched by the FPGA hardware. This approach, however, helps fitting substantially more query engines on the FPGA, a benefit accentuated as the number of variables and the variable region sets' size increase.

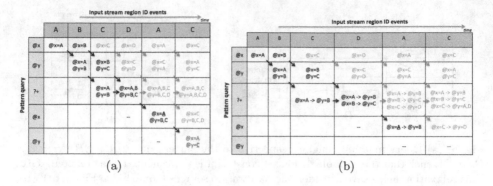

Fig. 6. Event-by-event matching of the pattern query $@x.@y.?^{+}.@x.@y$ $\{@x :$ $A, B, C, D\}$ $\{@y : A, B, C, D\}$. The resulting match in (a) is a false positive; whereas enough state is saved in (b) at the aggregator node $(?^{+})$ to eliminate that false positive.

4.5 Evaluating Patterns with a Single Variable and with Wildstar/Wildplus Predicates

The remainder of this discussion is applicable to both wildplus and wildstar query nodes. As detailed earlier (Fig. 3(a)), wildplus nodes act as **aggregator** nodes. When no variables are used, the only propagating information across nodes is a single bit value. In that case, a simple OR gate would suffice for aggregation (state saving).

When a wildplus predicate is located in between two instances of a variable, all values of the region-ID buffer should be stored, and forwarded to the next stages (nodes). Keeping that history is required in order to not result in false negatives. However, due to performance and resource utilization constraints, storing all that history is not desired. Using variable region set constraints, this limitation can be overcome by simply OR-ing the propagating buffer similarly to the match state buffer. This approach would store the information needed, and no history is lost. No false positives are generated, thus pattern evaluation is achieved at streaming mode.

4.6 Evaluating Patterns with Multiple Variables and with Wildstar/Wildplus Predicates

When more than one variable predicate is used in a pattern query, and with wildplus nodes in between instances of both these variables, the previous mechanism can lead to false positve matches, as even more state should be saved than discussed earlier. Figure 6(a) shows an event-by-event example of a pattern matching resulting in a false positive match. Each cell in the grid holds the values stored inside each respective variable buffer. Buffers for the variable $@x$ are used at each pattern node, whereas buffers for the variable $@y$ span from the second pattern node (i.e. the first $@y$ node), up to the last pattern node.

As described earlier, the wildplus node is the only node in the pattern query allowing horizontal propagation of *matched* states. This is due to the nature of wildplus nodes which hold a *matched* state. As the variable buffers are OR-ed at that wildplus node, they will store the information of the union of all variable buffers encountered at that node. Looking at the $?^+$ row in Figure 6(a), notice that the variable buffers for both @x and @y hold an increasing number of regions. That level of stored information is not sufficient, as it will be shortly shown to result in a false positive.

Upon the D event, both variable buffers did not propagate to the second instance of @x. That is because the @x variable buffer does not reflect that the previous instance of @x held the value of D (yet). However, on the next event A, the variable buffers propagated, and the @x variable buffer was masked with the event region. Hence, B was removed from the @x variable buffer. The @y variable buffer remains unmodified, since the @x node is not allowed to modify it.

Finally, at the last event C, focusing at the second instance of @y (i.e. the last pattern predicate), a match is shown for @$x=A$ and @$y=C$. While @x and @y did hold these values at some point, looking closer at the input stream, A and C were initially separated by B, though the query requires that the distance between @x and @y is 1 (back-to-back regions visited).

In order to not result in false positives, the level of history kept at the aggregator node has to be increased. Instead of only storing the union of all variable buffers, the information at the wildplus node should be the set of all variable buffers encountered. To reduce storage, that solution can be simplified such that, for each @x variable value, a list of all corresponding @y values are stored (as shown in Fig. 6(b)). Focusing on the aggregator row, every value of @x is associated with a list of @y values. These can be deduced from the propagating variable buffers into the wildplus node. Note that @$x=A$ is associated with @$y=B$. Therefore, the tuple @$x=A$, @$y=C$ cannot result in a match, as is the case in Fig. 6(a).

Nonetheless, implementing this solution in hardware is extremely costly in terms of resource utilization (and impact on the critical path/performance), especially with larger region sets and many variables per pattern. Furthermore, this solution does not scale with many variables, and does not hold with more aggregator nodes.

Another approach to eliminate false positives in such cases is a brute-force implementation of each query using all variable region-set combinations. For instance, the query $S = $ @$x.$@$y.?^+.$@$x.$@y {@$x : A, B$}{@$y : C, D$} can be implemented as four simpler queries, namely:

1. $S_1 = A.C.?^+.A.C$
2. $S_2 = A.D.?^+.A.D$
3. $S_3 = B.C.?^+.B.C$
4. $S_4 = B.D.?^+.B.D$

This approach is encouraging when the number of variables and the size of the region sets is relatively small. Otherwise, the implied resource utilization increases too much, even though each query is built using simple matching engines (no propagating variable buffers). Nonetheless, the common prefix (among similar pattern queries) optimization helps with the scalability.

In order to better evaluate the benefits of each of the above approach, a study on the resulting false positives versus resource utilization is performed in Section 5. In summary, when pattern queries make use of two or more variables, and with an aggregator node in between the occurrences of these variables, the proposed approaches are:

1. **Making use of propagating variable buffers:** this approach results in the least false positives;
2. **Implementing region set constraints as an OR:** the number of false positives here is a superset of the above case, and resource utilization is minimal. False positives are a superset, since the condition (OR check) to allow a match to propagate through a variable node is a superset of the first approach's variable node conditions (propagating buffers);
3. **A brute-force mapping approach:** this approach map each query as the combination of all variable region-sets. It has no false positives, but does not scale well with more variables and larger region sets.

5   Experimental Evaluation

We now present an extensive experimental evaluation of the proposed hardware architecture. We first describe the datasets used in the experiments, followed by the experimental setup. We then detail a thorough design space exploration on the proposed architecture, alongside a study on matching accuracy. Finally, we show the performance evaluation between the proposed architecture solutions with the CPU-based software approach.

5.1   Dataset Description

In our experimental evaluation, we use four real trajectory datasets. The first two datasets are the *Trucks* and *Buses* from [1]. Both datasets represent moving objects in the metropolitan area of Athens, Greece. The *Trucks* dataset has 276 trajectories of 50 trucks where the longest trajectory timestamp is 13,540 time units. The *Buses* dataset has 145 trajectories of school buses with maximum timestamp 992. The third dataset, *CabsSF*, consists of GPS coordinates of 483 taxi cabs operating in the San Francisco area [22] collected over a period of almost a month. The fourth dataset, *GeoLife*, contains GPS trajectory data generated from people that participated in the GeoLife project [34] during a period of over three years. This dataset has 17,621 trajectories with a total distance of about 1.2 million kilometers and duration of more than 48,000 hours.

5.2   Experiments Setup

For simplicity of the experimental evaluation, we partition the spatial domain in uniform grid sizes. These grid cells become the alphabet for our pattern queries.

In order to generate relevant pattern queries for each dataset, we randomly sample and fragment the original trajectories using a custom trajectory query generator. The length and location of each fragment are randomly chosen. These fragments are then concatenated to create a pattern query. We generate up to 2,048 pattern queries with different number of predicates, variables, and wildcards. The location of each variable and wildcard inside the query are randomly chosen.

Our FPGA platform consists of a Pico M-501 board connected to an Intel Xeon processor via 8 lanes of PCI-e Gen. 2 [21]. We make use of one Xilinx Virtex 6 FPGA LX240T, a low to mid-size FPGA relative to modern standards. The PCIc hardware interface and software drivers are provided as part of the Pico framework. The hardware engines communicate with the input and output PCIe interfaces through one stream each way, with dual-clock BRAM FIFOs in between our logic and the interfaces. Hence, the clock of the filtering engine is independent of the global clock. The PCIe interfaces incur an overhead of ≈8% of available FPGA resources.

The RAM on the FPGA board is not residing in the same virtual address space of the CPU RAM. Data is streamed from the CPU RAM to the FPGA. Since the proposed solution does not require memory offloading, RAM on the FPGA board is not used. Xilinx ISE 14 is used for synthesis and place-and-route. Default settings are set.

5.3   Design Space Exploration

Here we discuss the resource utilization and achievable performance (throughput) of the hardware engines. Figure 7(a) shows the resource utilization, and Fig. 7(b) shows the respective frequencies of the hardware engines, such that the number of queries (varying from 32, 64, 128, ... up to 2,048 queries), the query length (4 and 8 predicates), and number of variables in a pattern query (0 and 1 variable, in this last case a variable with a region set of 5 regions is assumed).

As the query compiler applies the common prefix optimization, and further resource sharing techniques are exercised by the synthesis/place-and-route tools, resource utilization does not double as the number of queries is doubled. Rather, a penalty of approximately 70% occurs.

Similarly, as the query length is doubled, an average increase of 80% in resources is found. However, adding one variable to each query results in, on average, doubling resource utilization. Note that the propagating buffer approach is employed for variable matching, and that these buffers propagate from the first occurrence of the variable to the last.

Overall, up to several thousands of query matching engines can fit on the target Xilinx V6LX240T FPGA, a mid- to low-size FPGA. While these numbers address the scalability of the proposed matching engines, Fig. 7(b) details the respective achievable performance in terms of:

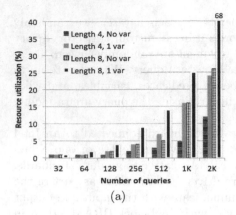
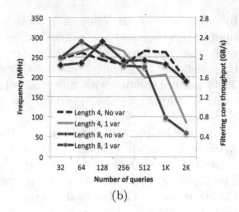

(a)   (b)

Fig. 7. (a) Resource utilization and (b) respective frequencies/throughput of the hardware engines, such that the number of queries is doubled, the query length is doubled, and variable predicate is present or not in the pattern query

1. **Operational frequency (MHz)**: measured as a function of the critical path, i.e., the longest wire connection of the FPGA circuit. This number is obtained post the place-and-route process of the FPGA tools;

2. **Throughput (GB/s)**: as the query matching engines process one ⟨region-ID, timestamp⟩ pair per hardware cycle, the FPGA throughput can be deduced from the circuit's operational frequency, given that the size of each input pair is 8 Bytes (2 integers). Nonetheless, this computed throughput is respective to the FPGA circuitry, and might not reflect the end-to-end (CPU-FPGA and back) performance, which is platform dependent. The end-to-end measurements are discussed in the sequence.

As the number of queries increases, frequency/throughput is initially around the 250MHz/2GBs mark. Fluctuations are due to the heuristic-based nature of the FPGA tools, though generally a trend is deduced. As the number of queries becomes too large, frequency drops considerably for queries with variables. The drop is not as steep for queries with no variables; the reason being that queries with variables can be thought of as longer queries (due to the propagating buffers). This drop in frequency occurs because of the large fan-out from the *region-ID* decoder to the many sinks, being the query nodes and propagating buffers.

Replicating the *region-ID* decoder (and event detector) helps reducing fan-out, and will potentially eliminate it. Each *region-ID* decoder is then connected to a set of queries. We refer to a *region-ID* decoder and its connected queries as a **cluster**. Note that each query belongs to exactly one cluster. The query compiler is developed to take as input parameter the cluster size, as a function of query nodes. Thorough experimentation shows that clusters need not hold less than 1,024 or even 512 query nodes (data omitted due to lack of space). Larger clusters result in performance deterioration; smaller clusters do not offer

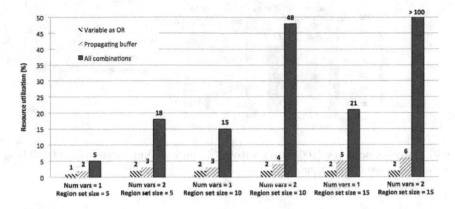

Fig. 8. Scalability of the each of the following three implementations of 100 queries of length 6 holding variables: **variable as OR**, **propagating buffer**, and **all combinations**

any benefits in performance, rather present an increase in resource utilization (due to the replication of the *region-ID* decoder and event detector per cluster).

5.4   Query Engine Implementations and False Positives

As described in previous sections, a query holding variables can be evaluated in one of three ways, namely:

1. **Variable as OR:** implementing the region set constraints as ORs (resulting in most false positives);
2. **Propagating buffer:** making use of propagating buffers (false positives arise only when using multiple variables alongside wildstar/wildplus nodes);
3. **All combinations:** brute-force mapping of each query as the combination of all variable region sets (no false positives).

Figure 8 illustrates the resource utilization of 100 queries of length 6 holding variofheaforementionedthreeapproaches.Thevaried
factors are the number of variables in each pattern query, and the respective region set size.

When implementing a *variable* as *OR*, each variable node is replaced with a simpler OR node. Thus, as expected (see Fig. 8), increasing the number of variables has almost no effect on resource utilization. The same applies to increasing the region set size. On the other hand, the *propagating buffer* technique starts off as utilizing slightly less than double the resources of the *variable* as the *OR* approach. Furthermore, doubling the region set size results in a 50% area penalty. Doubling the number of variables per pattern query exhibits similar behavior.

Finally, when transforming a query into a set of queries based on *all combinations* of the region sets, resource utilization starts off as more than double that of the *propagating buffer* technique. Doubling the number of variables naturally

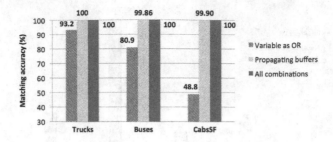

Fig. 9. Matching accuracy (100-false positives %) for each implementation of 100 long queries, over three datasets, namely *Trucks*, *Buses* and *CabsSF*

has a steeper effect than doubling the region set size on resource utilization. Note that the common prefix optimization helps with the scalability of this approach. Nonetheless, when using two variables with region set size of 15, the resulting circuitry did not fit on the FPGA. Practically, it is best to make use of this approach for critical pattern queries where false positives are not tolerated.

We now evaluate the number of false positive matches for each of the three query engine implementations previously discussed. In this experiment, as shown in Fig. 9, the matching accuracy (100-false positives %) is recorded for each implementation of 100 long queries, over three datasets, namely *Trucks*, *Buses* and *CabsSF* (the results for the *GeoLife* dataset follow the same pattern). Queries are generated using our query generator tool, where each query contains two variables, as well as one or more aggregator (?*/?+) nodes. Note that the *Propagating buffers* approach does not result in any false positives, unless multiple variables are used alongside aggregators.

As expected by its design, the *All combinations* approach results in no false positives. However, while the *Variable as OR* technique results in the most false positives (as expected), the matching accuracy varies from high (93.2%), to somewhat low (48.8%). On the other hand, matching accuracy is close to perfect (> 99.8%) for the *Propagating buffers* implementation, even as false positives increase as a result of the *Variable as OR* implementation. No false positives are recorded on the *Trucks* dataset when making use of the *propagating buffers*.

While the mileage of the *Variable as OR* implementation may vary, its scalability is key. Even when false positives are not tolerable, query matching engines can employ this technique, where the FPGA would be used as a pre-processing step with the goal of reducing the query set. The same applies for the *propagating buffers* implementation technique, where the query set would be reduced the most. Since the performance of CPU-based software approaches scales linearly with the number of pattern queries, reducing the query set has desirable advantages, especially that the time required for this pre-processing FPGA step is negligible.

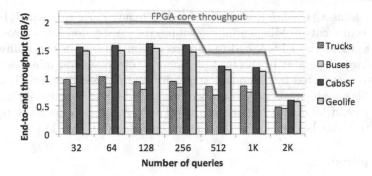

Fig. 10. End-to-end (CPU-RAM to FPGA and back) throughput of queries of length 4 with 1 variable. The throughput of the FPGA filtering core is drawn in red line.

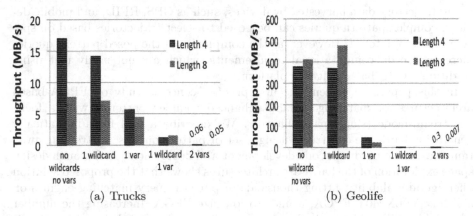

(a) Trucks   (b) Geolife

Fig. 11. FlexTrack (software) *IJP* throughput (MB/s) resulting from matching for 2,048 queries with varying properties on the (a) *Trucks* and (b) *GeoLife* datasets. Increasing query complexity (adding variables/wildcards) greatly decreases throughput.

5.5 Performance Evaluation

In the last set of experiments, we compare the performance evaluation between our proposed architecture solutions and the CPU-based software approach. Figure 10 shows the end-to-end (CPU-RAM to FPGA and back) throughput of length 4 queries with 1 variable. Throughput is lower from the FPGA filtering core for smaller trajectory files since steady state is not reached, and communication setup penalty is not hidden. For larger files, throughput is closer to the FPGA core's, given the physical limitations. Note that the throughput of the FPGA setup is independent of the trajectory file contents, as well as query structure (given a certain operational circuit frequency).

Figure 11 depicts the FlexTrack (software) *IJP* throughput (MB/s) resulting from matching for 2,048 queries with varying properties on the Fig. 11(a) *Trucks* and Fig. 11(b) *GeoLife* datasets. Pre-processing (index building) time is excluded. When considering simple queries, throughput is initially higher for the larger dataset (*GeoLife*), where processing steady-state is reached. Increasing query complexity (adding variables/wildcards) greatly decreases throughput. Note that where the FPGA end-to-end execution time is in the milliseconds range, software operates in the tens of seconds (up to several minutes) range, and is greatly affected by the query structure and dataset contents; hence the considerable speedup (over three orders of magnitude) and benefits of the FPGA setup. It should be noted that the proposed FPGA solution does not result in false positive matches for any of the queries considered in Fig. 11.

6   Conclusions

The wide and increasing availability of collected data in the form of trajectory has lead to research advances in behavioral aspects of the monitored subjects. Using trajectory data harvested by devices, such as GPS, RFID, and mobile devices, complex pattern queries can be posed to select trajectories based on specific events of interest. However, as the complexity of the posed pattern queries increases, so do computational requirements, which are not easily met using traditional CPU-based software platforms.

In this paper, we present the first proof-of-concept study on FPGA-based architectures for matching variable-enhanced complex patterns, with a focus on stream-mode (single pass) filtering. We describe a tool for automatically generating hardware constructs using a set of pattern queries, abstracting away ramifications of hardware code development and deployment. A thorough design space exploration of the hardware architectures shows that the proposed solution offers good scalability, fitting thousands of pattern query matching engines on a Xilinx V6LX240T FPGA, a mid- to low-size FPGA. Increasing the number of variables and wildcards is shown to have linear effect on the resulting circuit size, and negligible on performance. This behavior does not happen in CPU-based solutions, since performance is greatly affected from such pattern query characteristics.

When handling pattern queries with (a) no variables, (b) one variable, or (c) no wildcards with two or more variables, the proposed hardware architecture is able to process the trajectory data in a single pass. When two or more variables occur in a pattern query alongside wildcards, the proposed solution may have the drawback of resulting in false positive matches (though these are minimal in practice). Nonetheless, a no-false-positive solution is proposed, though being limited in scalability.

As part of our future research, we are working on enhancing the proposed framework to allow online pattern query updates. In this way, the deployed generic pattern query engines will support *any* pattern query structure and node values. A stream of bits forwarded to the FPGA will program the connections between deployed pattern query nodes. It should be noticed that this approach is different to the Dynamic Partial Reconfiguration (DPR), where the bit configuration of the FPGA itself is updated.

Acknowledgments. This work was partially supported by NSF grants IIS-1144158 and IIS-1161997.

References

1. Chorochronos (2013), http://www.chorochronos.org
2. Aggarwal, C., Agrawal, D.: On nearest neighbor indexing of nonlinear trajectories. In: Proc. ACM Symp. on Principles of Database Systems (PODS), pp. 252–259 (2003)
3. Cazalas, J., Guha, R.: GEDS: GPU Execution of Continuous Queries on Spatio-Temporal Data Streams. In: IEEE/IFIP Int'l Conf. on Embedded and Ubiquitous Computing (EUC), pp. 112–119 (2010)
4. du Mouza, C., Rigaux, P., Scholl, M.: Efficient evaluation of parameterized pattern queries. In: Proc. ACM Int'l Conf. on Information and Knowledge Management (CIKM), pp. 728–735 (2005)
5. Erwig, M., Schneider, M.: Spatio-Temporal Predicates. IEEE Trans. Knowl. Data Eng. 14(4), 881–901 (2002)
6. Fender, J., Rose, J.: A High-Speed Ray Tracing Engine Built on a Field-Programmable System. In: Proc. IEEE Int'l Conf. on Field-Programmable Technology (FPT), pp. 188–195 (2003)
7. Hadjieleftheriou, M., Kollios, G., Bakalov, P., Tsotras, V.J.: Complex Spatio-temporal Pattern Queries. In: Proc. Intl. Conf. on Very Large Data Bases (VLDB), pp. 877–888 (2005)
8. Hadjieleftheriou, M., Kollios, G., Tsotras, V.J., Gunopulos, D.: Indexing Spatiotemporal Archives. VLDB J. 15(2), 143–164 (2006)
9. Heckbert, P.S.: Graphics Gems IV, vol. 4. Morgan Kaufmann (1994)
10. Kim, S.-S., Nam, S.-W., Lee, I.-H.: Fast Ray-Triangle Intersection Computation Using Reconfigurable Hardware. In: Computer Vision/Computer Graphics Collaboration Techniques, pp. 70–81 (2007)
11. Knuth, D., Morris, J., Pratt, V.: Fast Pattern Matching in Strings. SIAM J. Comput. 6(2), 323–350 (1977)
12. Kumar, S., Dharmapurikar, S., Yu, F., Crowley, P., Turner, J.: Algorithms to Accelerate Multiple Regular Expressions Matching for Deep Packet Inspection. In: ACM SIGCOMM Conf. on Applications, Technologies, Architectures, and Protocols for Computer Communications, pp. 339–350 (2006)
13. Mitra, A., Najjar, W., Bhuyan, L.: Compiling PCRE to FPGA for Accelerating SNORT IDS. In: ACM/IEEE Symp. on Architecture for Networking and Communications Systems (ANCS), pp. 127–136 (2007)
14. Mokhtar, H., Su, J., Ibarra, O.: On Moving Object Queries. In: Proc. ACM Symp. on Principles of Database Systems (PODS), pp. 188–198 (2002)
15. Moussalli, R., Halstead, R., Salloum, M., Najjar, W., Tsotras, V.J.: Efficient XML Path Filtering Using GPUs. In: Workshop on Accelerating Data Management Systems, ADMS (2011)
16. Moussalli, R., Najjar, W., Luo, X., Khan, A.: A High Throughput No-Stall Golomb-Rice Hardware Decoder. In: IEEE Annual Int'l Symp. on Field-Programmable Custom Computing Machines, FCCM (2013)
17. Moussalli, R., Salloum, M., Najjar, W., Tsotras, V.: Accelerating XML Query Matching through Custom Stack Generation on FPGAs. In: Patt, Y.N., Foglia, P., Duesterwald, E., Faraboschi, P., Martorell, X. (eds.) HiPEAC 2010. LNCS, vol. 5952, pp. 141–155. Springer, Heidelberg (2010)

18. Moussalli, R., Salloum, M., Najjar, W., Tsotras, V.J.: Massively Parallel XML Twig Filtering Using Dynamic Programming on FPGAs. In: Proc. IEEE Int'l Conf. on Data Engineering (ICDE) (2011)

19. Mouza, C., Rigaux, P.: Mobility Patterns. Geoinformatica 9(4), 297–319 (2005)

20. Pfoser, D., Jensen, C., Theodoridis, Y.: Novel Approaches in Query Processing for Moving Object Trajectories. In: Proc. Intl. Conf. on Very Large Data Bases (VLDB), pp. 395–406 (2000)

21. Pico Computing M-Series Modules (2012), http://picocomputing.com/m-series/m-501

22. Piorkowski, M., Sarafijanovoc-Djukic, N., Grossglauser, M.: A Parsimonious Model of Mobile Partitioned Networks with Clustering. In: Int'l Communication Systems and Networks and Workshops (2009)

23. Sadoghi, M., Labrecque, M., Singh, H., Shum, W., Jacobsen, H.-A.: Efficient Event Processing Through Reconfigurable Hardware for Algorithmic Trading. Proc. VLDB Endow. 3(1-2), 1525–1528 (2010)

24. Attia Sakr, M., Güting, R.H.: Spatiotemporal Pattern Queries in SECONDO. In: Mamoulis, N., Seidl, T., Pedersen, T.B., Torp, K., Assent, I. (eds.) SSTD 2009. LNCS, vol. 5644, pp. 422–426. Springer, Heidelberg (2009)

25. Schmittler, J., Woop, S., Wagner, D., Paul, W.J., Slusallek, P.: Realtime Ray Tracing of Dynamic Scenes on an FPGA Chip. In: Proc. ACM Conf. on Graphics Hardware (HWWS), pp. 95–106 (2004)

26. Sidhu, R., Prasanna, V.K.: Fast Regular Expression Matching Using FPGAs. In: Proc. the Annual IEEE Symp. on Field-Programmable Custom Computing Machines (FCCM), pp. 227–238 (2001)

27. Tao, Y., Papadias, D.: MV3R-Tree: A Spatio-Temporal Access Method for Timestamp and Interval Queries. In: Proc. Intl. Conf. on Very Large Data Bases (VLDB), pp. 431–440 (2001)

28. Tao, Y., Papadias, D., Shen, Q.: Continuous Nearest Neighbor Search. In: Proc. Intl. Conf. on Very Large Data Bases (VLDB), pp. 287–298 (2002)

29. Tao, Y., Papadias, D., Sun, J.: The TPR*-Tree: An Optimized Spatio-Temporal Access Method for Predictive Queries. In: Proc. Intl. Conf. on Very Large Data Bases (VLDB), pp. 790–801 (2003)

30. Teubner, J., Müller, R., Alonso, G.: FPGA Acceleration for the Frequent Item Problem. In: Proc. IEEE Int'l Conf. on Data Engineering (ICDE), pp. 669–680 (2010)

31. Vieira, M.R., Bakalov, P., Tsotras, V.J.: Querying Trajectories Using Flexible Patterns. In: Proc. Int. Conf. on Extending Database Technology (EDBT), pp. 406–417 (2010)

32. Vieira, M.R., Bakalov, P., Tsotras, V.J.: FlexTrack: a System for Querying Flexible Patterns in Trajectory Databases. In: Proc. Int'l Symp. on Advances in Spatial and Temporal Databases (SSTD), pp. 475–480 (2011)

33. Woods, L., Teubner, J., Alonso, G.: Complex Event Detection at Wire Speed with FPGAs. Proc. VLDB Endow. 3(1-2), 660–669 (2010)

34. Zheng, Y., Xie, X., Ma, W.-Y.: GeoLife: A Collaborative Social Networking Service Among User, Location and Trajectory. IEEE Data Engineering Bulletin 33(2), 32–40 (2010)

Best Upgrade Plans for Large Road Networks

Yimin Lin and Kyriakos Mouratidis

School of Information Systems
Singapore Management University
80 Stamford Road, Singapore 178902
{yimin.lin.2007,kyriakos}@smu.edu.sg

Abstract. In this paper, we consider a new problem in the context of road network databases, named *Resource Constrained Best Upgrade Plan* computation (BUP, for short). Consider a transportation network (weighted graph) G where a subset of the edges are *upgradable*, i.e., for each such edge there is a cost, which if spent, the weight of the edge can be reduced to a specific new value. Given a source and a destination in G, and a budget (resource constraint) B, the BUP problem is to identify which upgradable edges should be upgraded so that the shortest path distance between source and destination (in the updated network) is minimized, without exceeding the available budget for the upgrade. In addition to transportation networks, the BUP query arises in other domains too, such as telecommunications. We propose a framework for BUP processing and evaluate it with experiments on large, real road networks.

1 Introduction

Graph processing finds application in a multitude of domains. Problems in transportations, telecommunications, bioinformatics and social networks are often modeled by graphs. A large body of research considers queries related to reachability, shortest path computation, path matching, etc. One of the less studied topics, which however is of large practical significance, is the distribution of available resources in a graph in order to achieve certain objectives. Here we consider road networks in particular, and the objective is to minimize the traveling time (shortest path distance) from a source to a destination by amending the weights of selected edges.

As an example, consider the transportation authority of a city, where a new hospital (or an important facility of another type) is opened, and the authority wishes to upgrade the road network to ease access to this facility from another key location (e.g., from the airport). While several road segments (network edges) may be amenable to physical upgrade, this comes at a monetary cost. The *Resource Constrained Best Upgrade Plan* problem (BUP) is to select some among the upgradable edges so that the traveling time between source and destination is minimized and at the same time the summed upgrade cost does not exceed a specific budget (resource constraint).

Another, more time-critical application example is an intelligent transportation system that monitors the traffic in the road network of a city, and schedules accordingly the traffic lights in road junctions in real-time. Assume that a major event is taking place in the city and heavy traffic is expected from a specific source (e.g., a sports stadium) to

M.A. Nascimento et al. (Eds.): SSTD 2013, LNCS 8098, pp. 223–240, 2013.

a specific destination (e.g., the marina). With appropriate traffic light reconfiguration, the driving time across some edges in the network can be reduced, at the cost of longer waits for walkers at affected pedestrian crosses. Assuming that along with each upgradable edge there is a cost associated to capture the burden imposed to pedestrians, a BUP could indicate which road segments to favor in the traffic light schedule so that (i) the traveling time from the stadium to the marina is minimized and (ii) the summed cost against pedestrian priority does not exceed a certain value.

Although we focus on transportation networks, BUP finds application in other domains too. Consider for example a communication network, where on-demand dynamic allocation of bandwidth and QoS parameters (e.g., latency) is possible for some links between nodes (routers). In the usual case of leased network infrastructure in the Internet Protocol (IP) layer, upgrading a link in terms of capacity or QoS access parameters would incur a monetary cost. When a large volume of time-sensitive traffic (e.g., VoIP) is expected between two nodes, BUP would indicate to the network operators which links to upgrade in order to minimize the network latency between the two nodes, subject to the available monetary budget.

Figure 1 shows an example of BUP query in a road network. The edges drawn as dashed lines are upgradable. Each upgradable edge is associated with a triplet of numbers (e.g., $\langle 9|10|16 \rangle$), indicating respectively the new weight (if the edge is chosen for upgrade), the cost for the upgrade, and the original weight of the edge. For normal, non-upgradable edges, the number associated with them indicates their (unchangeable) weight; weights are only illustrated for edges that affect our example (all the rest are assumed to have a weight of 15).

The input of the query is a source node s and a destination node t in the network, plus a resource constraint B. Let U be the set of upgradable edges. The objective in

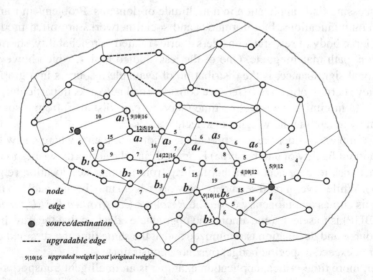

Fig. 1. Example of BUP query

BUP is to select a subset of edges from U which, if upgraded, will lead to the minimum possible shortest path distance from s to t, while the sum of their upgrade costs does not exceed B. Assuming a resource constraint $B = 20$, the output of BUP in our example includes edge (b_4, b_5) and leads to a shortest path distance of 58 via path $\{s, b_1, b_2, b_3, b_4, b_5, b_6, t\}$. The resource consumption in this case is 10, i.e., smaller than B, and thus permissible. Note that if B were larger, an even shorter distance could be achieved (namely, 53) by upgrading edges (a_1, a_2), (a_2, a_3) and (a_6, a_t). This however would incur a cost of 24 that exceeds our budget B.

There are several bodies of research that are related to BUP, such as methods to construct from scratch or modify the topology of a network to serve a specific objective [1–3]. Currently, however, there is no work on BUP, while algorithms for related problems cannot be adapted to address it.

In this paper, we formalize BUP and propose a suite of algorithms for its efficient processing. As will become clear in Section 3, the main performance challenge in BUP is the intractability of the search space and the need for numerous shortest path computations. We develop techniques that limit both these factors with the aim of efficient processing. To demonstrate the practicality of our framework, we conduct an extensive empirical evaluation using large, real road networks.

2   Related Work

There are several streams of work related to BUP query. In this section, we give a brief overview of these streams and indicate their differences from our problem.

2.1   Road Network Databases

There has been considerable research in the area of road network databases, including methods for network storage and querying (e.g., ranges and nearest neighbors) [4–6], the processing of queries that involve a notion of dominance based on proximity [7], continuous versions of proximity queries [8], etc. There have also been several studies on materialization with the purpose of accelerating shortest distance/path queries [9, 10]. All these techniques focus on data organization and querying mechanisms on a network that is used as-is, i.e., they do not consider the selective amendment of edge weights. The closest related piece in this area is [11], which considers shortest path computation over time-dependent networks, i.e., where the weight of each edge is given by a function of time. Again, there is no option to select edges for upgrade nor any control over edge weights.

2.2   Network Topology Modification

Another related body of work includes methods to modify the network topology in order to meet specific optimization objectives.

Algorithms on *network topology optimization* and *network design* compute/derive a topology for nodes and edges in a network (e.g., number of nodes, placement of nodes and edges, etc) to meet certain goals. Literature on this topic falls under wireless networks [12, 13], wired backbone networks [14–17] and service overlay networks [18–21]. These methods design the topology of the network, affecting its very structure. In BUP, instead, the topology is preserved and the question is which edges to upgrade within a specific budget (the budget not being a consideration among methods in this category).

In the *network hub allocation problem* the purpose is to locate hubs and allocate demand nodes to hubs in order to route the traffic between origin-destination pairs [22]. There are different lines of work: $p-$hub median techniques, $p-$hub center algorithms, hub covering, and hub allocation methods with fixed costs. These are essentially location-allocation problems, and far from the BUP setting.

2.3   Resource Allocation and Network Improvement

In this section, we review work which does not seek to modify the network topology, but is intended to allocate resources or select certain nodes/edges from the network to meet specific optimization objectives.

The *resource allocation problem in networks* is to efficiently distribute resources to users, such as bandwidth and energy, in order to achieve certain goals, like upholding QoS contracts. Most of the work in this field focuses on pricing and auction mechanisms [1, 23, 24]. Game theory is the main vehicle to address these problems, whose objectives differ from BUP.

Probably the closest related topic to BUP is the *network improvement problem*. The setting is similar to BUP, with an option to lower the weights of edges, but the objective is to reduce the diameter of the network, i.e., the maximum distance between any pair of nodes. In [2], the authors discuss the complexity of the problem with budget constraints. Budget constraints are also considered in [25], which proposes methods to minimize the diameter of a tree-structured network. [26] addresses the $q-$upgrading arc problem, where q edges are selected for upgrading to minimize the diameter of the graph. In [3] the problem is to identify the non-dominated paths in a space where each is represented by its upgrade cost and the overall improvement achieved in traveling time across a set of source-destination pairs if the path is upgraded. The problem definitions in this body of work are fundamentally different from BUP, and the proposed algorithms are inapplicable to it.

In a *resource constrained shortest path* problem, there are different types of resources required to cross each edge and the goal is to identify the shortest path that does not exceed the available budget in each resource type [27–30]. *Restricted shortest path* is another example where each edge is associated with a cost and a delay. The objective is to identify the path which incurs the minimum cost while the delay along the path does not exceed a specific time limit. Both exact and approximate methods have been proposed [31–33]. In both aforementioned problems, the choice is whether to pass through a certain edge or not, as opposed to choosing whether to upgrade it.

3   Problem Formalization

We first formalize the problem and identify the key challenges in its processing.

Let $G = \langle V, E \rangle$ be a road network (weighted graph), where V is the set of nodes (vertices) and E is the set of edges (arcs). Every edge $e = (v_i, v_j)$ in E is associated with a weight $e.w$ that models the traveling time from v_i to v_j via e. For simplicity, we assume an undirected network (but our methods can be easily extended to directed ones too). A subset U of the network's edges are *upgradable*. That is, every $e \in U$ is also associated with an upgrade cost $e.c$ (where $e.c > 0$), and a new weight value $e.w'$ (where $e.w' < e.w$), indicating that if $e.c$ is spent, the weight of e can drop to $e.w'$. In this graph G, the BUP problem is defined as follows.

Given a budget (resource constraint) B, a source node $s \in V$, and a destination node $t \in V$, the BUP problem is to select a subset R of the edges in U for upgrade so that (i) the summed upgrade cost in R does not exceed budget B and (ii) the shortest path distance between s and t in the updated network is minimized. Formally, the output R of BUP is the result of the following optimization problem:

$$\underset{R \subseteq U}{\operatorname{argmin}} \quad SP(s, t, R)$$

$$\text{subject to} \quad \sum_{e \in R} e.c \leq B$$

where $SP(s, t, R)$ is the (traveling time along the) new shortest path between s and t when edges in R are upgraded. If there are multiple subsets R that abide by the resource constraint and lead to the same smallest traveling time between s and t, BUP reports the one with the smallest total cost.

We now define some terms and establish conventions. Any subset $R \subseteq U$ is called a *plan*. The total cost of R is denoted by $C(R)$, i.e., $C(R) = \sum_{e \in R} e.c$. If $C(R)$ is no greater than B, we say that R is a *permissible* plan. For brevity, we call *length* of a path the total traveling time along it. For ease of presentation, we use $SP(s, t, R)$ to refer both to the shortest path (between s and t in the updated network) and to its length, depending on the context. Table 1 summarizes frequently used notation.

To provide an idea about the difficulty of the problem, and to identify directions to tackle it, we consider a straightforward BUP processing method. A naïve approach is to consider all possible subsets of U. For each subset R, we check whether it is permissible, and if so, we *evaluate* it. That is, we compute the shortest path $SP(s, t, R)$ when the edges in R (and only those) are upgraded. After considering all possible subsets, we report the permissible plan R that leads to the smallest $SP(s, t, R)$.

The number of all possible subsets of U is $2^{|U|}$ (technically, the set of all subsets of U is called *power-set*). The problem of the naïve approach is that it needs to examine an excessive number of alternative plans, and for each (permissible) of them, to perform a shortest path computation. To give an example, if $|U| = 20$, the number of possible plans is 1,048,576, which are too many to even enumerate, let alone to evaluate.

In the following, we center our efforts on a twofold objective, i.e., we aim to reduce (i) the number of evaluated plans and (ii) the cost to evaluate each of them. We refer to item (i) as the *search space* of the problem. Item (ii) is bound to the cost of shortest path search and is directly dependent on the size of the graph (which we aim to reduce). Towards this dual goal, we propose graph reduction techniques in Section 4 and elaborate algorithms in Section 5. Graph reduction techniques assist towards both our

Table 1. Notation

Symbol	Meaning
$G = \langle V, E \rangle$	Road network with node set V and edge set E
U	Set of upgradable edges ($U \subseteq E$)
$\lvert U \rvert$	The cardinality of U
B	Resource constraint (budget)
R	Edges chosen for upgrade ($R \subseteq U$)
$e.w$	Original weight of edge e
$e.c, e.w'$	Upgrade cost and upgraded weight of $e \in U$
s, t	Source and destination nodes
$C(R)$	Sum of upgrade costs (across all edges) in R
$SP(s, t, R)$	Shortest path (length) after upgrades in R
G_c	Concise graph, BUP-equivalent to G (Section 4)
R_{temp}	A permissible, heuristic BUP solution (Section 4)
$A(G_c)$	Augmented version of G_c (Section 5.1)
U_c	Set of upgradable edges in G_c (Section 5.2)
$length(p, R)$	Length of path p after upgrade R (Section 5.2)
R_{max}	Maximum improvement set (Section 5.3)
$I(R_{max})$	Total weight improvement in R_{max} (Section 5.3)

design goals, while our algorithms are targeted at search space reduction in particular (i.e., limiting the number of evaluated plans).

4   Graph-Size Reduction Techniques

In this section, we propose two orthogonal methods to reduce G into a concise graph G_c, which is *BUP-equivalent* to G, i.e., the BUP solution R on (the much smaller) G_c is guaranteed to be the solution of BUP on the original network G. The first method is graph shrinking by edge pruning. The second is a resource constraint preserver technique that abstracts the remaining part of G (after pruning) into a concise graph which is BUP-equivalent to the original G.

4.1   Graph Shrinking via Edge Pruning

Our intuition is that any edge (upgradable or not) that lies too far from s and t cannot affect BUP processing and, thus, is safe to *prune*, i.e., to remove from G. We start with two important lemmas.

Lemma 1. *Let R be the BUP result and $SP(s, t, R)$ be the achieved shortest path in the updated network. R includes only upgradable edges along the path $SP(s, t, R)$.*

Proof. The lemma is based on our problem definition, and specifically on the fact that among permissible plans that lead to the same (minimal) distance between s and t, BUP reports the lowest-cost one. We prove it by contradiction. Suppose that the BUP result R includes an upgradable edge e which is not along $SP(s, t, R)$. If we apply upgrade set $R - \{e\}$, the shortest path between s and t will pass through exactly the same edges as with R, and we will have achieved the same shortest path distance at a cost reduced by $e.c$, which contradicts the hypothesis that R is the BUP result.

Lemma 2. *Let R be the BUP solution in G. Any subgraph of G that includes all the edges along $SP(s,t,R)$ is BUP-equivalent to G.*

Proof. Consider the subgraph G_{sp} of G that comprises *only* the (upgradable and non-upgradable) edges along path $SP(s,t,R)$. A direct implication of Lemma 1 is that if we solved BUP on G_{sp}, we would derive the same result R. In turn, this means that any subgraph of G that is a supergraph of G_{sp} is BUP-equivalent to G.

Lemma 2 asserts that we can safely prune any edge, upgradable or not, that does not belong to $SP(s,t,R)$ (where R is the BUP solution). We show how edges can be pruned safely, without knowing R in advance.

Consider the *fully upgraded* network G, i.e., where all edges in U are upgraded. $SP(v_i, v_j, U)$ denotes the distance between a pair of nodes v_i, v_j in this graph. By definition, $SP(v_i, v_j, U)$ is the lower bound of the distance between v_i and v_j after any possible upgrade plan $R \subseteq U$.

Let T be the length of $SP(s,t,R)$, where R is the BUP solution, and assume that we *somehow* know T in advance. We will show that certain edges (be them upgradable or not) lie too far from s and t to belong to R, and are therefore safe to prune.

Lemma 3. *It is safe to prune every edge $e = (v_i, v_j)$ for which:*
(i) $SP(s, v_i, U) + w + SP(v_j, t, U) > T$ and
(ii) $SP(s, v_j, U) + w + SP(v_i, t, U) > T$
where w equals $e.w'$ if e is upgradable, or simply $e.w$ if e is not upgradable.

Proof. The shortest possible path between s and t that passes through edge e is the one that corresponds to a fully upgraded G. The length of that path is either $SP(s, v_i, U) + w + SP(v_j, t, U)$ or $SP(s, v_j, U) + w + SP(v_i, t, U)$, whichever is smaller. If that minimum value is greater than T, edge e could not possibly belong to $SP(s,t,R)$ (where R is the final BUP result) and, therefore, e can be safely pruned.

Value T is not known in advance. However, Lemma 3 can be applied if T is replaced by any number that is greater than or equal to T. The closer that number to T, the more effective the lemma. We use $SP(s,t,R_{temp})$ instead of T, where R_{temp} is a permissible (suboptimal) plan that leads to a sufficiently short distance from s to t.

Effective R_{temp} Selection: To derive R_{temp} quickly and effectively, we first compute the shortest path $SP(s,t,U)$ in the fully upgraded graph. Next, we form R_{temp} using only the upgradable edges included in $SP(s,t,U)$. If R_{temp} exceeds the resource constraint, we execute a knapsack algorithm [34] to derive the subset of R_{temp} that achieves the minimum sum of weights along path $SP(s,t,R_{temp})$ without violating B. Note that this is different from a BUP problem, because we essentially fix the path to $SP(s,t,R_{temp})$ and seek the permissible subset of its upgradable edges that minimizes the path length *along the specific path only*. The result of the knapsack algorithm is used as the R_{temp} for pruning.

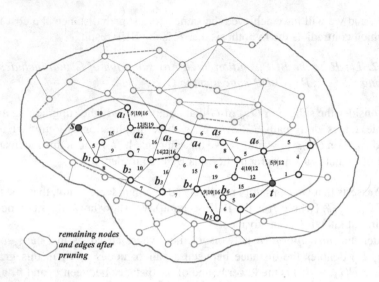

remaining nodes
and edges after
pruning

Fig. 2. Example of edge pruning

Figure 2 continues the running example of Figure 1. Assuming a weight of 15 units for every edge whose weight is not explicitly illustrated, and a $SP(s, t, R_{temp})$ distance of 60, Lemma 3 prunes every edge out of the inner (green-border) closed curve.

Implementation: To implement pruning, we perform two Dijkstra expansions [34] from s and t on the fully upgraded graph. Each expansion reaches up to distance $SP(s, t, R_{temp})$ from its start node (s and t, respectively). All edges that are not encountered or encountered by only one of the expansions, are pruned. Each of the remaining edges is checked against Lemma 3 and pruned (or not) accordingly.

4.2 Resource Constraint Preserver

In this section, we propose the *resource constraint preserver* technique, which transforms the remaining part of G (after pruning) into a concise graph G_c that is BUP-equivalent to G, i.e., a much smaller graph whose BUP solution (for the same s, t, B input) is guaranteed to be identical to the original road network. The concepts of *key nodes* and *plain paths* are central to this technique.

Definition 1. *Key node.* *A node* $v \in V$ *is a key node iff it is* s, t, *or end-node of an upgradable edge.*

Definition 2. *Plain path.* *A path is plain if it does not include any key nodes (except for its very first and very last nodes).*

We construct the network abstraction G_c as follows. First, we compute the *shortest plain path* for any pair of key nodes. The shortest plain path between key nodes v_i and v_j is the shortest among the plain paths that connect them. Computing this path can be done using any standard shortest path algorithm, by treating key nodes other than v_i and v_j as non-existent (thus preventing the reported path from including any intermediate key node). The second step to produce G_c is to replace each shortest plain path by a *virtual edge*, whose weight is equal to the length of the path. The edge set of G_c comprises only virtual and upgradable edges; the non-upgradable edges of the original graph are discarded. The node set of G_c includes only key nodes.

Lemma 4. G_c *is BUP-equivalent to the original network G.*

Proof. Let R be the BUP solution in G. Consider the sequence of key nodes in $SP(s,t, R)$ in order of appearance (from s to t). For every pair of consecutive key nodes v_i, v_j in this sequence, either v_i, v_j are the end-nodes of the same upgradable edge, or they are connected by a plain path. In the latter case, that plain path between v_i, v_j is also the shortest (by definition, every sub-path of a shortest path, is the shortest path between the intermediate nodes it connects). Thus, $SP(s, t, R)$ is a sequence of upgradable edges and shortest plain paths. Since G_c preserves the upgradable edges and includes all shortest plain paths between key nodes (in the form of equivalent virtual edges), it contains all edges comprising $SP(s, t, R)$. Hence, by Lemma 2, G_c is BUP-equivalent to G.

Further Shrinking: If the majority of edges are not upgradable, the preserver method will reduce the graph size. However, creating a fully connected graph among key nodes introduces many virtual edges, most of which unnecessary. To cure the problem, we apply Lemma 3 to each virtual edge before inclusion into G_c and prune it if the lemma permits.

Implementation: To accelerate the construction of G_c, we incorporate Lemma 3 into the computation of shortest plain paths. Specifically, for each key node v_i we perform a Dijkstra search (with source at v_i). When another key node v_j is encountered (i.e., popped by the Dijkstra heap), we add a virtual edge between v_i and v_j to G_c. However, we do not expand v_j (i.e., we do not push into the heap the adjacent nodes of v_j) so as to ensure plain paths. Let M be the smallest of $SP(s, v_i, U)$ and $SP(v_i, t, U)$ (both these values are known since the pruning stage). The Dijkstra search can safely terminate if it has reached up to distance $SP(s, t, R_{temp}) - M$ from v_i (any virtual edge longer than that threshold is useless according to Lemma 3), where R_{temp} is the heuristic (suboptimal) BUP solution from Section 4.1.

Figure 3 shows the G_c abstraction derived in our running example by the resource constraint preserver technique.

5 BUP Processing Algorithms

In this section, we present algorithms to compute the BUP solution on G_c, i.e., the graph resulting after the application of the edge pruning and resource preserver techniques from Section 4. G_c includes upgradable and virtual edges. For brevity, in the following we refer to virtual edges simply as edges. We denote by U_c the set of upgradable edges in G_c (since some edges in U have been pruned, $U_c \subseteq U$).

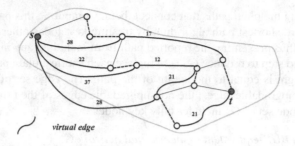

Fig. 3. The resulting graph G_c

Even with a smaller set of candidate edges for upgrade, the approach of evaluating arbitrary subsets of U_c is not only impractical, but not very meaningful either. That is, Lemma 1 suggests that the BUP solution R includes only upgradable edges along the shortest path from s to t (in the updated network). If candidate plans (subsets of U_c) were arbitrarily chosen for evaluation, in the majority of cases, their upgradable edges would fall at random and irrelevant locations, rather than on the shortest path from s to t. This observation motivates our processing methodology, which is path-centered.

Our approach is to iteratively compute alternative paths (from s to t) in increasing order of length, and evaluate them in this order. We distinguish three variants of this general approach, depending on which version of G_c is used for the incremental path exploration; it could be the original G_c, the fully upgraded G_c, or another version of G_c that we call *augmented*. Regardless of the underlying graph, we use the path ranking method of [35] to incrementally produce paths from s to t in increasing length order.

5.1 Augmented Graph Algorithm

Our first technique relies on the *augmented* version of G_c, denoted as $A(G_c)$, to which it also owes its name, i.e., *Augmented Graph* algorithm (AG). The augmented graph has the same node set as G_c, but its edge set is a superset of G_c. Specifically, every edge e in G_c becomes an edge in $A(G_c)$, retaining its original weight $e.w$ (be it upgradable or not). Additionally, for every $e \in U_c$, the augmented graph also includes a second edge e' between the same end-nodes as e, but with weight equal to $e.w'$ (i.e., the new weight if e is upgraded).

Figure 4 gives an example, showing the original G_c on the left and its augmented version $A(G_c)$ on the right. For the sake of the example, assume that non-upgradable edges have a unit weight. All edges of G_c appear in $A(G_c)$ with their original weights. Since edges e_1, e_2, e_3 are upgradable in G_c, graph $A(G_c)$ additionally includes e'_1, e'_2, e'_3 with the respective upgraded weights (shown next to the edge labels).

AG calls the path ranking algorithm of [35] in $A(G_c)$, and iteratively examines paths in increasing length order. In our example, assume that the budget is $B = 20$. The shortest path in $A(G_c)$ is $p_1 = \{s, d_1, d_2, f_1, t\}$ via e'_2 and e'_3 (both are upgraded links). The length of p_1 is 21. It passes via upgraded edges e'_2 and e'_3, thus requiring a total

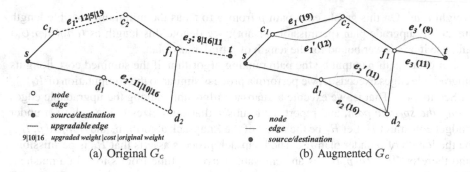

Fig. 4. Augmented graph example

cost of $e_2.c + e_3.c = 26$. That cost exceeds B and the path is ignored. The path ranking algorithm is probed again to produce the next best path, that is $p_2 = \{s, c_1, c_2, f_1, t\}$ via e'_1 and e'_3. Its length is 22, but its cost $e_1.c + e_3.c = 21$ exceeds B. Hence, this path is ignored too, and the path ranking algorithm is probed to produce the next best path, which is $p_3 = \{s, d_1, d_2, f_1, t\}$ via e'_2 and e_3 (upgraded and non-upgraded link, respectively). Its length is 24. The path passes via one upgraded edge, e'_2, which means that the total path cost is $e_2.c = 10$. That is within our budget, and AG terminates here with result $R = \{e_2\}$, achieving $SP(s, t, R) = 24$.

Observe that every path p output by path ranking in $A(G_c)$ corresponds to a specific upgrade plan, namely, to the plan that includes all upgraded links e' that p passes from. Consider a pair of paths p and p' in $A(G_c)$ that are identical, except that p passes via edge e, while p' passes from that edge's upgraded counterpart e'. For what path ranking in $A(G_c)$ is concerned, these are two different paths (since e and e' are modeled as different edges) corresponding to different upgrade plans.

Correctness: Path ranking, if probed enough times, will output *all possible paths* between s and t in $A(G_c)$. Hence, $SP(s, t, R)$ is guaranteed to be among them (where R is the BUP result), as long as AG does not terminate prematurely. AG stops probing the path ranking process when the latter outputs the first path p that abides by resource constraint B. Since path ranking outputs paths in increasing length order, p is guaranteed to be the shortest permissible path, i.e., to coincide with $SP(s, t, R)$.

Note that, in the worst case, path ranking will need to output all possible paths between s and t in $A(G_c)$. This is still preferable to evaluating all subsets of U_c, because AG essentially considers only combinations of upgradable edges *along acyclic paths* from s to t. For example, in Figure 4(b), path ranking would never output a path that includes both e'_1 and e'_2, while a blind evaluation of U_c subsets would consider (and waste computations for the evaluation of) plan $\{e'_1, e'_2\}$.

5.2 Fully Upgraded Graph Algorithm

In this section, we present the *Fully Upgraded Graph* algorithm (FG). FG runs path ranking on the fully upgraded G_c, where all edges $e \in U_c$ have their upgraded (reduced) weight $e.w'$. On this graph, each path p from s to t has the minimum possible length under any upgrade plan (permissible or not); we denote this length as $length(p, U_c)$ and use it as a lower bound for the length of p under any plan.

For every path p output by the path ranking algorithm, if the summed cost along its upgradable edges exceeds B, we perform a process similar to the computation of R_{temp} in Section 4.1. That is, we execute a knapsack algorithm among the upgradable edges *along the specific path*, and report their subset that minimizes the length of p under budget constraint B. Let R_p be the result of the knapsack algorithm, and $length(p, R_p)$ be the length of p under plan R_p. The knapsack process asserts that R_p is permissible, and therefore $length(p, R_p)$ is an achievable traveling time from s to t. In a nutshell, FG treats each path p output by path ranking as an umbrella construct representing all possible upgrade plans among the upgradable edges in p, and from these plans it keeps the best permissible plan, R_p.

While the path ranking algorithm iteratively reports new paths, we keep track of the one, say, p^* (and the respective R_{p^*} set) that achieves the smallest $length(p, R_p)$ among all paths considered so far.[1] Once path ranking reports a path p whose length (on the fully upgraded graph) is greater than $length(p^*, R_{p^*})$, FG terminates with R_{p^*} as the BUP result (achieving length $SP(s, t, R_{p^*}) = length(p^*, R_{p^*})$).

Referring to our example in Figure 4, the fully upgraded G_c would look like Figure 4(a) with weights 12, 11, and 8 for edges e_1, e_2, and e_3, respectively. Path ranking would first report path $p_1 = \{s, d_1, d_2, f_1, t\}$ with length 21. A knapsack algorithm on its set of upgradable edges (i.e., on set $\{e_2, e_3\}$) with resource constraint $B = 20$ reports $R_{p_1} = \{e_2\}$ and $length(p_1, R_{p_1}) = 24$. The next path output by path ranking is $p_2 = \{s, c_1, c_2, f_1, t\}$ with length 22. If that length were larger than $length(p_1, R_{p_1}) = 24$, FG would terminate. This is not the case, so a knapsack process on the upgradable edges along p_2 reports that $R_{p_2} = \{e_1\}$ with $length(p_2, R_{p_2}) = 25$. Path ranking is probed again, but reports NULL (i.e., all paths from s to t have been output) and FG terminates with result $R = R_{p_1} = \{e_2\}$.

Correctness: If probed enough times, path ranking in the fully upgraded G_c will report all possible paths from s to t on this graph. The paths are reported in increasing $length(p, U_c)$ order. Our termination condition guarantees that all paths not yet output by path ranking have $length(p, U_c)$ greater than $length(p^*, R_{p^*})$, and therefore could not lead to a shorter traveling time between s and t under any plan (permissible or not).

A note here is that every path output by path ranking in the augmented graph $A(G_c)$ (in Section 5.1) corresponds to an upgrade plan. In FG, instead, each path p output by path ranking in the fully upgraded G_c leads to the consideration of all possible upgrade plans along p (this is essentially what the knapsack-modeled derivation of R_p does).

5.3 Original Graph Algorithm

The *Original Graph* algorithm (OG) executes path ranking in the original G_c, i.e., assuming that no edge is upgraded. For every path p output by path ranking, it solves a knapsack problem to derive the subset R_p of the upgradable edges along p that achieves the minimum path length $length(p, R_p)$ without violating the resource constraint B.

---

[1] In case of tie between two alternative paths, we keep as p^* the one with the smallest $C(R_{p^*})$.

While new paths are being output by path ranking, OG maintains the path p^* (and the respective R_{p^*} set) that achieves the smallest $length(p, R_p)$ so far.

Regarding the termination condition of OG, we introduce the *maximum improvement set* R_{max}. Among all permissible subsets of U_c, R_{max} is the one that achieves the maximum total weight reduction (regardless of where the contained edges are located or whether they contribute to shorten the traveling time from s to t). We denote the total weight reduction achieved by R_{max} as $I(R_{max})$. The latter serves as an upper bound for the length reduction in *any* path under *any* permissible plan.

In the example of Figure 4(a), to derive R_{max} (and $I(R_{max})$), we solve a knapsack problem on $U_c = \{e_1, e_2, e_3\}$. Their individual weight reductions (i.e., values $e.w'$ − $e.w$) are 7, 5, 3 and their costs ($e.c$) are 5, 10, 16. The knapsack problem uses limit $B = 20$ for the total cost. The result is $R_{max} = \{e_1, e_2\}$ with total reduction $I(R_{max}) = 12$.

Returning to OG execution, let p be the next best path output by path ranking in the original (un-updated) G_c. Under any permissible upgrade plan, the length of p can be reduced at maximum by $I(R_{max})$, i.e., under any upgrade plan the new length of p cannot be lower than $length(p, \emptyset) - I(R_{max})$. If the latter value is greater than $length(p^*, R_{p^*})$, OG can safely terminate with BUP result $R = R_{p^*}$.

In the original G_c in Figure 4(a), edges e_1, e_2, and e_3 have weights 19, 16, and 11, respectively. Path ranking would first report path $p_1 = \{s, d_1, d_2, f_1, t\}$ with length 29. For p_1 we derive (via a knapsack execution on its upgradable edges) $R_{p_1} = \{e_2\}$ and $length(p_1, R_{p_1}) = 24$. The next path output by path ranking is $p_2 = \{s, c_1, c_2, f_1, t\}$ with length 32. Before we even solve a knapsack problem for p_2, we know that under any permissible plan its length cannot drop below $length(p_2, \emptyset) - I(R_{max}) = 32 - 12 = 20$. If that last value were greater than $length(p_1, R_{p_1}) = 24$, OG would terminate. This is not the case, so a knapsack process on p_2 reports $R_{p_2} = \{e_1\}$ with $length(p_2, R_{p_2}) = 25$. Path ranking is probed again, reports NULL (since all paths from s to t have been output), and OG terminates with BUP result $R = R_{p_1} = \{e_2\}$.

Correctness: The correctness of OG relies on similar principles to FG. First, path ranking, if probed enough times, will output all paths from s to t. For each of these paths p, OG computes the best permissible plan along its edges, i.e., R_p. Therefore, it will discover the optimal plan R at some point, unless terminated prematurely. Since path ranking in the original G_c outputs paths p in increasing $length(p, \emptyset)$ order, the termination condition of OG guarantees that all paths not yet output, even if improved to the maximum possible degree (i.e., $I(R_{max})$), cannot become shorter than p^*.

6   Experimental Evaluation

In this section, we first experimentally evaluate the effectiveness of our graph-size reduction techniques (from Section 4). Then proceed to compare the efficiency of our processing algorithms (from Section 5).

As default network G we use the road network of Germany, which has 28,867 nodes, 30,429 edges, and a diameter (i.e., maximum distance between any pair of nodes) of 14,383. The network is available at: www.maproom.psu.edu/dcw/. We study the impact of three parameters: (original) *path length* between source and destination; *upgrade ratio*; and *resource ratio*. The upgrade ratio indicates the ratio of upgradable edges over their total number (i.e., $|U|/|E|$). Upgradable edges are selected randomly from E. Their new weight is set to $e.w' = x \cdot e.w$, where x is a random number between 0.5 and 1. The upgrade cost is set to $e.c = y \cdot e.w$, where y is a random number from 0 to 1. The resource ratio indicates how strict the budget B is. Specifically, for each query we compute the sum of upgrade costs of all upgradable edges in the shortest path from s to t in the original network; let this sum be C. We set B to a fraction of this cost. The resource ratio equals B/C. Table 2 shows the parameter values tested and their default (in bold). In every experiment, we vary one parameter and set the other two to their default. Each measurement is the average over 20 queries. We use an Intel Core 2 Duo CPU 2.40GHz with 2GB RAM and keep the networks in memory.

Table 2. Experiment parameters

Parameter	Value Range
Path length	1000, 2000, **4000**, 6000
Upgrade ratio	0.04, 0.06, **0.08**, 0.1
Resource ratio	0.2, **0.4**, 0.6, 0.8

6.1   Evaluation of Graph-Size Reduction Methods

In this section, we leave aside BUP processing, and evaluate our graph-size reduction methods in three aspects: reduction of number of nodes, reduction of number of edges, and running time. We report results for pruning (from Section 4.1) when applied alone, and when applied in tandem with the preserver method (from Section 4.2).

Effect of Path Length: In Figure 5, we vary the path length and plot the number of remaining nodes/edges with each approach. We also present their running times; for each method ("Pruning" and "Pruning+Preserver") we include its full-fledged version (with all optimizations described in Section 4) and its version without the implementation optimization in the last paragraph of Section 4.1.

The original network has 28,867 nodes and 30,429 edges, out of which fewer than 500 nodes and 800 edges remain after pruning, achieving a vast reduction. The latter are

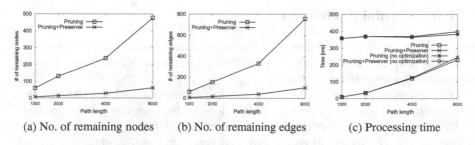

(a) No. of remaining nodes   (b) No. of remaining edges   (c) Processing time

Fig. 5. Effect of path length

further reduced by the preserver technique to fewer than 60 and 100, respectively, lowering down the problem size dramatically, even for the most distant source-destination pairs we tried. The number of remaining nodes/edges grows with the path length, because $SP(s, t, R_{temp})$ increases and, hence, Lemma 3 prunes fewer edges (recall that R_{temp} is a permissible, heuristic BUP solution, and $SP(s, t, R_{temp})$ is the length of the shortest path from s to t under plan R_{temp}). In terms of running time, both approaches take longer for larger path lengths, because the reduced graph is larger. The optimized versions of the algorithms are very efficient, requiring fewer than 250msec in all cases.

Effect of Upgrade Ratio: In Figure 6, we vary the upgrade ratio from 0.04 to 0.1, i.e., 4% to 10% of the network edges are upgradable. Lemma 3 is applied on the fully upgraded G, considering for each edge e its shortest possible distance from s and t, in order to guarantee correctness. Hence, a higher upgrade ratio implies looser pruning (equivalently, more remaining nodes and edges).

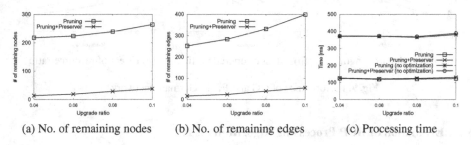

(a) No. of remaining nodes   (b) No. of remaining edges   (c) Processing time

Fig. 6. Effect of upgrade ratio

Effect of Resource Ratio: In Figure 7, we vary the resource ratio from 0.2 to 0.8 – that is, B ranges from 20% to 80% of C (described in the beginning of the experiment section). A greater ratio implies a larger budget B and, therefore, a smaller $SP(s, t, R_{temp})$. In turn, this means more extensive pruning by Lemma 3.

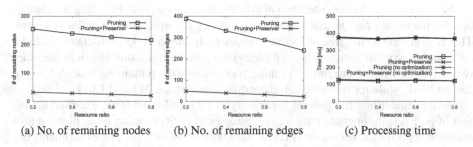

(a) No. of remaining nodes   (b) No. of remaining edges   (c) Processing time

Fig. 7. Effect of resource ratio

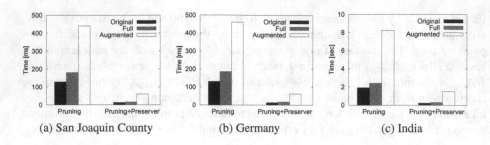

Fig. 8. Running time of BUP algorithms on different networks

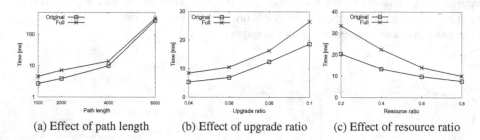

Fig. 9. Performance of OG and FG in Germany network

6.2 Evaluation of BUP Processing Algorithms

Given a reduced graph G_c (produced either by "Pruning" or "Pruning+Preserver"), in this section we evaluate the three BUP algorithms from Section 5. For understandability, in the plots we label AG as "Augmented", FG as "Full", and OG as "Original".

In Figure 8, in addition to Germany, we use two other real road networks; one smaller (San Joaquin County, with 18,263 nodes and 23,874 edges, from www.cs.utah.edu/~lifeifei/SpatialDataset.htm) and the other larger (India, with 149,566 nodes and 155,483 edges, from www.maproom.psu.edu/dcw/). For each road network, we present the processing time of all three BUP algorithms, assuming reduction by "Pruning" or "Pruning+Preserver", for the default parameter values in Table 2.

We observe that OG consistently outperforms alternatives, with FG being the runner-up. An interesting fact is that the running time of all algorithms in India is longer. This is irrelevant to the size of the network. A path length of 4,000 (default) in India corresponds to paths with much more edges than paths of the same length in the other two networks[2]. To see this, after "Pruning" in Germany the remaining nodes/edges are 238 and 331, while for India the corresponding numbers are 711 and 1,087.

Having established the general superiority of OG, in Figure 9 we examine the effect of path length, upgrade ratio and resource ratio on its running time, plotting also measurements for the runner-up (FG) for the sake of comparison. The experiments use our default network (Germany) after reduction by "Pruning+Preserver". We observe a

---

[2] We did not apply any normalization on edge weights across the three networks in order to retain their original distance semantics.

direct correlation between the running time of the BUP algorithms and the size of the reduced graph (investigated in Section 6.1) – for example, performance in Figure 9(a) follows the trends in Figures 5(a) and 5(b). This verifies that indeed the size of graph G_c is a major performance determinant and justifies our design effort in Section 4 to reduce it.

7   Conclusion

In this paper, we study the *Resource Constrained Best Upgrade Plan* query (BUP). In a road network where a fraction of the edges are upgradable at some cost, the BUP query computes the subset of these edges to be upgraded so that the shortest path distance for a source-destination pair is minimized and the total upgrade cost does not exceed a user-specified budget. Our methodology centers on two axes: the effective reduction of the network size and the efficient BUP processing in the resulting graph. Experiments on real road networks verify the effectiveness of our techniques and the efficiency of our framework overall. A direction for future work is the consideration of multiple concurrent constraints (on different resource types). Another is BUP processing when the optimization goal involves multiple source-destination pairs instead of a just one.

References

1. Jain, R., Walrand, J.: An efficient nash-implementation mechanism for network resource allocation. Automatica 46(8), 1276–1283 (2010)
2. Zhang, L.: Upgrading arc problem with budget constraint. In: 43rd Annual Southeast Regional Conference, vol. 1, pp. 150–152 (2005)
3. Nepal, K.P., Park, D., Choi, C.H.: Upgrading arc median shortest path problem for an urban transportation network. Journal of Transportation Engineering 135(10), 783–790 (2009)
4. Papadias, D., Zhang, J., Mamoulis, N., Tao, Y.: Query processing in spatial network databases. In: VLDB, pp. 802–813 (2003)
5. Shahabi, C., Kolahdouzan, M.R., Sharifzadeh, M.: A road network embedding technique for k-nearest neighbor search in moving object databases. GeoInformatica 7(3), 255–273 (2003)
6. Jensen, C.S., Kolárvr, J., Pedersen, T.B., Timko, I.: Nearest neighbor queries in road networks. In: GIS, pp. 1–8 (2003)
7. Deng, K., Zhou, X., Shen, H.T.: Multi-source skyline query processing in road networks. In: ICDE, pp. 796–805 (2007)
8. Stojanovic, D., Papadopoulos, A.N., Predic, B., Djordjevic-Kajan, S., Nanopoulos, A.: Continuous range monitoring of mobile objects in road networks. Data Knowl. Eng. 64(1), 77–100 (2008)
9. Kriegel, H.-P., Kröger, P., Renz, M., Schmidt, T.: Hierarchical graph embedding for efficient query processing in very large traffic networks. In: Ludäscher, B., Mamoulis, N. (eds.) SSDBM 2008. LNCS, vol. 5069, pp. 150–167. Springer, Heidelberg (2008)
10. Jung, S., Pramanik, S.: An efficient path computation model for hierarchically structured topographical road maps. IEEE Trans. Knowl. Data Eng. 14(5) (2002)
11. Ding, B., Yu, J.X., Qin, L.: Finding time-dependent shortest paths over large graphs. In: EDBT, pp. 205–216 (2008)
12. Hills, A.: Mentor: an algorithm for mesh network topological optimization and routing. IEEE Transactions on Communications 39(11), 98–107 (2001)

13. Amaldi, E., Capone, A., Cesana, M., Malucelli, F.: Optimization models for the radio planning of wireless mesh networks. In: Akyildiz, I.F., Sivakumar, R., Ekici, E., Oliveira, J.C., de McNair, J. (eds.) NETWORKING 2007. LNCS, vol. 4479, pp. 287–298. Springer, Heidelberg (2007)

14. Boorstyn, R., Frank, H.: Large-scale network topological optimization. IEEE Transactions on Communications 25(1), 29–47 (1977)

15. Ratnasamy, S., Handley, M., Karp, R.M., Shenker, S.: Topologically-aware overlay construction and server selection. In: INFOCOM (2002)

16. Kershenbaum, A., Kermani, P., Grover, G.A.: Mentor: an algorithm for mesh network topological optimization and routing. IEEE Transactions on Communications 39(4), 503–513 (1991)

17. Minoux, M.: Networks synthesis and optimum network design problems: Models, solution methods and applications. Networks 19, 313–360 (1989)

18. Li, Z., Mohapatra, P.: On investigating overlay service topologies. Computer Networks 51(1), 54–68 (2007)

19. Capone, A., Elias, J., Martignon, F.: Models and algorithms for the design of service overlay networks. IEEE Transactions on Network and Service Management 5(3), 143–156 (2008)

20. Fan, J., Ammar, M.H.: Dynamic topology configuration in service overlay networks: A study of reconfiguration policies. In: INFOCOM (2006)

21. Roy, S., Pucha, H., Zhang, Z., Hu, Y.C., Qiu, L.: Overlay node placement: Analysis, algorithms and impact on applications. In: ICDCS, p. 53 (2007)

22. Alumur, S.A., Kara, B.Y.: Network hub location problems: The state of the art. European Journal of Operational Research 190(1), 1–21 (2008)

23. Johari, R., Tsitsiklis, J.N.: Efficiency loss in a network resource allocation game. Math. Oper. Res. 29(3), 407–435 (2004)

24. Maillé, P., Tuffin, B.: Multi-bid auctions for bandwidth allocation in communication networks. In: INFOCOM (2004)

25. Ben-Moshe, B., Omri, E., Elkin, M.: Optimizing budget allocation in graphs. In: CCCG (2011)

26. Campbell, A.M., Lowe, T.J., Zhang, L.: Upgrading arcs to minimize the maximum travel time in a network. Networks 47(2), 72–80 (2006)

27. Handler, G.Y., Zang, I.: A dual algorithm for the constrained shortest path problem. Networks 10, 293–309 (1980)

28. Beasley, J.E., Christofides, N.: An algorithm for the resource constrained shortest path problem. Networks 19, 379–394 (1989)

29. Mehlhorn, K., Ziegelmann, M.: Resource constrained shortest paths. In: Paterson, M. (ed.) ESA 2000. LNCS, vol. 1879, pp. 326–337. Springer, Heidelberg (2000)

30. Ribeiro, C.C., Minoux, M.: A heuristic approach to hard constrained shortest path problems. Discrete Applied Mathematics 10(2), 125–137 (1985)

31. Hassin, R.: Approximation schemes for the restricted shortest path problem. Math. Oper. Res. 17(1), 36–42 (1992)

32. Lorenz, D.H., Raz, D.: A simple efficient approximation scheme for the restricted shortest path problem. Operations Research Letters 28(5), 213–219 (2001)

33. Dumitrescu, I., Boland, N.: Improved preprocessing, labeling and scaling algorithms for the weight-constrained shortest path problem. Networks 42, 135–153 (2003)

34. Cormen, T.H., Leiserson, C.E., Rivest, R.L., Stein, C.: Introduction to Algorithms, 2nd edn. The MIT Press and McGraw-Hill Book Company (2001)

35. Martins, E.Q.V., Pascoal, M.M.B.: A new implementation of yen's ranking loopless paths algorithm. 4OR 1(2) (2003)

Compact Representation of GPS Trajectories over Vectorial Road Networks

Ranit Gotsman and Yaron Kanza

Department of Computer Science,
Technion – Israel Institute of Technology
{ranitg,kanza}@cs.technion.ac.il

Abstract. Many devices nowadays record traveling routes, of users, as sequences of GPS locations. With the growing popularity of smartphones, millions of such routes are generated each day, and many routes have to be stored locally on the device or transmitted to a remote database. It is, thus, essential to encode the sequences, to decrease the volume of the stored or transmitted data. In this paper we study the problem of coding routes over a vectorial road network (map), where GPS locations can be associated with vertices or with road segments. We consider a three-step process of dilution, map-matching and coding. We present two methods to code routes. The first method represents the given route as a sequence of *greedy paths*. We provide two algorithms to generate a greedy-path code for a sequence of n vertices on the map. The first algorithm has $O(n)$ time complexity, and the second one has $O(n^2)$ time complexity, but it is optimal, meaning that it generates the shortest possible greedy-path code. Decoding a greedy-path code can be done in $O(n)$ time. The second method codes a route as a sequence of shortest paths. We provide a simple algorithm to generate a shortest-path code in $O(kn^2 \log n)$ time, where k is the length of the produced code, and we prove that this code is optimal. Decoding a shortest-path code also requires $O(kn^2 \log n)$ time. Our experimental evaluation shows that shortest-path codes are more compact than greedy-path codes, justifying the larger time complexity.

1 Introduction

Many devices, such as smartphones, contain a GPS receiver that allows users to record their locations, as they travel. Recording sequences of locations generates trajectories that can be used by various applications. Trajectories can be shared to recommend travel routes to users [1] or to find significant locations [2]. They can be used to determine similarity between users [3] or specify user behavior [4, 5]. They can be collected and analyzed to provide statistics about travels of individuals or of groups of people. Such statistics can be utilized by urban planners and policy makers in municipal, provincial and federal decision making.

An emerging problem is how to efficiently code these data sets in a world where millions of these trajectories are generated each day, and all have to be stored or transmitted for future processing in remote servers. Previous solutions

M.A. Nascimento et al. (Eds.): SSTD 2013, LNCS 8098, pp. 241–258, 2013.

were based on sampling and dilution [6–9]. In this paper, we consider the representation of trajectories over a road network, and we present a comprehensive approach that uses the topology of the road network to provide a compact representation of the traveled route—a representation that is much more compact than the mere result of the dilution.

We present in this paper a three-step process that starts by applying dilution of the trajectory using the standard Douglas-Peucker polyline-simplification algorithm [10]. Then, we apply map-matching to provide a route over the road network. Once a route is generated based on the GPS trajectory, it may be represented as a path in a planar graph, namely, a sequence of vertices in the graph. A compact representation of the route is computed using the topology and the geometry of the network. The proposed approach allows applying the dilution prior to the map matching, e.g., in cases where the dilution is conducted in a mobile device that does not hold a map of the area.

Our main contribution is two novel ways to compactly represent a path in a planar graph, and efficient algorithms to compute these compact representations. In both methods, we represent the path as a subsequence of vertices such that this path can be uniquely reconstructed from the vetrrtices by computing for each pair of consecutive vertices a well-defined path and concatenating these paths. For example, given a path, we seek to decompose it into the smallest possible sequence of shortest paths. Then, given the subsequence of vertices and the graph, the route may be recovered by generating a shortest path between every two consecutive vertices in the code.

In Section 2 we define the problem and provide an overview of the approach. The dilution phase is described in Section 3. The map-matching step is presented in Section 4. Computing compact codes for the paths produced by the map matching is presented in Section 5. Experimental evaluation over real data is provided in Section 6. In Section 7, we conclude and discuss future work.

2   Framework

A *vectorial road network* is a representation of a road map as a directed planer graph $G = (V, E)$ comprising a set V of vertices and a set E of edges, with a geometry X. The edges of the graph represent road segments and the vertices represent junctions. Each vertex v of G is associated with its real-world location, denote by $X(v)$. In this paper we consider recordings of travel routes over a vectorial road networks.

Devices with an embedded GPS allow recording user locations. Based on recorded locations, travel routes of users can be represented as sequences of points (locations). Each sequence has the form $(x_1, \ldots, x_n)$ where for each $i < j$, point x_i is a location that was visited and recorded prior to point x_j. We refer to such a sequence as a *trajectory*.

Trajectories are raw sequences of locations. Over a road network, our aim is to represent each sequence as a path on the graph. A path in G is a sequence of vertices $(v_1, \ldots, v_m)$ of V such that each two consecutive vertices are connected by an edge. To represent a sequence of points as a sequence of vertices, we first need to map the points of the sequence to the graph, namely, apply *map matching*. This produces the actual travel path on G. Then, we can compute a compact representation of the path. In this paper, we consider a *compact representation* of a path $P = v_1, \ldots, v_m$ to be a subsequence $C = (v_{i_1}, \ldots, v_{i_k})$ of P such that there is a known method to restore P from C.

Problem Definition: Given a trajectory as a sequence of points, the goal is to provide a compact representation, as short as possible, of the path of G that matches the given trajectory.

Our general approach is to apply the following three steps, for a given sequence of n location points. *(1)* Dilute the sequence, to remove unnecessary redundant points. *(2)* Apply map matching to associate the remaining points to vertices (junctions) of the road network. *(3)* Compute a compact representation of the sequence of vertices. In the following sections we describe these steps.

3   Trajectory Dilution

The first step of our method is to *dilute* (or *simplify*) the trajectory by removing redundant points. A redundant point is a point that is "almost" on the line connecting the points before and after it, as it does not add much new information about the location of the user. Since our map-matching step is not very sensitive to differences in the density of the GPS trajectory versus the density of vertices of the network, dilution does not reduce the accuracy of the matching.

Given a trajectory of points $X = (x_1, x_2, \ldots, x_n)$, removal of redundant points can be done using the Douglas-Peucker (DP) polyline-simplification algorithm [10] which has $O(n^2)$ time complexity. The DP algorithm is controlled by a single parameter—the distance a point is allowed to deviate from a straight line. The algorithm discards most of the points and marks just those to be kept. The algorithm proceeds recursively as follows: Initially it starts with the pair of indices $(1, n)$, representing the sequence of all the points $x_1, x_2, \ldots, x_n$ of the trajectory. It automatically marks the indices 1 and n to be kept. It then finds the index i of the point x_i that is furthest from the line segment between x_1 and x_n. If the point is closer than ε to that line segment, then all points with indices $2, \ldots, n-1$ may be discarded without the diluted trajectory being further than ε from the line segment, and the recursion terminates. If the point is further than ε, then index i is marked to be kept. The algorithm then calls itself twice recursively, first with the pair $(1, i)$ and then with the pair (i, n). When the procedure is complete, the generated trajectory consists of all (and only) those points whose indices have been marked to be kept.

Simplifying a trajectory can typically reduce the number of points significantly, say from 1,000 in an extremely dense trajectory to a mere 30 points while preserving the geometric integrity of the trajectory. A slightly better reduction can be achieved by taking into account the heading of the travel and the distances between adjacent points, as shown in [8]. The DP simplification

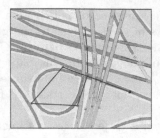
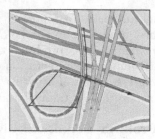

Fig. 1. Noisy GPS readings (green) and the associated polyline (blue). Black arrows show road matching options for the GPS points

Fig. 2. Each of the (green) GPS trajectory points is "snapped" to the closest map edge (orange points), leading to an incorrect map-match

Fig. 3. The orange points connected by the red polyline are the corresponding map-matched route computed by our algorithm

algorithm also helps in removing redundant trajectory points which accumulate while a vehicle stops in a traffic jam or at a traffic light. These points contain no additional information and just introduce noise because of GPS inaccuracy.

4   Map Matching

The second step after dilution is applying map matching. Map-matching has been studied for more than a decade, and the algorithms have evolved from very simple to quite sophisticated. Many papers studied this topic and it is not the focus of this paper, thus we do not present all the previous work in this area. Yet, so that the paper will be self contained, we present the map matching method we used, which is an adaptation of existing methods to handle well diluted trajectories. For a review of existing algorithms, we refer the reader to the comprehensive surveys of White, Bernstein, & Kornhauser [11], Quddus, Ochieng, Zhao, & Noland [12] and Quddus, Ochieng, & Noland [13].

4.1   Map Matching and HMM

Many recent map-matching algorithms are based on a Hidden-Markov Model (HMM) probabilistic approach [14]. Treating a GPS trajectory of edges $T = (t_1, t_2, \ldots, t_n)$ as a sequence of empirical observations (i.e. measurements), they attempt to compute the most likely sequence of map edges traversed given that sequence of observations.

A key principle in the HMM approach is that the algorithm must work simultaneously on the two inputs: the map and the GPS trajectory, hence operates in a state space consisting of states which are pairs of entities, one from the map and one from the GPS trajectory. Thus solving the HMM involves building a trellis, which is a replication of the map n times (one per each GPS trajectory point). Each replica is a layer of the trellis, containing all map edges and represents a trajectory edge. Thus, in this layered trellis graph, each trellis node represents a pair: an edge from the GPS trajectory and an edge from the map, and each trellis edge represents a connection between two map edges relevant to that edge of the trajectory. A trellis node (t_i, e_j) is connected to a trellis node (t_{i+1}, e_k) if and only if the two map edges e_j and e_k are relevant (i.e. sufficiently close) to the GPS trajectory edges t_i and t_{i+1} and connected one to the other. Note that trellis edges exist only between two adjacent layers of the trellis. Each trellis node (t_i, e_j) has an emission probability that estimates the correlation between the GPS measurement t_i and the edge e_j based on (Euclidean) distance between them. The trellis edge connecting node (t_i, e_j) to node (t_{i+1}, e_k) has a transition probability that estimates the distance between the two map edges e_j and e_k. In essence, the original HMM algorithm [14] proceeds monotonically along the temporal axis described by T, namely, along the horizontal dimension of the trellis, essentially traversing the map edges while traversing the trajectory, following the shortest weighted path through the trellis. The weight of a path is derived from the emission and transition probabilities of the vertices and edges along that path. The fact that there are no edges within layers allows efficient computation of this shortest path using the Viterbi dynamic programming algorithm [15]. The result is a list of map edges, which is the map-matched route.

The original HMM algorithm was designed primarily for the scenario of *dense* (but perhaps noisy) GPS trajectories. By "dense", we mean that, on the average, there are many GPS points per map edge. This means that the horizontal dimension of the trellis will be much larger than the vertical dimension, and there will be many edges in the shortest path computed through the trellis which will "march" along the same map edge. This precludes the opposite scenario—that of sparse GPS trajectories. In sparse trajectories, the trellis has a very small horizontal dimension, and many map edges should be traversed for a single trajectory edge. Since there are no edges within a trellis layer, this is not supported well, and the shortest path through the trellis is meaningless.

The variants of the HMM algorithm of Newson & Krumm [16] for map matching, attempts to modify the algorithm to deal also with the case of sparse GPS trajectories. For each trajectory edge, all the map edges in its vicinity—those that are not further away than some radius r are considered. An edge is added between two adjacent layers of the trellis corresponding to explicit shortest paths computed between any pair of map edges in adjacent vicinities. This way there are still no edges within trellis layers, but it is possible to move between layers, each layer corresponding to a GPS trajectory point, even if these points are quite far apart. While this modified HMM algorithm is now capable of map-matching sparse trajectories, the main problem is that it requires the computation of many shortest paths on the map, related to many of the trajectory edges, in order to construct the trellis in the first place. This can be time consuming.

4.2   Our Variation of the Map-Matching Algorithm

We now describe our map-matching algorithm, also based on a trellis graph, which deals correctly and naturally with sparse GPS trajectories. In contrast to the HMM algorithm of Newson & Krumm [16], it does not require to construct all the explicit shortest paths between map edges.

The key idea behind our algorithm is to allow the map and the GPS trajectory to play completely symmetric roles. The algorithm advances along the trajectory T and map edges in parallel, allowing each to advance at the correct speed, slowing down if necessary by staying put at a specific trajectory edge or map edge. This is ultimately formulated as a shortest path problem on the same type of trellis graph used by other HMM algorithms, whose nodes are pairs of edges— one from the GPS trajectory and one from the map. An edge exists between two trellis nodes, (i, j) and (k, l) (i and k are indices of GPS trajectory edges and j and l are indices of map edges) if and only if edge k is a successor of edge i in the trajectory and l is a neighboring edge of j on the map. The main difference between our trellis and the standard HMM trellis is that ours contains edges within layers. The weight of a trellis edge is a combination of the directionality of the comprised edges and the Euclidean distance between them. Note that the trellis graph is very sparse. A solution to the map-matching problem is the path with the minimal length among the following paths: the shortest paths between (t_1, e_i) and (t_n, e_j), where edge e_i is an edge within a radius r of the edge t_1 and edge e_j is an edge within radius r of the edge t_n (we found that $r = 20$m gives good results). If there are no edges within this radius r, then r will be increased, until there is some minimal number (typically 5) of edges to consider (both for the starting edges and for the ending edges).

Constructing the Trellis Graph. Given a map M with m edges and a GPS trajectory of edges $T = (t_1, t_2, \ldots, t_n)$, we build a trellis graph G, with $O(nm)$ nodes. As mentioned before, each node is a pair of edges, one (t) from T, and one from the edges in the vicinity of t in M. As we will see, G is very sparse since every node is connected to very few other nodes. Graph G has the same trellis structure as the graph used by the standard HMM algorithms, namely, can be viewed as n layers of the edges of the map M. Trellis edges within a layer correspond to neighboring edges (i.e. two edges where the target vertex of the first edge coincides with the source vertex of the second edge) within a single vicinity in the map, and edges between layers correspond to graph edges connecting between the vicinities of trajectory edges. Thus, movement within each layer corresponds to movement within the map at a given trajectory edge, and movement between layers corresponds to movement along the trajectory. Algorithm 4 describes this construction in detail.

The values dir_1 and dir_2 are the direction of edge t_i relative to edge e and the direction of edge x relative to edge y, respectively. The parameter d_1 is the minimum among *(1)* the distance from the source of t_i to e and *(2)* the distance from the source of e to t_i. The parameter d_2 is defined similarly— the minimum between *(1)* the distance from the source of x to y and *(2)* the distance from the source of y to x. The parameters $d_1, d_2, tLen_1, tLen_2, mLen_1$ and $mLen_2$ measure the distances between all the edges, as illustrated in Fig. 6. The dominant weight is the distance between the map edge and the trajectory edge, since if this distance is large, then there is a smaller chance that the true

---

Trellis-Graph Construction

Input: GPS trajectory $T = (t_1, t_2, \ldots, t_n)$,
a table *Neighbors* of map-edge adjacencies
Output: Trellis graph G

```
 1: for i = 1 to n do
 2:     J is the group of relevant edges from the map in the vicinity of ti
 3:     for each edge e ∈ J do
 4:        for each x ∈ {ti, ti+1} do
 5:           if x = ti then
 6:              N ← Neighbors(e)
 7:           else
 8:              N ← {e} ∪ Neighbors(e)
 9:           for each edge y ∈ N do
10:              add ē = ((ti, e), (x, y)) to G
```
11:              assign a weight of $\dfrac{(d_1+d_2)*(tLen_1+tLen_2+mLen_1+mLen_2)}{dir_1 * dir_2}$ to $\bar{e}$
```
12: return G
```

Fig. 4. Constructing the trellis graph G

route passed through that edge. Using these weights allows the algorithm to take into account how far the map edges and the trajectory edges are from each other. Fig. 5 shows a trellis graph constructed by the algorithm in Fig. 4.

After constructing the trellis graph G, we choose a couple of choices for the source edge on the map and a couple of choices for the target edge on the map. This is done by taking all the map edges that fall within a small radius r from the first and last point of the trajectory.

Computing the Matching. The last step of the algorithm is to find the weighted shortest path from a pair (t_1, e) to a pair (t_n, e'), where e is an optional starting edges and e' is an optional ending edge of G. The resulting path P will consist of pairs (t, e''), where $t \in T$ and e'' is an edge of the map. The map-matched route of the GPS trajectory to the map will be the ordered map edges of P after deleting consecutive duplicates of map edges. For example, in Fig. 5, P (the bold red path) is $((A, e_1), (B, e_3), (B, e_{10}), (C, e_{11}), (C, e_{12}))$, corresponding to the map-matched route $(e_1, e_3, e_{10}, e_{11}, e_{12})$.

The algorithm fails if no shortest path can be found. This usually means that either the map is not connected in the region we are working on, or that we did not extract enough map edges to support such a path during the extraction of relevant data. In such case, we may run the algorithm again on larger trajectory edge vicinities.

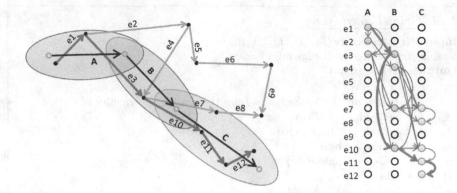

Fig. 5. Illustration of our map-matching algorithm. (Left) Sparse GPS trajectory. (Right) The trellis graph constructed by our algorithm from the map and trajectory. The bold blue path is the shortest path between e_1 and e_{12} through the trellis, corresponding to bold red path in the input graph, which is the resulting map-match of the GPS trajectory.

5   Path Codes

Once a route is generated based on a GPS trajectory, it may be represented as a path in a planar graph, namely, a sequence of vertices in the graph, implying edges between every two consecutive vertices, which translates to a sequence of vertex IDs. Thus, storing (or transmitting) long paths could be quite costly. In applications which involve building large databases of user paths, these costs could be prohibitive.

Thus, we present two novel ways to compactly represent a path in a planar graph, and efficient algorithms to compute these compact representations. Our methods represent the path as a subsequence of vertices from which the path can be uniquely reconstructed as a sequence of well-defined paths between each two consecutive vertices. In this representation, given the subsequence of vertices and the graph, the route may be recovered by generating the relevant paths between each two consecutive vertices of the code.

5.1   Greedy-Path Coding

Our first method of representing a path in a graph is as a sequence of consecutive *greedy paths*.

Definition 1 (Greedy Path). *Given a planar graph $G = (V, E)$ with geometry X (i.e., a mapping of vertices to geographic locations), a path $P = (i_1, i_2, \ldots, i_m)$ is a greedy path from vertex i_1 to vertex i_m when the sequence of Euclidean distances $\|X(i_1) - X(i_m)\|, \|X(i_2) - X(i_m)\|, \ldots, \|X(i_{m-1}) - X(i_m)\|$ is monotonically decreasing.*

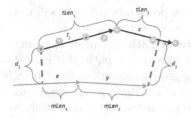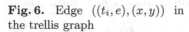

Fig. 6. Edge $((t_i, e), (x, y))$ in the trellis graph

Fig. 7. Greedy paths

Intuitively, a greedy path between vertex v and vertex u is one where each vertex w along the path is closer to u than $pred(w)$ (the predecessor of w). This defines a greedy path in a weak sense, and we add another condition to define a greedy path in a stronger sense.

Definition 2. *Given a planar graph $G = (V, E)$ with geometry X, a path $P = (i_1, i_2, \ldots, i_m)$ is a greedy path from vertex i_1 to vertex i_m in G iff the sequence of Euclidean distances $\|X(i_1) - X(i_m)\|, \|X(i_2) - X(i_m)\|, \ldots, \|X(i_m) - X(i_m)\|$ is monotonically decreasing and for all $1 \le k < m$, the following holds: $i_{k+1} = \text{argmin}_{j \in neighbors(i_k)}(\|X(j) - X(i_m)\|)$.*

The extra condition implies that not only is each vertex w along the path closer to u than $pred(w)$, but is the closest to u among all neighbors of $pred(w)$. A greedy path in the strong sense can be viewed as the discrete equivalent of a gradient descent path from v to u when considering the Euclidean distance function from u. The motivation for this extra condition is that under mild conditions on the graph, the greedy path in the strong sense will be unique, as opposed to the greedy path in the weak sense, which is typically not unique. As we will see later, uniqueness is important for the path coding application.

Note that a greedy path (in the weak sense, and certainly in the strong sense) between two given vertices in a planar graph is not always guaranteed to exist, even if the graph is connected. This can happen, for example, if a greedy walk from v to u gets stuck at a vertex w from which no neighbors are closer to u than w. This is the equivalent of getting stuck at a local minimum when performing gradient descent in the continuous case. For some specific planar graphs, the situation is better, for example, it is known that a greedy path in the weak sense exists between any two vertices of a Delaunay triangulation [17]. Such greedy paths are used extensively for routing in embedded networks, where messages are greedily forwarded towards their destination. Fig. 7 shows some examples of greedy paths in the weak and strong senses in a planar graph. In Fig. 7 (Left), the green path is a greedy path in the weak sense between A and B_1, and the orange path is the greedy path in the strong sense. In Fig. 7 (Right), a greedy path in the weak sense exists between A and B_2 (depicted in green), but no greedy path in the strong sense exists. This is evident from the fact that a greedy walk proceeds along the orange path and reaches a dead end (i.e. a local minimum of the Euclidean distance function from B_2). From this point onwards, we will use just the term greedy path to mean greedy in the strong sense.

It is easy to decide whether a given path is a greedy path by simply checking the definition. It is not too difficult either to compute a greedy path (if it exists) between vertex i_1 and vertex i_m using the following greedy algorithm. Start from vertex i_1. When at i_k, choose as i_{k+1} the neighbor of i_k which is the closest to the final destination i_m and also closer than i_k to i_m (if the latter condition is not satisfied, then the algorithm is stuck at a local minimum and fails). Then continue in the same manner from i_{k+1}.

Given a path $P = (i_1, i_2, \ldots, i_m)$, a greedy-path code of P is a subsequence $Q = (j_1, j_2, \ldots, j_k)$ of P such that $i_1 = j_1$, $i_m = j_k$, and P is identical to the concatenation of the greedy paths between j_t and j_{t+1} for $1 \leq t < k$, namely, if $j_t = i_r$ and $j_{t+1} = i_s$ then the sub-path $(i_r, \ldots, i_s)$ of P is a greedy path. An optimal greedy path code of P is a shortest possible Q (as measured by k). The objective is to produce a code such that greedy paths indeed exist between the code vertices. These greedy paths will be unique because of the extra (strengthening) condition.

We now describe two algorithms to compute a greedy path code of a path in a graph. The first is the simplest possible, running in linear time, but not necessarily generating an optimal greedy path code. The second algorithm is less efficient, but optimal. Note that in the worst case, the greedy path code of a path is the path itself.

Both algorithms take advantage of the fact that greedy paths have the *suffix property*, namely, any suffix of a greedy path is also a greedy path, which is a trivial consequence of the definition of a greedy path. It also means that given a graph G and a target vertex t, the uniqueness of the greedy paths implies that all greedy paths from all other vertices of G to t (if they exist) form a *greedy tree* rooted at t (after reversing the direction of the edges). This tree does not span the entire vertex set of G, rather only those vertices from which a greedy path to t exists.

Given a greedy path code of a path $(i_1, i_2, \ldots, i_m)$, it may be decoded in time complexity $O(m)$ by simply computing the greedy paths in the graph between each two consecutive vertices of the code. The uniqueness of the greedy path guarantees that the decoding is correct, i.e. indeed recovers the original path. The linear complexity assumes that all vertices have a bounded valence, thus computing the correct neighbor of a vertex in a greedy path requires $O(1)$ time.

5.2   Simple Greedy-Path Coding Algorithm

The simple greedy-path coding algorithm, presented in Fig. 8, starts from i_m, and proceeds checking backwards if the path is greedy. A codeword (an index of a vertex in the graph) is generated when the path ceases to be a greedy path, and the procedure repeats from there.

The suffix property of the greedy paths allows to check greediness in Line 4 by checking just the current s at each step, saving checking the greediness of the entire subpath between s and t. This algorithm has $O(m)$ time complexity,

---

Simple Greedy-Path Coding

Input: Path $P = (i_1, i_2, \ldots, i_m)$ in the planar graph $G = ((V, E), X)$,
Output: Greedy path code of the path P
1: $C \leftarrow (i_m)$
2: $t \leftarrow m,\ s \leftarrow m - 1$
3: **while** $t > 1$ **do**
4: **while** $s > 1$ and $i_s = \text{argmin}_{j \in neighbors(i_{s-1})}(\|X(j) - X(i_t)\|)$ and $\|X(i_s) - X(i_t)\| < \|X(i_{s-1}) - X(X_{i_t})\|$ **do**
5:    $s \leftarrow s - 1$
6: insert i_s at the beginning of C
7: $t = s$
8: **return** C

Fig. 8. Algorithm for computing a simple greedy-path coding

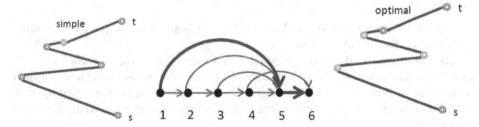

Fig. 9. Simple greedy-path code. The code is the 5 purple points.

Fig. 10. Graph R and the shortest path between 1 and 6, in purple, (see Fig. 12)

Fig. 11. Optimal greedy path code. The code is the 3 purple points.

where m is the number of vertices in the input path. The linear complexity assumes that all vertices have a bounded valence, thus checking the greediness of an edge in the path requires $O(1)$ time. Unfortunately, this algorithm is not guaranteed to find the shortest possible greedy path code. See Figures 9, 10 and 11 for an example of greedy-path coding in a graph G consisting of a single path. A path of 6 vertices (which is also the entire graph G) is coded into 5 points using the simple greedy path coding algorithm, but using the optimal algorithm to be described next results in a greedy path code of 3 points.

5.3 Optimal Greedy-Path Coding Algorithm

The optimal greedy-path coding algorithm, presented in Fig. 12, computes an optimal greedy-path code—a code with a minimal number of points. It is somewhat similar to the Imai-Iri algorithm [18] for simplifying a polyline. It starts by building a graph on the input points where an edge (v, u) represents the existence of a greedy path between v and u. Then, it computes a shortest path in

---

Optimal Greedy-Path Coding

Input: Path $P = (i_1, i_2, \ldots, i_m)$ in the planar graph $G = ((V, E), X)$,
Output: Optimal greedy-path code of the path P

1: create a graph R with m nodes and no edges
2: **for** $t = 2$ to m **do**
3:     $s \leftarrow t$
4:     **while** $s > 1$ and $i_s = \text{argmin}_{j \in neighbors(i_{s-1})}(\|X(j) - X(i_t)\|)$ and $\|X(i_s) - X(i_t)\| < \|X(i_{s-1}) - X(X_{i_t})\|$ **do**
5:         add the edge (s, t) to R
6:         $s \leftarrow s - 1$
7: add the edge (s, t) to R
8: Find the shortest path, S, from Node 1 to Node m in R
9: **return** S

Fig. 12. Algorithm for computing an optimal greedy-path coding

this graph between the first and last vertices. This generates a greedy-path code with the minimal number of vertices.

The time complexity of this algorithm is $O(m^2)$, since the outer loop (on t) iterates m times, and the inner loop can add up to t edges, resulting in a graph R containing m vertices and $O(m^2)$ edges. Thus the shortest path computation in Line 8 also requires $O(m^2)$ time when using Djikstra's algorithm with Fibonacci heaps [19].

The optimal greedy path coder relies on finding a shortest path in the graph R (Line 8 of Fig. 12). In order to guarantee a unique coding (e.g. in order to determine if two paths are identical based only on their codes), this shortest path of R must be unique, i.e. independent of the shortest-path algorithm (e.g. Dijkstra, Bellman-Ford) used by the encoder. Since a priori there is no reason that the shortest path should be unique, we achieve this by slightly modifying the content of the graph R in a way that guarantees uniqueness without compromising the true shortest path, as described by Mehlhorn [20]. Essentially, the weight of edge (i_r, i_s) will be $w_{rs} = 1 + m^{-2}(s - r)^2$, where m is the number of points in P.

Using these perturbed weights will have the effect of generating shortest paths with a similar number of edges. Among all such codes, it will prefer those whose greedy path segments have approximately the same number of edges. This is because all candidate codes have the same number k of greedy path segments, representing the same total number of edges m (as in the input path). Denoting by x_i the number of edges in the i-th greedy path segment, minimizing the sum of the squares $\sum_{i=1}^{k} (x_i)^2$ prefers uniform distribution of the x_i's, as the following lemma formalizes.

Lemma 1. *The solution to* $\min \sum_{i=1}^{k} (x_i)^2$ *subject to* $\sum_{i=1}^{k} x_i = m$ *(m is a positive constant) is* $x_i = m/k$ *for* $i = 1, \ldots, k$.

The proof of the lemma is straightforward using Lagrange multipliers.

5.4   Shortest-Path Coding

Greedy-path coding seeks to find the subsequence of points of P that segments P into a number of sub-paths, which are greedy paths between consecutive points of the subsequence. Greedy-path coding is relatively simple and decoding is extremely fast. It relies on the extrinsic geometry (i.e. coordinates of the embedding) of the graph. However, more compact codes are possible. In this section we explore shortest path coding, i.e. representing P as the subsequence of points of P which segments P into a number of sub-paths which are shortest paths between consecutive points of the subsequence. As we will see, these codes will be more difficult to compute and decoding them will be slower, but they will be more compact.

Define the length of a path to be the sum of the Euclidean lengths of the edges in the path. A shortest path between Vertex i and Vertex j is the path between the two vertices whose length is the shortest possible. This path can be computed using Dijkstra's algorithm and its many variants [21, 22]. As such, it relies only on the intrinsic geometry (edge lengths) of the graph.

In contrast with the greedy-path coding algorithms, shortest-path coding requires considering a larger portion of the graph than just the given path P and its neighboring edges—an entire bounding box of the path. Since the algorithm relies on computation of shortest paths between vertices, we need a much broader view of the region.

5.5   Optimal Shortest-Path Coding Algorithm

Shortest paths have the sub-path property, namely, any sub-path between vertex u and vertex v within a shortest path is necessarily also a shortest path between u and v. In particular, this implies the prefix property and the suffix property, that any prefix or suffix of a shortest path is a shortest path. The prefix property implies the well-known fact that given a graph G and a source vertex s all shortest paths from s to all other vertices form a spanning tree of G rooted at s. Using the suffix property, it is possible to prove that the following simple (i.e. greedy in the algorithmic sense) shortest-path coding algorithm is in fact optimal. The algorithm is presented in Fig. 13. Essentially, it is similar to the simple greedy-path coding algorithm, except that it proceeds in the forward direction, as opposed to the reverse direction. It checks incrementally whether sub-paths of the input path are shortest paths, taking advantage of the suffix property to save computations. We assume that all path lengths are different real numbers. This is needed to guarantee that the shortest path tree computed in Line 4 is unique, to allow the decoder to reconstruct the original path from the code. The optimality of the algorithm follows from the next proposition.

Proposition 1. *Any shortest-path code C' of a path P in graph G will have length greater than or equal to the length of C—the output of the algorithm.*

Proof. Let $C = (i_1, \ldots, i_k)$ be the output of the algorithm in Fig. 13 and $C' = (j_1, \ldots, j_r)$ be the output of any other shortest-path coding algorithm. It suffices

---

Optimal Shortest-Path Coding

Input: Path $P = (i_1, i_2, \ldots, i_m)$ in the planar graph $G = ((V, E), X)$,
Output: Optimal shortest-path code of P
  1: $C \leftarrow (i_1)$
  2: $s \leftarrow 1$
  3: **while** $s < m$ **do**
  4:    compute the shortest-path tree, rooted at i_s, whose leaves are all vertices i_t
       where $s < t \leq m$
  5:    $t \leftarrow s + 1$
  6:    let v_b be the vertex before i_t in the shortest path between i_s and i_t
  7:    **while** $t \leq m$ and $v_b = i_{t-1}$ **do**
  8:      $t \leftarrow t + 1$
  9:      set v_b to be the vertex before i_t in the shortest path between i_s and i_t
10:    append i_{t-1} to C
11:    $s \leftarrow t - 1$
12: **return** C

Fig. 13. Algorithm for computing an optimal shortest-path coding

Fig. 14. Illustration of the proof of Proposition 1. Purple points are the optimal coding C. Cyan points are some of the code C'. The path $(i_s, \ldots, i_{s+1})$ does not contain any element of C'. If the path $(j_p, \ldots, j_{p+1})$ is a shortest path, then the suffix property implies that $(i_s, \ldots, j_{p+1})$ is also a shortest path.

to prove that each of the $k-1$ segments $(i_s, \ldots, i_{s+1})$ contains at least one element of C' for all $1 \leq s \leq k-1$, since then $k \leq r$.

Note that the claim holds trivially for the first segment ($s = 1$) since $i_1 = j_1$. So, assume $1 < s < k$. Now assume by way of contradiction that the segment $(i_s, \ldots, i_{s+1})$ does not contain any element of C'. Let j_p be the largest element of C' such that $j_p < i_s$ and j_{p+1} the next element of C' (in the "worst case", $p = 1$). By the assumption, $j_{p+1} \geq i_{s+1}$. Now, by definition, $(j_p, \ldots, j_{p+1})$ is a shortest path, so the suffix property implies that $(i_s, \ldots, j_{p+1})$ is also a shortest path, in contradiction to the fact that $(i_s, \ldots, i_{s+1})$ is the longest possible shortest path starting at i_s. (See illustration in Fig. 14.)

Note that this proof does not hold for the simple greedy-path coding algorithm (Fig. 8), because the algorithm does not guarantee the final contradiction— that $(i_s, \ldots, i_{s+1})$ is the longest possible greedy path starting at i_s, since the algorithm operates in reverse.

The complexity of the algorithm is $O(k(n+n\log n+m))$ where n is the number of edges/nodes in the effective graph M (the path bounding box) and k is the number of points in the code. In general, n is $O(m^2)$, since this is the relationship between the number of edges in a one-dimensional path and the number of edges in a two dimensional region whose boundary length is $O(m)$, giving a complexity of $O(km^2\log m)$. The decoding also has $O(km^2\log m)$ time complexity due to the need to compute the shortest path between each consecutive pair in the code (there are $k-1$ pairs).

6   Experiments

To test the effectiveness of our methods, we implemented them and tested them experimentally. We implemented our map-matching algorithm in an interactive browser-based system, using the Google Maps Javascript API and the Open Street Map digital database. The system was written in Javascript for the client side and uses JSP/Servlets on the server side. The algorithms were implemented in MATLAB and compiled to run independently on the server by JSP/Servlet calls. The machine we used contained an Intel i7 CPU with 8GB RAM.

We used the dataset of GPS trajectories of the ACM SIGSPATIAL Cup 2012 contest (see http://depts.washington.edu/giscup/) and the GPS trajectory dataset used in [16], recorded in the Seattle area, to test our algorithms. These trajectories consist of GPS recording at a frequency of 1Hz through urban and rural areas (highways, small streets and intersections), which translates to a recording every 5-20 meters, depending on the vehicle velocity. These are considered dense recordings. The noise level was $\sigma = 10m$. A typical GPS trajectory contained 500 points. We also used a number of GPS trajectories we recorded ourselves using a smartphone application, while driving in the city of Haifa. These trajectory recordings were made such that at least 10 seconds and at least 10 meters elapsed between two successive recordings. These are quite sparse recordings. Here too the noise level was $\sigma = 10m$. In all the experiments, a grid-based spatial index was used for an efficient retrieval of road segments that are in a certain area or in the vicinity of a certain point.

Figures 15, 16 and 17 compare the different types of codes. They illustrate typical compact representations of a path. The coding points are depicted in purple and the other removed points appear in orange. Note that by using the optimal shortest-path code only 5 points are required to represent a path of 214 points. In general, the difference between the simple greedy-path code and the optimal greedy-path code is relatively small, but the shortest-path code is typically much more compact than the other two codes.

We ran statistics on a set of 33 routes that were map-matched (using our algorithm) from the GPS trajectories in the ACM SIGSPATIAL Cup 2012 dataset and the GPS trajectory dataset used in [16], to determine the average coding ratio and running time of the various algorithms. A typical path contains approximately 125 vertices after the dilution and the map-matching phases. The results are shown in Fig. 18 and Fig. 19. As evident there, the simple greedy-path coding algorithm reduces the number of vertices to 7.3% of the original on

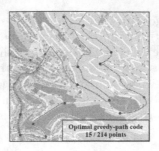

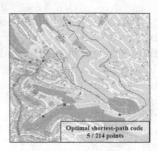

Fig. 15. Simple greedy-path code (19 points out of 214 original points).

Fig. 16. Optimal greedy-path code (15 points out of 214 original points)

Fig. 17. Optimal shortest path code (5 points out of 214 original points)

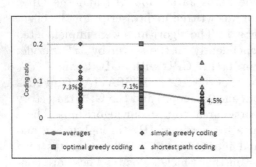

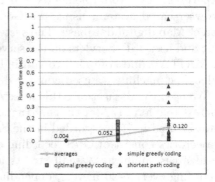

Fig. 18. Coding ratios of the three algorithms. Each data point is a path in the dataset.

Fig. 19. Running times. Each point is a path in the dataset.

the average, the optimal greedy-path coding algorithm reduces slightly more, to 7.1%. The shortest-path coding algorithm reduces to 4.5%, on the average.

Typically, it is important that the decoder will be efficient since the process of decoding is done many times (essentially every time a route is extracted from a database) and in real-time, as opposed to the encoding process which usually happens only once, and is typically done in an offline process. Decoding of the greedy path codes takes $O(m)$ time and decoding of the more compact shortest path code takes $O(km^2 \log m)$ time (where k is the length of the code).

In some applications it is important to code a path online (as it is being generated). This would seem to be impossible for the two greedy-path coding algorithms, since they operate in reverse. Nonetheless, it is possible to modify these algorithms to run in forward order, paying a penalty in time complexity. In contrast, the optimal shortest-path encoding algorithm can be executed online with a lag of just one path vertex, i.e. it is possible to decide whether a path vertex is part of the shortest path code only after the next route vertex has been seen. There will also be a running-time penalty to implement this in practice.

7 Conclusions

We study the problem of computing a compact coding of routes over a vectorial road network. Given a trajectory as a sequence of GPS measurements, it is shown how to represent it compactly, in a three-step process: (1) diluting the sequence, (2) applying map-matching to receive a sequence of map vertices, and (3) generating a compact representation of the traveled route.

For the classical problem of map-matching, the paper presents an adaptation of an HMM-based method. The aim is to handle effectively scenarios where the GPS measurements are sparse and noisy. This ability is lacking in many existing approaches. The result of the map-matching is a route in the form of a sequence of vertices of the road network. We present two approaches to represent a route compactly—as a sequence of greedy paths or as a sequence of shortest paths. We provide two algorithms for computing the sequence of greedy paths. One algorithms is simple and highly efficient, having $O(n)$ time complexity, over a sequence of n points, and the second algorithm has $O(n^2)$ time complexity, however, it computes the optimal greedy-path code. Decoding a greedy-path code can be done in $O(n)$ time. For generating the sequence of shortest paths, we provide an algorithm with $O(kn^2 \log n)$ time complexity, where k is the length of the (output) code. Decoding a shortest-path code also has $O(kn^2 \log n)$ time complexity. Experimentally, when applying our algorithm to real-world data sets, we observed that shortest-path codes are more compact than greedy-path codes but it takes more time to compute them. Evidently, our representation is more compact than merely applying dilution and map-matching.

Compact coding of routes on a map, coupled with a very fast decoding algorithm, is important for storage and transmission of this type of data from large (online) databases, especially as these databases become more and more widespread in the connected mobile world. An important related question is when is it possible to perform computations on routes in their coded form, i.e. without explicitly decoding them. For example, is it possible to intersect two routes by intersecting their greedy path or shortest path codes without decoding the two routes first? Similarly, is it possible to determine proximity of a given map vertex to a coded route, without decoding the route? These questions remain as future work. Future work also includes the question of how to use the timestamps of the GPS measurements, for improving the representation, and how to recover times when reconstructing a route.

References

1. Zheng, Y., Zhang, L., Ma, Z., Xie, X., Ma, W.Y.: Recommending friends and locations based on individual location history. ACM Trans. Web 5(1), 5:1–5:44 (2011)
2. Cao, X., Cong, G., Jensen, C.S.: Mining significant semantic locations from gps data. Proc. VLDB Endow. 3(1-2), 1009–1020 (2010)
3. Li, Q., Zheng, Y., Xie, X., Chen, Y., Liu, W., Ma, W.Y.: Mining user similarity based on location history. In: Proc. of the 16th ACM SIGSPATIAL GIS (2008)

4. Giannotti, F., Nanni, M., Pedreschi, D., Pinelli, F., Renso, C., Rinzivillo, S., Trasarti, R.: Unveiling the complexity of human mobility by querying and mining massive trajectory data. The VLDB Journal 20, 695–719 (2011)
5. Zheng, Y., Li, Q., Chen, Y., Xie, X., Ma, W.Y.: Understanding mobility based on gps data. In: Proceedings of the 10th International Conference on Ubiquitous Computing, pp. 312–321 (2008)
6. Meratnia, N., de By, R.A.: Spatiotemportal compression techniques for moving point objects. In: Proc. of the 9th International Conference on Extending Database Technology (2004)
7. Muckell, J., Hwang, J.H., Patil, V., Lawson, C.T., Ping, F., Ravi, S.S.: Squish: an online approach for gps trajectory compression. In: Proc. of the 2nd COM.Geo. COM.Geo 2011, pp. 13:1–13:8. ACM (2011)
8. Chen, Y., Jiang, K., Zheng, Y., Li, C., Yu, N.: Trajectory simplification method for location-based social networking services. In: Proc. of the 2009 International Workshop on Location Based Social Networks, LBSN 2009, pp. 33–40 (2009)
9. Zheng, Y., Zhou, X. (eds.): Computing with Spatial Trajectories. Springer (2011)
10. Douglas, D.H., Peucker, T.K.: Algorithms for the reduction of the number of points required to represent a digitized line or its caricature. Cartographica: Inter. Journal for Geographic Information and Geovisualization 10(2), 112–122 (1973)
11. White, C.E., Bernstein, D., Kornhauser, A.L.: Some map matching algorithms for personal navigation assistants. In: Transportation Research Part C: Emerging Technologies, vol. 8, pp. 91–108 (2000)
12. Quddus, M.A., Ochieng, W., Zhao, L., Noland, R.B.: A general map matching algorithm for transport telematics applications. GPS Solution 7(3) (2003)
13. Quddus, M.A., Ochieng, W.Y., Noland, R.B.: Current map-matching algorithms for transport applications: State-of-the art and future research directions. In: Transportation Research Part C: Emerging Technologies, pp. 312–328 (2007)
14. Hummel, B.: Map matching for vehicle guidance. In: Dynamic and Mobile GIS: Investigating Space and Time, pp. 437–438. CRC Press (2006)
15. Viterbi, A.J.: Error bounds for convolutional codes and an asymptotically optimum decoding algorithm. Transactions on Information Theory 13(2), 260–269 (1967)
16. Newson, P., Krumm, J.: Hidden markov map matching through noise and sparseness. In: Proceedings of the 17th ACM SIGSPATIAL International Conference on Advances in Geographic Information Systems, pp. 336–343 (2009)
17. Bose, P., Morin, P.: Online routing in triangulations. SIAM Journal of Computing 33, 937–951 (2004)
18. Imai, H., Iri, M.: Computational-geometric methods for polygonal approximations of a curve. Comp. Vision, Graphics, and Image Processing 36(1), 31–41 (1986)
19. Fredman, M.L., Tarjan, R.E.: Fibonacci heaps and their uses in improved network optimization algorithms. In: Proc. of the 25th Annual Symposium on Foundations of Computer Science, pp. 338–346. IEEE (1984)
20. Mehlhorn, K.: Unique shortest paths. In: Selected Topics in Algorithms: Course Notes (2009)
21. Bellman, R.: On a routing problem. Quarterly of Applied Mathematics 16(1), 87–90 (1958)
22. Dijkstra, E.W.: A note on two problems in connexion with graphs. Numerische Mathematik 1(1), 269–271 (1959)

Group Trip Planning Queries in Spatial Databases

Tanzima Hashem[1], Tahrima Hashem[2],
Mohammed Eunus Ali[1], and Lars Kulik[3]

[1] Department of Computer Science and Engineering
Bangladesh University of Engineering and Technology, Dhaka, Bangladesh
{tanzimahashem,eunus}@cse.buet.ac.bd

[2] Department of Computer Science and Engineering
Dhaka University, Dhaka, Bangladesh
tahrimacsedu14@gmail.com

[3] Department of Computing and Information System
University of Melbourne, VIC 3010, Australia
lkulik@unimelb.edu.au

Abstract. Location-based social networks grow at a remarkable pace. Current location-aware mobile devices enable us to access these networks from anywhere and to connect to friends via social networks in a seamless manner. These networks allow people to interact with friends and colleagues in a novel way, for example, they may want to spontaneously meet in the next hour for dinner at a restaurant nearby followed by a joint visit to a movie theater. This motivates a new query type, which we call a group trip planning (GTP) query: the group has an interest to minimize the total travel distance for all members, and this distance is the sum of each user's travel distance from each user's start location to destination via the restaurant and the movie theater. Formally, for a set of user source-destination pairs in a group and different types of data points (e.g., a movie theater versus a restaurant), a GTP query returns for each type of data points those locations that minimize the total travel distance for the entire group. We develop efficient algorithms to answer GTP queries, which we show in extensive experiments.

Keywords: Group nearest neighbor queries, Group trip planning queries, Location-based services, Location-based social networks, Spatial Databases.

1 Introduction

Location-based social networks such as Facebook [1], Google+ [2], and Loopt [3] enable a group of friends to remain connected from virtually anywhere at any time via location aware mobile devices. In these networks users can share their locations with others and interact with other users in a novel way. For example, a group of friends may want to spontaneously meet in the next hour for dinner at a restaurant nearby followed by a joint visit to a movie theater. The users are typically at different places, e.g., on a late afternoon all of them may still be at their offices and workplaces. Before going home in the evening, they would still like to organise a group dinner and movie event. Usually, all users would like to meet at a restaurant and movie theater that is nearby. To enable users to arrange a trip with a minimum total travel distance, we introduce a new type of query called a *group trip planning (GTP) query*.

M.A. Nascimento et al. (Eds.): SSTD 2013, LNCS 8098, pp. 259–276, 2013.
© Springer-Verlag Berlin Heidelberg 2013

Specifically, for a set of user source-destination pairs in a group and different types of data points (e.g., a movie theater versus a restaurant), a GTP query returns for each type of data points those locations that minimize the total travel distance for the entire group. The total travel distance of a group is the sum of each user's travel distance from source to destination via the data points. Figure 1 shows an example, where the pair of data points (p_1'', p_2') minimizes the total travel distance for the group trip. The group may fix the order of the planned locations (e.g., first the movie theater, then the restaurant) or keep the order flexible (e.g., the group is happy to visit the movie theater and the restaurant in any order). We call the former type *ordered GTP queries* and the latter *flexible GTP queries*. In this paper, we develop algorithms to evaluate both type of GTP queries.

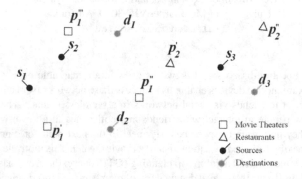

Fig. 1. An example scenario

Query processing for a group of users instead of a single user involves high computational overhead because this query has to be evaluated with respect to a set of locations instead of a single location. Existing studies [4,5] show that iteratively applying algorithms that are designed for a single user incur large query processing overheads if they are used to evaluate group queries. Thus, the development of efficient algorithms for processing group nearest neighbor (GNN) queries [4,5] and its variants [6,7] has recently evolved into an important research area. A GNN query finds the data point that minimizes the aggregate distance with respect to the locations of the users in the group. In this paper, we address group trip planning queries. Though trip planning queries for a single user have been addressed [8,9] in the literature, GTP queries have not yet been explored.

A GTP query can be mapped to a GNN query, if a group meets at a single location during their trip; existing GNN algorithms can be applied to the GTP query to find the data point that minimizes the total travel distance for a set of source and destination locations of the group members. However, if the group stops at least at two different types of locations, we show that applying GNN algorithms [4,5] to evaluate GTP queries increases the query processing cost significantly. Such a technique evaluates GNN queries multiple times and requires multiple independent searches on the database. We develop an efficient algorithm for processing GTP queries that finds the data points with a single traversal on the database. The key idea of our algorithm is to reduce the search space, i.e., to avoid the computation of total travel distances for data points that cannot be part of the query answer. We develop a set of pruning techniques to eliminate the data points from the search while evaluating GTP queries.

In summary, we make the following contributions:

- We propose a new type of query, the group trip planning (GTP) query and propose the first solution to process the query.
- We develop an efficient algorithm to evaluate GTP queries. Our algorithm can evaluate both ordered and flexible GTP queries.
- We perform an extensive experimental study to show the efficiency of our proposed algorithms, including a detailed comparative analysis.

The rest of the paper is organized as follows. Section 2 presents the problem setup and Section 3 reviews existing works related to this problem. In Section 4, we propose our algorithms to evaluate GTP queries and in Section 5, we present extensive experiments to validate the efficiency of our algorithms. Section 6 concludes the paper with future research directions.

2 Problem Overview

In a group trip planning (GTP) query, a group of users specify their source and destination locations, the types of data points that they want to visit together while traveling from their source to destination locations. For a group of n users let S represent the set of source locations $\{s_1, s_2, \ldots, s_n\}$ and D the set of destination locations $\{d_1, d_2, \ldots, d_n\}$. The source and destination locations of a user u_i are denoted as s_i and d_i, respectively. A set of data points of type t in a 2-dimensional space is denoted by D_t. If a group plans a trip with m different types of data points along their trip, the GTP query returns a set of data points $\{p_1, p_2, \ldots, p_m\}$, $p_t \in D_t$, such that an aggregate function $f(S, D, p_1, p_2, \ldots, p_m)$ is minimal. The function $f(S, D, p_1, p_2, \ldots, p_m)$ represents the total travel distance for the group trip as $\left(\sum_{i=1}^{n} Dist(s_i, p_1) + n \times \sum_{i=1}^{m-1} Dist(p_i, p_{i+1}) + \sum_{i=1}^{n} Dist(d_i, p_m) \right)$, where $Dist(.,.)$ is the Euclidean distance between two locations. In general, a group trip planning (GTP) query is formally defined as follows:

Definition 1. *(Group Trip Planning (GTP) Queries). Given a set of source locations S, a set of destination locations D, sets of m types of data points $\{D_1, D_2, \ldots, D_m\}$, and an aggregate function f the GTP query returns a set of data points $\{p_1, p_2, \ldots, p_m\}$, $p_t \in D_t$, that minimizes f.*

For a set of data points $\{p_1, p_2, \ldots, p_m\}$, in an ordered GTP query, the group determines the order in which the group plans to visit the m data points. The parameters $p_1, p_2, \ldots, p_m$ in the aggregate function f are passed in the order specified by the group. In a flexible GTP query, however f is evaluated for every possible order of the data points $p_1, p_2, \ldots, p_m$.

A group may be interested in k sets of data points $\{p_1^1, p_2^1, \ldots, p_m^1\}$, $\{p_1^2, p_2^2, \ldots, p_m^2\}$, $\ldots$, $\{p_1^k, p_2^k, \ldots, p_m^k\}$ that have the k smallest total travel distances for the group trip. The group then select one set by considering other factors such as cost and recommendations. If a group queries for k sets of data points for a group trip then the query is called a *k group trip planning (kGTP) query*. In Section 4, we develop algorithms to evaluate *k*GTP queries.

The symbols used in this paper are summarized in Table 1.

Table 1. Symbols

Symbol	Meaning
s_i and d_i	Source and destination locations of a user u_i
$S = \{s_1, s_2, \ldots, s_n\}$	A set of source locations of n users in the group
$D = \{d_1, d_2, \ldots, d_n\}$	A set of destination locations of n users in the group
$\{p_1^k, p_2^k, \ldots, p_m^k\}$	A set of data points that has k^{th} minimum distance for the group trip
$Dist(.,.)$	The Euclidean distance between two locations
$f(S, D, p_1, p_2, \ldots, p_m)$	A function that returns the total travel distance as $\{\sum_{i=1}^{n} Dist(s_i, p_1) + n \times Dist(p_1, p_2) + \cdots + n \times Dist(p_{m-1}, p_m) + \sum_{i=1}^{n} Dist(d_i, p_m)\}$

3   Related Work

Most of the existing techniques for processing spatial queries assume that data points are indexed, e.g., using an R-tree [10] or its variant R^*-tree [11]. In an R-tree, nearby objects are grouped together with minimum bounding rectangles (MBRs). These MBRs are organized in a hierarchical way such that the root MBR covers the whole data space and the MBR of a parent node covers the MBRs of all of its children. In this paper, we use separate R^*-trees, to index different types of data points.

The most popular query type in spatial database is nearest neighbor queries that return the nearest data point with respect to a given location. A well known approach to evaluate the k NNs is to traverse the R^*-tree in a best-first (BF) [12] manner, which is called the best-first search (BFS). In BFS, the search starts from the root of the tree, and the child nodes are recursively accessed in the increasing order of their distances from the query point. The search process terminates as soon as the k nearest data points are retrieved from the tree.

A group nearest neighbor (GNN) query [4,5,13], a variant of the NN query, finds the nearest data point with respect to all user locations of the group. A GNN query minimizes the aggregate distance for the group, where an aggregate distance is measured as the total, minimum or maximum distance of the data point from the group. Papadias et al. [4,5] have proposed three techniques: multiple query method (MQM), single point method (SPM), and minimum bounding method (MBM), to evaluate GNN queries. All of these three variants use the BFS technique to traverse the R^*-tree. Among these three techniques, MBM performs the best as it traverses the R^*-tree once and takes the area covering the users' location into account. We use MBM for evaluating GNN queries in our approach. In [13], Li et al. have developed exact and approximation algorithms for GNN queries that minimize the maximum distance of the group. In this paper, we require GNN algorithms that minimize the total travel distance.

Recently, some variants of GNN queries have been proposed [6,7]. In [6], Deng et al. have proposed a group nearest group (GNG) query that finds a subset of data points, as opposed to a single data point in a GNN query, from the dataset such that the aggregate distance from the group to the subset is minimum. The aggregate distance is computed as the summation of all Euclidean distances between a query point and the nearest data point in the subset. In [7], Li et al. have proposed a flexible aggregate similarity search that finds the nearest data point and the corresponding subgroup for a fixed subgroup size; for example, a group may query for the nearest data point to 50% of group members. In this paper, we introduce a group based trip planning query, which is different from the above mentioned variants.

Trip planning queries [9,8,14,15] have been studied in the literature with respect to a single user. In [9], Li et al. have developed approximation algorithms for finding a set of data points that minimize the total travel distance for the trip. The travel distance starts from the source location, passes through the set of data points, and ends at the destination location of the user, where each data point corresponds to a type (e.g., a restaurant) specified by the user. In [8], Sharifzadeh et al. have developed algorithms to evaluate a optimal sequenced route (OSR) query that returns a route with the minimum length passing through a set of data points in a particular order from the source location of a user, where both order and type of data points are specified by the user. Chen et al. [14] have proposed a generalization of the trip planning query, called multi-rule partial sequenced route (MRPSR) query. A MRPSR query provides a uniform framework to evaluate both of the above mentioned variants [9,8] of trip planning queries. All these existing techniques assume a single user, and thus are not suitable for a group trip.

A large body of research works focus on developing algorithms for processing a route planing query and its variants [16,17,18,19]. A route planning query finds a suitable route that minimizes the desire function such as travel time, shortest path, cost, etc. for *a single source and destination pair*. Route planing queries do not include data points in the route and are applicable for a single user. On the other hand, we propose an efficient solution for a kGTP query that finds different types of data points for a group trip that minimize the total travel distance of the group.

4   Algorithms

In this section, we present algorithms to process kGTP queries. A straightforward way to evaluate kGTP queries would be applying a trip planning algorithm [14] for every user in the group independently and determining the group travel distance for every computed set of data points. The process continues to incrementally determine the sets of data points that minimize the total travel distance of every user in the group until the set of data points with the minimum group travel distance have been identified. However, this straightforward solution is not scalable and incurs very high query processing overhead as same data points are accessed by multiple users separately. Instead, we first propose an iterative algorithm as baseline method that does not evaluate the trip planing query independently for every user but still requires to access same data multiple times.

To avoid the limitation of the baseline method, we develop an efficient hierarchical algorithm that evaluates kGTP queries with a single traversal of the database and incurs less processing overhead. The input to the algorithm for a group of n users are source locations $S = \{s_1, s_2, \ldots, s_n\}$, destination locations $D = \{d_1, d_2, \ldots, d_n\}$,

m types of data points for $m > 0$, and the number k of required sets of data points. The output of the algorithm is A that consists of k sets of data points $\{p_1^1, p_2^1, \ldots, p_m^1\}, \{p_1^2, p_2^2, \ldots, p_m^2\}, \ldots, \{p_1^k, p_2^k, \ldots, p_m^k\}$ having the k smallest total travel distances for the group trip. We assume that each type of data points are indexed using a separate R^*-tree [11] in the database. Let $R_1, R_2, \ldots, R_m$ represent the R^*-trees to index the set of data points in $D_1, D_2, \ldots, D_m$, respectively. Note that, our algorithms can also be adopted if a single R^*-tree is used to index all types of data points. We present our two approaches, i.e., iterative and hierarchical, for processing kGTP queries in Section 4.1 and Section 4.2, respectively.

4.1   Iterative Approach

The basic idea of our iterative approach to evaluate a kGTP query is to use group nearest neighbor (GNN) queries. A kGNN query returns the locations of k data points that have the k smallest total travel distances for the group. For ease of understanding, we start the discussion of the iterative algorithm for the number of data type, $m = 1, m = 2$, and then generalize the algorithm for any value of m.

For $m = 1$, the iterative approach can evaluate the kGTP query answer with a single iteration using any existing kGNN algorithm [4,5]. The iterative approach determines the kGTP query answer $p_1^1, p_1^2, \ldots, p_1^k$ as k GNNs (group nearest neighbors) from D_1 with respect to $\{S \cup D\}$.

Consider the case $m = 2$, where the group specifies the order of visit as $\{1, 2\}$, i.e., the group visits the data point of type 1 before the data point of type 2. The iterative approach determines the 1st GNN as p_1 from D_1 with respect to S and the 1st GNN as p_2 from D_2 with respect to $\{p_1 \cup D\}$. For the pair of data points p_1 and p_2, the iterative algorithm determines the total travel distance for the group trip. Then the algorithm determines the 2nd GNN as p_2' from D_2 with respect to $\{p_1 \cup D\}$ and computes the total travel distance for the group trip for p_1 and p_2'. The process continues until the kth group nearest data point as p_2 from D_2 with respect to $\{p_1 \cup D\}$ has been determined.

In the next iteration, the algorithm determines the 2nd GNN as p_1 from D_1 with respect to S and repeats the same procedure mentioned above for the 1st GNN from D_1. The iteration continues to incrementally determine the GNNs from D_1 with respect to S until the algorithm finds k pairs of data points that have k smallest total travel distances for the group trip.

Algorithm 1 shows the pseudocode to iteratively evaluate the ordered kGTP query for 2 stops, i.e., $m = 2$. Function $FindGNN(l, S)$ is used in Algorithm 1 to compute the lth GNN that has l^{th} smallest total distance with respect to a set of points S.

In each iteration ($l = 1, 2, \ldots$), the algorithm computes k pairs of data points using $FindGNN$ (see Lines 1.5 - 1.12). For each pair p_1 and p_2, p_1 is evaluated as the lth GNN of S and p_2 is evaluated as j^{th} GNN of $\{p_1 \cup D\}$, where j ranges from 1 to k. For each of these k pairs of data points, the algorithm determines the total travel distance for the group trip as $CurrentDist$. We use an array $MinDist[1..k]$ of k entries to prune the pair of data points that cannot be the part of the answer set A and to check the termination condition of the algorithm. Each entry $MinDist[i]$ represents the ith smallest distance of total travel distances for the group trip computed so far. The entries of $MinDist$ are initialized to ∞ and then checked for update after every computation of $CurrentDist$.

Algorithm 1. 2S-Ordered-kGTP-IA(S, D, k)

    Input : $S = \{s_1, s_2, \ldots, s_n\}$, $D = \{d_1, d_2, \ldots, d_n\}$, and k.
    Output: $A = \{\{p_1^1, p_2^1\}, \{p_1^2, p_2^2\}, \ldots, \{p_1^k, p_2^k\}\}$.

1.1 $A \leftarrow \emptyset$
1.2 $MinDist[1..k] \leftarrow \{\infty\}$
1.3 $l \leftarrow 1$
1.4   **repeat**
1.5 $p_1 \leftarrow FindGNN(l, S)$
1.6 $j \leftarrow 1$
1.7   **while** $j \leq k$ **do**
1.8 $p_2 \leftarrow FindGNN(j, \{p_1 \cup D\})$
1.9 $CurrentDist \leftarrow f(S, D, p_1, p_2)$
1.10   **if** $CurrentDist \leq MinDist[k]$ **then**
1.11 $Update(CurrentDist, MinDist, A)$
1.12 $j \leftarrow j + 1$
1.13 $l \leftarrow l + 1$
1.14   **until** $\sum_{i=1}^{n} Dist(s_i, p_1) \leq MinDist[k]$
1.15   **return** A

If $CurrentDist \leq MinDist[k]$, p_1 and p_2 become one of the k pairs that have k smallest total travel distance among the explored pairs of data points so far and the function $Update$ is called to update $MinDist$ and A (Line 1.11).

The parameter l is initialized to 1 and is incremented by 1 at every iteration until k pairs of data points that minimize the total travel distances for the group trip have been determined. We use the following lemma to check the terminating condition of Algorithm 1.

Lemma 1. *Let p_1 be the lth group nearest neighbor with respect to the set of source locations S and $MinDist[k]$ be the kth smallest total travel distances for the group trip computed so far. The k pairs of data points that minimize the total travel distance for the group trip have been found by the algorithm if $\sum_{i=1}^{n} Dist(s_i, p_1) > MinDist[k]$.*

Proof. For any t^{th} group nearest neighbor p_1' with respect to S, where $t > l$, we have $\sum_{i=1}^{n} Dist(s_i, p_1') > \sum_{i=1}^{n} Dist(s_i, p_1) > MinDist[k]$. Thus, if the condition $\sum_{i=1}^{n} Dist(s_i, p_1) > MinDist[k]$ becomes true then there can be no other unexplored pair of data points that can further minimize the total travel distance for the group trip. $\square$

The proof of Lemma 1 also shows the correctness of our proposed nested algorithm, i.e., the answer set, A includes k pairs of data points that have k smallest total travel distances with respect to S and D of a group.

Algorithm 1 also works for a flexible GTP query with slight modifications. For a flexible GTP query we have to execute Algorithm 1 twice considering both orders, first p_1 then p_2 and the reverse order, i.e., first p_2 and then p_1. The evaluation of kGTP queries for each order computes k group trips; among these $2k$ group trips we have to select k group trips with k smallest total travel distances.

Algorithm 2. 3S-kGTP-IA(S, D, k)

    Input : $S = \{s_1, s_2, \ldots, s_n\}$, $D = \{d_1, d_2, \ldots, d_n\}$, and k.
    Output: $A = \{\{p_1^1, p_2^1, p_3^1\}, \{p_1^2, p_2^2, p_3^2\}, \ldots, \{p_1^k, p_2^k, p_3^k\}\}$.
2.1 $A \leftarrow \emptyset$
2.2 $MinDist[1..k] \leftarrow \{\infty\}$
2.3 $l \leftarrow 1$
2.4 **repeat**
2.5 $p_1 \leftarrow FindGNN(l, S)$
2.6 $u \leftarrow 1$
2.7   **repeat**
2.8 $p_2 \leftarrow FindNN(u, p_1)$
2.9 $v \leftarrow 1$
2.10   **while** $v \leq k$ **do**
2.11 $p_3 \leftarrow FindGNN(j, \{p_2 \cup D\})$
2.12 $CurrentDist \leftarrow f(S, D, p_1, p_2, p_3)$
2.13   **if** $CurrentDist \leq MinDist[k]$ **then**
2.14 $Update(CurrentDist, MinDist, A)$
2.15 $v \leftarrow j + 1$
2.16 $u \leftarrow v + 1$
2.17   **until** $\sum_{i=1}^{n} Dist(s_i, p_1) + n \times Dist(p_1, p_2) \leq MinDist[k]$
2.18 $l \leftarrow l + 1$
2.19 **until** $\sum_{i=1}^{n} Dist(s_i, p_1) \leq MinDist[k]$
2.20 **return** A

Consider the case for $m = 3$. Assume that the group specifies the order of visit as $\{1, 2, 3\}$. Algorithm 2 shows the pseudocode for evaluating ordered kGTP queries for $m = 3$. Algorithm 1 works in a similar way to Algorithm 1. The only difference for Algorithm 2 is the addition of an intermediary step using existing nearest neighbor algorithms [20,12]. We use function $FindNN$ to evaluate nearest neighbors (Line 2.8). Similarly, Algorithm 2 can be generalized for ordered kGTP queries with m types of data points by adding $m - 2$ intermediary steps using $FindNN$. For flexible kGTP queries with m types of data points, the query evaluation requires to consider all possible combinations of orderings of the data points and adds additional query processing overheads.

For $m \geq 2$, the limitation of the iterative approach is its high computational overhead. To determine the sets of data points that minimize the total travel distance for the group trip, the iterative approach requires to traverse the same data sets multiple times and the overhead increases with the increase of m.

4.2 Hierarchical Approach

In this section, we present an efficient hierarchical approach to evaluate a kGTP query in a single traversal of R^*-trees $R_1, R_2, \ldots, R_m$. The base idea of our hierarchical algorithm, GTP-HA, is to exploit the hierarchical properties of R^*-trees and use a modified best first search (BFS) on $R_1, R_2, \ldots, R_m$ to find k sets of data points that minimize the total travel distances for the group trip.

Algorithm 3. GTP-HA(S, D, k)

    Input : $S = \{s_1, s_2, \ldots, s_n\}$, $D = \{d_1, d_2, \ldots, d_n\}$, and k.
    Output: $A - \{\{p_1^1, p_2^1, \ldots, p_m^1\}, \{p_1^2, p_2^2, \ldots, p_m^2\}, \ldots, \{p_1^k, p_2^k, \ldots, p_2^m\})\}$.

3.1 $end \leftarrow 0$
3.2 $A \leftarrow \emptyset$
3.3 $MinDist[1..k] \leftarrow \{\infty\}$
3.4 $i \leftarrow 1$
3.5 $Enqueue(Q_p, root_1, root_2, \ldots, root_m, 0, d_{max}(root_1, root_2, \ldots, root_m))$
3.6   **while** Q_p *is not empty and end* $= 0$ **do**
3.7 $\{r_1, r_2, \ldots, r_m, d_{min}(r_1, r_2, \ldots, r_m), d_{max}(r_1, r_2, \ldots, r_m)\} \leftarrow Dequeue(Q_p)$
3.8   **if** $r_1, r_2, \ldots, r_m$ *are data points* **then**
3.9 $\{p_1^i, p_2^i, \ldots, p_m^i\} \leftarrow \{r_1, r_2, \ldots, r_m\}$
3.10   **if** $i - k$ **then**
3.11 $end \leftarrow 1$
3.12 $i \leftarrow i + 1$
3.13   **else**
3.14 $W \leftarrow FindSets(r_1, r_2, \ldots, r_m)$
3.15   **for** *each* $(w_1, w_2, \ldots, w_m) \in W$ **do**
3.16   Compute $d_{min}(w_1, w_2, \ldots, w_m)$ and $d_{max}(w_1, w_2, \ldots, w_m)$
3.17   **if** $d_{min}(w_1, w_2, \ldots, w_m) \leq MinDist[k]$ **then**
3.18 $Enqueue(Q_p, w_1, w_2, \ldots, w_m, d_{min}(w_1, w_2, \ldots, w_m), d_{max}(w_1, w_2, \ldots, w_m))$
3.19   **if** $d_{max}(w_1, w_2, \ldots, w_m) \leq MinDist[k]$ **then**
3.20 $Update(MinDist, d_{max}(w_1, w_2, \ldots, w_m))$

3.21 **return** A

Algorithm 3 shows the steps for GTP-HA. The notations that we have used for hierarchical approach are summarized below:

- r_j: a data point or a minimum bounding rectangle of a node of R_j.
- $\{w_1, w_2, \ldots, w_m\}$: A set of entities, where each entity w_j represents a data point r_j or a minimum bounding rectangle of a child node of r_j.
- $Dist_{min}(.,.)(Dist_{max}(.,.))$: a function that returns the minimum (maximum) distance between two parameters, where a parameter can be either a point or a minimum bounding rectangle of a R^*-tree node.
- $d_{min}(w_1, w_2, \ldots, w_m)$: the distance computed as $\sum_{i=1}^{n} Dist_{min}(s_i, w_1) + n \times Dist_{min}(w_1, w_2) + \cdots + n \times Dist_{min}(w_{m-1}, w_m) + \sum_{i=1}^{n} Dist_{min}(d_i, w_m)$

- $d_{max}(w_1, w_2, \ldots, w_m)$: the distance computed as $\sum_{i=1}^{n} Dist_{max}(s_i, w_1) + n \times Dist_{max}(w_1, w_2) + \cdots + n \times Dist_{max}(w_{m-1}, w_m) + \sum_{i=1}^{n} Dist_{max}(d_i, w_m)$
- $MinDist[k]$: The kth smallest distance of already computed $d_{max}(w_1, w_2, \ldots, w_m)$s.

The algorithm starts the search from the root nodes of $R_1, R_2, \ldots, R_m$. The algorithm inserts the root nodes of $R_1, R_2, \ldots, R_m$ together with their d_{min} and d_{max} into a priority queue Q_p. The elements of Q_p are ordered in order of d_{min}. In each iteration of the search, the algorithm dequeues $r_1, r_2, \ldots, r_m$ from Q_p. If all $r_1, r_2, \ldots, r_m$ are data points then the data points are added to A (Line 3.9). Otherwise, the algorithm computes W using the function $FindSets$. The input parameters of $FindSets$ are $r_1, r_2, \ldots, r_m$. $FindSets$ determines all possible sets $\{w_1, w_2, \ldots, w_m\}$s, where w_j represents either one of the child node of r_j or the data point r_j. In case of an ordered GTP query, $FindSets$ only computes ordered set of data points/R^*-tree nodes, whereas in case of a flexible GTP query, $FindSets$ considers all combination of data points/R^*-tree nodes. Consider an example, where r_1, r_2, r_3 represent data points. If a group specifies the order of visiting types of data points as $\{3, 1, 2\}$ then the computed set is $\{r_3, r_1, r_2\}$, i.e., $w_1 = r_3$, $w_2 = r_1$, and $w_3 = r_2$. If the group is flexible then the computed sets are $\{r_1, r_2, r_3\}$, $\{r_1, r_3, r_2\}$, $\{r_2, r_1, r_3\}$, $\{r_2, r_3, r_1\}$, $\{r_3, r_1, r_2\}$, and $\{r_3, r_2, r_1\}$.

For each set $\{w_1, w_2, \ldots, w_m\}$ in W, the algorithm computes $d_{min}(w_1, w_2, \ldots, w_m)$ and $d_{max}(w_1, w_2, \ldots, w_m)$. Similar to the iterative approach, we use an array $MinDist[1..k]$ to check whether R^*-tree nodes/data points can be pruned using the following lemma:

Lemma 2. *A set of data points or R^*-tree nodes $\{w_1, w_2, \ldots, w_m\}$ can be pruned if $d_{min}(w_1, w_2, \ldots, w_m) > MinDist[k]$.*

If the condition of Lemma 2 is satisfied for any set $\{w_1, w_2, \ldots, w_m\}$, then $\{w_1, w_2, \ldots, w_m\}$ or any set computed from the child nodes of $w_1, w_2, \ldots, w_m$ can never be part of A. If the condition of Lemma 2 is not satisfied for $\{w_1, w_2, \ldots, w_m\}$, i.e., $d_{min}(w_1, w_2, \ldots, w_m) \leq MinDist[k]$, then the algorithm inserts $(w_1, w_2, \ldots, w_m)$ into Q_p and updates $MinDist$ if $d_{max}(w_1, w_2, \ldots, w_m) \leq MinDist[k]$. The search continues until k sets of data points have been added to A.

The following theorem shows the correctness of the hierarchical algorithm.

Theorem 1. *Let A be the answer set returned by $GTP - HA$. A includes k sets of data points that have k smallest total travel distances with respect to S and D of a group.*

Proof. (By contradiction) Assume that a set of data points, $\{p_1^j, p_2^j, \ldots, p_m^j\}$, have j^{th} ($1 \leq j \leq k$) minimum total travel distance with respect to S and D of a group. The set of data points $\{p_1^j, p_2^j, \ldots, p_m^j\}$, may not have been included in A for two reasons: (i) the set of data points $\{p_1^j, p_2^j, \ldots, p_m^j\}$ or R^*-tree nodes $\{w_1, w_2, \ldots, w_m\}$ containing the set of data points $\{p_1^j, p_2^j, \ldots, p_m^j\}$ have been pruned or (ii) the algorithm has terminated before the set of data points $\{p_1^j, p_2^j, \ldots, p_m^j\}$ have been included in A.

We know that the total travel distance of a set $\{p_1^j, p_2^j, \ldots, p_m^j\}$ is within $d_{min}(w_1, w_2, \ldots, w_m)$ and $d_{max}(w_1, w_2, \ldots, w_m)$, where $w_1, w_2, \ldots, w_m$ represent $p_1^j, p_2^j, \ldots, p_m^j$ or a minimum bounding rectangle of R^*-tree node containing $p_1^j, p_2^j, \ldots, p_m^j$, respectively. In the algorithm, $Mindist[k]$ represents the current

k^{th} smallest distance of already computed $d_{max}(w_1, w_2, \ldots, w_m)$s and remains same or decreases with the execution of the algorithm. According to the assumption, $d_{min}(w_1, w_2, \ldots, w_m) \leq Mindist[k]$. However, in the algorithm, $(w_1, w_2, \ldots, w_m)$ is only pruned if $d_{min}(w_1, w_2, \ldots, w_m) > Mindist[k]$, which contradicts the assumption.

Further, the algorithm terminates when k sets of data points have been dequeued from Q_p, where elements of Q_p are maintained in order of d_{min}s. If $\{p_1^j, p_2^j, \ldots, p_m^j\}$ is not included in A then according to our algorithm $\{p_1^j, p_2^j, \ldots, p_m^j\}$ has not been dequeued as one of k sets of data points and has total travel distance larger than the kth smallest total travel distance, which again contradicts the assumption.   □

To further improve the pruning capabilities of our hierarchical algorithm, we can first determine the upper bound of the kth minimum total travel distance for the group trip and use it to prune data points/R^*-tree nodes while computing the optimal total travel distance. We use the following heuristics to determine the upper bound of the kth minimum total travel distance for the group trip.

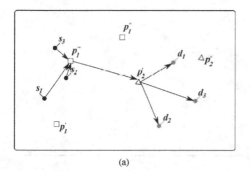 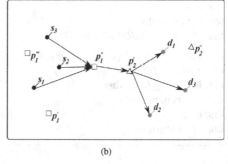

(a)   (b)

Fig. 2. Two example scenarios (a) and (b) for $k = 1$ and $m = 2$, where Heuristics 1 and Heuristic 2 are applied, respectively, to compute the upper bound

Heuristic 1. *Let p_1 be the GNN with respect to S, p_j be the nearest neighbor from p_{j-1} for $2 \geq j < m$, and p_m be the kth GNN with respect to $p_{m-1} \cup D$. The upper bound of the kth minimum total travel distance for the group trip is computed as $f(S, D, p_1, p_2, \ldots, p_m)$.*

Heuristic 2. *Let p_1 be the GNN with respect to $S \cup D$, p_j be the GNN with respect to $p_{j-1} \cup D$ for $2 \geq j < m$, and p_m be the kth GNN with respect to $p_{m-1} \cup D$. The upper bound of the kth minimum total travel distance for the group trip is computed as $f(S, D, p_1, p_2, \ldots, p_m)$.*

In Heuristic 1, the upper bound is computed based only on the distance, whereas in Heuristic 2, the direction of D from S in addition to the distance are used for upper bound computation. Depending on the distribution of locations of sources, destinations and data points, Heuristic 1 or Heuristic 2 can provide the smaller upper bound of the kth total travel distance. Figure 2 shows example scenarios for $k = 1$ and $m = 2$, where Heuristics 1 and Heuristic 2 provide smaller upper bound in Figure 2(a) and 2(b), respectively. In Figure 2(a), the upper bound is computed for the pair (p_1''', p_2') using Heuristics 1 and in Figure 2(b), the upper bound is computed for the pair (p_1'', p_2') using Heuristics 2.

Algorithm 4. UpperBound_MinDist(S, D, k)

Input : $S = \{s_1, s_2, \ldots, s_n\}$, $D = \{d_1, d_2, \ldots, d_n\}$, and k.
Output: The upper bound of the kth minimum total travel distance.
```
// Heuristic 1
```
4.1 $p_1 \leftarrow FindGNN(1, S)$
4.2 $i \leftarrow 2$
4.3 **for** $i < m$ **do**
4.4 $\quad\lfloor\ p_i \leftarrow FindNN(1, p_{i-1})$
4.5 $p_m \leftarrow FindGNN(k, p_{m-1} \cup D)$
4.6 $Dist_1 \leftarrow f(S, D, p_1, p_2, \ldots, p_m)$
```
// Heuristic 2
```
4.7 $p_1 \leftarrow FindGNN(1, S \cup D)$
4.8 $i \leftarrow 2$
4.9 **for** $i < m$ **do**
4.10 $\quad\lfloor\ p_i \leftarrow FindGNN(1, p_{i-1} \cup D)$
4.11 $p_m \leftarrow FindGNN(k, p_{m-1} \cup D)$
4.12 $Dist_2 \leftarrow f(S, D, p_1, p_2, \ldots, p_m)$
4.13 **if** $Dist_1 \leq Dist_2$ **then**
4.14 $\quad\mid$ **return** $Dist_1$
4.15 **else**
4.16 $\quad\lfloor$ **return** $Dist_2$

Algorithm 4 shows the steps for computing the upper bound of the kth minimum total travel distance for the group trip. The algorithm determines the upper bound using Heuristic 1 and Heuristic 2 separately (Line 4.3 and Line 4.6, respectively) and returns the smaller upper bound of the kth total travel distance. Note that the functions $FindGNN$ and $FindNN$ used in Algorithm 4 to compute the GNN and the nearest neighbor are the same function used in Algorithm 2.

After computing the upper bound, in Algorithm 3, $MinDist[k]$ is initialized with the upper bound of the kth minimum total travel distance for the group trip instead of ∞ (Line 3.3) and used to prune data points/R^*-tree nodes using Lemma 2 (Line 3.17).

5 Experiments

In this section, we evaluate the performance of our proposed algorithm kGTP-HA with the base line algorithm kGTP-IA through an extensive set of experiments. In our experiments, we use two variants of our hierarchical algorithms: one without upper-bound of kth minimum total travel distance, called kGTP-HA1, and the other with pre-computed upper-bound, called kGTP-HA2. We use both real and synthetic data sets in our experiments. The real data set C is taken from 62,556 points of interests (i.e., data points) of California. We generate synthetic data sets using a uniform (U) and a Zipfian (Z) distribution, respectively. We vary the size of U and Z as 5000, 10,000, 15,000, and 20,000 point locations. For all data sets, the data space is normalized into a span of $10,000 \times 10,000$ square units. Since we need m categories of data points, we equally divide the set of data points of a dataset and index them using R*-trees. We run the experiments on a desktop with a Intel Core 2 Duo 2.40 GHz CPU and 4 GBytes RAM.

In realistic scenarios where a group of users plans a trip that involves different types of data objects, the number of different data points is typically limited to 2 or 3. A group may want to go to dinner, enjoy a movie at a cinema afterwards, and possibly go later to a pub, but it is unlikely that a group trip will be planned in advance for a larger number of places, i.e., data points. Thus we set the number of data types (m) to 2 and 3 in our experiments. A group will usually know in advance in which order they will visit the different places. This also motivates to consider ordered kGTP queries in our experiments. Note that our algorithm works with any number of data types for both ordered and flexible kGTP queries.

We vary different parameters: the group size (n), the number of required sets of data points (k), the query area, i.e., the minimum bounding rectangle covering the source and destination locations (M), and the dataset size in different sets of experiments. In all experiments, we measure IO costs and the query processing time to measure the efficiency of our algorithms. Table 2 summarizes the values used for each parameter in our experiments and their default values.

Table 2. Experiment Setup

Parameter	Range	Default
Group size	4, 16, 64, 256	64
Query area M	2%, 4%, 8%, 16%	4%
k	2, 4, 8, 16	4
Data set size (Synthetic)	5K, 10K, 15K, 20K	-

We first present the experimental results for two types of data points (i.e., $m = 2$), which we call a 2 stops GTP (2S-kGTP) query. Then we present the results for three types of data points (i.e., $m = 3$), which we call a 3 stops GTP (3S-kGTP) query.

5.1   2S-kGTP Queries

In this section, for each set of experiments, we execute 100 2S-kGTP queries and show the average results. These queries are generated randomly in the total space.

Effect of Group Size. Figures 3(a) and 3(b) show the IOs and the processing time, respectively, required by 2S-GTP-IA, 2S-GTP-HA1, and 2S-GTP-HA2 for different group sizes. Both IOs and the processing time increase with the increase of group size for all three approaches. Figure 3(a) shows that 2S-GTP-IA requires on average two order of magnitude more IOs than that of both 2S-GTP-HA1 and 2S-GTP-HA2. We observe in Figure 3(b) that the processing time of 2S-GTP-IA is on average an order of magnitude higher than that of 2S-GTP-HA. We also observe that 2S-GTP-HA2 takes on average 20% less IOs and 27% less processing time than 2S-GTP-HA1, which is expected due to a tighter upper-bound of the termination condition of 2S-GTP-HA2.

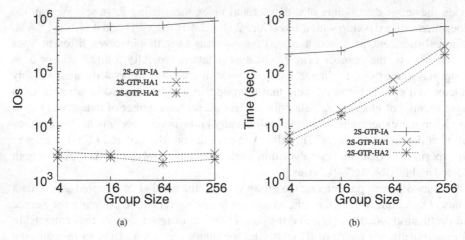

Fig. 3. Effect of group size (data set C)

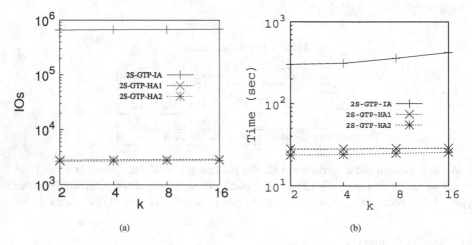

Fig. 4. Effect of k (data set C)

Effect of k. In this set of experiments, we vary the value of k as 2, 4, 8, and 16. In Figure 4(a), we observe that the IO cost slightly increases with the increases of k for 2S-GTP-IA and remains almost constant for 2S-GTP-HA1 and 2S-GTP-HA2. From the experimental results, we find that 2S-GTP-IA requires at least two orders of magnitude times more IOs than that of both 2S-GTP-HA1 and 2S-GTP-HA2, and 2S-GTP-HA2 takes on average 4% less IOs than that of 2S-GTP-HA1.

Figure 4(b) shows that the processing time of 2S-GTP-IA is at least one order of magnitude higher than that of both 2S-GTP-HA1 and 2S-GTP-HA2. On the other hand, the processing time of 2S-GTP-HA2 is on average 13% lower than that of 2S-GTP-HA1, which is expected as 2S-GTP-HA2 uses a tighter upper-bound to facilitate more pruning capabilities.

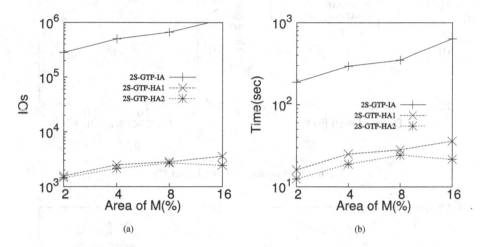

(a)   (b)

Fig. 5. Effect of query area M(data set C)

Effect of Query Area (M). In this set of experiments, we vary the query area M as 2%, 4%, 8%, and 16% of the data space. Figures 5(a) and 5(b) show the IO cost and the processing time, respectively, required by 2S-GTP-IA, 2S-GTP-HA1, and 2S-GTP-HA2 for different M. The IO cost and the processing time increase with the increase of M area for all three algorithms as for a larger M we need retrieve to access more data points from R*-trees than that of a smaller M.

Figure 5(a) shows that 2S-GTP-IA requires at least two orders of magnitude more IOs than that of both 2S-GTP-HA1 and 2S-GTP-HA2, and 2S-GTP-HA2 takes 14% less IOs than that of 2S-GTP-HA1. Similarly, Figure 5(b) shows that the processing time of 2S-GTP-IA is on average one order of magnitude higher than that of both 2S-GTP-HA1 and 2S-GTP-HA2, and 2S-GTP-HA2 takes on average 25% less processing time than that of 2S-GTP-HA1.

Effect of Data Set Size. In this set of experiments, we vary the dataset size as 5000, 10000, 15000, and 20000 for both uniform (U) and Zipfian (Z) distributions.

Figures 6(a) and 6(b) show the IO cost and processing time, respectively, for different data set sizes with U distribution. The experimental results show that both 2S-GTP-HA1 and 2S-GTP-HA2 outperform 2S-GTP-IA in a greater margin for a larger dataset in terms of both IOs and processing time. Figures 7(a) and 7(b) show the IO cost and processing time, respectively, for different data set sizes with Z distribution and we observe that the experimental results for Z distribution follow similar trends to U distribution.

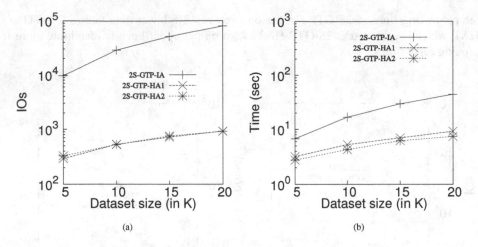

Fig. 6. Effect of dataset size (dataset U)

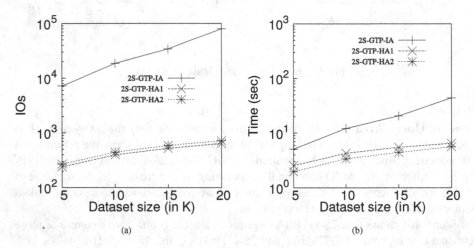

Fig. 7. Effect of dataset size (dataset Z)

5.2   3S-kGTP Queries

From the experimental results of 2S-kGTP queries, we observe that both of our hierarchical approaches significantly outperform the iterative approach. Thus, in the next set of experiments for 3S-kGTP queries, we only evaluate the performance of two variants of our hierarchical approaches, 3S-GTP-HA1 and 3S-GTP-HA2. In these experiments, we vary different parameters such as group size, k, M, and dataset size. Figures 8(a) and 8(b) show the effect of varying group size on IO cost and processing time, respectively, for California dataset. The experimental results show that 3S-GTP-HA2 always outperforms 3S-GTP-HA1 in terms of both IOs and processing time for any group size. We observe in Figures 8(a) that the processing time of 3S-GTP-HA2 is on average 55% less than that of 3S-GTP-HA1. Similarly, 3S-GTP-HA2 requires on average 64% less

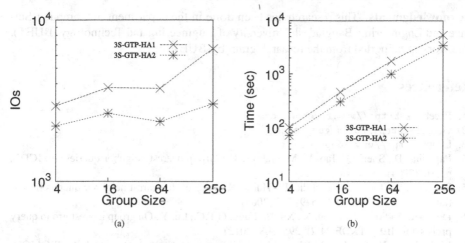

Fig. 8. Effect of group size (data set C)

IOs than that of 3S-GTP-HA1. The results also show that 3S-GTP-HA2 outperforms 3S-GTP-HA1 in a greater margin for a larger group size.

We omit the experimental results for other parameters (e.g., k, M, and dataset sizes) for space constraint. The results show similar behavior as of 2S-GTP queries evaluation presented in the previous section. However, the processing time and IOs of 3S-GTP queries are higher than those of 2S-GTP queries, which is expected. Since in realistic scenarios, the number of different data points is typically limited to 2 or 3 for a kGTP query, the trend of increased processing overheads for a larger value of m would not effect the applicability of our algorithms.

6   Conclusion

In this paper, we introduced a new query type: the k group trip planning (kGTP) query. This query has many real world applications, in particular with the increased use of location-based social networks. We proposed an efficient hierarchical algorithm to evaluate kGTP queries. Our hierarchical algorithm evaluates the query with a single search on the database. We also developed an iterative approach as baseline method, which transforms kGTP queries into group nearest neighbor queries. We performed extensive experiments to show the efficiency of our algorithms and benchmark our hierarchical approach against the baseline method. The hierarchical algorithm performs on average an order of magnitude time faster and requires on average two orders of magnitude less IOs than the iterative approach. In the future, we plan to evaluate GTP queries in road networks. We also aim to protect the location privacy [21,22] of users while evaluating GTP queries, i.e., we will study scenarios where the group of users does not reveal their locations among each other.

Acknowledgments. This research has been done in the department of Computer Science and Engineering, Bangladesh University of Engineering and Technology (BUET). The work is supported from the research grant by BUET.

References

1. Facebook, http://www.facebook.com
2. Google+, http://plus.google.com
3. Loopt, http://www.loopt.com
4. Papadias, D., Shen, Q., Tao, Y., Mouratidis, K.: Group nearest neighbor queries. In: ICDE, p. 301 (2004)
5. Papadias, D., Tao, Y., Mouratidis, K., Hui, C.K.: Aggregate nearest neighbor queries in spatial databases. TODS 30(2), 529–576 (2005)
6. Deng, K., Sadiq, S.W., Zhou, X., Xu, H., Fung, G.P.C., Lu, Y.: On group nearest group query processing. IEEE TKDE 24(2), 295–308 (2012)
7. Li, Y., Li, F., Yi, K., Yao, B., Wang, M.: Flexible aggregate similarity search. In: SIGMOD, pp. 1009–1020 (2011)
8. Sharifzadeh, M., Kolahdouzan, M., Shahabi, C.: The optimal sequenced route query. The VLDB Journal 17(4), 765–787 (2008)
9. Li, F., Cheng, D., Hadjieleftheriou, M., Kollios, G., Teng, S.-H.: On trip planning queries in spatial databases. In: Medeiros, C.B., Egenhofer, M., Bertino, E. (eds.) SSTD 2005. LNCS, vol. 3633, pp. 273–290. Springer, Heidelberg (2005)
10. Guttman, A.: R-trees: a dynamic index structure for spatial searching. In: SIGMOD, pp. 47–57 (1984)
11. Beckmann, N., Kriegel, H.P., Schneider, R., Seeger, B.: The R*-tree: an efficient and robust access method for points and rectangles. SIGMOD Rec. 19(2), 322–331 (1990)
12. Hjaltason, G.R., Samet, H.: Ranking in spatial databases. In: International Symposium on Advances in Spatial Databases, pp. 83–95 (1995)
13. Li, F., Yao, B., Kumar, P.: Group enclosing queries. IEEE TKDE 23(10), 1526–1540 (2011)
14. Chen, H., Ku, W.S., Sun, M.T., Zimmermann, R.: The multi-rule partial sequenced route query. In: GIS, pp. 10:1–10:10(2008)
15. Ohsawa, Y., Htoo, H., Sonehara, N., Sakauchi, M.: Sequenced route query in road network distance based on incremental euclidean restriction. In: Liddle, S.W., Schewe, K.-D., Tjoa, A.M., Zhou, X. (eds.) DEXA 2012, Part I. LNCS, vol. 7446, pp. 484–491. Springer, Heidelberg (2012)
16. Malviya, N., Madden, S., Bhattacharya, A.: A continuous query system for dynamic route planning. In: ICDE, pp. 792–803 (2011)
17. Chen, Z., Shen, H.T., Zhou, X.: Discovering popular routes from trajectories. In: ICDE, pp. 900–911 (2011)
18. Geisberger, R., Kobitzsch, M., Sanders, P.: Route planning with flexible objective functions. In: ALENEX, pp. 124–137 (2010)
19. Chen, Z., Shen, H.T., Zhou, X., Zheng, Y., Xie, X.: Searching trajectories by locations: an efficiency study. In: SIGMOD, pp. 255–266 (2010)
20. Roussopoulos, N., Kelley, S., Vincent, F.: Nearest neighbor queries. In: SIGMOD, pp. 71–79 (1995)
21. Hashem, T., Kulik, L., Zhang, R.: Privacy preserving group nearest neighbor queries. In: EDBT, pp. 489–500 (2010)
22. Mokbel, M.F., Chow, C.Y., Aref, W.G.: The new casper: query processing for location services without compromising privacy. In: VLDB, pp. 763–774 (2006)

Reverse-k-Nearest-Neighbor Join Processing*

Tobias Emrich, Hans-Peter Kriegel, Peer Kröger, Johannes Niedermayer,
Matthias Renz, and Andreas Züfle

Institute for Informatics, Ludwig-Maximilians-Universität München
{emrich,kriegel,kroeger,niedermayer,renz,zuefle}@dbs.ifi.lmu.de

Abstract. A reverse k-nearest neighbour (RkNN) query determines the objects
from a database that have the query as one of their k-nearest neighbors. Processing such a query has received plenty of attention in research. However, the
effect of running multiple RkNN queries at once (join) or within a short time
interval (bulk/group query) has only received little attention so far. In this paper, we analyze different types of RkNN joins and discuss possible solutions for
solving the non-trivial variants of this problem, including self and mutual pruning
strategies. The results indicate that even with a moderate number of query objects
($|R| \approx 0.0007|S|$), the performance (CPU) of the state-of-the-art mutual prun-
ing based RkNN-queries deteriorates and hence algorithms based on self pruning
without precomputation produce better results. During an extensive performance
analysis we provide evaluation results showing the IO and CPU performance of
the compared algorithms for a wide range of different setups and suggest appropriate query algorithms for specific scenarios.

1 Introduction

A Reverse k-Nearest Neighbor (RkNN) query retrieves all objects from a multidimensional database having a given query object as one of their k nearest neighbors. Various algorithms for efficient RkNN query processing have been studied under different
conditions due to the query's relevance in a wide variety of domains — applications
include decision support, profile-based marketing and similarity updates in spatial and
multimedia databases.

An important problem in database environments that has not received much attention
so far is the scenario where the query does not consist of a single point but instead of a
whole set of points, for each of which an RkNN query has to be performed. This setting
is often referred to as *group query*, *bulk query* or simply *join* of two sets R and S. This
problem frequently arises in the strategic decision making process of companies that
supply products to clients which are typically shops (cf. Figure 1). Consider a supplier
(e.g. supplying video stores) that has a set of products R (e.g. videos each described
by a given set of features like genre, length, etc.). Each client (e.g. video store) also
has a portfolio S (e.g. a set of videos) which typically include different groups of products (videos) satisfying different preferences and, thus, different groups of customers.
In order to judge which products should be offered by the supplier to a given client,
the supplier needs information about which objects in R fit to the characteristics of the
client's portfolio S and/or would be a good supplement to extend this portfolio. From

---

* Part of this work was supported by the DFG under grant number KR 3358/4-1.

M.A. Nascimento et al. (Eds.): SSTD 2013, LNCS 8098, pp. 277–294, 2013.

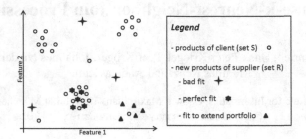

Fig. 1. Application of R*k*NN join between two sets of products R and S for product (set) recommendation

the supplier's point of view it is particularly important to know about the data characteristics. Thus, many companies rely on the following process in order to recommend updates for their clients: First, for each product s in S the kNNs are computed. Second, for each product r in R, it is examined whether or not r is amongst the kNN of which product s. In other words, for each r, an R*k*NN query in S is launched. If r has a lot of R*k*NNs in S, this indicates that r fits well to a corresponding group within the data distribution of s (although this usually needs additional inspection). If r has no R*k*NNs in S, r obviously does not fit well. In addition, if many products r_i have the same s as their R*k*NN and s is so far an outlier in S, the products r_i may be a good addition to extend the portfolio (probably depending on the current success of s). Analogously, R*k*NN joins can also be employed for solving *inverse* queries [1] where the task is to find for a given set of query objects the set of database objects, having all (/most) query objects in their kNN set. Furthermore, the R*k*NN join operation plays a key role in updating patterns derived by almost all data mining algorithms that rely on kNN information after changes to the database, e.g. shared-neighbor clustering [2,3] and kNN-based outlier detection [4,5].

For evaluating single R*k*NN queries, two groups of algorithms have evolved over time. *Self pruning* approaches (e.g. [6,7,8]) have to perform costly precomputations in order to materialize kNN-spheres for all database objects. These kNN-spheres are used for pruning candidates during query execution. In contrast, *mutual pruning* approaches (e.g. [9,10,11]) do not perform any precomputations. This results in more flexibility in terms of updates and the choice of k because materialized results need not to be updated each time the database changes. Furthermore, in contrast to self pruning approaches, mutual pruning approaches do not require the parameter k to be known prior to index generation. However, mutual pruning introduces costly refinement of candidates, resulting in higher overall cost.

Recently, we sketched a mutual pruning approach for R*k*NN join processing in [12]. In this paper, we discuss how self pruning approaches adapt to R*k*NN joins and we compare their performance to the existing mutual pruning approach. We will show that the overhead of performing a traditional R*k*NN-query for each point in the query set R separately cannot be justified, even if R is small. Additionally, we will see that with increasing size of R self pruning approaches that compute kNN spheres on the fly become more useful than approaches based on mutual pruning.

Beside an overview of general related work (cf. Section 3), the key contributions of this paper are as follows:

- We provide a formal overview over variants of the RkNN join, showing that the monochromatic RkNN join addressed within this paper is a non-trivial instance of the RkNN join problem (cf. Section 2).
- We suggest an algorithm for performing monochromatic RkNN joins based on self-pruning in Section 4. It does not rely on materialized information but computes necessary information on the fly. Thus, this approach features great flexibility in terms of database updates and the choice of k.
- A systematic comparison in Section 5 shows that the performance of classic algorithms for single RkNN queries deteriorates even for relatively small query sets ($|R| \approx 0.0007|S|$). Furthermore, the proposed solution outperforms the mutual-pruning solution of [12] by orders of magnitude.

Section 6 concludes the paper.

2   Problem Definition

In this section, we recap the definition of RkNN queries and formally define the RkNN join and important variants.

2.1   Background

Given a finite multidimensional data set $S \subset \mathbb{R}^d$ ($s_i \in \mathbb{R}^d$) and a query point $r \in \mathbb{R}^d$, a k-nearest neighbor (kNN) query returns the k nearest neighbors of r in S:

$$kNN(r, S) = \{s \in S : |\{s' \in S : dist(s', r) < dist(s, r)\}| < k\}$$

A monochromatic RkNN query, where r and $s \in S$ have the same type, can be defined by employing the kNN query:

$$RkNN(r, S) = \{s \in S | r \in (k + 1)NN(s, S \cup \{r\})\}$$

Thus, an RkNN query returns all points $s_i \in S$ that would have r as one of its nearest neighbors. In Figure 2 (a) an R2NN query is shown. Arrows denote a subset of the 2NN relationships between points from S. Since r is closer to s_2 than its 2NN s_1, the result set of an R2NN query with query point r is $\{s_2\}$. s_3 is not a result of the query since its 2NN s_2 is closer than r. Note that the RkNN query is not symmetric, i.e. the kNN result $kNN(r,S) \neq RkNN(r, S)$, because the 2NNs of r are s_2 and s_3. Therefore the result of an RkNN(r,S) query cannot be directly inferred from the result of a kNN query $kNN(r,S)$.

Although similar, the bichromatic RkNN query is slightly different. In this case, two sets R and S are given. The goal is to compute all points in S for which a query point $r \in R$ is one of the k closest points from R [13]:

$$BRkNN(r, R, S) = \{s \in S | r \in kNN(s, R)\}$$

Both variants of RkNN queries vary in the data set on which the kNN-query is performed: for an RkNN query the kNN-query is (with some modifications) performed on S, whereas for a BRkNN-query it is performed on R.

.

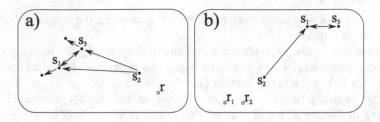

Fig. 2. Monochromatic R2NN Query (a), Monochromatic R1NN Join(b)

2.2 The R*k*NN Join

In this paper, we address the problem of R*k*NN joins. Given two sets R and S, the goal of a monochromatic R*k*NN join is to compute, for each point $r \in R$ its monochromatic R*k*NNs in S.

Definition 1 (Monochromatic R*k*NN join [12]). *Given finite sets $S \subset \mathbb{R}^d$ and $R \subset \mathbb{R}^d$, the monochromatic R*k*NN join $R \overset{MRkNN}{\bowtie} S$ returns a set of pairs containing for each $r \in R$ its R*k*NN from S:$R \overset{MRkNN}{\bowtie} S = \{(r, s)|r \in R \wedge s \in S \wedge s \in RkNN(r, S)\}$*

An example for $k = 1$ can be found in Figure 2 (b). The result for both objects from R in this example is $R1NN(r_1) = R1NN(r_2) = \{s_2\}$, i.e. $R \overset{MRkNN}{\bowtie} S = \{(r_1, s_2), (r_2, s_2)\}$. Note that the elements r_1 and r_2 from R do not influence each other, i.e., r_1 cannot be a result object of r_2 and vice versa. This follows directly from the definition of the MR*k*NN join. A variant of the R*k*NN join is the bichromatic join $R \overset{BRkNN}{\bowtie} S$ where for all $r \in R$ a bichromatic R*k*NN query is performed:

Definition 2 (Bichromatic R*k*NN join).*Given two finite sets $S \subset \mathbb{R}^d$ and $R \subset \mathbb{R}^d$, the bichromatic R*k*NN join $R \overset{BRkNN}{\bowtie} S$ returns a set of pairs containing for each $r \in R$ its BR*k*NN from S:$R \overset{BRkNN}{\bowtie} S = \{(r, s)|r \in R \wedge s \in S \wedge s \in BRkNN(r, R, S)\}$*

A BR*k*NN join can be expressed with a *k*NN join. Based on the definition of the BR*k*NN query, the BR*k*NN join can be converted:

$$R \overset{BRkNN}{\bowtie} S = \{(r, s) : r \in R \wedge s \in S \wedge r \in kNN(s, R)\}$$

The *k*NN join can be defined as:

$$S \overset{kNN}{\bowtie} R = \{(s, r) : s \in S \wedge r \in R \wedge r \in kNN(s, R)\}$$

$$\Leftrightarrow R \overset{BRkNN}{\bowtie} S = S \overset{kNN}{\bowtie} R$$

Hence, a BR*k*NN join can be performed by reusing the techniques from *k*NN research, such as [14,15]. Furthermore, this problem has already been solved for an incremental setting in [16].

Last but not least let us analyze the special case where $R = S$. The monochromatic R*k*NN-self-join can be defined as follows:

Definition 3 (Monochromatic RkNN Self Join). *Given a finite set $S \subset \mathbb{R}^d$, the mono-chromatic RkNN self join is defined as $S \overset{MRkNN}{\bowtie} S$.*

Performing a monochromatic RkNN-self-join is trivial, since it is possible to perform a self-kNN-join on S:

$$S \overset{MRkNN}{\bowtie} S = \{(r,s) | r \in S \wedge s \in S \wedge r \in (k+1)NN(s, S \cup \{r\})\}$$

$$= \{(r,s) | r \in S \wedge s \in S \wedge r \in kNN(s,S)\}$$

The resulting pairs (r,s) of the kNN-join just have to be inverted to produce the result of the RkNN join. Notice that with a self RkNN join, a query point always returns itself as one of its RkNNs.

In summary, we observe that the monochromatic RkNN join where $R \neq S$ cannot be matched to existing well-addressed problems. Therefore, we will address the processing of this problem in this paper.

3 Related Work

The problem of performing multiple RkNN queries at a time, i.e., a RkNN join, has hardly been addressed. The authors of [16] deal with incremental bichromatic RkNN joins as a by-product of incremental kNN joins, aiming at maintaining a result set over time instead of performing bulk evaluation of large sets. Since it does not address the problem of a monochromatic join, it solves a different problem. In [12] we employed a mutual pruning approach for RkNN joins. In this paper however, we will investigate self pruning approaches for RkNN joins.

Contrary to RkNN joins, the problem of efficiently supporting RkNN queries has been studied extensively in the past years. Existing approaches for Euclidean RkNN search can be classified as self pruning approaches or mutual pruning approaches. Since these approaches are the foundations of RkNN join processing, we review them in the following.

Self pruning approaches like the RNN-Tree [6] and the RdNN-Tree [7] are usually designed on top of a hierarchically organized tree-like index structure. They try to conservatively or exactly estimate the kNN distance of each index entry e. If this estimate is smaller than the distance of e to the query q, then e can be pruned. Thereby, self pruning approaches do not usually consider other entries (database points or index nodes) in order to estimate the kNN distance of an entry e, but simply pre-compute kNN distances of database points and propagate these distances to higher level index nodes. The major limitation of these approaches is that the pre-computation and update (in case of database changes) of kNN distances is time consuming, the storage of these distances wastes memory and, thus, these methods are usually limited to one specific or very few values of k. Approaches like [8,17] try to overcome these limitations by using approximations of kNN distances but this in turn yields an additional refinement overhead during query processing – or only approximate results.

Mutual pruning approaches such as [9,10,11,18] use other points to prune a given index entry e. The most general and efficient approach called TPL is presented in [11].

It uses any hierarchical tree-based index structure such as an R-Tree to compute a nearest neighbor ranking of the query point q. The key idea is to iteratively construct Voronoi hyper-planes around q w.r.t. to the points from the ranking. Points and index entries that are beyond k Voronoi hyper-planes w.r.t. q can be pruned and do not have to be considered for Voronoi construction anymore.

A *combination of self- and mutual pruning* is presented in [19]. It obtains conservative and progressive distance approximations between a query point and arbitrarily approximated regions of a metric index structure. A further specialization of this approach to Euclidean data is proposed in [20] exploiting geometric properties to achieve a higher pruning power.

Beside solutions for Euclidean data, solutions for *general metric spaces* (e.g. [8,17]) usually implement a self pruning approach. Typically, metric approaches are less efficient than the approaches tailored for Euclidean data because they cannot make use of the Euclidean geometry.

Furthermore, there exist approximate solutions for the RkNN query problem that aim at reducing the query execution time at the cost of accuracy (e.g. [10,21]).

4   A Self Pruning Approach

4.1   General Idea

As the key contribution of this paper, we developed a self pruning approach that does not rely on materialized information as existing self pruning techniques but computes kNN distances on the fly. For single RkNN queries, such a self pruning approach obviously suffers from this overhead. However, intuitively, these on the fly computations may amortize for a large number of queries, i.e., a large outer set R (and we will see in the experiments that the break even point is surprisingly small). Thus, the idea of the following solution provides a dedicated algorithm to perform an RkNN join using the techniques of self pruning. In the following, we still assume that $\mathcal{R}$ and $\mathcal{S}$ are aR^*-tree representations of R and S, respectively. Let us first decompose the definition of the RkNN join into smaller pieces. The RkNN query returns all pairs of points $(r \in R, s \in S)$ such that r is located in the kNN-spheres of s. Therefore, as a first step we simply compute the kNN-spheres of all points in S. This can be done by performing a self-$(k+1)$NN-join on S.[1] Then, for each of the resulting kNN-spheres, the set of points in R that is enclosed by this sphere has to be returned. This can be done by performing an ϵ-range query for each kNN-sphere on the set R. Note that the later query does not correspond to an ϵ-range-join, since the kNN-spheres of each $s \in S$ will usually have different radii. Rather, a "varying-range-join" needs to be applied. This introduces some interesting possibilities of optimization when pruning subtrees in $\mathcal{R}$. We will address this problem in Section 4.3. To summarize, an RkNN join can be performed by combining a self-$(k+1)$NN-join of S with a varying-range-join, a generalization of the ϵ-range-join. We will now introduce algorithms for computing the kNN-join and the varying-range-join.

---

[1] Each point will have itself in its kNN set, thus we need a $k + 1$NN-join to find the kNNs of each point in S.

4.2   Implementing the Self-kNN-Join

A variety of researchers addressed the problem of efficiently computing the result of a kNN-Join. In our implementation we decided to evaluate two different approaches. One of them is a sequential second-order nested-loop join, that does not directly facilitate the structure of the underlying index, however it exploits the spatial proximity of points in leaf nodes of the tree. The other one is a hierarchical join that fully utilizes the index to reduce the number of unnecessary comparisons. Both of them shall be introduced in this section. In Section 5 we will also investigate which of the suggested algorithms fits best for specific data sets. For the sake of clarity, we assume to have two virtual copies S_l and S_r of S in the following. An entry from S_l is denoted by e_l^S and an entry from S_r as e_r^S.

Sequential Self-kNN-Join. For the sequential self-kNN-join we implemented a cache-aware nested-block loop join that takes the specific properties of the self-join, and the spatial proximity of values in the leaf nodes of a tree-structured index into account. It is based on the implementation in the ELKI-framework [22]. The algorithm basically takes a leaf from the index S, performs a self-join within the same leaf first in order to initialize its maximum kNNDist ($MaxKNNDist_{temp}$) as its pruning distance. The pruning distance is used to prune whole leaf pages later on without processing the points contained in them. Then it sequentially accesses all leaf nodes of the index and joins them with the currently processed node. This version is not only easy to implement, it also enables to return partial results at each time where a page e_l^S has been processed such that memory can be freed after directly performing a varying-range-join on the partial result.

For each leaf node e_l^S from S_l, the algorithm proceeds as follows. First, a sequential self-kNN-join of e_l^S is performed in order to initialize the $MaxKNNDist_{temp}()$ of e_l^S. The kNN-join basically collects for each point in $p \in e_l^S$ the k points from e_r^S that are closest to p. Note that the actual kNN's are not relevant, just the distances of current kNN-candidates have to be tracked and hence the memory consumption can be reduced significantly compared to traditional kNN join processing. In a second step, each leaf $e_r^S \neq e_l^S$ of S_r is traversed in arbitrary, but deterministic, order. The page e_r^S can be pruned if $MINDIST(e_r^S, e_l^S) > MaxKNNDist_{temp}(e_l^S)$. If the page cannot be pruned, each point from e_l^S is joined with each point from e_r^S. After joining the contained points, the pruning distance $MaxKNNDist_{temp}(e_l^S)$ is updated when necessary. After finishing the join of e_l^S, the kNN-distances of points in this page can be returned as a partial result. The join of the following leaf node $e_{l'}^S$ of S_l is again joined with itself first, however the join of remaining nodes is performed in opposite order. The intention of this proceeding is that some nodes from the last traversal are still stored in the page cache such that they do not have to be reloaded, reducing the number of page accesses.

Hierarchical Self-kNN-Join. The hierarchical kNN-Join is based on the best-first kNN search algorithm from [15], however with some minor adaptions to fit the special properties of our setting, a self-kNN-join. Actually, this join is only semi-hierarchical since it does not join two trees but rather joins each leaf from one tree with another tree, yielding an asymptotic complexity of $O(|S| \log(|S|))$ for a self join in contrast to $O(|S|^2)$ for the sequential self-join described in the section before. In contrast to a fully hierarchical approach this algorithm enables the possibility to return partial results each time a page e_l^S has been processed; we will exploit this property when processing the varying-range-join.

For each leaf node e_l^S taken from $\mathcal{S}_l$, the tree $\mathcal{S}_r$ is traversed starting with the root node in best-first order according to the $MINDIST$ of the current group and the currently processed sub-tree. For this purpose, the root entry of $\mathcal{S}_r$ is inserted into a priority queue. The priority queue is ordered by increasing $MINDIST$ to e_l^S. $MINDIST$ between two overlapping entries is always zero, however a bigger overlap is usually better than a smaller one since this can reduce the $MaxKNNDist_{temp}$ of the currently processed page more effectively. Therefore e_l^S that overlap e_r^S more than other pages are preferred. The algorithm pops the first entry e_r^S from the priority queue and checks if $MINDIST(e_r^S, e_l^S) \leq MaxKNNDist_{temp}$. Entries with $MINDIST(e_r^S, e_l^S) > MaxKNNDist_{temp}$ are pruned as in the sequential approach. If an entry cannot be pruned, the corresponding node has to be accessed. If the node is an intermediate node, all its children that cannot be pruned are inserted into the priority queue, i.e., the page is resolved. If the node is a leaf node, the page is joined. Joining is similar to the proceeding of the sequential join, however it employs some additional improvements from [15]: points $r_i \in e_r^S$ with $MINDIST(e_l^S, r_i) > kNNDist_{temp}(e_l^S)$ and all points $l_i \in e_l^S$ with $NNDist(l_i) < MINDIST(l_i, e_r^S)$ can be pruned during an initial scan of the processed pages. This preprocessing reduces the quadratic cost of comparing all entries from one node with all entries from the other node. After finishing the join $MaxKNNDist_{temp}$ is updated when necessary.

4.3   Implementing the Varying-Range-Join

A trivial solution for performing a varying-range-join would be to sequentially search the set R for each $s \in S$ to find $\{r \in R | dist(r, s) < kNNDist(s)\}$. This operation returns the subset of R that would contain s as one of its RkNN-results. As a first improvement, the tree structure of $\mathcal{R}$ can be exploited to reduce the asymptotic complexity to a logarithmic one by performing a depth-first traversal: Only nodes e^R of $\mathcal{R}$ that intersect the kNN-sphere of s need to be considered, i.e., subtrees with $MINDIST(e^R, s) > kNNDist(s))$ can be pruned. For reducing the number of disc accesses, e.g. if a slow HDD is used instead of a fast SSD, it would be a good idea to traverse $\mathcal{R}$ less often with a larger subset of S. Since both self-kNN-join algorithms join pages e^S of objects, partial results show spatial proximity. This spatial proximity can be used to prune subtrees from $\mathcal{R}$ that cannot contain result candidates for any point in e^S, greatly speeding up the range query. This technique can even be extended. The straightforward approach for computing an RkNN join-result would be to materialize all kNN-spheres first and then perform a range query for each of the spheres on $\mathcal{R}$. However, in order to keep the memory consumption for storing kNN-spheres low, a readily kNN-joined page can be directly range-joined with $\mathcal{R}$ such that the corresponding kNN-spheres of the page do not have to be stored further.

This varying-range-join is implemented in form of a depth-first index traversal. It receives a set e^S containing points from S in combination with their kNN-distances and $\mathcal{R}$. The algorithm first checks each child of the root node of $\mathcal{R}$ whether or not it has to be accessed, i.e., if this page could contain a varying-range-join result of any

---

Algorithm 1. VaryingRangeJoin(Entry e^R, Set e^S)

  if R is LeafNode **then**
    for all $r \in e^R$ **do**
      for all $s \in e^S$ **do**
        if $dist(r,s) \leq kNNDist(s)$ **then**
          report (r,s) as a result tuple
        end if
      end for
    end for
  else
    for all $e_1^R \in e^R$ **do**
      if PossibleCandidate(e^S, e_1^R) **then**
        VaryingRangeJoin(e_1^R, e^S)
      end if
    end for
  end if

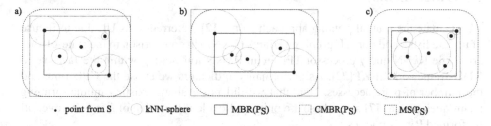

  • point from S   ◯ kNN-sphere   ☐ MBR(P_S)   ☐ CMBR(P_S)   ☐ MS(P_S)

Fig. 3. Comparison of $MS(e^S)$ and $CMBR(e^S)$ in an (a) average, (b) worst, and (c) best case

point in e^S. If the child has to be accessed, it proceeds in a depth-first order, accessing its children and testing them recursively. The pseudo code is depicted in Algorithm 1. Checking if any child from e^R might have a point from e^S as an RkNN is evaluated in the function PossibleCandidate(e^S, e_1^R). Checking can be achieved in two ways, by employing the Minkowski sum, or by laying an MBR around all kNN-spheres contained in e^S:

- Use the Minkowski sum $MS(e^S)$, following the idea of [7]: This method checks whether the MBR of $e^{\mathcal{R}}$ has a distance of more than $\max_{s \in e^S} kNNDist(s)$ to the MBR of the points in e^S and can be evaluated by checking whether $MINDIST(MBR(e_1^R), MBR(e^S))$ is less than or equal to $\max_{s \in e^S} kNNDist(s)$

- Build a bounding box around all kNN-spheres, following [6]: Let $CMBR(e^S) = MBR(\bigcup_{s \in e^S} kNNSphere(s))$. e_1^R has to be evaluated if $CMBR(e^S) \cap MBR(e_1^R) \neq \emptyset$. The MBR $CMBR(e^S)$ can be calculated as:

$$CMBR(e^S).min_i = \min_{p_j \in e^S} \{p_j[i] - kNNDist(p_j)\}$$

$$CMBR(e^S).max_i = \max_{p_j \in e^S} \{p_j[i] + kNNDist(p_j)\}$$

where $0 \leq i \leq d$.

While the first check is quite simple to perform and for traditional range-joins even very restrictive, the second approach is usually better when the diameter of kNN-spheres differs. An example for both bounding boxes is shown in Figure 3 (a). However, note that none of the checks is the best solution for all scenarios. There exist special cases where the Minkowski sum fits the contained spheres tighter than $CMBR(e^S)$. For example if $|e^S| = 1$, the volume covered by a Minkowski sum is smaller than $CMBR(e^S)$ (cf. Figure 3 (b)). More general, the worst case for $CMBR(e^S)$ happens if at each face of $MBR(e^S)$ there is a point p with $kNNDist(p) = \max\limits_{s \in e^S} kNNDist(s)$. In this case there is $MS(e^S) \subset CMBR(e^S)$ and hence the Minkovski sum shows higher pruning power. The best case (Figure 3 (c)) happens if for each point $p \in e^S$ there is $kNNSphere(p) \setminus MBR(e^S) = \emptyset$, i.e., points with large kNN-spheres are close to the center of $MBR(e^S)$. In this scenario, $CMBR(e^S)$ performs better than the Minkowski sum approach.

5   Experiments

We evaluate the mutual pruning approach from [12] (referred to as UL due to its use of update lists), and our self pruning variants (kNN$_*$) in comparison to the state-of-the-art single RkNN query processor TPL in an RkNN join setting within the Java-based KDD-framework ELKI [22]. As performance indicators we chose the CPU time and the number of page accesses. Note that we did not evaluate existing mutual-pruning techniques such as [7] since these are only applicable if the value of k is fixed over all performed RkNN queries.

It should be noted that, for estimating the *CPU time* of a particular algorithm, we measured the thread time of the corresponding Java thread and performed dry runs before executing the actual simulations in order to keep the impact of garbage collection and Just-In-Time compilation on the results low. A test run was aborted if it lasted at least 20 times longer than executing a self pruning based RkNN join with the same set of variables. For measuring the number of *page accesses*, we assumed that a given number of pages fit into a dedicated cache. If a page has to be accessed but is not contained in the page cache, it has to be reloaded. If the cache is already full and a new page has to be loaded, an old page is kicked out in LRU manner. The page cache only manages data pages from secondary storage, remaining data structures have to be stored in main memory. In order to avoid everything being stored in memory, we employed a restrictive approach, assuming that only node- and entry-IDs can be stored in main memory. Therefore, each time information about an entry, e.g. an MBR or a data point, has to be accessed, the corresponding data page has to be reloaded into the cache. We chose this restrictive strategy because all algorithms employ different methods on processing data. While a sequential kNN-join processes only two data pages at once, mutual pruning based approaches like TPL and UL process the whole tree at a time by facilitating lists of entries, e.g. the TPL entry heap. These data structures are principally unbounded, such that storing whole entries in memory could lead to a situation where the whole tree can be accessed without reloading any page from disk. This would bound the number of page accesses to the tree size, which is unrealistic in large database environments. We set the cache size to fit about 5% of all nodes in the default setting.

We chose the underlying synthetic data sets from R and S to be normally distributed with equivalent mean and a standard deviation of 0.15. We set the default size of R to $|R| = 0.01|S|$, since the performance of TPL degenerates with increasing $|R|$. For each of the analyzed algorithms we used exactly the same data set given a specific set of input variables in order to reduce skewed results. As an index structure for querying we employed an aggregated R*-tree (aR*-Tree [23]). During performance analysis, we analyzed the impact of k, the number of data points in R and S, the dimensionality d, the overlap o between the data sets R and S, the page size p and the cache size c on the performance of the evaluated algorithms keeping all but one variable at a fixed default value while varying a single independent variable. Input values for each of the analyzed independent variables can be found in Table 1. Furthermore, we investigated empirically for which sizes of the sets R and S TPL and $k\text{NN}_*$-algorithms show equivalent performance. This experiment is of great practical relevance since it gives hints on which specific algorithm to use in a specific setting.

Table 1. Values for the evaluated independent variables. Default values are denoted in bold.

Variable	Values	Unit
k	5, **10**, 100, 500, 1000	points
d	2, **4**, 6, 8, 10	dimensions
$|R|$	10, **100**, 1000, 10000, 20000, 40000	points
$|S|$	10, 1000, **10000**, 20000, 40000, 80000	points
o	**0.0**, 0.2, 0.4	$|\mu_S - \mu_R|$
p	512, **1024**, 2048, 4096, 8192	bytes
c	4096, 16364, **32768**, 65536, 131072	bytes

TPL was implemented as suggested in [11], however we did not incorporate the clipping step since this technique increased the computational performance of the algorithm and had only a marginal effect on the page accesses (especially when $d > 2$). Instead we implemented the decision criterion from [24] to enable pruning on intermediate levels of the indexes.

Concerning the nomenclature of the algorithms we use the following notation. UL is the mutual pruning based algorithm from [12]. The additional subscript S (Single) means that every *single* point of R was queried on its own. With UL_G (Group), a whole set of points, a leaf page, was queried at once. UL_P (Parallel) traversed both indexes for R and S in parallel. For the $k\text{NN}$ algorithms we employ a similar nomenclature $k\text{NN}_{ABC}$:

- $A \in \{H, S\}$ denotes whether a hierarchical or sequential nearest neighbor join is used(compare Section 4).
- $B \in \{S, G\}$: S denotes whether the whole $k\text{NN}$ join is processed first and then each resulting $k\text{NN}$ sphere is used to perform a *single* range query on R. G (*group*) denotes that after the $k\text{NN}$s of a single page from S is computed, the corresponding page is used to perform a varying-range-query on the tree of R. The second method can be employed to keep the number of intermediate results low and avoid swapping these intermediate results to disk. Furthermore both methods vary in the algorithm used for the range-join, i.e., whether each point is queried separately or groups of points are queried together.

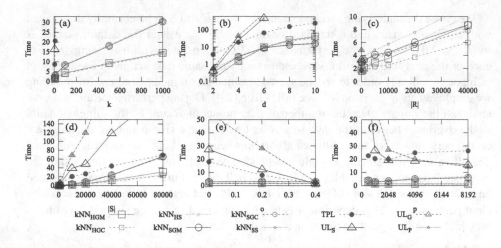

Fig. 4. Performance (CPU time), synthetic dataset. Time is measured in seconds.

– $C \in \{C, M\}$ indicates how the MBR is computed when performing a group range-join (C: CMBR, M: Minkovski sum, compare Section 4).

5.1   Experiments on Synthetic Data

Varying k. In a first series of experiments, we varied the parameter k. While the execution time of the self pruning based $k\text{NN}_*$ increases moderately with k, the remaining mutual pruning approaches become already unusable with low values for k (cf. Figure 4 a). The runtime of TPL increases considerably fast. The runtime of the UL algorithms degenerates similar to TPL; the main problem with this family of algorithms is their use of update lists whose size increases with k, compare [12]. The runtime behaviour of all $k\text{NN}$-based algorithms increases more moderate with increasing k. The performance of the hierarchical join is more stable w.r.t. different values of k, on the one hand side because the hierarchical version of the join can prune whole subtrees on the index level, and on the other hand because the hierarchical join employs further optimizations when joining leaf pages. Furthermore, note that the effect of performing a single/group varying-range join and using different MBRs is quite small. The more sophisticated approach $k\text{NN\_HGC}$ is indeed the best one, however the difference in execution time compared to the simpler solutions $k\text{NN\_HGM}$ and $k\text{NN\_HS}$ is quite low because in this setting $|R|$ is too small.

Concerning the number of page accesses, the picture is quite similar (cf. Figure 5 a). TPL and UL show a worse performance than the $k\text{NN}$-based solutions. However, interestingly the curves of the sequential and hierarchical $k\text{NN}$-join intersect if k becomes rather large, making the sequential join a better choice. This shows that with large values of k the pruning of whole subtrees of the index-based approach fails such that the overhead for traversing the index nodes of the tree cannot be justified any more.

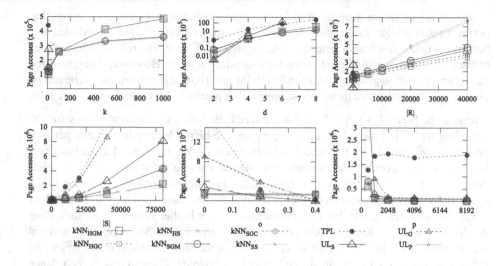

Fig. 5. Performance (page acesses), synthetic dataset

Varying the Dimensionality (d). Taking a look at the performance of the different algorithms with varying dimensionality offers other interesting results. Note that the scale of
these graphs is logarithmic (cf. Figure 4 b and 5 b). The UL approaches scale worse than
the other approaches, because the pruning power of index-level pruning decreases with
increasing dimensionality. TPL and the hierarchical join variants scale similarly, however, in this specific setting, the TPL based join performs by over a magnitude worse than
the hierarchical join. Comparing sequential and hierarchical join, the performance of the
index used by the hierarchical join degenerates with higher dimensionality, such that the
sequential join is able to outperform all remaining approaches if the number of dimensions becomes higher than 6. This behaviour is most likely well explained by the "curse
of dimensionality" and the degradation of spatial index structures in high dimensions.

The results in terms of the number of Disk accesses look very similar, therefore
they shall not be further investigated. However note that the UL approaches show much
better performance in terms of the number of disk accesses if the dimensionality is low.
Therefore, we recommend to use UL for spatial applications (i.e. 2D data).

Varying the Size of R ($|R|$). Varying $|R|$ shows a negative effect on the mutual pruning
approaches TPL and UL (cf Figure 4 c). Especially TPL and UL_S scale linearly with
$|R|$. Although the execution time of the join-based algorithms grows with $|R|$ as well,
this increase is moderate. Note that with a larger $|R|$ the difference between the different
range-join algorithms becomes more obvious. If $|R|$ is very large, the $CMBR$ method
become by about 30% better than the more simple Minkowsky MBRs. Furthermore, the
epsilon-range-join where each kNN-sphere is queried on its own is outperformed by its
group-based counterparts. Recall that we set the default size of R to $|R| = 100$. In this
case even linearly scanning R for a kNN-sphere from S would not lead to a significant
loss in performance. However, if $|R|$ becomes larger, the logarithmic performance of a
hierarchical join becomes visible, and pruning subtrees earlier can further speed up the
query such that the choice of an MBR falls clearly on the $CMBR$s.

Taking a look at the number of disk accesses (illustrated in Figure 5 c) shows that performing an epsilon-range query for each kNN-sphere from S is not a good idea if R becomes large. In this scenario, many of the pages from R have to be reloaded during each of the $|S|$ queries, putting high load on the secondary storage. However, if neighboured values are queried in groups as done in the HGM and HGC algorithms, the cache can be used more efficiently even though these approaches mutually load R and S into the cache.

Varying the Size of S ($|S|$). Next we analyzed the effect of different values for $|S|$ regarding the CPU time (cf Figure 4 d). Again, the hierarchical join variants perform best, followed directly by the sequential join. On the other hand, the UL variants perform worst. However, the UL variant that queries a single point from R during each iteration performs best since this enables highest pruning power. Most importantly the shape of the TPL algorithm looks totally different. It increases faster than the curve of the kNN-join variants first, but then intersects them for higher values of $|S|$, indicating the different asymptotic complexity of the different algorithms. Not taking into account the epsilon-range join, the asymptotic complexity of the sequential join is $O(|S|^2)$ while the complexity of the hierarchical join is about $O(|S|\log(|S|))$ which is in accordance to the empirical results. To the best of our knowledge the theoretical runtime complexity of the TPL algorithm has not been analyzed so far, however the structure of the algorithm suggests a logarithmic complexity for a single query point. For the number of disk accesses the picture is similar which can be observed in Figure 5 d. However the break even point is shifted towards larger values of $|S|$.

Thinking further, the point of intersection between TPL and the kNN-join variants is determined by the size of R since a new value in R introduces a new TPL-query – which is expensive – but only a single more point as a possible range-query result for the kNN algorithms – which is cheap. Specifically given a fixed set $|S|$, there exists a set R_e for which TPL and the hierarchical join algorithms show similar performance. In a practical setting the size of $|R_e|$ for a given set $|S|$ is of great value, since this knowledge can be used to decide whether to use TPL or a kNN join for performing the query. Therefore we ran an additional experiment showing the size of R_e for a given size of $|S|$ where both algorithms have an equal runtime. The results can be found in Figure 6(a). Due to the quadratic complexity of the sequential join, the size of R_e increases superlinear with the size of S. In contrast the size of R_e increases sublinear for the hierarchical kNN-join, suggesting this algorithm for large databases. Given a database with 1 million points, for the hierarchical join there is $R_e \approx 0.0007|S|$ and for the sequential join $R_e \approx 0.003|S|$. Thus as a rule of thumb it is strongly recommendable to use a self pruning based approach if $|R| > 0.01 \cdot |S|$.

Varying the Overlap between R and S (o). Until now we assumed that the normally distributed sets of values R and S overlap completely, i.e. both sets have the same mean. This assumption is quite intuitive for example if we assume that R and S are drawn from the same distribution. In this experiment however, we aim at analyzing what happens if R and S are taken from different distributions. We do so by varying the mean of R and S, i.e. by decreasing the overlap of the two sets ($o = mean(R) - mean(S)$). First of all, the runtime performance of the join-based algorithms stays pretty much constant (cf Figure 4 e). Although this might seem counter-intuitive in the beginning, recall that the kNN-join is only performed on the set S. Therefore the distance between R and S

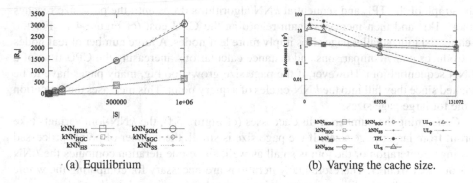

(a) Equilibrium   (b) Varying the cache size.

Fig. 6. Given a set S of size $|S|$, the left figure visualizes the size of set R_e for which TPL and the kNN-based joins have the same computational performance. The right figure visualizes the results for varying cache size. Note the logarithmic scale on the y-Axis.

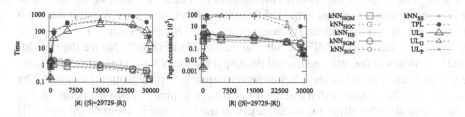

Fig. 7. Performance (CPU time in seconds, page accesses), real dataset (HSV)

does not affect this cost of the RkNN join. Furthermore, the set R is small in our setting, hence the cost for the varying-range-join can be neglected. However only this varying-range-join is affected by a different overlap between R and S. All remaining variants can take severe profit from lower overlap between the sets R and S. All of them employ pruning to avoid descending into subtrees that do not have to be taken into account to answer the query. If the overlap decreases, subtrees can be pruned earlier, reducing the CPU-time and number of page accesses (cf Figure 5 e).

Varying the Cache Size (c). For analyzing the cache size (see Figure 6(b)) we only provide an insight into the impact on the number of page accesses since the cache does not affect the CPU time. Both kNN-join variants are barely affected by the size of the cache, the gain of all of these approaches is about 30%. In contrast, the mutual pruning based algorithms behave by over an order of magnitude (TPL) and two orders of magnitude (UL) better for larger cache sizes. The results show that in a scenario where the accessible memory is large (about 10% of the index), the UL algorithms can be a useful choice. Although they show a higher computational complexity, this disadvantage is compensated by the low number of disc accesses performed by this group of algorithms. If the cache size is relatively small, e.g. due to a multi-user environment, the kNN-based algorithms are the matter of choice.

Varying the Page Size (p). The effect of an increasing page size is twofold. While the UL and hierarchical kNN-join approaches take mainly profit from a larger page size, the graph of the TPL and sequential kNN algorithms drops until the page size reaches about 1kB and then increases again regarding the CPU cost (cf Figure 4 f). For the sequential join, smaller page sizes imply more leaf nodes. A large number of leaf nodes introduces many comparisons and distance calculations, increasing the CPU time of a kNN sequential join. However, if the page size grows too big, many pages have to be visited since they fall into the kNN-circles of a query point. This increases the execution time for large page sizes.

Concerning the number of disk accesses (cf Figure 5 f), the kNN-join variants take profit from larger page sizes. If the page size is small, the number of points processed during one iteration of the join is small as well, since one iteration computes the kNN for one leaf from S. Therefore many iterations are necessary for computing the whole join result, and each of these iterations involves reloading pages into the cache.

5.2 Real Data Experiments

Now let us take a look at experiments driven with real data. As an input, we employed 3D HSV color feature vectors extracted from the caltech data set[2]. We split the input data set containing 29639 feature vectors into two sets R and S such that $|R| + |S| = 29639$, varying the size of R. The results can be found in Figure 7. Notice the two different behaviours of the self- and mutual pruning techniques. Concerning the CPU time, the self pruning approaches show better performance if R is large and S is small. The reason for this behaviour is that a self-join is always a quite expensive operation. For the hierarchical kNN self-join the complexity in the best case is about $O(|S| \log(|S|))$ while for the sequential join the complexity is about $O(|S|^2)$. Combined with the varying-range-join we have an overall complexity of $O(|S| \log(|S|) + |S| \log(|R|))$ or $O(|S|^2 + |S| \log(|R|))$. Now if $|S|$ is large (and $|R|$ can be neglected), this leads to a complexity of $O(|S| \log(|S|))$ and $O(|S|^2)$, respectively. On the other hand, if $|S|$ can be neglected, we have a complexity of $O(\log(|R|))$ in both cases. This also explains why the hierarchical and sequential techniques show the same performance if $|S|$ becomes small. For TPL, the behaviour is different: its complexity in $|S|$ is sublinear but for each $r \in R$ a separate query has to be performed, such that its complexity increases linear with $|R|$. Therefore the performance of TPL is better for small R (and large S), but worse for large R (and small S). For the UL approaches the results can be explained equivalently. Further note that the the the UL_S approach in this scenario shows a significantly lower number of page accesses than UL_G and UL_P

5.3 Comparing CPU-Cost and IO-Cost

Last but not least let us shortly analyze whether the techniques are IO- or CPU-bound. For this purpose, we reuse the results from the experiment where we varied the cache size. From the number of IO operations we computed the resulting IO time, assuming an SSD with page access time of $0.1ms$ (given a HDD, the IO time is higher)[25]. The graph in Figure 8 shows the amount of IO time and CPU time for sample approaches of the self- and mutual pruning approaches. Note that in this setting, most algorithms are IO-bound. For the self pruning approaches this is the case even if the cache size is very high. In contrast, the UL approaches (which have been optimized to greatly reduce

---

[2] http://www.vision.caltech.edu/Image_Datasets/Caltech256/

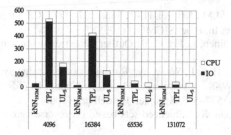

Fig. 8. CPU and IO time in seconds when varying the cache size

the number of page accesses), these approaches become CPU-bound if the cache size becomes very large.

6   Conclusions

In this paper, we addressed the problem of running multiple RkNN-queries at a time, a.k.a RkNN join. For this purpose, we formally classified variants of the RkNN join, including monochromatic and bichromatic scenarios as well as self joins. We proposed a dedicated algorithm for RkNN join queries based on the well-known self pruning paradigm for single RkNN queries. In our experiments, we evaluated the proposed approach against existing work on RkNN joins and RkNN queries. To summarize the contribution of these experiments, we suggest different scenarios for RkNN query processing: *First*, if the database is relatively static and many RkNN queries are run expecting low latency, preprocessing as used in [6,7] is a useful choice since self pruning is usually more selective than mutual pruning. This avoids computing the kNN-spheres each time a query is performed as done with our join algorithms. Note that precomputation usually can only be used if k is fixed. *Second*, if the database is dynamic and RkNN queries are performed in a bulk, our proposed join algorithms, especially the self pruning variant clearly performs best. The technique is also preferable if an RkNN join of intermediate query results has to be computed. Furthermore, our self-pruning algorithm could be easily adapted to performing RkNN joins when each of the objects in R has another value of k. For low-dimensional data, such as 2D geo-spatial data, the mutual pruning RkNN join algorithm from [12] is the matter of choice. *Third*, if the database is highly dynamic and single RkNN-queries have to be performed immediately, single RkNN queries based on mutual pruning like TPL [11] are the method of choice. This is also the case where the database is static but the self join of the whole set S cannot be justified by a low number of RkNN queries.

References

1. Bernecker, T., Emrich, T., Kriegel, H.-P., Mamoulis, N., Renz, M., Zhang, S., Züfle, A.: Inverse queries for multidimensional spaces. In: Pfoser, D., Tao, Y., Mouratidis, K., Nascimento, M.A., Mokbel, M., Shekhar, S., Huang, Y. (eds.) SSTD 2011. LNCS, vol. 6849, pp. 330–347. Springer, Heidelberg (2011)
2. Jarvis, R.A., Patrick, E.A.: Clustering using a similarity measure based on shared near neighbors, vol. C-22(11) (1973)

3. Ankerst, M., Breunig, M.M., Kriegel, H.-P., Sander, J.: OPTICS: Ordering points to identify the clustering structure. In: Proc. SIGMOD (1999)
4. Hautamäki, V., Kärkkäinen, I., Fränti, P.: Outlier detection using k-nearest neighbor graph. In: Proc. IPCR (2004)
5. Jin, W., Tung, A.K.H., Han, J., Wang, W.: Ranking outliers using symmetric neighborhood relationship. In: Ng, W.-K., Kitsuregawa, M., Li, J., Chang, K. (eds.) PAKDD 2006. LNCS (LNAI), vol. 3918, pp. 577–593. Springer, Heidelberg (2006)
6. Korn, F., Muthukrishnan, S.: Influenced sets based on reverse nearest neighbor queries. In: Proc. SIGMOD (2000)
7. Yang, C., Lin, K.-I.: An index structure for efficient reverse nearest neighbor queries. In: Proc. ICDE (2001)
8. Achtert, E., Böhm, C., Kröger, P., Kunath, P., Pryakhin, A., Renz, M.: Efficient reverse k-nearest neighbor search in arbitrary metric spaces. In: Proc. SIGMOD (2006)
9. Stanoi, I., Agrawal, D., Abbadi, A.E.: Reverse nearest neighbor queries for dynamic databases. In: Proc. DMKD (2000)
10. Singh, A., Ferhatosmanoglu, H., Tosun, A.S.: High dimensional reverse nearest neighbor queries. In: Proc. CIKM (2003)
11. Tao, Y., Papadias, D., Lian, X.: Reverse kNN search in arbitrary dimensionality. In: Proc. VLDB (2004)
12. Emrich, T., Kriegel, H.-P., Kröger, P., Niedermayer, J., Renz, M., Züfle, A.: A mutual-pruning approach for RKNN join processing. Proc. BTW (2013)
13. Wu, W., Yang, F., Chan, C.-Y., Tan, K.: FINCH: Evaluating reverse k-nearest-neighbor queries on location data. In: Proc. VLDB (2008)
14. Böhm, C., Krebs, F.: The k-nearest neighbor join: Turbo charging the KDD process. In: KAIS, vol. 6(6) (2004)
15. Zhang, J., Mamoulis, N., Papadias, D., Tao, Y.: All-nearest-neighbors queries in spatial databases. In: Proc. SSDBM (2004)
16. Yu, C., Zhang, R., Huang, Y., Xiong, H.: High-dimensional KNN joins with incremental updates. Geoinformatica 14(1) (2010)
17. Tao, Y., Yiu, M.L., Mamoulis, N.: Reverse nearest neighbor search in metric spaces. IEEE TKDE 18(9) (2006)
18. Cheema, M.A., Lin, X., Zhang, W., Zhang, Y.: Influence zone: Efficiently processing reverse k nearest neighbors queries. In: Proc. ICDE (2011)
19. Achtert, E., Kriegel, H.-P., Kröger, P., Renz, M., Züfle, A.: Reverse k-nearest neighbor search in dynamic and general metric databases. In: Proc. EDBT (2009)
20. Kriegel, H.-P., Kröger, P., Renz, M., Züfle, A., Katzdobler, A.: Reverse k-nearest neighbor search based on aggregate point access methods. In: Winslett, M. (ed.) SSDBM 2009. LNCS, vol. 5566, pp. 444–460. Springer, Heidelberg (2009)
21. Xia, C., Hsu, W., Lee, M.L., Joxan, J., Xia, C., Hsu, W.: Erknn: efficient reverse k-nearest neighbors retrieval with local knn-distance estimation. In: Proc. CIKM (2005)
22. Achtert, E., Hettab, A., Kriegel, H.-P., Schubert, E., Zimek, A.: Spatial outlier detection: Data, algorithms, visualizations. In: Pfoser, D., Tao, Y., Mouratidis, K., Nascimento, M.A., Mokbel, M., Shekhar, S., Huang, Y. (eds.) SSTD 2011. LNCS, vol. 6849, pp. 512–516. Springer, Heidelberg (2011)
23. Papadias, D., Kalnis, P., Zhang, J., Tao, Y.: Efficient OLAP operations in spatial data warehouses. In: Jensen, C.S., Schneider, M., Seeger, B., Tsotras, V.J. (eds.) SSTD 2001. LNCS, vol. 2121, p. 443. Springer, Heidelberg (2001)
24. Emrich, T., Kriegel, H.-P., Kröger, P., Renz, M., Züfle, A.: Boosting spatial pruning: On optimal pruning of mbrs. In: Proc. SIGMOD (2010)
25. Emrich, T., Graf, F., Kriegel, H.-P., Schubert, M., Thoma, M.: On the impact of flash SSDS on spatial indexing. In: Proc. DaMoN (2010)

DART: An Efficient Method for Direction-Aware Bichromatic Reverse k Nearest Neighbor Queries

Kyoung-Won Lee[1], Dong-Wan Choi[2], and Chin-Wan Chung[1,2]

[1] Division of Web Science Technology,
Korea Advanced Institute of Science & Technology, Korea
[2] Department of Computer Science,
Korea Advanced Institute of Science & Technology, Korea
{kyoungwon.lee,dongwan}@islab.kaist.ac.kr, chungcw@kaist.edu

Abstract. This paper presents a novel type of queries in spatial databases, called *the direction-aware bichromatic reverse k nearest neighbor(DBRkNN)* queries, which extend the bichromatic reverse nearest neighbor queries. Given two disjoint sets, P and S, of spatial objects, and a query object q in S, the DBRkNN query returns a subset P' of P such that k nearest neighbors of each object in P' include q and each object in P' has a direction toward q within a pre-defined distance. We formally define the DBRkNN query, and then propose an efficient algorithm, called *DART*, for processing the DBRkNN query. Our method utilizes a grid-based index to cluster the spatial objects, and the B$^+$-tree to index the direction angle. We adopt a filter-refinement framework that is widely used in many algorithms for reverse nearest neighbor queries. In the filtering step, DART eliminates all the objects that are away from the query object more than the pre-defined distance, or have an invalid direction angle. In the refinement step, remaining objects are verified whether the query object is actually one of the k nearest neighbors of them. From extensive experiments, we show that DART outperforms an R-tree-based naive algorithm in both indexing time and query processing time.

Keywords: reverse nearest neighbor, direction-aware, query optimization.

1 Introduction

Recently, with the rapid dissemination of mobile devices and location-based services(LBSs), various applications have started utilizing spatial databases for mobile users. *Bichromatic reverse nearest neighbor(BRNN)* queries extended from *reverse nearest neighbor(RNN)* queries are one of the most popular and important queries for spatio-temporal information, and widely used in many applications. For example, in the case of mobile advertising, an advertiser can promote a product to specifically targeted customers who are close to the advertiser based on each customer's location by searching BRNNs of the advertiser. Many researches addressed that one of the future challenges of location-based services is

M.A. Nascimento et al. (Eds.): SSTD 2013, LNCS 8098, pp. 295–311, 2013.

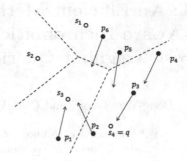

Fig. 1. An example of BRNN query with a direction constraint

personalization [5,12,15], which provides more customized services, based on the user's implicit behaviour and preferences, and explicitly given details. In order to achieve personalization in LBSs, considering only location is not sufficient to retrieve more accurate results in terms of the user's intention.

In that respect, the direction is another important feature that represents user's intention as there exist extensive researches that consider the direction to predict moving object's future location [17]. Each mobile user can have a certain direction with respect to his/her movement or sight, and the direction can be easily obtained by a mobile device with GPS and a compass sensor [18]. However, there are only a few researches that considers a direction-aware environment, and existing studies only focus on user-centric query processing, not objective-centric query processing.

Considering the above, BRNN queries without the direction constraints can be ineffective in many applications to find targeted users in the sense that users looking(or moving) in the opposite direction are less influenced by the query objects even if they are close to the query object. For example, there are many customers in a marketplace, and they are moving around and looking for some products they need. In this situation, a restaurant manager may want to find potential customers who have an intention to enter the marketplace, and hang around the restaurant, because the manager wants to reduce the advertisement budget and does not want to be regarded as a spammer to customers. There are also other kinds of applications where a direction property needs to be considered such as providing a battle strategy to a moving military squad during the war. In these applications, the direction as well as the location are important properties to obtain more accurate results in terms of the targets' intention.

Fig. 1 shows an example of the BRNN query with a direction constraint. Consider a set $P = \{p_1, p_2, p_3, p_4, p_5, p_6\}$ of customers and a set $S = \{s_1, s_2, s_3, s_4\}$ of advertisers. Given a querying advertiser s_4, the usual BRNN query returns p_2, p_3 and p_4 since their closest advertiser is q (i.e., s_4). However, the customers whose directions (represented by arrows) are not toward q do not need to be considered because they are not effective advertising targets. Thus, although customer p_2 has q as its nearest advertiser, p_2 should be discarded from the final result in the direction-aware environment since the direction of p_2 is toward s_3, not q.

Furthermore, in order to maximize the effectiveness of advertising, it is better to consider the distance. If we adjust a maximum distance on p_4, it also can be discarded depending on the maximum distance, even if their nearest advertiser is q. Therefore, only p_3 can be an answer for the BRNN query with the direction constraint.

There have been extensive algorithms studied for processing RNN queries [1,10,19,22] and BRNN queries [14,21,25,26], based on various effective pruning techniques using objects' locations. However, the straightforward adaptation of these algorithms are inefficient to solve the problem of finding BRNNs with the direction constraint. This is because each object has an arbitrary direction, which does not have any correlation with its location.

In this paper, we present a novel type of queries, called *direction-aware bichromatic reverse k nearest neighbor queries(DBRkNN)*, in spatial databases, which extends the previous BRNN query by considering the direction as well as the location. Moreover, we propose an efficient algorithm, called *DART*, for our DBRkNN queries to overcome the difficulties of pruning in a direction-aware environment. DART attempts to minimize pre-processing time by using only a grid-based index to access the set of spatially clustered objects and the B^+-tree to index objects' directions. In common with many previous studies, we follow a filter-refinement framework. In specific, in the filtering step, DART returns a set of candidates, each of which has q as one of its k nearest neighbors and a direction toward q within a pre-defined distance, while the refinement step removes false hits from the set of candidates.

The contributions of this paper are as follows:

- We propose a novel type of query, the direction-aware bichromatic reverse k nearest neighbor (DBRkNN) query, which is an interesting variant of the bichromatic reverse nearest neighbor query. The DBRkNN query is useful in many applications which require to process a large amount of spatial objects with arbitrary directions.
- We propose an efficient algorithm, namely DART, to process DBRkNN queries specially focusing on a direction-aware pruning technique. To effectively prune unnecessary objects, DART uses simple index structures and yet significantly reduces the pre-processing time.
- We experimentally evaluate the proposed algorithm by using synthetic datasets. Experimental results show that our proposed algorithm is on the average 6.5 times faster for the indexing time, and 6.4 times faster for the query processing time than an R-tree-based naive algorithm.

The rest of the paper is organized as follows. Section 2 surveys related work. Section 3 presents the formal definition of the DBRNN query. The proposed algorithm for the DBRNN query is explained in Section 4. Section 5 experimentally evaluates the proposed algorithm. Finally, Section 6 concludes the paper with some directions for future work.

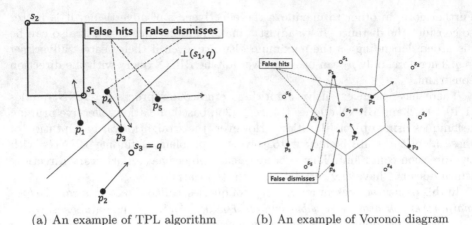

(a) An example of TPL algorithm   (b) An example of Voronoi diagram

Fig. 2. An example of the false hits/false dismisses of the previous works

2   Related Work

We first examine existing studies [1,2,4,9–11,14,19–24,26] about the RNN query, which has received considerable attention due to its importance and effectiveness in many applications. The first algorithm for processing the RNN query was proposed by Korn, F. et al [10]. However, this algorithm requires to index all data points, and to pre-compute their nearest neighbors which is inefficient in dynamic database environments. Stanoi et al. [19] proposed "60-degree-pruning", which maintains only an index tree without any pre-processing structure. They divide the space around the query point into six equal regions having 60° at the query point, and the answers are retrieved by selecting a candidate point from each region. The TPL algorithm was proposed by Tao et al [22], which utilizes the perpendicular bisector between the query point and each point to maximize the pruning area.

The above algorithms for RNN queries, however, are inefficient to process the DBRkNN query since they do not consider the direction constraint for the query processing. For example, Fig. 2(a) shows the false hits/false dismisses of the TPL algorithm for the DBRkNN query. There is a bisector between the query object q (i.e., s_3) and object s_1, so p_2, p_3, and p_5 are selected as candidates since they reside in the half-plane containing q. Although the object p_3 and p_5 are the BRNNs of q, these are not the DBRNNs of q because their directions are not toward q. Moreover, the pruned object p_4 that is located in the opposite half-plane to q can be an answer in the DBRNN query, because its direction is toward q across the bisector.

The traditional RNN queries have been further branched out into the bichromatic RNN (BRNN) query, which is the closest to our DBRNN query. Given two disjoint sets, P and S, of points, and a query point which is one of the points in S, the BRNN query retrieves a set of points in P that have q as their nearest neighbor. There are some researches [9,14,20,25] on the BRNN query, and all their solutions are basically focusing on finding the Voronoi polygon that contains the query point by using a Voronoi diagram.

However, the above methods for the BRNN query are not efficient in our environment. For example, Fig. 2(b) shows an example of using Voronoi diagram to solve BRNN query. There are seven Voronoi polygons each of which is generated by an object in S. In the case of p_2 and p_6, they are the BRNN of q (i.e., s_7), but the directions of them are toward the opposite side of q, which makes them to be false hits/false dismisses. Similarly, although p_4 and p_1 are not the BRNN of q, they are the DBRNN of q because their directions are toward q.

For other types of RNN queries, there has been many researches for processing the *continuous RNN (CRNN)* [2–4,6,9,21] query and the *stream RNN* [11] query. The goal of each type of queries is basically to find the RNN with regard to the query object in a specific environment. However, the solutions for these queries are neither applicable nor relevant to the DBRkNN query due to the properties of the query.

By focusing on the influence of obstacles on the visibility of objects, there are works on a different type of RNN queries [7, 16, 27]. Gao et al. [7] first introduced the *visible reverse nearest neighbor (VRNN)* search, which considers the visibility and the obstacle that significantly affect the result of RNN queries. However, the visibility is defined only for the query object to verify objects that are not influenced by the presence of obstacles while the direction in the DBRNN query is defined for each object to represent its movement or sight. Furthermore, we also adjust the maximum distance to give a flexibility on the spatial environment.

Recently, Li et al. [13] proposed the *direction-aware spatial keyword search*, called *DESKS*, that finds the k nearest neighbors satisfying both keyword and direction constraints to the query. They assumed that the direction is given and addressed that the existing methods on the spatial keyword search are inefficient to solve the spatial keyword search with a direction constraint. However there is a big difference between this work and ours in the sense that we focus on RkNN queries (not kNN), and every object has a direction in our environment while only the query object has a direction in DESKS.

3   Problem Formulation

In this section, we formally define the DBRNN query along with the DBRkNN query. Table 1 summarizes the notations frequently used.

3.1   Problem Definition

We consider two disjoint sets, P and S, of spatial objects, and a query object q in S. Each object in P, called a *customer* object, includes its location and a direction which is represented by a counterclockwise angle from the positive x-axis (i.e., the direction of 3 o'clock is $0°$), and has a fan-shaped region, called a *valid area*, based on its *direction's angle*. On the other hand, objects in S, called

Table 1. The notation of the DBRkNN

Symbol	Description
$P = \{p_1, ..., p_n\}$	the set of customer objects with directions
$S = \{s_1, ..., s_m\}$	the set of advertiser objects
p	a customer object with a direction in P
s	an advertiser object in S
q	the query object selected from advertiser objects in S
r	the maximum distance
d	the direction angle ($0° \sim 359°$)
θ	the valid angle range

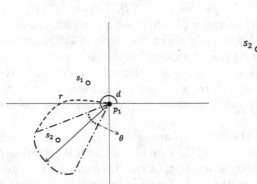 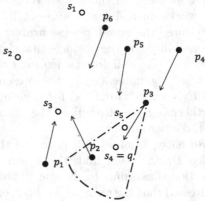

(a) An illustration of the DBRNN query   (b) An illustration of the DBRkNN query ($k = 2$)

Fig. 3. An illustration of the DBRNN query and the DBRkNN query

advertiser objects, have only locations, and one of the advertiser objects can be a query object. We first define the notion "valid area" which is an important concept for selecting candidates and verifying answers as follows:

Definition 1 (Valid Area). *Let p denote an object in P. Then the valid area of p is represented by a fan-shaped region, which has the following properties:*

- *A valid area consists of angle θ and radius r, both of which are pre-defined by a system.*
- *θ is a viewing angle, and r is the maximum distance to discard an object the distance of which is greater than r.*

Now, based on Definition 1, we define the DBRNN query as follows:

Definition 2 (Direction-aware Bichromatic Reverse Nearest Neighbor Query). *Given two disjoint sets, P and S, the direction-aware bichromatic reverse nearest neighbor query retrieves the subset P' of P such that each object in P' has $q \in S$ as its nearest neighbor, and contains q in its valid area.*

Fig. 3(a) illustrates the basic concept of the DBRNN query. The object p_1 has a valid area based on the maximum distance r, the direction's angle d and the valid angle range θ. If the query is invoked on s_2, p_1 is the DBRNN of s_2 even though s_1 is closer than s_2 since only s_2 is located within the valid area of p_1. Similar to Definition 2, the DBRkNN query can be defined as follows:

Definition 3 (Direction-aware Bichromatic Reverse k Nearest Neighbor Query). *Given two disjoint sets, P and S, the direction-aware bichromatic reverse k nearest neighbor query retrieves the subset P' of such P that each object in P' has $q \in S$ as one of its k nearest neighbors, and contains q in its valid area.*

Fig. 3(b) shows an illustration of the DBRkNN query. Let us assume that k is 2, and the query is invoked on s_4. If the value of k is 1, there is no answer for the query. However, in this case, although s_5 is the nearest neighbor of p_3 as well as a DBRkNN of q, because q is the second-nearest neighbor of p_3.

Problem Statement. In this paper, our goal is to find an efficient method that gives the set of exact answers for the DBRNN query and the DBRkNN query. Specifically, we focus on minimizing both the indexing time and the query time.

4 The DART Algorithm

In this section, we present our proposed algorithm, called DART, that solves DBRNN queries. First, we overview our proposed method, and then explain the details of DART.

4.1 Overview

Essentially, our solution is based on a grid-based index to access the spatially clustered objects and the B$^+$-tree to index the direction's angles. We adopt a filter-refinement framework that is widely used in many algorithms for RNN queries. In the filtering step, DART eliminates all the objects that are more than the maximum distance away from the query object or have an invalid direction's angle. After that, in the refinement step, the remaining objects are verified to check whether the query object is actually the nearest neighbor of each object. The important features of DART are explained as follows:

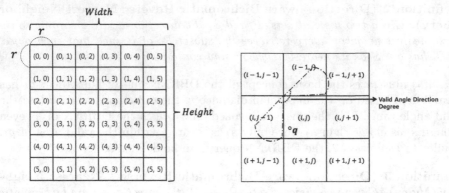

(a) An example of the grid-based space partitioning

(b) An example of the Valid Direction Angle Degree

Fig. 4. The key elements for DART

The Grid-Based Space Partitioning. The whole space is divided into a grid of the equal-sized cells that are represented by rectangles of $r \times r$ size (recall that r is the maximum distance). The number of rows and columns depends on the *width* and *height* of the space. Each cell has a unique id number that represents its location. Fig. 4(a) shows an example of our space partitioning scheme using grid cells. For each cell, we not only maintain two lists of objects (advertiser object and customer object) that are located in the area of the cell but also index the direction's angle of each customer object by using the B$^+$-tree. Note that this space partitioning takes just linear time while an R-tree takes at least $O(n \log n)$ time complexity for indexing n spatial objects [8].

Direction Angle Index. The directions' angles are indexed by the B$^+$-tree to reduce unnecessary checks for the objects that are toward the wrong direction. In this structure, there are at most 360 keys which represent degrees of the directions' angles. For each key, we maintain a list of the objects that have the same direction angle degree as the key value. Similar to the grid-based space partitioning, the construction of this B$^+$-tree can be done in linear time, since each insertion requires only $O(\log 360)$ time (i.e., a constant time).

Valid Direction Angle Range. Each grid cell has a static valid direction angle range (hereafter called "valid angle range") that guarantees, if the direction angle of an object is not within the valid angle range, the object cannot have an appropriate direction toward the query object. Fig. 4(b) shows an example of the valid angle range. When the query is posed, we first figure out which grid cell contains the query object, and then retrieve neighboring cells around the grid cell that has the query object. For each neighboring cell, we define the valid angle range accordingly. As we discussed in Section 3, we use counterclockwise angles; the 3 o'clock position is 0° and the 6 o'clock position is 270°.

Let us first consider the valid angle range of cell $(i-1, j-1)$. In an extreme case, an object in P in cell $(i-1, j-1)$ can be located at the bottom right corner of the cell and the query object can be located at the top right corner or the bottom left corner of cell (i, j). In this case, the valid angle range of an object in P should cover the top or left boundary of cell (i, j) to have the query object within its valid area. Therefore, the valid angle range of cell $(i-1, j-1)$ should be $(0°, \frac{\theta}{2}°)$ and $(270° - \frac{\theta}{2}°, 360°]$ as depicted in Fig. 4(b). The valid angle range of other three corner cells (i.e., $(i-1, j+1), (i+1, j-1)$, and $(i+1, j+1)$) are defined in a similar way. On the other hand, the cells on the cross line (i.e., $(i-1, j), (i, j-1), (i, j+1), (i+1, j)$) are defined in a different manner due to the positional characteristics. For example, in an extreme case of an object in P in cell $(i-1, j)$ can be located at the bottom left or bottom right corner of the cell and the query object can be located at the opposite side of the object in the cell (i, j). In this case, the valid angle range of an object in P should cover the top boundary of cell (i, j) to have the query object within its valid area. Therefore, the valid angle range of cell $(i-1, j)$ should be $(0°, \frac{\theta}{2}°)$ and $(180° - \frac{\theta}{2}°, 360°]$. Similar to the corner cells, the other cells on the cross line have similar valid angle ranges. The specific ranges of the valid angle ranges are shown in Table 2.

Table 2. The Valid Angle Range of each cell

Cell No.	Valid Angle Degree
$(i-1, j-1)$	$(0°, \frac{\theta}{2}°)$ and $(270° - \frac{\theta}{2}°, 360°]$
$(i-1, j)$	$(0°, \frac{\theta}{2}°)$ and $(180° - \frac{\theta}{2}°, 360°]$
$(i-1, j+1)$	$(180° - \frac{\theta}{2}°, 270° + \frac{\theta}{2}°)$
$(i, j-1)$	$(0°, 90° + \frac{\theta}{2}°)$ and $(270° - \frac{\theta}{2}°, 360°]$
(i, j)	$(0°, 360°]$
$(i, j+1)$	$(90° - \frac{\theta}{2}°, 270° + \frac{\theta}{2}°)$
$(i+1, j-1)$	$(0°, 90° + \frac{\theta}{2}°)$ and $(180° - \frac{\theta}{2}°, 360°]$
$(i+1, j)$	$(0°, 180° + \frac{\theta}{2}°)$ and $(360° - \frac{\theta}{2}°, 360°]$
$(i+1, j+1)$	$(90° - \frac{\theta}{2}°, 180° + \frac{\theta}{2}°)$

4.2 DART for DBRNN Query Processing

Based on the above key features, DART follows a two-step framework, where a set of candidate objects are returned and then false hits are verified to retrieve the exact solution. Our algorithm does not index the exact locations of the objects, but use a grid-based structure to access the set of spatially clustered objects efficiently. Moreover, our method inserts the object's direction's angle degree into the B$^+$-tree to maintain the object's direction's angle.

Algorithm 1. The construction of the basic structure

Input: Sets of objects P and S
Output: A set G of grid cell's list and the B^+-tree

```
 1  G ← ∅;
 2  foreach obj in P ∪ S do
 3      cellId ← Assign(obj.x, obj.y) ;
 4      list ← lists[cellId].;
 5      list.add(obj);
 6      G.add(list);
 7      if obj ∈ P then
 8          d ← obj.d;
 9          Btree ← Btrees[cellId];
10          Btree.insert(d, obj);
11      end
12  end
13  return G;
```

Index Construction Step. Algorithm 1 shows the process of constructing the basic structures. When an object is inserted, the assignment algorithm determines which grid cell contains the object (Line 3). In addition, the method just adds the object into the proper list and stores the list for the corresponding cell (Lines 4-6). For the two types of objects, we maintain two lists of objects separately. In addition, DART also maintains a B^+-tree to index direction's angle degree for each cell (Lines 8-10). As mentioned earlier, the entire construction process can performed in linear time.

Filtering Step. In the filtering step, DART eliminates unnecessary objects by considering the maximum distance and the valid angle range based on the location and angle of the object. The overall algorithm flow is shown in Algorithm 2.

First, the method gets the grid cell number that contains the query object by using *assign* function. In this function, it is easy to retrieve the neighboring cells by using the *width* and *height* of the space (Lines 2-3). Because the size of each cell is determined by the maximum distance, we do not have to consider other cells except for the neighboring cells. By doing this, we can prune numerous objects whose distances from the query object are larger than the maximum distance.

Next, for each cell, DART selects the candidate set by processing the range search on the B^+-trees on directions' angles of objects located in the cell(Lines 5-7). Note that this range search requires only a constant time since there are at most 360 keys. As we discussed the valid angle range in Section 4.1, we can easily find the objects whose directions' angles are within the valid angle ranges of the corresponding cells (See Table 2). Before we put an object into the candidate set, we double check the actual distance and the angle degree between the query object and the object (Lines 8-13). The reason for doing this step is to guarantee

---

Algorithm 2. The filtering step of DART

Input: The query object q
Output: A set of candidates

1 $candidate \leftarrow \emptyset$;
2 $CellId \leftarrow$ `Assign(query.x, query.y)`;
3 $neighbor[\ ] \leftarrow$ `getNeighbor`($CellId$);
4 **for** $i \leftarrow 0$ **to** *size of neighbor[ ]* **do**
5      $Btree \leftarrow neighbor[i].Btree$;
6      $range \leftarrow neighbor[i].ValidAngleRange$;
7      $list[] \leftarrow Btree.$`rangequery`($range$);
8   **for** $j \leftarrow 0$ **to** *size of list[ ]* **do**
9          $angle \leftarrow$ `getAngle`(q, $list[j]$);
10   **if** `getDistance`($list[j], q$) $\leq r\ AND\ |(angle - list[j].d)| \leq \frac{\theta}{2}$ **then**
11             $candidate.add(list[j])$;
12   **end**
13   **end**
14 **end**
15 **return** $candidate$

---

that the candidate set contains only objects whose valid area cover the query object. If the distance between the two objects is within the maximum distance and the difference of the two directions' angles (angle degree between two objects and the object's direction angle degree) is less than $\frac{\theta}{2}$, then we finally add the object to the candidate set.

Refinement Step. After the termination of the filtering step, we have a candidate set that contains all the objects whose valid area contain the query object. In the refinement step, we verify that the actual nearest neighbor of each candidate object is the query object. Algorithm 3 shows the flow of the refinement step. Basically, the method confirms the answer set by checking the nearest neighbor of each candidate. For candidate object p, if there is an advertiser object s_i closer than the query object, it is possible that s_i is the nearest neighbor of p and is within the valid area of p. To determine this, for all the advertiser objects in S that are contained in neighboring cells, the method calculates the distance between the candidate object and each advertiser object (Line 7). Moreover, if there is an advertiser object closer than the query object, we check the actual angle degree (Lines 7-13). Similar to the above procedure, if the difference is smaller than half of θ, it means that the direction's angle is also facing the object s_i, and the nearest neighbor of the candidate object is not the query object (Lines 9-12). Otherwise, the candidate object can be an answer of the DBRNN query.

Algorithm 3. The refinement step of DART for the DBRNN query

Input: A candidate set C
Output: a set of answers *Answer*
1 $neighbor[] \leftarrow$ getNeighborGrid();
2 $list \leftarrow$ getObjectS($neighbor[]$);
3 $Answer \leftarrow C$;
4 **for** $i \leftarrow 0$ **to** $size\ of\ C$ **do**
5 | $distance \leftarrow$ getDistance(q, $C[i]$);
6 | **for** $j \leftarrow 0$ **to** $size\ of\ list[]$ **do**
7 | | **if** $distance >$ getDistance($C[i]$, $list[j]$) **then**
8 | | | $angleS \leftarrow$ getAngle($C[i]$, $list[j]$);
9 | | | **if** $|(angleS - C[i].d)| \leq \frac{\theta}{2}$ **then**
10 | | | | $Answer.$remove($C[i]$);
11 | | | | break;
12 | | **end**
13 | **end**
14 **end**
15 **end**
16 **return** *Answer*

4.3   DART for the DBR*k*NN Query Processing

In this section, we extend the algorithm for DBRNN queries to process DBR*k*NN queries for an arbitrary value of k, which means the query result should be all the customer objects that have q within k nearest neighbors (k is a positive integer, typically small). For processing DBR*k*NN queries, although the arbitrary value k is added, the overall flow is almost the same. The filtering step does not need to be changed, because we prune the unnecessary objects only considering the distance and angle constraints. In the refinement step, DART should be slightly modified so that k advertiser objects can be checked when finding advertiser objects closer than the query object (Lines 10-11 in Algorithm 3).

5   Experiments

In this section, we evaluate the performance of our proposed algorithm for the DBRNN query and the DBR*k*NN query by using four synthetic datasets. In particular, we generate synthetic datasets for both spatial object sets, P and S, under the uniform distribution. We set the size of the dataset for spatial objects in P to be from 10,000 to 10,000,000 and that in S to be $|P|/100$ on the 2-dimensional 1,000 × 1,000 euclidean space. For experimental parameters, we vary the valid angle range, the maximum distance, and the cardinality of the dataset. The values of parameters are presented in Table 3.

The experiment investigates the index time and query time for varying values of parameters such as the valid angle range, the maximum distance, and the cardinality. For comparison, we also implement a naive method that utilizes an

Table 3. The values of parameters

Parameter	The range of values
Valid angle range	30 - 90 degree
	(60 degree by default)
Maximum distance	50 - 200
	(100 by default)
Cardinality of P	10,000 - 10,000,000
	(1,000,000 by default)

R-tree on the objects' locations because the existing methods for RNN queries and BRNN queries cannot guarantee the exact solution for DBRNN queries, which implies that the methods are not applicable to DBRNN queries and the DBR*k*NN queries (See Fig. 2). Similar to the proposed algorithm, the naive method also prunes the objects outside of the circular range that has a radius equal to a maximum distance. After that, this method conducts the same procedure as the refinement step to remove false hits. The only difference from DART is that the direction pruning is not performed in the filtering step. All algorithms are implemented in Java, and the experiments are conducted on a PC equipped with Intel Core i7 CPU 3.4GHz and 16GB memory.

5.1   Experimental Results of DBRNN Query

Fig. 5 shows the performance of DART and the naive method when processing DBRNN queries with varying values of experimental parameters. Fig. 5(a), 5(c), and 5(e) represent the index time of each experiment, and Fig. 5(b), 5(d), and 5(f) represent the query time. For all results on the index and query processing, DART shows a superior performance compared to the naive method.

First, we conduct an experiment with varying the valid angle range varying from 30 degree to 90 degree. According to Fig. 5(a), the grid-based clustering and the B$^+$-tree indexing on the direction's angles do not need heavy indexing time while the R-tree based indexing is time consuming. We can observe that the grid-based clustering takes less time than the R-tree to maintain the object's location. Moreover, the direction angle indexing time is not a big issue in the whole pre-processing step because we have at most 360 angle degrees so that there are at most 360 keys as we mentioned earlier. On the other hand, Fig. 5(b) shows an increasing gap between the query time of DART and that of the naive method. The reason that the naive method shows an increasing curve as the valid angle range increases is due to the total candidates of answer objects. For instance, DART filters irrelevant objects with its valid angle range, and then conducts refinement on a candidate set. However, the naive method just filters objects which have a longer distance than the maximum distance by conducting range search on the R-tree, and checks for the direction's angle for each object.

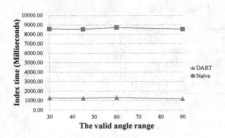

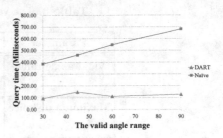

(a) Index time for varying valid angle range

(b) Query time for varying valid angle range

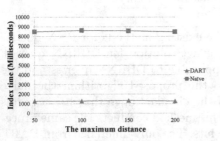

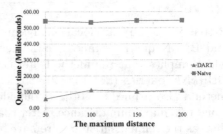

(c) Index time for varying maximum distance

(d) Query time for varying maximum distance

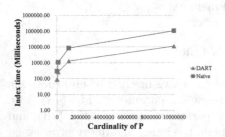

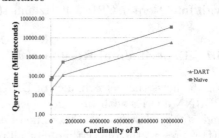

(e) Index time for varying cardinality

(f) Query time for varying cardinality

Fig. 5. Experimental Results of DBRNN Query

In this step, the naive method generates more candidates as the valid angle range gets wider, and hence the number of total candidates increases.

According to Fig. 5(c), indexing time is almost similar to the result of the valid angle range, and Fig. 5(d) indicates that the query processing time shows a steady gap between DART and the naive method. There is a little increasing line for DART for the maximum distance from 50 to 100, because the number of total candidates is quite small for the data3 dataset.

In Fig. 5(e) and Fig. 5(f), we conduct experiments with varying the cardinality by using all the datasets. Both index time and query time show similar increasing curves as the cardinality becomes bigger, however, the difference of the performance is upto 10 times between DART and the naive method. In our

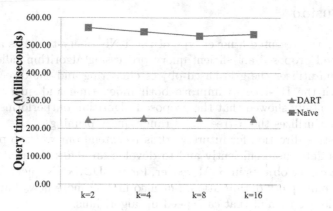

Fig. 6. Experimental Results of DBRkNN Query

observation, DART can handle bigger sized datasets more efficiently than the
naive method, even though the dataset reaches 10 millions of objects.

5.2   Experimental Results of DBRkNN Query

Fig. 6 shows the performance of DART and the naive method for the DBRkNN
queries. For the arbitrary k value, we start from $k = 2$ and exponentially increase
k until 16. The experimental results show that the query times for the cases are
almost uniformly distributed. In our observation, this is due to the maximum
distance and the direction constraint. Only a small computation is increased
because the constraints limit the boundary for the DBRkNN search. From this
result, we can claim that our proposed algorithm is also much more efficient for
processing the DBRkNN query than the naive method.

5.3   Summary

In summary, we have shown through extensive experiments that DART outperforms the naive method in both indexing time and query processing time. We
conducted several experiments by changing the values of parameters, namely the
valid angle area, the maximum distance, and the cardinality to show the effect
of those parameters on the performances. The results indicate that DART can
handle more than 10 million objects within a minute. Therefore, DART is suitable for a snapshot query with at most one minute time interval to secure index
time and query time. In addition, although DART approximately prunes irrelevant objects by using a grid-based space partitioning(while the naive method
prunes certain objects whose distance are longer than the maximum distance),
its direction angle pruning technique makes up the time of double checking for
the maximum distance.

6   Conclusion

In this work, we presented a novel type of the RNN query that has a direction constraint, and proposed an efficient query processing algorithm called DART. Our algorithm utilizes the grid-based object clustering and the direction angle indexing with the B^+-tree to improve both index time and query time. We also experimentally showed that the proposed algorithm outperforms the naive algorithm that utilizes the R-tree based range query pruning.

An interesting direction for future work is to extend our work to process the larger size of data more efficiently and to develop an algorithm that efficiently prunes unnecessary objects in S. Moreover, for the DBRkNN query, we plan to work on not only query optimization, but also finding more new characteristics of the direction constraint that can speed up algorithms.

Acknowledgments. This work was supported in part by WCU (World Class University) program under the National Research Foundation of Korea funded by the Ministry of Education, Science and Technology of Korea (No. R31-30007), and in part by the National Research Foundation of Korea grant funded by the Korean government (MSIP) (No. NRF-2009-0081365).

References

1. Achtert, E., Böhm, C., Kröger, P., Kunath, P., Pryakhin, A., Renz, M.: Efficient reverse k-nearest neighbor search in arbitrary metric spaces. In: Proceedings of the 2006 ACM SIGMOD International Conference on Management of Data, SIGMOD 2006, pp. 515–526. ACM, New York (2006)
2. Benetis, R., Jensen, S., Karciauskas, G., Saltenis, S.: Nearest and reverse nearest neighbor queries for moving objects. The VLDB Journal 15(3), 229–249 (2006)
3. Cheema, M.A., Zhang, W., Lin, X., Zhang, Y.: Efficiently processing snapshot and continuous reverse k nearest neighbors queries. VLDB Journal 21(5), 703–728 (2012)
4. Cheema, M.A., Zhang, W., Lin, X., Zhang, Y., Li, X.: Continuous reverse k nearest neighbors queries in euclidean space and in spatial networks. The VLDB Journal 21(1), 69–95 (2012)
5. Dhar, S., Varshney, U.: Challenges and business models for mobile location-based services and advertising. Commun. ACM 54(5), 121–128 (2011)
6. Emrich, T., Kriegel, H.P., Kröger, P., Renz, M., Xu, N., Züfle, A.: Reverse k-nearest neighbor monitoring on mobile objects. In: Proceedings of the 18th SIGSPATIAL International Conference on Advances in Geographic Information Systems, GIS 2010, pp. 494–497. ACM, New York (2010)
7. Gao, Y., Zheng, B., Chen, G., Lee, W.C., Lee, K.C.K., Li, Q.: Visible reverse k-nearest neighbor query processing in spatial databases. IEEE Trans. on Knowl. and Data Eng. 21(9), 1314–1327 (2009)
8. Guttman, A.: R-trees: a dynamic index structure for spatial searching. In: Proceedings of the 1984 ACM SIGMOD International Conference on Management of Datam, SIGMOD 1984, pp. 47–57. ACM, New York (1984)
9. Kang, J.M., Mokbel, M.F., Shekhar, S., Xia, T., Zhang, D.: Continuous evaluation of monochromatic and bichromatic reverse nearest neighbors. In: ICDE (2007)

10. Korn, F., Muthukrishnan, S.: Influence sets based on reverse nearest neighbor queries. In: Proceedings of the 2000 ACM SIGMOD International Conference on Management of Data, SIGMOD 2000, pp. 201–212. ACM, New York (2000)
11. Korn, F., Muthukrishnan, S., Srivastava, D.: Reverse nearest neighbor aggregates over data streams. In: Proceedings of the 28th International Conference on Very Large Data Bases, VLDB 2002, pp. 814–825. VLDB Endowment (2002)
12. Krumm, J.: Ubiquitous advertising: The killer application for the 21st century. IEEE Pervasive Computing 10(1), 66–73 (2011)
13. Li, G., Feng, J., Xu, J.: Desks: Direction-aware spatial keyword search. In: Kementsietsidis, A., Salles, M.A.V. (eds.) ICDE, pp. 474–485. IEEE Computer Society (2012)
14. Lian, X., Chen, L.: Monochromatic and bichromatic reverse skyline search over uncertain databases. In: Proceedings of the 2008 ACM SIGMOD International Conference on Management of Data, SIGMOD 2008, pp. 213–226. ACM, New York (2008)
15. Mokbel, M.F., Levandoski, J.J.: Toward context and preference-aware location-based services. In: Proceedings of the Eighth ACM International Workshop on Data Engineering for Wireless and Mobile Access, MobiDE 2009, pp. 25–32. ACM, New York (2009)
16. Nutanong, S., Tanin, E., Zhang, R.: Incremental evaluation of visible nearest neighbor queries. IEEE Trans. on Knowl. and Data Eng. 22(5), 665–681 (2010)
17. Qiao, S., Tang, C., Jin, H., Long, T., Dai, S., Ku, Y., Chau, M.: Putmode: prediction of uncertain trajectories in moving objects databases. Applied Intelligence 33, 370–386 (2010)
18. Qin, C., Bao, X., Roy Choudhury, R., Nelakuditi, S.: Tagsense: a smartphone-based approach to automatic image tagging. In: Proceedings of the 9th International Conference on Mobile Systems, Applications, and Services, Mobisys 2011, pp. 1–14. ACM, New York (2011)
19. Stanoi, I., Agrawal, D., Abbadi, A.E.: Reverse nearest neighbor queries for dynamic databases. In: ACM SIGMOD Workshop on Research Issues in Data Mining and Knowledge Discovery, pp. 44–53 (2000)
20. Stanoi, I., Riedewald, M., Agrawal, D., Abbadi, A.E.: Discovery of influence sets in frequently updated databases. In: Proceedings of the 27th International Conference on Very Large Data Bases, VLDB 2001, pp. 99–108. Morgan Kaufmann Publishers Inc., San Francisco (2001)
21. Taniar, D., Safar, M., Tran, Q.T., Rahayu, W., Park, J.H.: Spatial network rnn queries in gis. Comput. J. 54(4), 617–627 (2011)
22. Tao, Y., Papadias, D., Lian, X.: Reverse knn search in arbitrary dimensionality. In: Proceedings of the Thirtieth International Conference on Very Large Data Bases, VLDB 2004, vol. 30, pp. 744–755. VLDB Endowment (2004)
23. Tao, Y., Papadias, D., Lian, X., Xiao, X.: Multidimensional reverse knn search. The VLDB Journal 16(3), 293–316 (2007)
24. Tao, Y., Yiu, M.L., Mamoulis, N.: Reverse nearest neighbor search in metric spaces. IEEE Trans. on Knowl. and Data Eng. 18(9), 1239–1252 (2006)
25. Tran, Q.T., Taniar, D., Safar, M.: Bichromatic reverse nearest-neighbor search in mobile systems. IEEE Systems Journal 4(2), 230–242 (2010)
26. Vlachou, A., Doulkeridis, C., Kotidis, Y., Norvag, K.: Monochromatic and bichromatic reverse top-k queries. IEEE Trans. on Knowl. and Data Eng. 23(8), 1215–1229 (2011)
27. Wang, Y., Gao, Y., Chen, L., Chen, G., Li, Q.: All-visible-k-nearest-neighbor queries. In: Liddle, S.W., Schewe, K.-D., Tjoa, A.M., Zhou, X. (eds.) DEXA 2012, Part II. LNCS, vol. 7447, pp. 392–407. Springer, Heidelberg (2012)

User-Contributed Relevance
and Nearest Neighbor Queries

Christodoulos Efstathiades[1] and Dieter Pfoser[2,3]

[1] Knowledge and Database Systems Laboratory
National Technical University of Athens, Greece
`cefstathiades@dblab.ece.ntua.gr`
[2] Research Center "Athena", Maroussi, Greece
`pfoser@imis.athena-innovation.gr`
[3] George Mason University, Fairfax, VA, USA
`dpfoser@gmu.edu`

Abstract. Novel Web technologies and resulting applications have lead
to a participatory data ecosystem that when utilized properly will lead to
more rewarding services. In this work, we investigate the case of Location-
based Services and specifically of how to improve the typical location-
based Point-Of-Interest (POI) request processed as a k-Nearest-Neighbor
query. This work introduces Links-of-interest (LOI) between POIs as a
means to increase the relevance and overall result quality of such queries.
By analyzing user-contributed content in the form of travel blogs, we establish the overall popularity of a LOI, i.e., how frequently the respective
POI pair is mentioned in the same context. Our contribution is a query
processing method for so-called k-Relevant Nearest Neighbor (k-RNN)
queries that considers spatial proximity in combination with LOI information to retrieve close-by and relevant (as judged by the crowd) POIs.
Our method is based on intelligently combining indices for spatial data
(a spatial grid) and for relevance data (a graph) during query processing. An experimental evaluation using real and synthetic data establishes
that our approach efficiently solves the k-RNN problem when compared
to existing methods.

1 Introduction

Location-based Services have been at the forefront of mobile computing as they
provide an answer to the simple question as to what is around me. A lot of effort
has been dedicated to improving such services typically by improving the selectivity of each request. Rating sites augment POIs with quality criteria. Preferences, when available, add further user specific parameters to a request. Context
limits the available information based on situational choices. However, what has
not been captured yet are user experiences per se, i.e., assessing what people
want in terms of what people in the same situation have done in the past. Our
objective is to provide location-based services, specifically relevant k-NN search
based on the crowdsourced choices and experiences other users had in the past.

M.A. Nascimento et al. (Eds.): SSTD 2013, LNCS 8098, pp. 312–329, 2013.
© Springer-Verlag Berlin Heidelberg 2013

This work focusses on semantically enriching k-RNN search by taking user experience into account. Specifically, we introduce the concept of a Link-of-interest (LOI) between two POIs to express respective *relevance*, i.e., find related nearest POIs to my location. Relevance is inferred by extracting pairs of POIs that are frequently mentioned together in the same context by users. In our work, we extract this information (co-occurrence of POI pairs) by parsing travel blogs and using the page structure to derive relevance. This relevance information can be represented by means of a graph in which POIs represent nodes and LOIs are links. At the same time, we need to capture the spatial properties of POIs. The challenge in this work will be as to how we efficiently combine relevance captured by a graph and the spatial properties captured by a spatial index.

While existing work addresses spatio-textual search, i.e., introducing a spatial aspect to (Web) search, thus, enabling it to index and retrieve documents according to their geographic context, to the best of our knowledge, the exact problem of combining user experience (expressed as relevance) with spatial proximity in the form of k-RNN search (R stands for Relevance) has not been studied in literature. In this work, we are trying to solve the problem of how to combine searching a graph structure with a spatial index so as to efficiently process k-RNN queries. The spatial aspect is indexed by a static spatial grid, while the relevance is captured by a graph. We propose two methods, (i) GR-Sync (GR = Grid/Graph), which expands the two indexes separately, but synchronizes their search at certain steps and (ii) GR-Link, which uses a tighter integration in that spatial search results are seeded to the graph search so as to minimize costly (and most often unnecessary) expansions. Experimentation shows that the performance of GR-Link is best since it examines smaller portions of the data.

The outline of the remainder of this work is as follows. Section 2 discusses related work. Section 3 provides background on the data (spatial+relevance), queries (k-RNN), and the basic access methods that we use. The respective k-RNN query processing approaches are outlined in Section 4. Section 5 details the experimental evaluation and Section 6 concludes and provides directions for future research.

2   Related Work

To the best of our knowledge, the exact problem of combining user experience (expressed as relevance) with spatial proximity has not been studied in literature. Our methods are somewhat related to the research on Spatio-Textual Search. According to [16], studies in this area try to make web search geographically-aware, thus enabling it to index and retrieve documents according to their geographic context. Current research focusses on combining spatial and textual indexes and proposes hybrid methods to support geographical awareness. Current approaches are dealing with merging R-trees, regular grids, and space-filling curves with a textual index such as inverted files or signature files. One of the first works in Spatio-Textual indexing was conducted in the context of the SPIRIT search engine [16]. SPIRIT facilitates the use of regular grids as spatial indexes and inverted files for the indexing of the documents. Following this idea, Zhou et al. [17] present hybrid indexing approaches comprising of R*-trees [1] for spatial indexing and inverted files for text indexing. Several approaches following the combination of R*-trees and inverted files followed [17,4,15]. Chen et al. [3] use space-filling curves for spatial indexing, whereas in [5], R-trees are combined with signature files in the internal nodes of the tree, so that a combined index is created. The algorithm in [9] is used for incremental nearest neighbor queries. Cong et al. [4] propose the IR-tree index, creating an inverted file for each node of an R-tree, again using [9]. Here, linear interpolation is used [13], in order for the spatial distance to be combined with the textual distance and therefore produce a combined score. In our work, we use a similar approach when combining the relevance score with spatial distance. A very related approach to [4] is presented in [11]. Recently, Rocha-Junior et al. [15] propose the hybrid index S2I, which uses aR-Trees [14]. The authors claim to outperform all other proposed approaches in this problem domain. Our work differs significantly in that we do not consider textual similarity, but use the co-occurrence of POIs in text as an indicator for relevance. This relevance information (and no actual textual information) is then combined with the spatial aspect of the POIs to retrieve relevant nearest neighbors.

To the best of our knowledge, the most related work that combines user experiences and spatial data is [2], which introduces the notion of prestige to denote the textual relevance between nearby to the query point objects. It also uses a graph, but the way it is constructed is based on the spatial distance between the objects as opposed to our method of using relevance. The prestige from a given query point is propagated through the graph, an information that is later used to extend the IR-tree [4]. Compared to our work, our index does not include a factor of randomness in our relevance score calculation as this is the case with PageRank-based algorithms. In general, the notion of relevance in [2] is based on different measures that are query dependent (textual relevance), a fact that is not being considered in our work. Our query considers only the location of a query point. Relevance is only derived from the collected dataset.

The problem of combining relevance and spatial distance is closely related to the problem of computing top-k queries that are based on different subsystems such as studied by Fagin et al. [7,8]. The difference to our approach is that we have a specific focus on spatial data and on how to combine the result sets. Also, in our final k-RNN list we have the exact scores, compared to the approximate scores of the NRA algorithm.

Finally, in this work relevance is computed by counting the co-occurrences of POIs in the same paragraphs of texts. In spite of the simplicity of this metric, recent results show that POIs co-occur in documents when they are spatially close, have similar properties, or interact with each other [12]. Essentially, these observations are a confirmation of Tobler's first law of geography that states that "everything is related to everything else, but near things are more related than distant things."

3 Data and k-RNN Queries

With the proliferation of the Internet as the primary medium for data publishing and information exchange, we have seen an explosion in the amount of online content available on the Web. In addition to professionally-produced material being offered free on the Internet, the public has also been encouraged to make its content available online to everyone as User-Generated Content (UGC). We, in the following describe the data we want to utilize and how to exploit it to provide k-RNN queries.

3.1 Data

Web-based services and tools can provide means for users through attentional (e.g., geo-wikis, geocoding photos) or un-attentional efforts (e.g., routes from their daily commutes) to create vast amounts of data concerning the real world that contain significant amounts of information ("crowdsourcing"). The simplest possible means to generate content is by means of *text* when (micro) blogging. Any type of text content may contain geospatial data such as the mentioning of POIs, but also data characterizing the relationship between two POIs, e.g., the Red Cross hospital is next to the St. Basil church.

In this work, what we try to discover in these data sources are collections of POIs. Consider the example of Figure 1a. In the specific text snippet of a travel blog, three distinct spatial objects are mentioned, $O_1 = \{\text{Acropolis}\}$, $O_2 = \{\text{Plaka}\}$, and $O_3 = \{\text{Ancient Agora}\}$. We introduce the concept of *Link-of-Interest* (LOI) as a means to express relevance between two POIs. Assuming $O = \{O_1, \ldots, O_n\}$ is the set of all discovered POIs, then a text paragraph $P_x \subset O$, i.e., contains a set of POIs. It also holds that $O = \bigcup P$. We state that there exists a LOI $L_{i,j}$ between two POIs O_i and O_j, if both POIs are mentioned in the same text paragraph P_x. The set of all LOIs is defined as follows.

$$L = \{L(O_i, O_j) | \exists P_x \in P : O_i \in P_x \wedge O_i \in P_x\} \tag{1}$$

Do note that the definition of relevance is a very simple one, i.e., co-occurrence in the same paragraph. In future work, we intend to exploit the entire document structure as well as use sentiment information. However, any of these considerations will only affect the way we compute relevance and not the presented techniques for computing the specific type of query.

3.2 k-RNN Queries

Combining relevance information with spatial distance will allow us to provide better query results, i.e., POIs that are close-by and relevant.

Let D be a spatial database that contains spatial objects O and Links of Interests L. Each spatial object is defined as $O_i = (O_i.id, O_i.loc)$ and each LOI as $L_k = (O_i.id, O_j.id, r)$, $O.id$ is a unique object identifier, $O.loc$ captures the object location in two-dimensional space, and r provides the relevance (score)

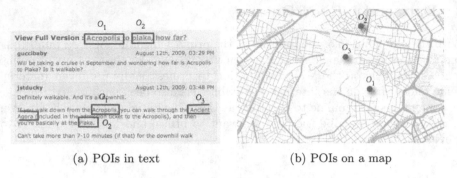

(a) POIs in text   (b) POIs on a map

Fig. 1. "Closeby" Points Of Interest

of a LOI existing between two POIs o_i and o_j. The relevance r is computed as the number of times a specific POI pair appears in different paragraphs, i.e., the more often, the higher r and the higher the relevance of this specific LOI.

We introduce the k-RNN query as follows. Given a query point that is represented as a POI, find the k relevant nearest neighbors by taking into account, both, spatial proximity and relevance. An example is given in Figure 2 for a 1-RNN neighbor. Essentially, the query takes into account how relevant a close-by POI is to the query point and combining this information with the spatial proximity between the two POIs computes a combined k-RNN score denoting the Spatio-Relevance Distance between the two points, i.e., the smaller the distance (score), the better.

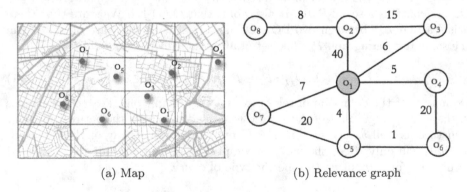

(a) Map   (b) Relevance graph

Fig. 2. Relevance Graph and spatial "map" data

Equation 2 provides a means to compute a combined k-RNN score s_{rnn} that considers, both, spatial distance and relevance.

$$s_{rnn} = \frac{\alpha * s_r(Q,O)}{s_r.max} + \frac{(1-\alpha) * s_d(Q,O)}{s_d.max} \qquad (2)$$

with $0 \leq \alpha \leq 1$ and where Q is the query point, O is the spatial object for which we compute the score with respect to Q, s_r is a score that takes into account the relevance score between two points and can be any kind of metric, $s_r.max$ is the maximum possible relevance score, s_d is the Euclidean distance between Q and O and $s_d.max$ is the maximum possible value the Spatial Distance can take and depends of course on the query space. Parameter α is used to denote the importance of each distance function (Relevance or Spatial) and can be tuned according to the user's needs.

In the next sections, detailed explanations are given as to the calculation of the Relevance Score and the computation of k-RNN queries.

3.3 Access Methods

Given a query and trying to define an efficient method for solving it in a data management context, typically access methods are used to speed up processing. In the following, we will discuss some methods that, either on their own, or, by combining them, efficiently solve the given query processing problem. When considering efficiency, we will also argue for the simplicity of a method, as the more complex a proposed access method is, the bigger is the challenge in implementing it in a given data management infrastructure.

Indexing Space and Relevance. Following this approach, we try to utilize spatial indexing methods with graph data structures. A regular spatial grid is used for indexing the locations of our POIs. The reason for using a regular grid instead of other types of access methods such as the R-tree is that (i) it can be updated in $O(1)$ time, (ii) there are no limitations to the index size (whole planet), (iii) k-NN queries can be processed incrementally by radiating out from the query point (see in the following), and (iv) it is a data structure that serves as a simple and elegant way of showing how a spatial index can be combined with others to index, e.g., space + relevance. The disadvantages of the method are that (a) space is divided into grid cells, which are considered to be disk pages and because of the static and non-uniform nature of the spatial data, pages will be underutilized and (b) the query time is not competitive when compared to other methods such as the R-tree and its more efficient variants.

However, in this work we do not consider the efficiency of our approach based merely on the use of the spatial index, but on how to create a hybrid method to answer k-RNN queries. Therefore, any type of spatial index can be used to replace the regular grid without having to change the algorithms used in our approach.

The *Relevance Graph* is defined as a graph $G(V, E)$, where V is the set of vertices that correspond to the POIs found in the set of documents, and E is the set of edges that correspond to links-of-interest (LOIs) between the POIs. The edge weights denote the relevance score r between a pair of vertices.

We consider a pair of POIs to be related if there is at least one co-occurrence in a paragraph and the relevance is derived from the number of co-occurrences. For example, in Figure 2, we can see that the *Acropolis Metro Station* (O_1) and

Parthenon (O_3) appear 6 distinct times in the same paragraphs, therefore their relevance score is 6.

Index Creation and Maintenance. Both indices are built incrementally during a pre-processing phase. In general, we consider that the data to be inserted concern one document/description at a time. Therefore, the input to the insert and update procedures are POIs and LOIs.

In the case of the Spatial Grid, considering its static nature, the corresponding grid cell is located based on the POI's coordinates and it is added to the respective node (corresponding to a cell). Since we aim for having a one-to-one correspondence between nodes and pages, should the node reach is maximum capacity, an overflow page is added.

For the Relevance Graph, we have to update the graph (add new points/vertices) and also update the relationships expressed as edge weights. Therefore, with the input of a list of points, as well as the identified LOIs, if a POI has not been previously added to the graph, we add it and also add the possible relations to other POIs. To perform this operation efficiently, a separate data structure is used to store the edges of the graph, so that they can be updated or added in $O(1)$ time.

4   *k*-RNN Query Processing

Having outlined structures to index the data, the major contribution of this work is how to integrate those methods so as to efficiently support the processing of *k*-RNN queries.

4.1   Index Synchronization

In order to tackle efficient *k*-RNN processing, we first define a basic query processing method, termed *GR-Sync* (derived from Grid/Graph synchronization) that combines the results of the two separate indices in order to answer *k*-RNN queries. GR-Sync consists of two separate methods for identifying the *k*-RNN candidates in the Spatial Grid and in the Relevance Graph, respectively. The intuition behind this approach is an intermixed, stepwise execution of the search in both indexes. After each step of the so-called *expansion process* in both data structures, the results (current list of respective *k*-NN neighbor candidates) are combined. If the neighbors found are guaranteed to be the *k*-RNN neighbors, then the procedure stops, otherwise it continues until the *k*-RNN neighbors are guaranteed to have been found.

Spatial Search. The *Spatial Expansion* algorithm uses a Spatial Grid that divides the surface of the earth into equal-sized grid cells as shown in Figure 2a. For each inserted POI we store four distances to the respective sides of the cell. Figure 3 shows an example of how the algorithm behaves for eight expansions beginning from a query POI. The algorithm first locates the grid cell of the query point. Given that this algorithm tries to establish the relevance and proximity between POIs ("Where to go next?"), the query point is recruited from the set of POIs. The objective of the spatial expansion is to discover the nearest-neighbor

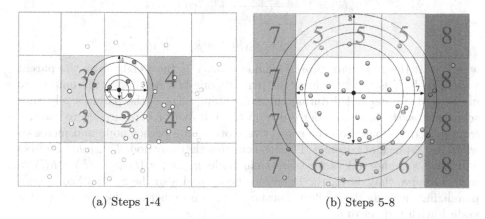

(a) Steps 1-4   (b) Steps 5-8

Fig. 3. Spatial expansion

POIs of the query point. We do so, by retrieving close-by POIs in neighboring cells in a step-wise fashion. As we will see, this results in *snail-like expansion* . The order in which the cells are retrieved depends on the location of the query point within its containing cell. Consider the example of Figure 3a. The query point is closest to the bottom side of the cell (arrow labeled "1"). To guarantee that all POIs have been examined that are within distance 1, only the cell of the query point needs to be loaded (indicated also by the circle around the query point and of radius "1"). However, for the case of distance "2" also the points of the bottom cell (labeled 2) need to be retrieved. This *snail-like expansion* next retrieves two cells labeled "3" and then two cells labeled "4". While we also retrieve points that are further away than d_4, we are also certain that we have not missed any candidates.

The points that are not within the maximum distance are added to a *retrievedPoints* list, whereas the points that are within the maximum distance are added to the k-NN list based on Euclidean distance.

With increasing distance, the search expands to neighboring cells as shown in the example of Figure 3b. As expected, the larger the distance from the query point, the more points are retrieved in each step, e.g., 4 cells in Step 8.

Computing the Relevance Score. To retrieve the *relevance score* s_r, we have to examine the Relevance Graph. Our method orients itself on the Breadth-First graph traversal. It starts with the query point and in a first step examines all adjacent nodes in the Relevance Graph. Subsequent steps examine all neighbors of the initially-visited nodes and so on.

To compute a Relevance Score, we use the following recursive formula that computes the score of a node k based on the score of its predecessor, or, parent node p. For the initial expansion $p = q$.

$$s_r(k) = 1 - \frac{w_{p,k}}{\sum_{i=1}^{N} w_{p,i}} + s_r(p) + s_h(k) \tag{3}$$

$$s_h(k) = s_h(p) + h(k) \tag{4}$$

where s_r is the *Relevance Score* for node k, $w_{p,k}$ is the weight from the parent node p to k, N is the number of the parent's one-hop neighbors, the number of nodes that are expanded during the same step, $s_r(p)$ is the parent's relevance score, and $s_h(k)$ is a score derived from the number of hops needed to reach k from q. $s_h(k)$ is based on the respective score of the parent node and increases with the distance of k from d. This also ensures that any node l with $h(l) > h(k)$ will have a higher relevance score than node k, i.e., $s_r(l) > s_r(k) : h(l) > h(k)$. Observe that, both, the Relevance Score and also the Spatial Score are a penalizing score as they reflect distance, i.e., the higher s_r, the less relevant a node k with respect to q.

Figure 4 depicts an example of the calculation of the Relevance Score for the neighbors of a query point. First, the sum of the weights of the edges that connect the query point to its neighbors is calculated. In this example $\Sigma(0) = 28$. To compute each one-hop neighbor's Relevance Score s_r, we need to get the intermediate score of each neighbor node based on the edge weight, i.e., $1 - (w_{p,k} / \sum_{i=1}^{N} w_{p,i})$. For example, node B has an intermediate score of $1 - 3/28 = 0.89$. Applying the rest of Equation 3, with $s_r(p) = 0$, since $p = q$ and $h = 0$ with the hop count starting at 0, $s_r(B) = 0.89$. Continuing the expansion, i.e., retrieving the two-hop neighbors, for example for node F, $s_r(F) = (1 - 2/6) + 0.89 + 1 = 2.56$ (the edge that connects the neighbor to its parent is not taken into consideration). It should be noted that in the case of computing the score of a node, we consider the closest path as far as the number of hops from the query point is considered. Additionally, should a node be reachable by the same number of hops through multiple nodes, we consider the lowest scoring node as a parent node.

Synchronizing Expansion. To compute the k-RNN score, both, the spatial search and the relevance graph search need to be synchronized and their scores combined. The following approach computes combined scores and evaluates the status of the search at fixed intervals determined by the number of expansions performed in each index. As described earlier, "expansions" are performed to retrieve k-RNN candidates. In the Spatial Grid, this results in a snail-like expansion and retrieval of cells surrounding the query point, and in the Relevance Graph, a BFS-like expansion of increasing distance to the query point is used. After each expansion step, the two lists contain the closest in terms of spatial and relevance score neighbors, respectively. The two scores in both lists need to be combined to assess k-RNN candidates. To synchronize the spatial expansion in the grid with the expansion of the Relevance Graph, we define a respective

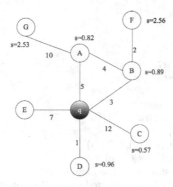

Fig. 4. Graph expansion and Relevance Score

rate of expansion steps. The expansion ratio χ determines the ratio of spatial to relevance graph expansions. χ will typically be in the range of 4 to 16, as spatial expansions are considerably cheaper. Each search maintains a list of expanded POIs and their respective score. After each expansion cycle (considering the expansion ratio), the two lists are checked and the common POIs, i.e., appearing in both lists, are identified. Using Equation 2, their combined score s_{rnn} is computed. All POIs with such a score are added to a queue sorted by s_{rnn} essentially containing all top k-RNN neighbors that have been identified at this point, i.e., POIs that have been examined in both data structures.

To define a termination criterion for the search, we have to guarantee that all k-RNN POIs have been found, i.e., further search will not reveal any POIs with a better score than the ones already identified. Each search keeps an *open list* for each set of respective POIs found so far recording the corresponding k-RNN score. If a POI is found in only one index, to compute its score, we use a best-case estimate for the missing score, i.e., that it will be found during the next expansion. For example, assuming a POI was retrieved in the spatial search but not yet in the relevance search, the relevance score will be the lowest possible score after the next expansion (relevance score is dominated by number of hops = expansion steps). Similarly, should the spatial score be missing, we assume the best case, i.e., that the POI will be discovered during the next round of expansion with a distance from the query point just beyond the current search distance. As the searches progress, the scores of the POIs, and thus the open lists, will be updated based on current number of hops and search distance. The GR-Sync algorithm is shown in Figure 5. GR-Sync uses s_{rnn}^{d} and s_{rnn}^{r} as the minimum predicted combined k-RNN scores for nodes with no valid spatial, or relevance score, respectively. If $s_{rnn}^{*} = min(s_{rnn}^{d}, s_{rnn}^{r})$, the minimum score that still could be found, is less than $max(s_{rnn})$, i.e., the maximum score of identified rnn candidates, it means that the POIs in the (top-k) result list are guaranteed to be the top k-RNN neighbors (Line 17). This condition can be used as termination criterion.

GR-SYNC(q, k)

```
1    RG ▷ Relevance Graph
2    SG ▷ Spatial Grid
3    rg ▷ Discovered POIs in RG
4    sg ▷ Discovered POIs in SG
5    rnn ▷ k-RNN result list
6    n ▷ Relevance Graph step count
7    χ ▷ Expansion ratio Spatial Grid
8    while ¬ complete
9        SPATIALEXPANSION(q, k, (n × χ), SG, sg)
10       SEMANTICEXPANSION(q, k, n, RG, rg)
11       rnn = sg ∩ rg ▷ NN results with complete score
12       s^r_{rnn} = MINSCORE(rg) ▷ Min. predicted score, Relevance Graph
13       s^d_{rnn} = MINSCORE(sg) ▷ Min. predicted score, Spatial Grid
14       s'_{rnn} = MIN(s^d_r nn, s^r_r nn) ▷ Min. expected score
15       s^*_r = MAXSCORE(rnn) ▷ Max. score in current result list
16       ▷ if the max k^{th} RNN distance found so far is less than the min predicted distance
17       complete = (|rnn| ≥ k ∧ s^*_r < s'_r nn)
```

Fig. 5. GR-Sync k-RNN algorithm

4.2   Index Linking

The k-RNN query processing algorithm presented so far does not actually combine the two indices (spatial and relevance) in any way, but only evaluates the results at times with the expansion steps used as a means of synchronization. An inherent problem with this method is the search in the Relevance Graph, which after the first hops becomes very costly. Big expansions in the Relevance Graph (empirically observed for hops > 3) retrieve a lot of data and, hence, incur disk activity. To address this problem, we have devised a variation of our query processing technique that manages to bound the expansion in the Relevance Graph, i.e., to compute the Relevance Scores without expanding large potions of the graph. In this *GR-Link* method, termed as such since it combines the two indexes (Grid/Graph linking), we use the results of the spatial search as seed elements to the search in the Relevance Graph. The intuition is that many 1-hop expansions (spatial search seeds) are "cheaper" than a single n-hop expansion of the query point. POIs retrieved by the spatial search (and not discovered yet in the relevance search) are expanded in the Relevance Graph (cf. Figure 6) The intuition is that the graph searches of (i) the seeded POIs and (ii) the query point will meet eventually and, thus, we can compute a POI's relevance score. Without *seeding the search*, we would have to wait until the query point expansion reaches a POI. The simplified example of Figure 6 should illustrate that the POIs expanded are much fewer than the POIs that would have been retrieved if a second expansion step from the query point would have been performed.

The GR-Link approach also allows us to better bound the estimates of missing scores. Placing a POI on the Relevance Graph, we know for sure that if

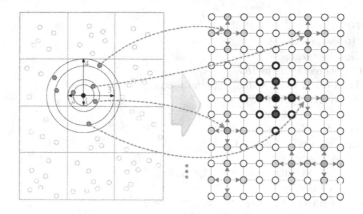

Fig. 6. GR-Link method using Grid POIs in Graph search

their expansions (POI and query point) do not meet, the base for their score computation is at least the sum of the two expansions plus one hop (since they did not meet). Therefore, the predicted s_r score for such a point is larger than what it would have been in the non-hybrid case.

The pseudo code for this method is shown in Figure 7. The difference from the GR-Sync algorithm are the statements in Lines 11-14, which seed POIs to the Relevance Search.

As the experimental section will show, the intuition of choosing many small Relevance Graph expansions over one large expansion pays off and the GR-Link method shows superior performance in terms IO when compared to the GR-Sync solution.

5   Experimental Evaluation

Did the previous sections define our approach to the processing of k-RNN queries, so will we in the following establish its efficiency. An empirical study using real and synthetic datasets will assess the performance of the query processing methods in terms of accessed data (disk I/O operations) under varying parameter settings. Overall, we compare three methods, (i) the GR-Sync method (naive method), (ii) the GR-Link method and (iii) an hypothetic ideal method (see below for an explanation). What we expect from the experiments is to see a well-performing GR-Link method that comes close to the performance of the ideal method.

5.1   Data

The experimentation relies on real and synthetic datasets. The real data consists 120k POIs and 670k LOIs extracted from a corpus of 120k documents. The

GR-LINK(q, k)

1   RG ▷ Relevance Graph
2   SG ▷ Spatial Grid
3   rg ▷ Discovered POIs in RG
4   rg' ▷ Discovered POIs in RG - spatial seeds
5   sg ▷ Discovered POIs in SG
6   rnn ▷ k-RNN result list
7   n ▷ Relevance Graph step count
8   χ ▷ Expansion ratio Spatial Grid
9   **while** ¬ complete
10   SPATIALEXPANSION($q, k, (n \times \chi), SG, sg$)
11   **for** each $poi \in sg$
12   SEMANTICEXPANSION($poi, 1, RG, rg'$)
13   SEMANTICEXPANSION(q, n, RG, rg)
14      $rg = $ CONNECT(rg, rg') ▷ Combine graph expansions
15      $rnn = sg \cap rg$ ▷ NN results with complete score
16      $s_{rnn}^r = $ MINSCORE(rg) ▷ Min. predicted score, Relevance Graph
17      $s_{rnn}^d = $ MINSCORE(sg) ▷ Min. predicted score, Spatial Grid
18      $s_{rnn}' = $ MIN($s_r^d nn, s_r^r nn$) ▷ Min. expected score
19      $s_r^* = $ MAXSCORE(rnn) ▷ Max. score in current result list
20   ▷ if max k^{th} RNN distance found so far is less than min predicted distance
21   complete $= (|rnn| \geq k \wedge s_r^* < s_r' nn)$

Fig. 7. GR-Link k-RNN algorithm

documents were collected from three different travelblog sites (travelblog, traveljournals, travelpod) as described in [6]. The texts were pre-processed to collect the necessary information for the index: (i) identification of POIs, (ii) geocoding of POIs (spatial position), and (iii) location within the document (paragraph id and offset).

To improve the significance of the experiments, we also generate a larger-in-size synthetic dataset. The procedure for generating the data follows simple heuristic rules derived from the characteristics of the real data, i.e., generating hotspots (cities) with POIs and links between these hotspots. We generate center points (cities, attractors) in an area of 30×30 degrees (approximately $3300 \times 3300 km$) using a uniform distribution. For each center point, using a normal (gaussian) distribution, we generate neighboring points (POIs). We then generate relationships (LOIs) between POIs using a normal distribution, i.e., the neighbors that are spatially closer to a POI in question are more likely to be linked to it than the ones further away (Tobler's spatial bias). On average, we generate 30 LOIs per POI (maximum $= 50$). The edge weights (denoting the number of common paragraphs) are generated randomly based on observations from our real dataset and vary between 0 and 500. Our synthetic dataset consists of 1 million POIs that form a Relevance Graph with a total of 6.5 million edges (LOIs). It is considerably larger than the Relevance Graph derived from the real data and, thus, should allow us to provide a more conclusive experimental evaluation.

5.2 Experimental Setup

We evaluate the various query processing methods not only with different datasets but also a varying set of parameters including (i) the number of RNN neighbors k and (ii) the preference parameter α emphasizing either spatial distance or relevance.

The performance is assessed in terms of number of page accesses (I/O) performed to retrieve the data from disk for each query. In each case, we performed 100 queries and computed the average I/O shown in the respective charts. An important aspect is the spatial grid used to index the data. We use a regular grid with a spacing of 0.02 degrees (approximately 2km) in longitude and latitude. The synthetic dataset consists of 1 million POIS covering an area of 30 degrees longitude and latitude, respectively, an extent somewhat comparable to Europe. The 1M POIs are grouped into 160k cells amounting to an average space utilization of 6, but not exceeding 30. Keep in mind that we only consider occupied cells in our index. The real dataset contains 120k points that are scattered all over the globe. Here, 80k cells contain at least one POI. This amounts to an average space utilization of < 2, with few cells containing more than 5 POIs. Essentially, we used this dataset as a template for the synthetic data generation. Still, we wanted to present the results of the performance study, to see whether the trends with respect to the indexing methods persist.

The expansion ratio χ was set to 12, i.e., for each expansion in the Relevance Graph, 12 expansions in the Spatial Grid are performed.

5.3 k-RNN Query Performance

In our experimental setup, we first vary the number of sought nearest neighbors k to see how scalable our algorithms are in terms of I/O. In this experiment, the parameter α is fixed to 0.5, i.e., considering the spatial and relevance score equally important.

As we can see in Figure 8a, the GR-Link approach outperforms the GR-Sync method by an order of magnitude. This is due to the fact that GR-Sync needs to search large parts of, both, the Spatial Grid and the Relevance Graph in order to guarantee the result. Therefore, when POIs are found in one of the two indices, it needs to keep searching the other to find the combined k-RNN score. This causes the algorithm to access a lot of unnecessary pages. On the other hand, the GR-Link method uses the results of the Spatial Search to limit the expansions in the Relevance Graph. Hence, this method has a considerably better performance.

While a comparison to the naive GR-Sync approach does not pose a challenge, so does the next experiment relate the performance of GR-Link to an "ideal" approach (cf. Figure 8b). The ideal approach simulates a Relevance Graph search that terminates as soon as the k-RNN POIs are found, i.e., we have a-priori knowledge of the results and are only expanding the graph until "discovered". As such this method is unrealistic, but should just serve as a lower-bound for the performance of the hybrid index. Figure 8b shows that the GR-Link method examines more data than the ideal approach (albeit little data overall). Still, in terms of comparison to the baseline approach, the hybrid index's performance is close to that of the ideal method.

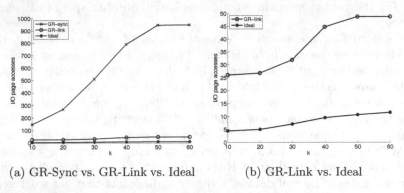

(a) GR-Sync vs. GR-Link vs. Ideal   (b) GR-Link vs. Ideal

Fig. 8. Naive, Hybrid and Ideal Index I/O performance

5.4   Spatial Grid vs. Graph Partitioning

All POI data is kept on disk. The assumption so far is that each populated Spatial Grid cell constitutes a page on disk. This approach has the disadvantage of taking into consideration only the spatial characteristics without taking into account that the nearest-neighbor search explores not only the spatial grid but the Relevance Graph as well. The more the points discovered in the Relevance Search (graph expansion) are spatially spread out, the more pages need to be retrieved, thus, increasing the I/O cost. Our approach in extending the current disk layout is now to group POIs together based on the proximity in the Relevance Graph. This is achieved by partitioning the Relevance Graph using the METIS graph partitioning tool [10]. However, the grid-based partitioning is essential for spatial search and the underlying expansion mechanism. Using (relevance) graph-based partitioning, we mapped the spatial grid cells to graph partitioning. I.e., does the spatial search request a specific cell, the respective one or more graph partitions are fetched. Figure 9a shows that graph-partitioning does not provide an advantage over the static spatial partitions. What is interesting to see is that the performance advantage of the ideal method when compared to GR-Link is diminished for the case of graph partitioning. An explanation here could be that the expansion solely relies on the graph and, hence, a respective partitioning would provide some (respective) advantage for this method. Overall however, the ideal method performs best in the case of the Spatial Grid, which is evident when comparing Figures 8b and 9b.

5.5   Score Balance

The preference parameter α balancing the effect of spatial distance vs. relevance on the query result is not only a critical factor for the quality of the result, but

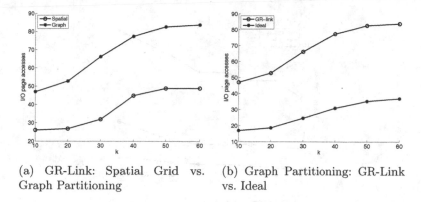

(a) GR-Link: Spatial Grid vs. Graph Partitioning

(b) Graph Partitioning: GR-Link vs. Ideal

Fig. 9. Partitioning: Spatial Grid vs. Relevance Graph

also affects the query processing cost. With $\alpha = 0.1$ it favors the Spatial Score and with $\alpha = 0.9$ it favors the Relevance Score. The two charts of Figure 10 show that the query cost does not change significantly with α. It seems that with an emphasis on Relevance ($\alpha > 0.5$), the cost slightly decreases due to fewer spatial expansions.

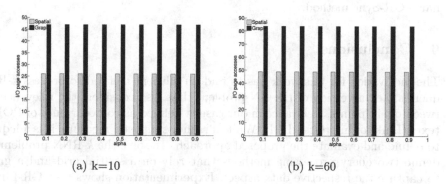

(a) k=10

(b) k=60

Fig. 10. Varying preference parameter α

5.6 Real Dataset Experiments

The number of page accesses when compared to the experiments with the synthetic dataset appear to be orders of magnitude greater in number because of the many spatial expansions that need to be performed in order for the algorithms to guarantee the k-RNN result in a sparse dataset. Overall, this experiment shows the same trends observed for synthetic data.

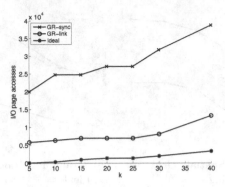

Fig. 11. Real Dataset: varying k

5.7 Summary

The experiments showed that one can provide an efficient indexing mechanism and query processing method for combining spatial and graph-based search by interlinking index traversal in both "spaces". The GR-Link method performs very close to an "ideal" method and an order of magnitude better than the naive GR-Sync method.

6 Conclusions

The motivation for this work was to find an efficient method to process k-RNN queries, i.e., a version of the NN problem that also considers the relevance between query points. Relevance in our case is defined as co-occurrence of POIs in texts (Links of Interests - LOI). While a rather simplistic measure, it is adequate to define and evaluate the proposed approach. To solve the k-RNN problem, we define two query processing methods that rely on a spatial grid and a graph to capture the respective data aspect. Experimentation shows that GR-Link is an efficient method that performs significantly better than a naive method and comes close to the performance of a hypothetical ideal method. This work outlines a first approach to combining spatial and graph search and consequently directions for future work are plenty. They include using more efficient spatial access methods, optimizing the search in the graph, and adding graph information to the spatial index. We aim also in performing an "interestingness" evaluation in order to measure the usefulness of the k-RNNquery to the users.

Acknowledgements. The research leading to these results has received funding from the European Union Seventh Framework Programme - Marie Curie Actions, Initial Training Network GEOCROWD (http://www.geocrowd.eu) under grant agreement No. FP7-PEOPLE-2010-ITN-264994.

References

1. Beckmann, N., Kriegel, H.-P., Schneider, R., Seeger, B.: The r*-tree: an efficient and robust access method for points and rectangles. In: Proc. SIGMOD Conf., pp. 322–331 (1990)
2. Cao, X., Cong, G., Jensen, C.S.: Retrieving top-k prestige-based relevant spatial web objects. PVLDB 3(1-2), 373–384 (2010)
3. Chen, Y.-Y., Suel, T., Markowetz, A.: Efficient query processing in geographic web search engines. In: Proc. SIGMOD Conf., pp. 277–288 (2006)
4. Cong, G., Jensen, C.S., Wu, D.: Efficient retrieval of the top-k most relevant spatial web objects. PVLDB 2(1), 337–348 (2009)
5. De Felipe, I., Hristidis, V., Rishe, N.: Keyword search on spatial databases. In: Proc. 24th ICDE Conf., pp. 656–665 (2008)
6. Drymonas, E., Pfoser, D.: Geospatial route extraction from texts. In: Proc. 1st Workshop on Data Mining for Geoinformatics, pp. 29–37 (2010)
7. Fagin, R.: Combining fuzzy information from multiple systems. J. Comput. Syst. Sci. 58(1), 83–99 (1999)
8. Fagin, R., Lotem, A., Naor, M.: Optimal aggregation algorithms for middleware. J. Comput. Syst. Sci. 66(4), 614–656 (2003)
9. Hjaltason, G.R., Samet, H.: Distance browsing in spatial databases. ACM Trans. on Database Syst. 24(2), 265–318 (1999)
10. Karypis, G., Kumar, V.: A fast and high quality multilevel scheme for partitioning irregular graphs. SIAM J. Sci. Comput. 20, 359–392 (1998)
11. Li, Z., Lee, K.C.K., Zheng, B., Lee, W.-C., Lee, D., Wang, X.: IR-tree: An efficient index for geographic document search. IEEE Trans. on Knowl. and Data Eng. 23(4), 585–599 (2011)
12. Liu, Y., Wang, F., Kang, C., Gao, Y., Lu, Y.: Analyzing relatedness by toponym co-occurrences on web pages. Transactions in GIS (to appear, 2013)
13. Martins, B., Silva, M.J., Andrade, L.: Indexing and ranking in geo-ir systems. In: Proc. Geographic Information Retrieval Workshop, pp. 31–34 (2005)
14. Papadias, D., Kalnis, P., Zhang, J., Tao, Y.: Efficient OLAP operations in spatial data warehouses. In: Jensen, C.S., Schneider, M., Seeger, B., Tsotras, V.J. (eds.) SSTD 2001. LNCS, vol. 2121, pp. 443–459. Springer, Heidelberg (2001)
15. Rocha-Junior, J.B., Gkorgkas, O., Jonassen, S., Nørvåg, K.: Efficient processing of top-k spatial keyword queries. In: Pfoser, D., Tao, Y., Mouratidis, K., Nascimento, M.A., Mokbel, M., Shekhar, S., Huang, Y. (eds.) SSTD 2011. LNCS, vol. 6849, pp. 205–222. Springer, Heidelberg (2011)
16. Vaid, S., Jones, C.B., Joho, H., Sanderson, M.: Spatio-textual indexing for geographical search on the web. In: Medeiros, C.B., Egenhofer, M., Bertino, E. (eds.) SSTD 2005. LNCS, vol. 3633, pp. 218–235. Springer, Heidelberg (2005)
17. Zhou, Y., Xie, X., Wang, C., Gong, Y., Ma, W.-Y.: Hybrid index structures for location-based web search. In: Proc. 14th CIKM Conf., pp. 155–162 (2005)

Using Hybrid Techniques for Resource Description and Selection in the Context of Distributed Geographic Information Retrieval

Stefan Kufer, Daniel Blank, and Andreas Henrich

University of Bamberg, D-96047, Germany
{stefan.kufer,daniel.blank,andreas.henrich}@uni-bamberg.de
http://www.uni-bamberg.de/minf/

Abstract. The amount of media items on the web is increasing tremendously, especially regarding personal media items. To effectively collaborate over and share these massive amounts of media objects, there is a strong need for adequate indexing and search techniques. Trends like social networks, large-storage mobile devices and high-bandwidth networks make peer-to-peer (P2P) information retrieval systems of deep interest.

Hence, resource selection based on compact resource descriptions is used to efficiently determine promising peers w.r.t. a query. To design effective media search applications, multiple search criteria need to be addressed. Subsequently, besides text or visual media content, geospatial data is frequently used.

We propose techniques to summarize and select collections of georeferenced media items in P2P systems. Generally, these summarization techniques can be divided into geometric and space partitioning approaches. This paper presents and evaluates techniques of a third category, hybrid approaches that combine features of geometric and space partitioning techniques.

1 Introduction

During the last years, the amount of (especially personal) media data accessible via the World Wide Web has vastly increased. People write blogs, twitter about events or their lifes, use remote photo or video communities and share media content in social networks. Therefore, people not only store these personal media objects, but also interact with each other through them, for example by collaboratively tagging or commenting on various items. Consequently, online resources varying in size, media type and update frequency have to be administered [1].

Additionally, the availability—and with it the usefulness—of geospatial metadata has increased dramatically in recent times. Nowadays, digital cameras as well as mobile phones are often equipped with GPS sensors at reasonable costs. Hence, these devices are able to capture georeferenced information to enrich media data in many situations, like shooting videos or taking pictures. Supplementary, geo-tagging tools with rich user interfaces have emerged in several domains and there are large geo-tagging initiatives attempting to georeference

M.A. Nascimento et al. (Eds.): SSTD 2013, LNCS 8098, pp. 330–347, 2013.

textual resources such as Wikipedia. Taken en masse, the increased importance of geospatial information in the context of searches can be recognized.

Obviously, geospatial information is not the only search criterion. Other criteria such as textual content, timestamps and (low-level) audio and visual content information can be used when searching for media items as well. An integrative combination of these criteria with spatial filter or ranking conditions can facilitate an effective retrieval of text, image, audio and video documents.

Our search scenario assumes a P2P system maintaining personal media archives. P2P systems are formed by computers (potentially) distributed all over the world, the peers, which can act as both clients and servers. By applying a scalable P2P IR protocol, a service of equals for the administration of media items can be established, without the requirement to maintain expensive infrastructure. In our scenario, the media items administered in a personal archive are stored locally on the peer (that is to the user's personal device), without the need to store media items on remote servers hosted by service providers such as Flickr or YouTube, reducing the dependency on service providers as informational gatekeepers. To facilitate retrieval, media items can be described by four criteria: 1) textual content, 2) low-level content features, 3) timestamps and 4) a geographic footprint. Hence, personal media archives can be represented by four corresponding resource descriptions (summaries). Each summary represents a feature aggregation in terms of the media items a resource (peer) maintains, for example an aggregation over all the geographic coordinates of all the media items a peer administers. As a scalable P2P protocol, Rumorama [2] is applied in our scenario, and establishes hierarchies of PlanetP-like [3] networks. In a PlanetP-like network, every peer knows all the resource descriptions of every other peer inside the network, enabling routing decisions while query processing (that is contacting the most promising peers, according to the summaries and with respect to a certain query, first). The distribution of resource descriptions in the network is assured by randomized rumor spreading [3].

The present paper studies novel techniques with respect to resource description and resource selection considering geographic metadata and thus continues work presented in [4]. There we examined techniques falling into either the category of geometrical approaches or the category of space partitioning approaches (cf. Sect. 2). The current paper introduces a third category, hybrid approaches combining features of the two previous approaches, and evaluates them with respect to [4].

2   Resource Description and Selection for Geographic Queries

Generally, in our scenario every peer maintains a chunk of images as media items, where every image is described by a single pair of lat/long-coordinates. These geocoordinates are basically treated as point data in a plane for resource description and selection. Consequently, distances are approximated using the Euclidean distance, since investigations in [5] showed that the usage of distance

Fig. 1. Visualization of the simple MBR approach (on the left) and the RecMAR$_k$ technique with k=3 (on the middle) respectively k=6 (on the right) as examples for geometric approaches. The blue points denote the data points of the sample peer.

measures better suited for distance calculations between two points on earth (like Haversine distance [6] or Vincenty distance [7]) do not conduct noticeable changes in case of our data collection (cf. Sect. 3.1).

Resource selection is performed by ranking peers based on their resource descriptions, the query location and maybe some additional information such as reference points. The peer ranking defines the order in which peers get contacted during query processing. When searching for the k closest images with respect to a query location, the peer ranking should reflect that peers with a higher probability of administering a bigger fraction of the top-k images receive a higher rank than peers maintaining a smaller fraction of the top-k images. Our scenario requires a reasonable trade-off between the expressiveness of a resource description (allowing better selectivity) and its storage requirements for its representation. For the techniques presented in the following, differences in evaluation time are negligible compared to the time needed for accessing peers and will therefore not be considered.

The remainder of this section shortly presents the two classes of summary types evaluated in [4] and also describes the two most promising approaches found in these studies (cf. Sect. 2.1 and Sect. 2.2). Next, the class of hybrid approaches and the specific techniques examined in this paper (cf. Sect. 2.3) get introduced. Finally, the resource ranking algorithms will be described in Sect. 2.4.

2.1 Geometric Approaches

The first class of summary types computes a single or multiple geometrical shapes to enclose the set of point data a peer administers. Calculating approximated, concise representations of complex forms (a peer's point cloud in our case) is a standard problem in many computer science domains [8], therefore plenty of computational geometry algorithms exist and are applicable for this category of summaries. Figure 1 (on the left) shows one of the most basic techniques in this field, as it simply encloses all of a peer's data points into a minimum bounding rectangle (MBR) to describe its data, requiring two pairs of lat/long-coordinates to be stored.

In [4], we found the most promising technique of this class to be an approach where a peer's point cloud is described by several so-called minimum area rectangles (MARs). The computation algorithm is based on work from Becker et. al. [8] to summarize a set of bounding boxes by two bounding boxes which minimize the area that is covered. This algorithm has been adjusted to point data and was transformed into a recursive version, continuing the disassembly until a predefined maximum of k MARs has been computed or a certain threshold $dist$ has been undercut, which is taking the distance between the center of a MAR and the most distant of its associated data points into account. The threshold needs to be adjusted with respect to the underlying data collection to achieve appealing results. This technique, denoted $RecMAR_k$, shows excellent selectivity (that is the fraction of peers contacted to retrieve the relevant images) while keeping summary sizes at a reasonable level, since the algorithm allocates more MARs for peers with "complex" point data and less MARs for peers with less "complex", spatially narrowed point clouds. See Fig. 1 on the middle and on the right for a visualization of $RecMAR_k$.

2.2   Space Partitioning Approaches

The second category of summary types globally segments the data space into a certain number of subspaces. Thus, the segmentation is the same for all peers. This allows storing the information if a peer maintains data points for a certain subspace (or not) in the very peer's summary. Basically, there are three approaches to store information for a subspace: storing how many data points are contained in a cell (using integer values), storing whether at least one data point is contained in a subspace or not (using bits), or storing some kind of distance information concerning a peer's data points in a subspace (for example, the maximum distance between any data point being located in a certain subspace and the subspace's center, using float values). Figure 2 on the left shows the simplest space partitioning technique, mapping the lat/long-coordinates to a uniform grid (which results in non-uniform grid cell sizes on the sphere). The yellow highlighted cells contain data points, the related information can be stored in any of the aforementioned ways in a peer's summary.

Our experiments in [4] reveal the so-called Ultra Fine-grained Summaries (UFS_n) to be best when taking both description selectivity and summary sizes into account. The data space gets segmented based on n predetermined reference points invoking a Voronoi-like space partitioning (see Fig. 2 on the right). Therefore, a data point is assigned to the cell of the reference point being closest to it. To attain convincing results, we found data space segmentation has to be adjusted with respect to the underlying data collection (yielding smaller subspaces for areas with high global point density and bigger subspaces for low density areas). This can be achieved by randomly choosing reference points out of the underlying collection, or (simulating cases where this is not possible) from an external source where data points are similarly distributed as in the data collection. Using UFS_n, for each peer and each cell there is only binary information (1 or 0) stored depending on whether there is at least one data point inside a certain cell or not, allowing compression techniques to greatly reduce summary

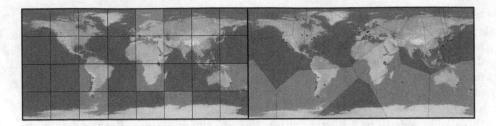

Fig. 2. Visualization of the simple grid (on the left) and the UFS_n technique (on the right), invoking a Voronoi-diagram-like space partitioning with 32 subspaces, as examples for space partitioning approaches. The black crosses on the right denote the reference points of the Voronoi diagram.

sizes. We compress the summaries using Java's standard gzip implementation[1]. To gain selectivity, the number of data cells can be increased. In [4], this approach results in very selective resource descriptions while keeping the average description size at a very low level.

2.3   Hybrid Approaches

This subsection introduces a third category of summary types, the hybrid approaches. These techniques combine characteristics of both geometric and space partitioning approaches, using geometric shapes as well as space segmenting to describe the location(s) of a peer's data. Basically, this category can be further distinguished into two subclasses, whether the techniques are using the geometric approach as first data description tool and affiliate some space partitioning method, or if they are doing it the other way round—some space segmentation followed by the usage of geometric shapes—. The first two techniques presented in the following are using geometric shapes as primary and space partitioning as secondary technique, while the subsequent two approaches are utilizing space partitioning primary and geometric shapes secondary.

MBRGrid$_r$. The first technique combines the basic geometric and space segmentation approaches presented earlier in the present paper. Initially, an MBR containing all of a peer's data points is computed. In a second step, the enclosed area gets segmented into subspaces by utilizing a uniform grid to partition the corresponding space. The number of subspaces can be adjusted by a parameter r, representing the number of rows on the grid. For this local segmentation, we did likewise as in [4] or [9] for global space segmentation, setting the number of

---

[1] The summaries of the hybrid summary types have been compressed the same way. For $RecMAR_k$, compression resulted in higher average summary sizes, therefore for this technique uncompressed summary sizes are taken into account when comparisons between the different techniques are drawn.

Fig. 3. Visualization of the MBRGrid$_r$ summaries with r=2 (on the left), respectively r=8 (on the right) for the interior grid

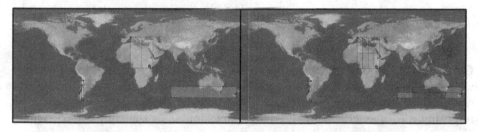

Fig. 4. Visualization of the KMARGrid$_k^r$ summaries with k=3 (computation of up to three enclosing rectangles) and r=1 (on the left), respectively r=2 (on the right) for the particular interior grids

columns twice as big as the number of rows. Adjusting the number of rows and columns according to, for example, the height/width-ratio of the enclosing MBR is conceivable and could be part of future work. See Fig. 3 for a visualization of an MBRGrid$_r$ summary.

A peer's summary is represented by a bit vector. First, the values encoding the two lat/long-coordinate pairs of the enclosing MBR are incorporated. Originally, the MBR bounds are captured as float values and get converted into binary information for summary inclusion ($4 \cdot 32$ bits). The summary's remainder consists of values 1 or 0, indicating whether the corresponding subspace contains at least one data point or not. If all of a peer's data points share (exactly) the same lat/long-coordinates, only the values specifying the MBR are encoded.

KMARGrid$_k^r$. Unsurprisingly, KMARGrid$_k^r$ takes the RecMAR$_k$ algorithm as a starting point to compute up to k MARs containing all of a peer's data points. The second step is similar to MBRGrid$_r$, except that in *each* MAR there is a Grid to be invoked. Again, grid granularity is determined by a parameter r, yielding r rows and $2 \cdot r$ columns for a MAR's grid. See Fig. 4 for a visualization of a KMARGrid$_k^r$ summary.

A bit vector represents a peer's summary as well. The grid-divided MARs are encoded one after another, with each using $4 \cdot 32$ bits for the rectangle extents and $r \cdot 2r$ bits to indicate cell occupancy for the respective interior grid.

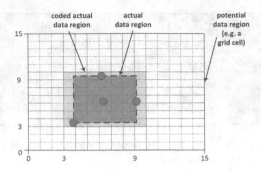

Fig. 5. Coding of an actual data region. The blue circles are the data points to be described (adapted from [14]).

GridMBR$_r^b$. GridMBR$_r^b$ is the first of two approaches doing it the alternative way, that is first segmenting the (whole) data space and afterwards using geometric shapes in a second step. GridMBR$_r^b$ initially segments the data space by imposing a regular grid onto the data space similar to the simple partitioning approach presented earlier. In the second step, for each occupied cell, an approximated MBR containing all the cell's data points is computed, pursuing an idea presented in [14].

Here a distinction between a potential data region (being the cell of a spatial access structure) and an actual data region is made, the latter being an MBR containing all the data points being located in the potential data region (that is a certain grid cell in our case). In [14], to reduce storage spent on encoding these interior MBRs, a technique originally introduced with the buddy-tree [15] is used, exploiting the presence of potential data regions. For encoding a MBR, generally four values need to be stored, specifying the lower left and upper right corner. Let's say b bits shall be used to encode one of these values. With this, we can distinguish 2^b different positions on an axis of a data cell. These positions can be used to encode an approximated MBR (also called the coded actual data region), being larger than the true MBR (the actual data region), but requiring significantly less storage than encoding the true MBR with float values. Figure 5 illustrates this for two-dimensional data with $b=4$.

For GridMBR$_r^b$, the parameter b is used to determine how much storage shall be used to encode one of the four required MBR bounding values, using b bits to encode a value (for example if $b=3$, eight different positions on each cell axis can be distinguished). Likewise as for example MBRGrid$_r$, a parameter r specifies the number of rows (r) and the number of columns ($2 \cdot r$) of the global grid.

As a peer's summary, a bit vector is used. Grid cells which do not contain any data point are encoded with 0, cells that contain at least one data point are encoded with 1. After an 1 representing an occupied cell, there follow $4 \cdot b$ bits encoding the four required values specifying lower left and upper right of the interior MBR. See Fig. 6 for an illustration of GridMBR$_r^b$ summaries for the sample peer.

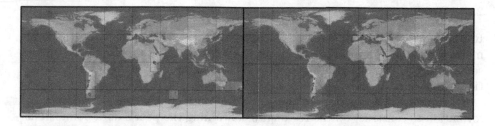

Fig. 6. Visualization of the GridMBR$_r^b$ summaries with r=4 (resulting in 32 grid cells) and b=2 (on the left) respectively b=4 (on the right), resulting in four respectively 16 possibilities on each axis to encode the interior MBRs

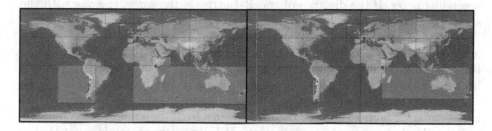

Fig. 7. Visualization of the non-uniform data space partitioning of K-D-MBR$_n^b$ with n=32 subspaces and b=2 (on the left) respectively b=4 (on the right) for encoding the cell-interior MBRs

K-D-MBR$_n^b$. The last hybrid approach takes the k-d-tree-like data space partitioning, called GFBu$_n$ in [4], as starting point. Using this technique, the data space gets segmented into rectangular cells of different sizes. Training data is used to learn a segmentation adjusted to the underlying data collection, while sources for training data are the same as for the reference points of the UFS$_n$ summaries, that is directly out of the data collection or from an external source with similar point distribution. At the beginning of the training process, the data space consists of only one cell or bucket. Training data points are sequentially inserted into this bucket, until a bucket-overflow occurs followed by splitting the bucket into two parts. Afterwards, further data points are inserted into the data structure. The whole process is repeated until the desired amount of n buckets has been reached.

As a second step after space partitioning, an approximated MBR is encoded for each cell containing at least one data point. The MBR computation is accomplished the same way as for GridMBR$_r^b$, that is there is a parameter b again, to adjust the amount of storage used to encode the MBR. Figure 7 illustrates this for the sample peer. For the summaries, bit vectors are used in the same way as for GridMBR$_r^b$.

2.4   Ranking

Ranking of peers is conducted based on the information supplied by the summaries. For the hybrid approaches, all the ranking algorithms operate on the same principle. Note that all these approaches at the very end encode information of rectangular areas in which a peer's data points are located.

To rank peers, for each peer the algorithm extracts the peer's summary information to construct all the rectangles containing the peer's data points. This is done in an offline phase. At query time, the minimum distance between each rectangle and the query location is calculated (a query point located inside a rectangle results in a distance of 0 for this rectangle). For each rectangle, distance information and the area covered by the rectangle are stored in a so-called R-Entry. All R-Entries of a peer are arranged in a queue, sorted by distance in ascending order. If the distance of two R-Entries is the same, the one with the smaller area covered is favored.

To determine a ranking between two peers, the sorted R-Entries are compared one after another. If the first R-Entry of peer p_a is closer to the query location than the first R-Entry of peer p_b, p_a is ranked higher than p_b and vice versa. If both R-Entries yield the same distance, p_a is ranked higher than p_b, if p_a's R-Entry covers a smaller area than p_b's R-Entry and vice versa. If the area covered is the same for both R-Entries, the next entries from the queues are compared, etc. (until a decision can be made)[2]. If the R-Entry comparison does not lead to a decision, a random ranking choice is made.

For RecMAR$_k$ evaluated in [4], ranking works the exact same way. The UFS$_n$ ranking shows a little variation due to the computational complexity for calculating Voronoi cell boarders. There, the reference points c_j ($j \in \{0; n-1\}$) are sorted in ascending order with respect to the distance to the query location. The first element of the sorted list L corresponds to the cluster center being closest to the query (called query cluster). If peer p_a administers documents in this query cluster (that is 1 is set for the cluster in the peer's summary) while peer p_b does not (that is 0 is set), p_a is ranked higher than p_b and vice versa. If both peers feature the same value for the query cluster, the next element out of L is chosen and both peers are ranked according to their summary values for this very cluster. This procedure continues until a decision favoring one of the peers can be made or the end of L is reached, resulting in a random decision.

3   Evaluation

In this section, we will give an intital brief description of the data collection used for the evaluation (cf. Sect. 3.1). For comparability with previous evaluated approaches, we use the experimental setting conducted in [4], but only apply

---

[2] Obviously, not all the peers hold the same number of rectangular areas containing data points. In this case, for the peer represented by fewer rectangular areas, dummy entries are generated, whose values have the most unfavorable impact (that is infinite distance and infinite area) on the ranking for the respective peer.

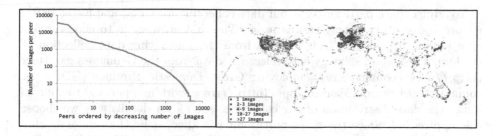

Fig. 8. Number of images per peer (left) and geographic distribution of images locations (right) for the evaluated data collection

one query mode (cf. Sect. 3.2), since results are very similar for different query point sources in [4]. Afterwards, the experimental results will be displayed and discussed in Sect. 3.3.

3.1 Data Collection

During 2007, a large amount of publicly available, georeferenced images uploaded to Flickr (http://www.flickr.com) was crawled. In our scenario, every Flickr user operates a peer of its own. Thus, we assign images to peers by means of the Flickr user ID. After some data cleansing, the collection consisted of 406,450 geo-tagged images from 5,951 different users/peers. The distribution of the number of images per peer (see Fig. 8) is very skewed which is typical for many P2P settings [3]. Few peers administer large amounts of the collection ("big peers"), while there are also many peers which store only few images ("small peers"). See [9] for a more detailed analysis. On the right, Fig. 8 illustrates the geographic distribution of the image locations, revealing an uneven spread with image hotspots in North America, Europe and Japan.

3.2 Experimental Setting

We use 500 image locations chosen in a two-step process as queries. First a random peer is selected, second we choose a random geo-location from this peer. This ensures the same likelihood that a query originates from an arbitrary peer regardless of its size, thus not favoring big peers in the ranking.

For parameterization, k for RecMAR$_k$ and KMARGrid$_k^r$ is set to 3, 6 and 9. For space partitioning approaches and hybrid techniques relying primarily on space partitioning, we choose the number of subspaces to be 512, 2,048 and 8,192, resulting in r being set to 16, 32 and 64 for MBRGrid$_r$ and GridMBR$_r^b$. For interior grids (used by MBRGrid$_r$ and KMARGrid$_k^r$), r is set to 8, 16, 32 and 64. To encode approximated interior MBRs for GridMBR$_r^b$ and K-D-MBR$_n^b$, we vary b from 2 to 4 to 6, corresponding to the bits used to encode one of the four required MBR values, respectively. These parameters are generally applicable to our or similarly distributed data collections. For significantly differing data sets, the parameters will have to be reconsidered.

To adjust space partitioning to our data collection for UFS_n and $K\text{-}D\text{-}MBR_n^b$, we use the same two strategies as in [4]. The first strategy is to choose reference points, respectively training data from the underlying data collection at random. The second strategy is to select the data from the Geonames gazetteer (http://www.geonames.org), employing Gross Domestic Product (GDP) per country statistics from Worldmapper (http://www.worldmapper.org) to approximate the data distribution of our collection (see [9] for details why we choose this approach)[3]. Since the space partitioning is affected by the randomly chosen training data, we run ten experiments with different seeds for the random number generators to minimize the effects of outliers.

For $RecMAR_k$ and $KMARGrid_k^r$, we use the same *dist* value (cf. Sect. 2.1) deployed in [4] for $RecMAR\_Q_k$, meaning the 0.75-quantile of the top-50 data point distances determined for 500 queries.

Space efficiency of different resource descriptions is measured by analyzing summary sizes (cf. Sect. 3.3). Remember we apply Java's gzip implementation with default parameters if summary compression is beneficial (which is the case for all summary types except $RecMAR_k$). The measurements include 27 byte serialization overhead necessary in order to distribute the resource descriptions within the network. To assess the selectivity of different approaches, we calculate the fraction of peers that needs to be contacted on average in order to retrieve the 50 image locations closest with respect to a given lat/long-pair as query location. Summary sizes are strongly dependent on the techniques used, so it is not possible to report the selectivity for specific given summary sizes.

To determine the 50 nearest neighbors, a k-nearest neighbors (kNN) algorithm, implemented as a range query with decreasing query radius, is used. First, all peers are ranked according to the ranking algorithms (cf. Sect. 2.4). For each of the ten best ranked peers, the 50 image locations closest to the query location are requested. Out of this set, the 50 closest image locations are determined to form the current intermediate top 50 results[4]. Afterwards, the already considered peers are removed from the set of peers yet to look at. The distance of the fiftieth closest image location is the query radius for the next round, in which ten further peers will be contacted. Peers which can be pruned from search, due to their resource description, the query location and the query radius, are removed from the set of peers still to consider. A renewed ranking is not conducted, meaning the set of peers is only ranked once in the first query round. In each round, the ten best ranked peers of the remaining set are enquired for their 50 closest image locations according to the query location, possibly leading to a substitution of (some of) the current top 50's images. Afterwards, these peers

---

[3] For differentiation, we expand _e to the respective technique acronym if data was chosen from the Geonames gazetteer as external source, for example UFS_n_e.

[4] The consideration of ten peers at once is done to exploit the parallelism of our scenario. Nevertheless, the determination of the top 50 image location's retrieval time (meaning the fraction of peers which had to be visited in order to retrieve the top 50 image locations) is captured on a single peer basis, meaning the contacting position of the appropriate peer is captured on a single step base and not on a ten step base.

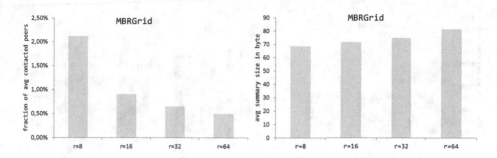

Fig. 9. Development of selectivity (on the left) and summary sizes (on the right) when varying parameter r for MBRGrid$_r$

are removed from the set of peers to consider. This procedure continues until the set of peers to consider is empty, meaning the 50 nearest neighbors with respect to the query location have been determined.

3.3   Experimental Results

In this section, we will evaluate the different techniques. For our 500 queries, an average of 8.234 peers maintain relevant image loactions, resulting in a fraction of 0.138% of the 5,951 peers. As a naive baseline it is interesting to note, that if all peers would directly transfer a (zipped) byte array containing all of their data points, average summary sizes would be 265.85 byte (cf. Table 3.3 at the end of this chapter).

In Fig. 9, retrieval performance and summary sizes are depicted for MBRGrid$_r$ with varying parameter r. There is a degressive improvement in selectivity with the increase of r, as the areas containing a peer's data points are described more precisely. At the same time, compression keeps summary size growth moderate. If r is altered from 8 to 64, selectivity is more than four times better while summaries are about 19% bigger on average. Also, selectivity growth is still disproportionate to summary size growth when increasing r from 32 to 64, making MBRGrid$_{64}$ the best choice when pondering between the two superordinate goals.

For KMARGrid$_k^r$, selectivity can obviously be enhanced by both increasing either one or both parameters k and r. Figure 10 on the left shows a degressive improvement as greater values for k and r are chosen. Combining geometric shapes with space partitioning seems very beneficial with respect to selectivity, since KMARGrid$_3^8$ is almost as good as RecMAR$_9$ (cf. Table 3.3), though it has to be admitted that the former requires about 13 byte extra storage on average. Generally, it can be seen that a decomposition into more MARs yields greater benefits for selectivity than a finer resolution of the interior grids. The utmost gains are attained by increasing k from 3 to 6. Afterwards, a further increase of k from 6 to 9 is not as beneficial, as our stopping criterion for further MAR decomposition comes into play, resulting in few peers to be described by the

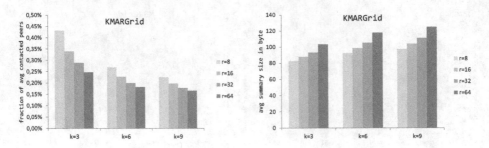

Fig. 10. Development of selectivity (on the left) and summary sizes (on the right) when varying parameters k and r for KMARGrid$_k^r$

maximum of nine or a matching amount of MARs. On the other hand, altering k from 6 to 9 leads to a rather moderate average summary size growth. The required average storage space is depicted in Fig. 10 on the right. Increasing k results in degressive summary size growth, while increasing r results in progressive summary size growth. When searching for the best compromise between selectivity and average summary sizes, we compare percentual selectivity gains and percentual average summary size growth when increasing our parameters k and r[5]. As long as percentual selectivity gains are higher than percentual summary size growth, we say it is beneficial to raise the parameters. Taking this into account, KMARGrid$_9^{32}$ results as best parameterization for this technique. Generally, at the beginning of the parameter rise (that is from $k=3$ and $r=8$ on), selectivity gains are vigorously disproportionate compared to average summary size growth, flattening during the further procedure until it is only slightly disproportionate or even slightly underproportionate.

The results for GridMBR$_r^b$ are shown in Fig. 11. A relatively precise encoding of interior MBRs in coarse grids results in high selectivity gains in comparison to increasing the number of subspaces. Taking for example GridMBR$_{32}^2$ as a basis and increasing parameters r, respectively b to the next level, GridMBR$_{32}^4$ shows a better selectivity compared to GridMBR$_{64}^2$, since GridMBR$_{32}^4$ only contacts a peer fraction of 0.94% while GridMBR$_{64}^2$ contacts 1.19% of the peers. At the same time, increasing b also results in more space efficient summary sizes (for example 58.2 byte for GridMBR$_{32}^4$ vs. 60.7 byte for GridMBR$_{64}^2$). Generally, it is beneficial to increase both r and b since the selectivity gain is still heavily disproportionate to summary size growth when comparing for example GridMBR$_{64}^6$ to both GridMBR$_{32}^6$ and GridMBR$_{64}^4$.

Results for both variants of K-D-MBR$_n^b$ are depicted in Fig. 12. Generally, selectivity increases most when raising b from 2 to 4, but also raising n from 512 to 2048 yields vast selectivity gains, both clearly disproportionate compared to summary size growth. A further parameter increment ($b=6$ and $n=8{,}192$) results in only slightly disproportionate or even underproportionate selectivity

---

[5] Obviously, the importance of selectivity and storage requirements has to be weighted for a concrete application scenario.

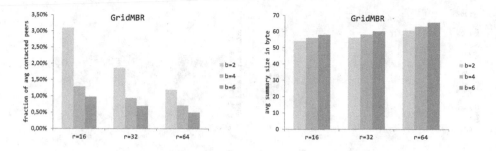

Fig. 11. Development of selectivity (on the left) and summary sizes (on the right) when varying parameters r and b for GridMBR$_r^b$

gains. Overall, it is very promising to encode interior MBRs for space partitioning with uneven cell sizes, as the larger subspaces often contain data points in only a narrow area. Thus, the data point occurrence areas can be described much more precisely. Likewise for MBRGrid$_r^b$, increasing b is more beneficial than increasing the number of subspaces (parameter n here). Generally, when comparing K-D-MBR$_n^b$ to K-D-MBR$_n^b$_e, the same behavior observed for UFS$_n$ and UFS$_n$_e in [4] can be noted. The _e-variant shows worse selectivity but lower required storage space, since space partitioning is less fitted to the underlying data collection, resulting in less subspaces to be occupied with a peer's data points on average. This reduces the amount of interior MBRs which need to be encoded. Overall, selectivity gains area slightly better for the _e-variant when increasing n and b, as the inferior base accuracy causes a higher gain potential. When comparing percentual selectivity gain with percentual summary size growth, different parameterizations are best for the respective variants. For training data right from the underlying data collection, K-D-MBR$_{2048}^6$ emerges as best compromise, while for training data from an external source, K-D-MBR$_{8192}^6$_e arises as most reasonable solution. It is worth mentioning that K-D-MBR$_{2048}^6$ outperforms K-D-MBR$_{8192}^6$_e both with respect to selectivity and summary size.

Comparing the different techniques on their respective best parameterization (cf. Fig. 12 and Table 3.3), both MBRGrid$_{64}$ and GridMBR$_{64}^6$ are significantly worse with respect to selectivity in comparison to the more complex hybrid approaches (and RecMAR$_9$ and UFS$_{8192}$ as well), even though for both the most precise examined parameterization has been chosen (cf. Fig. 13). Only UFS$_{8192}$_e with its just approximately fitted space partitioning is outperformed by MBRGrid$_{64}$ and GridMBR$_{64}^6$. However, GridMBR$_{64}^6$ requires significantly less storage than MBRGrid$_{64}$ and less storage than any other technique except UFS$_{8192}$_e, which is clearly outperformed by GridMBR$_{64}^6$ with respect to selectivity. Thus, GridMBR$_r^b$ could be worthwhile if somehow there is no possibility to adjust techniques with respect to the underlying data collection (which all of the other techniques except MBRGrid$_r$ to some extent do, may it be the use of stopping criteria or adapted space partitioning), as selectivity is still neat with very small average summary sizes.

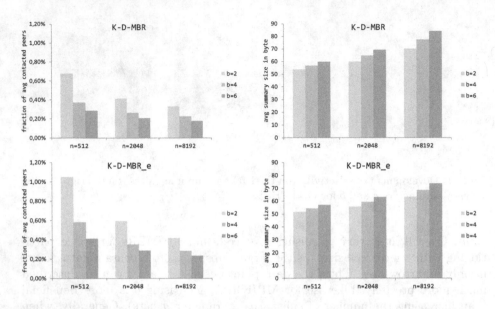

Fig. 12. Development of selectivity (on the left) and summary sizes (on the right) when varying parameters n and b for K-D-MBR$_n^b$(_e)

The K-D-MBR$_{2048}^6$(_e)-variants both significantly outperform their respective UFS$_{8192}$(_e)-variants in terms of selectivity, despite a four times lower amount of subspaces, due to their usage of cell-interior MBRs. This is achieved by only slightly bigger average summary sizes. Even more surprisingly, K-D-MBR$_{2048}^4$ (not shown in Fig. 13) can outperform UFS$_{8192}$ both on selectivity *and* summary sizes, making K-D-MBR$_n^b$ the overall superior technique compared to UFS$_n$. In terms of selectivity, K-D-MBR$_{2048}^6$ is yet slightly topped by KMARGrid$_9^{32}$, which is closing up to 0.04% with respect to the theoretical optimum (cf. Table 3.3). On the other hand, average summary sizes are almost 60% greater compared to K-D-MBR$_{2048}^6$. Generally, it can be seen that for the hybrid techniques, segmenting space first and computing geometric shapes second, results in far smaller summary sizes while selectivity is about equal compared to the alternative way (when matching GridMBR$_r^b$ to MBRGrid$_r$ and K-D-MBR$_n^b$ to KMARGrid$_k^r$).

Taking both selectivity and average summary sizes into account, K-D-MBR$_n^b$ with its training data directly chosen from the underlying data collection seems best as it offers almost prime selectivity combined with still very small average summary sizes, clearly outperforming the former state-of-the-art approaches UFS$_n$ and RecMAR$_k$. Furthermore, K-D-MBR$_n^b$ is much more insensitive in case training data origins from an external source than UFS$_n$. Considering pure selectivity, KMARGrid$_k^b$ constitutes an advance as well, though at the cost of big summary sizes.

Interestingly, selectivity is not directly related to the data space area spanned on average by the different summarization techniques (cf. Table 3.3). This can be seen as RecMAR$_9$ and both K-D-MBR$_{2048}^6$ and K-D-MBR$_{8192}^6$ offer much better

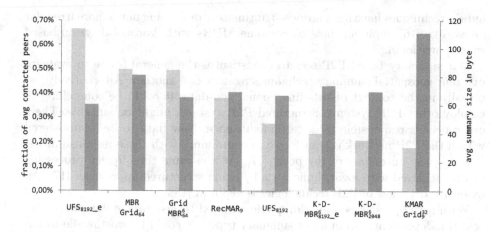

Fig. 13. Overview of evaluated techniques sorted by selectivity (light gray bars) and additionally showing average summary sizes (dark gray bars)

Table 1. Result table for our experiments (summary size values S_x in bytes), uncompressed summary size values are marked with *

Approach	Ø frac. of peers in %	$S_\emptyset$	S_{min}	S_{max}	Ø area covered
UFS_{8192_e}	0.665	60.5	49.8	295.3	not calculated
$MBRGrid_{64}$	0.498	81.3	53	414	0.984
$GridMBR_{64}^6$	0.495	65.6	55	329	0.630
$RecMAR_9$	0.386	69.3*	43*	171*	2.933
UFS_{8192}	0.276	66.88	48	467.4	not calculated
$K\text{-}D\text{-}MBR_{8192\_e}^6$	0.234	73.7	48.1	947.5	0.900
$K\text{-}D\text{-}MBR_{2048}^6$	0.208	69.5	46.45	577.5	1.015
$KMARGrid_9^{32}$	0.178	111.2	53	994	0.024
Baselines	0.138	265.8	53	43064	–

selectivity, while the area on average overlaid by the respective descriptions is (in case of $RecMAR_9$ clearly) larger compared to $MBRGrid_{64}$ and $GridMBR_{64}^6$. It seems that in areas with low point density, taller delineated descriptions are acceptable if in high density areas the descriptions are subsequently more precise.

4   Related Work

This paper introduced and evaluated techniques to summarize geospatial data. The techniques presented predominantly orientate themselves towards approaches known from spatial index structures (cf. [10]), like for example the R-tree [11], the k-d tree [12], Voronoi-diagram-based techniques (for example [13]) or the LSD^h-tree [14].

At the very end our hybrid techniques are all based on describing rectangular areas containing data points. In the context of spatial index structures, other hybrid techniques have been proposed, ultimately describing much more irregular areas, like [16] depicting how to combine MBRs with Voronoi-diagram-based space partitioning.

Our summary based P2P scenario constitutes the general frame to evaluate different geospatial summary techniques against one another. Alternatively, especially in the context of two-dimensional geo-data, it could be compelling to employ other P2P systems. Structured P2P systems might be suitable. There every peer is responsible for a certain subspace. New data to be administered within the P2P network is transferred in compliance with these responsibilities, reducing the autonomy of the peers [17]. At the same time, query processing can be achieved with logarithmic costs by using structured approaches [17]. An extensive overview of P2P technologies is given in [18].

Within the presented scenario, queries related to geographic data were processed independently from other summary types. It could be worthwhile to not only consider summary techniques in an isolated way, but also in cooperation. Effective searches of, for example, an image showing a sunset at the Grand Canyon, could be achieved this way. Currently, summaries for both—low level visual content and geographic information—would lead to two independent resource rankings. Both rankings would have to be combined into one with appropriate techniques (for example [19]). In [20], an approach aggregating several summary types into one summary to consider interdependency is presented. A further consideration of these combinations would be of value.

5   Conclusion and Outlook

This paper focuses on resource selection based on geographic information. It introduces a novel category of sumarization techniques besides the geometric approaches and space partitioning approaches in this context. Hybrid approaches combine features of the two former approaches. In terms of selectivity, the two more complex hybrid techniques (KMARGrid$_k^r$ and K-D-MBR$_n^b$) presented in this paper can outperform the best techniques from the geometric and space partitioning approaches. While KMARGrid$_k^r$ achieves this at the cost of significantly greater summary sizes, K-D-MBR$_n^b$ even outperforms the former state-of-the-art approaches with respect to summary size. Future work will mainly adress the evaluation of all these techniques on collections which are not extracted from a social network with its typical long tail distribution and therefore offer a more uniform allocation with respect to the number of image locations per resource. Furthermore, we plan to investigate the adaption of the resource summarization techniques for other application fields, such as index structures.

References

1. Thomas, P., Hawking, D.: Server selection methods in personal metasearch: a comparative empirical study. Information Retrieval 12(5), 581–604 (2009)

2. Müller, W., Eisenhardt, M., Henrich, A.: Scalable summary based retrieval in P2P networks. In: Intl. Conf. on Information and Knowledge Management, pp. 586–593 (2005)
3. Cuenca-Acuna, F., Peery, C., Martin, R.P., Nguyen, T.D.: PlanetP: Using gossiping to build content addressable peer-to-peer information sharing communities. In: IEEE Intl. Symp. on High Performance Distributed Computing, pp. 236–246 (2003)
4. Kufer, S., Blank, D., Henrich, A.: Techniken der Ressourcenbeschreibung und -auswahl für das geographische Information Retrieval. In: Proceedings of the IR Workshop at LWA 2012, pp. 1–8 (2012)
5. Blank, D., Henrich, A.: Describing and Selecting Collections of Georeferenced Media Items in Peer-to-Peer Information Retrieval Systems. In: Diaz, L., Granell, C., Huerta, J. (eds.) Discovery of Geospatial Resources: Methodologies, Technologies, and Emergent Applications, pp. 1–20 (2012)
6. Sinnott, R.: Virtues of the haversine. Sky and Telescope 68(2), 159 (1984)
7. Vincenty, T.: Direct and inverse solutions of geodesics on the ellipsoid with application of nested equations. Suvery Review 22(176), 88–93 (1975)
8. Becker, B., Franciosa, P.G., Gschwind, S., Ohler, T., Thiemt, G., Widmayer, P.: An Optimal Algorithm for Approximating a Set of Rectangles by Two Minimum Area Rectangles. In: Bieri, H., Noltemeier, H. (eds.) CG-WS 1991. LNCS, vol. 553, pp. 22–29. Springer, Heidelberg (1991)
9. Blank, D., Henrich, A.: Description and Selection of Media Archives for Geographic Nearest Neighbor Queries in P2P Networks. In: Inf. Acc. for Pers. Media Archives at ECIR 2010, pp. 22–29 (2010)
10. Samet, H.: Foundations of Multidimensional and Metric Data Structures. Morgan Kaufmann Publishers Inc., San Francisco (2005)
11. Guttman, A.: R-trees: a dynamic index structure for spatial searching. SIGMOD Rec. 14(2), 47–57 (1984)
12. Bentley, J.L.: Multidimensional binary search trees used for associative searching. Comm. ACM 18(9), 509–517 (1975)
13. Kolahdouzan, M., Shahabi, C.: Multidimensional binary search trees used for associative searching. In: Proc. of the 30th Intl. Conf. on Very Large Data Bases, pp. 840–851 (2004)
14. Henrich, A.: The LSD^h-tree: An Access Structure for Feature Vectors. In: Proc. of the 14th Intl. Conf. on Data Engineering, pp. 262–369 (1998)
15. Seeger, B., Kriegel, H.-P.: The buddy-tree: an efficient and robust access method for spatial data base systems. In: Proc. of th 13th Intl. Conf. on VLDB, pp. 590–601 (1990)
16. Sharifzadeh, M., Shahabi, C.: VoR-tree: R-trees with Voronoi diagrams for efficient processing of spatial nearest neighbor queries. Proc. VLDB Endow. 3(1-2), 1231–1242 (2010)
17. Doulkeridis, C., Vlachou, A., Nrvag, K., Vazirgiannis, M.: Part 4: Distributed Semantic Overlay Networks. Handbook of Peer-to-Peer Networking, 1st edn. Springer Science+Business Media (2009)
18. Shen, X., Yu, H., Buford, J., Akon, M.: Handbook of Peer-To-Peer Networking, 1st edn. Springer Publishing (2009)
19. Belkin, N.J., Kantor, P., Fox, E.A., Shaw, J.A.: Combining the evidence of multiple query representations for information retrieval. Inf. Processing and Management 31(3), 431–448 (1995)
20. Hariharan, R., Hore, B., Mehrotra, S.: Discovering gis sources on the web using summaries. In: Proc. of the 8th ACM/IEEE Joint Conf. on Digital Libraries, JCDL 2008, pp. 94–103 (2008)

Location-Based Sponsored Search Advertising

George Trimponias[1], Ilaria Bartolini[2], and Dimitris Papadias[1]

[1] Department of Computer Science and Engineering,
Hong Kong University of Science and Technology
{trimponias,dimitris}@cse.ust.hk
[2] Department of Computer Science and Engineering,
University of Bologna
i.bartolini@unibo.it

Abstract. The proliferation of powerful mobile devices with built-in navigational capabilities and the adoption in most metropolitan areas of fast wireless communication protocols have recently created unprecedented opportunities for location-based advertising. In this work, we provide models and investigate the market for location-based sponsored search, where advertisers pay the search engine to be displayed in slots alongside the search engine's main results. We distinguish between three cases: (1) advertisers only declare bids but not budgets, (2) advertisers declare budgets but not bids, and (3) advertisers declare both bids and budgets. We first cast these problems as game theoretical market problems, and we subsequently attempt to identify the equilibrium strategies for the corresponding games.

Keywords: Location-based advertising, Game theory, Nash equilibrium.

1 Introduction

The growing popularity of powerful and ubiquitous mobile devices has recently created an immense potential for location-based advertising (LBA) [5]. Smartphone use is rapidly increasing in all parts of the world; in the US only, its penetration is currently approaching 50% of all mobile subscribers, while around 60 percent of the new phones in 2011 were smartphones[1]. This development has certainly been facilitated by the adoption of broadband wireless protocols, e.g., 3G/4G networks, and the prevalence of Wi-Fi hotspots. Moreover, modern mobile devices possess built-in navigational functionalities using variations of sophisticated technologies such as triangulation, GPS, and cell-ID [5]. Advertisers can utilize this positional information to send advertising material to relevant consumers, which has in turn created an exciting market for LBA with companies such as AdMob (acquired by Google) and Quattro Wireless (acquired by Apple) leading the charge.

Location-based advertising, especially in its mobile form, is poised for tremendous growth because of its special characteristics [1]. First, it enables personalization:

---

[1] See http://www.informationweek.com/news/mobility/business/231602163

M.A. Nascimento et al. (Eds.): SSTD 2013, LNCS 8098, pp. 348–366, 2013.

a mobile device is associated with the identity of the user so the advertising material can be individually tailored. For example, users can state their preferences, or even specify the kind of advertising messages they are interested in. Second, it is context-aware, i.e., the advertising messages can take into account the context such as time and location. Third, mobile devices are portable and allow instant access: users carry their device most of the time, and advertisers can target interesting consumers any time of the day. Finally, mobile advertising can be interactive since it is possible to engage the user to discussions with the advertiser; this can also serve as a means of market research. As a result of the aforementioned reasons, marketers can reach their audience of interest in a much more targeted, personal and interactive manner, and thus increase their advertising campaign's success.

On the other hand, currently the most profitable and thriving business model for online advertising is *sponsored search advertising*; Google's total revenue alone in fiscal year 2010 was $29.3 billion and mainly came from advertising[2]. Sponsored search consists of three parties [12]: (i) *users* pose keyword queries with the goal of receiving relevant material; (ii) *advertisers* aim at promoting their product or service through a properly designed ad, and target relevant users by declaring to the search engine a set of keywords that capture their interest; (iii) the *search engine* mediates between users and advertisers, and facilitates their interaction. As several advertisers may match a given user query, an *auction* is run by the search engine every time a user poses a query to determine the winners as well as the *price per click*. Concretely, each advertiser declares to the engine a priori its *bid* for a given keyword, so the auction assigns ad slots to advertisers based on their bids.

In this work, inspired both by the success of sponsored search advertising and the immense potential for LBA, we study the promising area of *location-based sponsored search advertising*. In particular, we examine how the spatial component can be incorporated into the current sponsored search models, and investigate algorithms for selling advertising opportunities to advertisers. Similar to prior literature on conventional sponsored search advertising [8], in order to model the advertisers we distinguish between the three following cases: (1) the advertisers declare a maximum amount of money that they are willing to pay per click, but are not bounded by a total daily budget, (2) the advertisers have a maximum daily budget at their disposal, but do not have an upper bound on the amount of money that they are willing to pay per click, and (3) the advertisers have both a total daily budget and a maximum amount of money that they are willing to pay per click. We will explicitly show how the introduction of the spatial component affects the underlying sponsored search auction in each of the cases above by using tools and techniques from game theory.

The rest of the paper is organized as follows. Section 2 surveys related work. Section 3 provides a general model for location-based sponsored search. Sections 4-6 investigate three interesting settings for location-based sponsored search: (1) advertisers declare only bids (Section 4), (2) advertisers declare only budgets (Section 5), and (3) advertisers declare both bids and budgets (Section 6). Using tools from game theory, we analyze the three different cases and provide the Nash equilibrium strategies when possible. Finally, Section 7 concludes the paper providing interesting directions for future research.

---

[2] See http://investor.google.com/financial/2010/tables.html

2   Related Work

2.1   Location-Based Advertising

Location-based advertising (LBA) involves delivering advertising material to users based on their location. It follows two different modes of operation [3]: *pull-based* (also termed *query-based*) and *push-based*. The former provides advertising information only upon specific request by the user, e.g., when a car driver asks for the nearest gas stations. The latter delivers marketing material to users within a specified geographical area, without their explicit request; for instance, shops in a mall seeking to promote their new product may target all shoppers by delivering the corresponding advertising information. Moreover, push-based advertising is further divided into two types: *opt-in*, where users receive relevant advertising material by determining in advance the kind of ads they are interested in, and *opt-out*, where users receive marketing messages until they explicitly declare they do not wish to receive any further material.

LBA presents immense opportunities for higher return on investment compared to other traditional advertising avenues because it enables contextually relevant advertising [5]. Moreover, the ability to instantaneously connect users to places or resources of interest in their immediate vicinity can offer an unrivaled user experience and satisfaction. Interestingly, LBA can also serve as a subtle tool for market research: consumers constantly provide information about their behavior through their mobile activity, which can be subsequently used to increase the effectiveness of a marketing campaign. Despite its obvious benefits, consumer's privacy is still a major cause of concern for LBA. Advertisers need to be very clear about how they utilize, process, and store user information; data breaches, for instance, can be especially detrimental to the advertiser's reputation and long-term success, since they can reveal personal information. A second major concern stems from the intrusive nature of some forms of LBA, in particular push-based that occurs without the user's explicit request. Among the two modes, opt-out is associated with a higher intrusion risk and is thus used more rarely; opt-in, in contrast, is permission-based advertising and may be used to effectively rule out unsolicited marketing messages (i.e., spamming) [18].

2.2   Sponsored Search Advertising

Sponsored search advertising is the most profitable form of online advertising. It constitutes a large and rapidly growing source of revenue for search engines. Currently, the most prominent players in the sponsored search market are Google's AdWords [22], and Bing Ads [21]. In sponsored search advertising, advertisers place properly designed ads to promote their product or service. They target interesting users by declaring to the search engine a list of keywords that a relevant user may search for. For each keyword, they additionally specify their *maximum cost per click* (*maximum CPC*), also known as maximum bid, which corresponds to the maximum amount of money they are willing to spend to appear on the results page for a given keyword. Note that bidding takes place continuously. Moreover, advertisers may be limited by budget constraints, so they may declare a maximum daily budget as well. Every time a user enters a query, a limited number of *paid* (also, *sponsored*) links appears on top or to the right side of the *unpaid* (also, *organic* or *algorithmic*) search results. In order to determine the winning advertisers as well as the price they need to pay, an auction occurs in an automated fashion. In practice, large search engines also compute a *quality score* (*QS*) for every advertiser which measures how relevant the keyword, ad text and landing page are to a user.

Concretely, sponsored search advertising consists of three stages. (i) *Ad retrieval* returns all ads that are relevant to the user's query, and is usually performed by sophisticated machine learning algorithms. An ad's relevance is measured by several metrics, such as ad-query lexical and semantic similarity. To match an ad against a query, the search engine needs to take into account all ad information, including the bid phrase it is associated with, its title and description, the landing page it leads to, its URL, etc. Moreover, query substitution and query rewriting are frequently used to find relevant ads. As the ad pool may consist of millions of ads, efficient indexing techniques have been proposed to improve the performance of the first stage. (ii) After retrieving relevant ads, the search engine performs *ad ranking*. Ads are sorted in decreasing order of their rank, where the ad rank is determined by both the bid placed by the advertiser on the keyword, and the quality of the ad. The ad with the highest rank appears in the first position, and so on down the page, until all slots have been filled. Google AdWords[3], for instance, defines the ad rank as the product CPC·QS. (iii) The last stage is *ad pricing* through properly designed auctions to determine the price per click that the advertiser will be charged whenever the user clicks on their ad (pay per click model). The natural method would be to make bidders pay what they bid (i.e., *generalized first-price auction*), but that leads to several instabilities and inefficiencies. Instead, all large search engines currently employ a *generalized second-price auction* (*GSP*) [7][19]. A GSP auction charges an advertiser the minimum amount required to maintain their ad's position in search results, plus a tiny increment. For instance, suppose that ranking is based on Google's AdRank and that K slots are available, and are numbered $1, \ldots, K$, starting from the top and going down. Moreover, let the advertiser i at position i have a maximum bid b_i and a quality score QS_i. In GSP, the price for a click for advertiser i is determined by the advertiser $i+1$, and given by $b_{i+1} \cdot QS_{i+1}/QS_i$, which is the minimum that i would have to bid to attain its position. Note that in this pricing scheme, a bidder's payment does not take into consideration its own bid. Also, prices per click can be computed in linear time in the number of advertisers $O(N)$ for a fixed number of slots K.

Despite its prevalence as the standard auction format, GSP is not *truthful* (also known as *incentive-compatible*): advertisers have no incentive to declare their true valuations to the search engine. Stated equivalently, reporting the true bids may not constitute a Nash equilibrium [7]. As a result, advertisers may devote considerable resources to manipulate their bids, potentially paying less attention to ad quality and other campaign goals. Interestingly, we can alleviate this shortcoming by altering the

---

[3] See https://adwords.google.com/support/aw/bin/answer.py?hl=en&answer=6111

payment scheme: instead of paying the minimum amount of money required to win its position, an advertiser is requested to pay an amount of money equal to the externalities that it imposes on the others, i.e., the decreases in the valuations of other bidders because of its presence. This payment scheme yields the Vickrey-Clarke-Groves (VCG) auction, named after William Vickrey [20], Edward H. Clarke [4], and Theodore Groves [11]. Contrary to GSP, VCG gives bidders an incentive to bid their true value, and is *socially optimal*, i.e., the bidder with the highest valuation acquires the slot at the highest position, the bidder with the second-highest valuation receives the slot at the second-highest position, etc. Note that GSP rather than VCG is used in practice, even though the latter would (at least theoretically) diminish incentives for strategizing and facilitate the advertisers' task. We believe that the introduction of the ad quality score QS has also played a role in the wide adoption of GSP. Indeed, ad quality scores are now an integral part of both the ranking and pricing protocols; even if advertisers manipulate their bids, it is very difficult to game the system as they have no control over the ad quality scores.

3   Models for Location-Based Sponsored Search

Assume N advertisers and K slots 1, ..., K, where 1 is the top slot, 2 the second, and so on. There is ample evidence in the literature that higher slots are associated with higher revenues. There are numerous ways to model this; perhaps the easiest way is to characterize each slot with the click through rate, which denotes the probability that a user will actually click on an ad that is placed in that slot. In this work, we follow the same approach by assuming that whenever an ad is displayed in slot l, $1 \leq l \leq K$, it has a probability c_l, $0 \leq c_l \leq 1$, of being clicked. To incorporate the fact that higher slots are more valuable, we further assume that $c_l > c_{l'}$ whenever $l < l'$. Whether a user clicks on an ad or not depends on numerous factors including the other ads (*ad externalities*), but for the sake of simplicity we do not consider them here; i.e., an ad located at slot l is clicked with probability c_l independent of the rest of the slots [15]. To keep the model simple, we also do not consider quality scores for advertisers.

A salient feature of our work is that advertisers value users according to their location. To model this, we assume that the space is partitioned with a grid of L cells. There are (in expectation) M_j queries per day in cell j, which can be estimated based on historical data. Advertisers have different valuations for the different grid cells. For instance, a typical advertiser would have high valuations for cells nearby and lower valuations for more distant cells. We denote with $w_{i,j}$ the valuation of advertiser i per click inside cell j. Calculating the valuation is a difficult marketing/operational research problem, beyond the scope of our work. Finally, advertiser i may be bounded by a maximum daily budget B_i. We can assume that the advertisers are only aware of their own budget and valuations, which they declare to the search engine. Besides the budgets and valuations per click, the search engine has knowledge of relevant statistical information such as number of queries per cell, or percentage of total clicks that a slot receives, etc. We consider that advertisers are interested in exactly the same (unique) keyword; how keyword interactions affect our market is an interesting research topic in its own right, and can be explored in future work.

Finally, note that the valuation of an advertiser for a given cell is fixed for all points inside the cell. The grid granularity involves an inherent trade-off between valuation expressivity and search engine revenue. On the one hand, small cells allow advertisers to better capture their cells of interest, as opposed to coarse grid granularities that would force an advertiser to declare interest for the entire cell even if they were interested in just a small part. On the other hand, small cells may take a serious toll on the search engine's revenue because the expected number of advertisers expressing interest in a given cell decreases as the grid granularity becomes finer. In the worst case scenario, a cell could attract interest from just a single advertiser and would yield poor income for the search engine. For instance, assume a cell that attracts only one advertiser. The commonly used GSP protocol when advertisers only declare bids will then assign any query inside the cell to that advertiser for a price equal to 0, compromising the search engine's revenue goals. Determining the proper grid granularity is thus a critical factor of success for location-based sponsored search.

In the following sections, we discuss location-based sponsored search advertising focusing on three cases [8], depending on the advertiser input and constraints. In the first *bids-only* case, each advertiser i is not bounded by a daily budget, i.e., $B_i = \infty$, and is willing to pay up to its valuation per click $w_{i,j}$ in cell j, i.e., its maximum bid per click for cell j is equal to $w_{i,j}$. In the second *budgets-only* case, each advertiser i is bounded by a finite daily budget B_i, but is indifferent to the price per click that it is asked to pay, i.e., its maximum bid per click for any cell is unbounded. Finally, in the third *bids-and-budgets* case, each advertiser i is bounded by a finite daily budget B_i, and is willing to pay up to its valuation per click $w_{i,j}$ in cell j. We first cast all three cases as game theoretical problems, and we subsequently attempt to identify the equilibrium strategies for the corresponding games. Table 1 illustrates common symbols used in the rest of the paper.

Table 1. Frequent symbols

Symbol	Meaning
N, L, K	Number of advertisers, grid cells, and ad slots
B_i	Total daily budget of advertiser i
$B_{i,j}$	Part of total budget B_i that advertiser i allocates into cell j
$w_{i,j}$	Valuation per click of advertiser i for cell j
M_j	Expected number of queries per day for cell j
c_l	Probability that an ad located at slot l will get clicked
$U_{i,j} / U_i$	Total daily utility of advertiser i from cell j / from all cells
C_i	Set of cells where advertiser i has the highest valuation per click
$\tilde{U}$	Upper concave envelope of U
Y_j	Sum of budgets that have been allocated to cell j by all advertisers
p_j	Price per click in cell j (for case 3)
s_j	Permutation of advertisers such that $w_{s_j(i),j}$ is decreasing in i
$w_j^{(2)}$	Second-highest valuation per click in cell j (for case 3)

4   Bids-Only Case

Bids-only is the simplest case, as it constitutes a straightforward generalization of the conventional sponsored search framework. Whenever the search engine receives a query from a cell, it runs an auction where each advertiser is assumed to bid an amount of money equal to its valuation per click for that particular cell. We can utilize any auction format, such as the GSP or the VCG (see Section 2.2), to determine the K winners that will fill the slots, as well as the prices per click that they have to pay. These two auctions have been extensively studied in the literature, and as mentioned earlier truthfully reporting the bids constitutes a Nash equilibrium for the VCG auction, but is in general not an equilibrium for the GSP procedure. Note that the actual number of queries per cell does not matter: every single time a user issues a query, a new auction will play out in an automated way; cells with high workload will simply involve more auctions compared to cells with lower traffic.

Next, we discuss some useful metrics focusing on the GSP framework. A very interesting notion in auction theory concerns an advertiser's payoff, which refers to the net utility the advertiser receives from being advertised. In sponsored search auctions, an advertiser's payoff is defined in terms of a *quasi-linear model*: the payoff per click is equal to the valuation/utility v_i per click the advertiser i gets minus the price per click p_i that it must pay, i.e., $v_i - p_i = v_i - b_{i+1}$, where the bids b_i are in descending order. We can also define the expected payoff per day if we know the average number of queries per day M. Since slot l receives a c_l percentage of the total clicks, the expected payoff per day for slot l is $M \cdot c_l \cdot (v_i - b_{i+1})$. Non-winning advertisers get a payoff equal to 0. Finally, we define the search engine's (cumulative) profit per click simply as $p_1 + \ldots + p_K = b_2 + \ldots + b_{K+1}$ (similarly for the cumulative profit per day). It is now straightforward to generalize the above metrics in the location-based framework. For instance, the expected payoff per day that advertiser i gets in cell j if it gets assigned to slot l is $M_j \cdot c_l \cdot (w_{i,j} - w_{i+1,j})$ (where $w_{i,j}$ are in decreasing order for cell j).

5   Budgets-Only Case

In case 2, advertiser i declares a maximum daily budget B_i, as well as its valuations per click $w_{i,j}$ for each cell j. As opposed to case 1, where $w_{i,j}$ is the maximum amount that i is willing to pay per click, for case 2 the payments per click are bounded only by B_i, and the cell valuations are used just to determine the relative importance of cells. For simplicity, we initially consider a single slot ($K=1$) with probability of being clicked $c_1=1$, and deal with more slots later. Since case 2 only involves budget constraints, it is convenient to assume a *Fisher market model* [2]: under this model, money does not bear any intrinsic value and every advertiser is willing to burn their entire budget; note that this is different from the quasi-linear model that we assumed in Section 4. Our goal is to assign to every advertiser a probability that their ad will be displayed in any given cell, whenever a user in that cell issues a relevant query. Therefore, no auction takes place and we do not have a winner selection and price determination phase.

Based on the declared budgets and cell valuations, the system computes for each cell the probability that any advertiser will be chosen as a response to user query.

In conventional sponsored search with only one slot, the optimal solution to this problem displays an advertiser with a probability that is proportional to its budget [8][14][13]. So, the advertiser with the highest budget has the highest probability of being displayed, which is equal to its budget divided over the sum of all budgets; and so on for the rest of the advertisers. This rule is called *proportional sharing*, and, intuitively, it guarantees fairness.

In location-based sponsored search, on the other hand, advertisers declare a total daily budget for all cells, but do not specify how this budget should be allocated among the various cells. Now, assume that the advertiser (somehow) decides how to allocate its budget into the cells, so that each cell has a non-negative budget and the sum of budgets over all cells does not exceed the advertiser's total budget. If such an allocation were known for every advertiser, then we could simply apply the proportional sharing rule: in a given cell, an advertiser is advertised with a probability proportional to its budget for this specific cell. But then a natural question arises: how should every advertiser allocate its budget?

To answer this question, we will resort to the proportional-share allocation market by Feldman et al. [10]. Concretely, assume a budget allocation for advertiser i such that it assigns $B_{i,j} \geq 0$ to cell j and the sum of its allocations over all cells does not exceed B_i. The probability that i will be displayed in cell j is $B_{i,j}/Y_j$, where Y_j is the sum of budgets that have been allocated to cell j by all advertisers. The utility for advertiser i in cell j is then $U_{i,j} = w_{i,j} \cdot M_j \cdot B_{i,j}/Y_j$, since it gets a value $w_{i,j}$ for every query in j when displayed with a probability $B_{i,j}/Y_j$, and there are M_j queries in total in cell j. We assume *additive* utilities, so i's total utility U_i is the sum of its utilities $U_{i,j}$ over all cells: $U_i = \sum_j w_{i,j} M_j \frac{B_{i,j}}{\sum_k B_{k,j}}$. Note that the payoff of advertiser i is equal to its utility, because of the Fisher market model assumption (money bears no intrinsic value to the advertiser).

A given set of budget allocations will give rise to different corresponding utilities for the advertisers. So, how should advertisers allocate their budget? Ideally, we would like to allocate every individual budget in a way that maximizes the advertiser's utility. But since the advertisers compete against each other, one's gain may translate into another's loss. To come up with proper budget allocations, we thus utilize the notion of Nash equilibrium. The set of agents consists of the advertisers, while the strategy space for advertiser i is the convex, bounded and closed set $\{(B_{i,1}, ..., B_{i,L}) \mid B_{i,j} \geq 0 \text{ and } \sum_{1 \leq j \leq L} B_{i,j} = B_i\}$, i.e., the set of all valid budget allocations. From advertiser's i perspective, a best response strategy is simply a strategy $B_i = (B_{i,1}, ..., B_{i,L})$ that maximizes its utility given the other advertisers' budget allocations, i.e., the solution to the following optimization problem:

$$\text{maximize } U_i(B_{1,1}, ..., B_{1,L}, ..., B_{N,1}, ..., B_{N,L})$$
$$\text{subject to } \sum_{1 \leq j \leq L} B_{i,j} = B_i \text{ and } B_{i,j} \geq 0.$$

A Nash equilibrium then corresponds to the stable state where no advertiser has an incentive to deviate from their strategy given that the other advertisers stick to their strategy as well. Stated equivalently, every advertiser plays a best response strategy to the rest of the advertisers. Formally, a set of valid strategies $B_1^*, ..., B_N^*$ form a Nash equilibrium if for any other valid strategy B_i, $1 \leq i \leq N$, we have:

$$U_i(B_1^*, ..., B_i^*, ..., B_N^*) \geq U_i(B_1^*, ..., B_i, ..., B_N^*).$$

It turns out that the above game does not always accept a Nash equilibrium. To demonstrate this, consider two advertisers 1 and 2 with budgets $B_1, B_2 > 0$, and two cells 1 and 2 with expected number of queries per day $M_1, M_2 > 0$. Advertiser 1 is interested in both cells, whereas advertiser 2 is only interested in cell 1. For player 2, the best strategy would obviously be to allocate its entire budget B_2 to cell 1 to gain the maximum possible proportion of ads. For advertiser 1, on the other hand, the best strategy would be to allocate a tiny amount $\varepsilon > 0$ to cell 2 (and win all advertising opportunities in 2) and spend the rest $B_1 - \varepsilon$ on cell 1 (and maximize its share in cell 1 as well). Unfortunately, there is no optimal value of ε, since (1) it must be positive to ensure 1 gets all ads in cell 2, and (2) as small as possible so that 1 wins the largest possible share in cell 1. Alternatively, consider the simpler case with a single player 1 with $B_1 > 0$, interested in a single cell 1 with $M_1 > 0$. As before, player 1 should allocate the smallest possible positive $\varepsilon > 0$ on cell 1, but such an ε does not exist.

The root of the non-existence of a Nash Equilibrium in the examples above is due to the discontinuity of the utility functions at point 0. This problem can be circumvented in two different ways. First, we can enforce a *reserve price*, which is defined as the minimum possible price that an advertiser must pay per click. Indeed, a reserve price means that the advertiser cannot buy any click with an arbitrarily small budget, and the discontinuity at 0 ceases to exist. Second, we can restrict our attention to so called *strongly competitive* games [10], i.e., games where for a given cell there are at least two advertisers with positive valuations. Indeed, strong competition implies that if only one advertiser would allocate a tiny budget on a given cell, then any other advertiser who has non-zero valuation for that cell will have an incentive to also allocate (a tiny) budget in that cell to guarantee a percentage of ads [10].

Computing the Nash equilibrium is the next source of concern. There are 2 classes of algorithms for this purpose. The *best response algorithm* iteratively updates the budget allocations of every player to reflect the other players' current strategies. This algorithm simulates the best response dynamics of the game and thus has a very natural interpretation. We describe it in Figure 1; the interested reader is referred to [10] for further details. Note that its time complexity is dominated by the sorting procedure, so it is $O(N \log N)$. Theoretically, the best-response dynamics does not necessarily converge to a Nash equilibrium of the game; nevertheless, in practice the algorithm performs very well.

Repeat for each advertiser i, $1 \le i \le N$

1. Sort the cells according to $\dfrac{w_{i,j}}{\sum_{i' \ne i} B_{i',j}}$ in decreasing order, where $1 \le j \le L$

2. Compute the largest k such that

$$\frac{\sqrt{w_{i,k} M_k \sum_{i' \ne i} B_{i',k}}}{\sum_{m=1}^{k} w_{i,m} M_m \sum_{i' \ne i} B_{i',m}} \left(B_i + \sum_{m=1}^{k} \sum_{i' \ne i} B_{i',m} \right) - \sum_{i' \ne i} B_{i',k} > 0$$

3. Set $B_{i,j} = 0$ for $j > k$, and for $1 \le j \le k$ set

$$B_{i,j} = \frac{\sqrt{w_{i,j} M_j \sum_{i' \ne i} B_{i',j}}}{\sum_{m=1}^{k} w_{i,m} M_m \sum_{i' \ne i} B_{i',m}} \left(B_i + \sum_{m=1}^{k} \sum_{i' \ne i} B_{i',m} \right) - \sum_{i' \ne i} B_{i',j}$$

until convergence.

Fig. 1. Best-response dynamics for $K=1$ and strong competition

The alternative to best response dynamics is the *local greedy adjustment method* [10]. Under this algorithm, we first identify for every advertiser the two cells that provide the highest and lowest marginal utilities. We then move a fixed small amount of money from the cell with the lowest marginal utility to the cell with the highest one. This strategy aims to adjust the budget allocations so that the marginal values in each cell are the same. For concave utility functions (as ours), this is a sufficient condition for an optimal allocation. However, the method suffers from lower convergence rates.

As a last remark, note that contrary to case 1, the actual query distribution is now important. To understand why, assume the advertiser has a high valuation for cell 1 and a low valuation for cell 2. However, a small number of queries are issued in cell 1, whereas several queries are issued in cell 2. In bids-only sponsored search, a separate auction occurs every time a query is issued, so the advertiser can bid high for cell 1 and low for cell 2; since there are far more queries in cell 2 the advertiser will obviously participate in the auction for cell 2 far more times, but has no reason not to bid high for cell 1 and low for cell 2. In the budgets-only setting, however, query distribution has a profound effect on the budget allocation. In the above example, the advertiser may have to allocate a large part of its budget to cell 2 just because there are far too many queries in that cell.

Multiple Slots: We can generalize the above discussion in the case of several slots, by assuming for simplicity that a given advertiser may appear with non-zero probability in more than one slots (as opposed to the bids-only case). This assumption is necessary for a straightforward and simple generalization. Indeed, the idea is that every advertiser allocates part of its budget into all slots in every cell. The utility that advertiser i extracts from being advertised at slot l in cell j is $w_{i,j} c_l M_j \frac{B_{i,j,l}}{\sum_k B_{k,j,l}}$, where $B_{i,j,l}$ the amount of money that i allocates in slot l of cell j. Similar to before, we can assume additive utilities, so that the total utility of advertiser i the sum of its utilities over all slots and over all cells. Using the above techniques, we can then find budget allocations that constitute a Nash equilibrium.

6   Bids-and-Budgets Case

In this setting, advertiser i declares a maximum daily budget B_i as before, but contrary to case 2, i is now not willing to spend more than $w_{i,j}$ per click in cell j. Stated equivalently, the price that advertiser i pays per click in a given cell j cannot exceed its declared valuation $w_{i,j}$ for that cell. The valuations thus act as maximum bids per click, and we also refer to case 3 as bids-and-budgets case. We only deal with the case of a single slot, i.e., $K=1$ with $c_l=1$, and we assume again that money bears no intrinsic value to the advertisers (Fisher market model). The case of several slots is more complex, and can be investigated in future work.

Before dealing with the location-based setting, we first explore how conventional sponsored search addresses the case where both budgets and maximum bids per click are declared. In particular, we will attempt to highlight how this setting is inherently more complex than the budgets-only case. We focus on cell j with M_j queries per day and budget allocations in it $B_{1,j}, ..., B_{N,j}$. First, assume that every advertiser receives a share of the total ads proportional to its budget. Then, the price per click would be equal to $p_j = (B_{1,j} + ... + B_{N,j})/M_j$. As long as this quantity is not greater than all valuations per click $w_{1,j}, ..., w_{N,j}$, no problem occurs. But if an advertiser i exists with $w_{i,j} < p_j$, this advertiser would not be willing to pay as much as p_j per click, so the proportional allocation framework of Section 5 cannot be directly applied. To alleviate this problem, we need to come up with a price p_j^* such that all advertisers who can afford that price have sufficient budgets to purchase all the advertising opportunities. Figure 2 presents the price-setting mechanism by Feldman et al. [9][8] that determines that price p_j^*. It is essentially a price-descending mechanism: the price keeps falling until p_j^* is reached. Moreover, it has the desired property of being truthful.

---

1. Assume w.l.o.g. that $w_{1,j} > w_{2,j} > ... > w_{N,j} \geq 0$.

2. Let k^* be the first bidder such that $w_{k^*+1,j} \leq \frac{\sum_{i=1}^{k^*} B_{i,j}}{M_j}$. Set price $p_j^* = \min\left\{\frac{\sum_{i=1}^{k^*} B_{i,j}}{M_j}, w_{k^*,j}\right\}$.

3. Allocate $B_{i,j}/p_j^*$ ads to each advertiser $i \leq k^* - 1$. Allocate $M_j - \sum_{i=1}^{k^*} B_{i,j}/p_j^*$ ads to advertiser k^*. Allocate 0 ads to the rest of the bidders.

Fig. 2. The price-setting mechanism in cell j for $K=1$ slot in the bids-and-budgets case

Now, recall that in the case where only budgets are available, the price per query in cell j would be equal to $p_j = \frac{\sum_{i=1}^{N} B_{i,j}}{M_j}$. Obviously, p_j is linear in its arguments $B_{i,j}$ ($1 \leq i \leq N$) and continuous. On the other hand, the price-setting mechanism in Figure 2 yields prices that are clearly more complex. First, we notice the price p_j for a given cell j will again be an argument of only the budget allocations for that cell $B_{1,j}, ..., B_{N,j}$. However, it does not have the simple linear form as in the case of only budgets. To get a flavor of the price function, consider a setting with only 2 advertisers 1 and 2 with maximum bids $w_{1,j}$ and $w_{2,j}$ (with $w_{1,j} > w_{2,j}$) for cell j that has M_j queries per day. Figure 3 depicts how the price varies according to the budgets $B_{1,j}$ and $B_{2,j}$ that the advertisers allocate in cell j. In particular, if $B_{1,j} \geq M_j \cdot w_{2,j}$, then $k^* = 1$ and the price is determined as the minimum of $B_{1,j}/M_j$ and $w_{1,j}$. When $B_{1,j} \geq M_j \cdot w_{1,j}$ then the price is equal to $w_{1,j}$ (region I), while when $B_{1,j} < M_j \cdot w_{1,j}$, the price is equal to $B_{1,j}/M_j$ (region II). On the other hand, when $B_{1,j} < M_j \cdot w_{2,j}$, then $k^* = 2$ and the price is the minimum of $w_{2,j}$ and $(B_{1,j}+B_{2,j})/M_j$; for $B_{1,j}+B_{2,j} \geq M_j \cdot w_{2,j}$ the price is $w_{2,j}$ (region III), while for $B_{1,j}+B_{2,j} < M_j \cdot w_{2,j}$, the price is $(B_{1,j}+B_{2,j})/M_j$ (region IV). Inside a region, the price can be either constant or linear. We first observe that the price function is everywhere continuous; the boundaries of the regions are carefully chosen so that the price is continuous as we move from one region to the other. Note also that the price function for the price-setting mechanism is bounded: it achieves a minimum value of 0 at the origin $(0,0)$, and it can never get larger than $w_{1,j}$. On the contrary, the price per click in the budgets-only case is unbounded: it can get arbitrarily large as the budgets that the advertisers allocate grow larger.

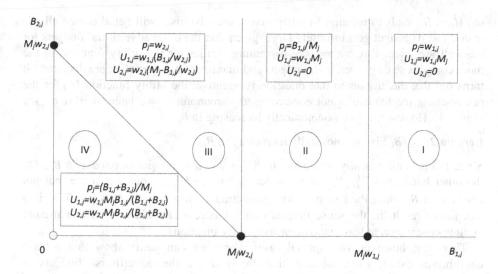

Fig. 3. Price p_j and utilities $U_{1,j}$ and $U_{2,j}$ in cell j when the number of advertisers is $N=2$

The above example captures some important properties of the price function in the case of both maximum bids and budgets. The price setting mechanism decomposes the budget space into N regions (one for each of the N possible k^*), and then further divides that region into two subregions: the price is constant inside one of them and linear in the other. In the following, we state several results. The proofs of all results are available in our technical report [17]. For every cell j, we consider the permutation s_j that reorders the bids in decreasing order, i.e., $w_{s_j(1),j} > w_{s_j(2),j} > \cdots > w_{s_j(N),j} > 0$ for every cell j. Moreover, all budget allocations $B_{i,j}$ are non-negative: $B_{i,j} \geq 0$.

Lemma 1: The price function $p_j(B_{1,j}, ..., B_{N,j})$ is continuous in $(B_{1,j}, ..., B_{N,j})$.

Let's now try to formalize the location-based setting where advertisers have valuations over the various cells. Similar to the previous case, we will be looking for a budget allocation $\boldsymbol{B_i} = (B_{i,1}, ..., B_{i,L})$ for every advertiser i, $1 \leq i \leq N$. For a given allocation, denote $z_{i,j}$ the share of ads that advertiser i gets in cell j. Then, its utility from cell j is $w_{i,j} \cdot z_{i,j}$; its total utility from all cells simply is $U_i = \sum_j w_{i,j} z_{i,j}$. In order to compute the share $z_{i,j}$, we will exploit the price-setting mechanism, assuming that $w_{s_j(1),j} > w_{s_j(2),j} > \cdots > w_{s_j(N),j} > 0$. If $w_{s_j(2),j} \leq B_{s_j(1),j}/M_j$, then $k_j^*=1$ and the price p_j^* is $\min\{B_{s_j(1),j}/M_j, w_{s_j(1),j}\}$. On the other hand, if $w_{s_j(2),j} > B_{s_j(1),j}/M_j$, we continue by checking whether $w_{s_j(3),j} \leq (B_{s_j(1),j} + B_{s_j(2),j})/M_j$. If the latter is true, then $k_j^*=2$ and the price $p_j^* = \min\{(B_{s_j(1),j} + B_{s_j(2),j})/M_j, w_{s_j(2),j}\}$. If it is false, we proceed in exactly the same way, until we come up with the proper k_j^*, and subsequently compute the price p_j^*. Figure 3 depicts the utility functions in cell j in the case of $N=2$ advertisers.

Next, we compare the above utility function with the simpler utility function in the case where only budgets are declared: $U_{i,j} = w_{i,j} M_j \frac{B_{i,j}}{\sum_k B_{k,j}}$. Clearly, the latter function is concave in $B_{i,j}$, gets a minimum value of 0 for $B_{i,j}=0$, and asymptotically converges to

$w_{i,j}M_j$ as $B_{i,j}$ tends to infinity. In other words, the advertiser will get allocated all ads in cell j as its budget gets infinitely large, given that the other advertisers' budgets for this cell are fixed. But can we say something similar for the utility function in the more complex setting when both budgets and maximum bids per click are declared? It turns out that the answer to that question is negative: the utility function $U_{i,j}$ for the price-setting mechanism is not concave in $B_{i,j}$ anymore, as we later show (e.g., see Figure 4). However, $U_{i,j}$ is monotonically increasing in $B_{i,j}$:

Lemma 2: $U_{i,j}(B_{i,j})$ is monotonically increasing in $B_{i,j}$.

Since $U_{i,j}$ is monotonically increasing in $B_{i,j}$, it will also be *quasi-concave* in $B_{i,j}$. On the other hand, $U_i = \sum_{j=1}^{L} U_{i,j}$. It turns out that when $U_{i,j}$ are quasi-concave, but not concave in $B_{i,j}$, then their sum is not quasi-concave in $(B_{i,1}, \ldots, B_{i,L})$ [6]. This is a worrisome result, in the sense that existence theorems for Nash equilibria usually assume concave or, at least, quasi-concave utility functions.

There are, however, two special cases where we can easily show that a Nash equilibrium exists. First, assume that the sum of the advertisers' budgets is sufficiently small, i.e., $\sum_{i=1}^{N} B_i \le M_j w_{N,j}$, for every cell j. In this case, independent of the budget allocation, we have in any cell j that $\sum_{i=1}^{N} B_{i,j} \le M_j w_{N,j}$, so the price setting mechanism will allocate to every advertiser a percentage of advertising opportunities proportional to the budget that they allocate in every cell. But this is identical to case 2, and it thus always admits a Nash equilibrium if 1) there is a reserve price, or 2) there is strong competition. Second, assume that every advertiser has sufficiently large budget, and that there is strong competition in every cell. For any advertiser i, consider the set of cells C_i where i has the highest valuation per click among all advertisers, i.e., $C_i = \{j | w_{i,j} = \max_{1 \le i' \le N}\{w_{i',j}\}\}$ (for some advertisers this set may be empty). For advertiser i, we then define the following budget allocation strategy: allocate 0 to cell j if $j \in C_i$, else allocate an amount of money equal to or greater than $M_j w_j^{(2)}$, where $w_j^{(2)} > 0$ the second highest valuation per click in cell j (it is positive because of the strong competition assumption). This is always possible if $B_i \ge \sum_{j \in C_i} M_j w_j^{(2)}$, for every advertiser i. It is easy to verify that the above sets of budget allocations correspond to Nash equilibria, since any advertiser cannot increase its utility by deviating to a different budget allocation. Indeed, with the previous budget allocation every advertiser i wins all ads for the cells that belong to C_i. Obviously, i cannot gain a higher utility by changing its budget allocation for cells $j \in C_i$. On the other hand, even if i allocates a positive budget in cells $j \notin C_i$, it will still gain 0 advertising opportunities, since the first advertiser has adequate budget and valuation to buy all ads in that cell. In fact, a Nash equilibrium in the case of sufficiently large budgets can be given by the following rule: in every cell the advertiser with the highest valuation per click pays a price per click equal to the valuation per click of the second highest advertiser, and wins all ads for that cell. But what we have just described is the GSP procedure. Stated equivalently, the GSP auction for sufficiently large budgets results in a Nash equilibrium.

We have thus observed how the bids-and-budgets case encompasses the simpler bids-only and budgets-only cases for sufficiently small or large budgets, respectively.

On the other hand, when only one advertiser has a positive valuation for a cell j, then using the same line of arguments as in Section 5 we can see that its utility function is discontinuous at 0, and the game accepts no Nash equilibrium. It is however possible to slightly modify the game in a way that makes the discontinuity at 0 disappear, similar to [10][13]. In this direction, we will introduce a *fictitious* advertiser $N+1$ who allocates a tiny budget $B_\varepsilon > 0$ in every cell, but has an arbitrarily large valuation per click for every cell. We call the perturbed game with the additional player G. So, what is the impact of the additional player $N+1$ on the game structure? Essentially, the arbitrarily large valuation per click for every cell implies that advertiser $N+1$ will have the highest valuation per click in every cell and will thus be able to pay any price that the price mechanism sets. On the other hand, we set B_ε to be very small so that player $N+1$ has a negligible impact. Note that the introduction of the fictitious player serves the same purpose as the reserve price of the budgets-only case, namely to smooth out the utility function and tackle the discontinuity at 0.

In the general case, we are currently not aware whether game G always accepts a Nash equilibrium since each advertiser's utility function is not quasi-concave. Although we cannot answer whether a Nash equilibrium exists, we can nevertheless find a budget allocation such that the maximum utility that an advertiser can gain by deviating is known.

In this direction, we will consider the *upper concave envelope* $\tilde{U}_{i,j}$ of the utility $U_{i,j}$, for any advertiser i and any cell j. Formally, we will be looking for the infimum of all functions that are concave and are greater than or equal to $U_{i,j}$ for any $B_{i,j}$. This is, in general, not an easy task, but as we shall see the upper concave envelope for the utility functions that arise in the bids-and-budgets setting has a relatively simple form.

We focus on advertiser i and cell j, $1 \leq i \leq N$ and $1 \leq j \leq L$. Assume the rest of the advertisers' budgets for cell j are fixed and equal to $B_{1,j}, ..., B_{i-1,j}, B_{i+1,j}, ..., B_{N,j}$. Also, w.l.o.g. assume that $w_{1,j} > ... > w_{N,j}$. We are interested in the first advertiser k^* such that $w_{k^*+1,j} \leq \frac{\sum_{i=1}^{k^*} B_{i,j}}{M_j}$ as $B_{i,j}$ varies. Let $k^* = k^0$ when $B_{i,j} = 0$. If $k^0 < i$, then no matter how much budget i allocates, the price setting mechanism allocates no advertising opportunities to them, because advertisers $1, ..., k^0$ have sufficient budget to buy all ads at a price that is higher than what i can afford; thus $U_{i,j} = 0$ and, subsequently, $\tilde{U}_{i,j} = U_{i,j} = 0$. If, on the other hand, $k^0 \geq i$, then the utility function $U_{i,j}$ will have the form that we depict in Figure 4. In particular, we can form the $i-k^0+1$ regions R_k, $i \leq k \leq k^0$, such that that the first advertiser in region R_k with the property that $w_{k+1,j} \leq \frac{\sum_{i=1}^{k} B_{i,j}}{M_j}$ is k. In particular, when $B_{i,j} = 0$ then $k^* = k^0$ and we get the leftmost region R_{k^0}; as $B_{i,j}$ grows larger k^* eventually becomes i and remains so thereafter. Points P, P_1, P_2, and P_3 in Figure 4 correspond to budget allocations $B_{i,j}$ equal to B, $B_1 = w_{k+1,j} \cdot M_j - S_{k,j} - B_{k,j}$, $B_2 = w_{k,j} \cdot M_j - S_{k,j} - B_{k,j}$, and $B_3 = w_{k,j} \cdot M_j - S_{k,j}$, respectively.

We will now determine the upper concave envelope of $U_{i,j}$ by focusing on regions R_k, with $i \leq k \leq k^0$. Define $S_{k,j} = \sum_{i'=1}^{i-1} B_{i,j} + \sum_{i'=i+1}^{k-1} B_{i,j}$ (for $k=i$ this expression gives $S_{i,j} = \sum_{i'=1}^{i-1} B_{i,j}$). Region R_i (rightmost region in Figure 4) consists of a concave part which corresponds to the utility function $U_{i,j}(B_{i,j}) = w_{i,j} M_j \frac{B_{i,j}}{\sum_{i'=1}^{i} B_{i',j}}$ for $B_{i,j} \in [w_{i+1,j} \cdot M_j - S_{i,j} - B_i, w_{i,j} \cdot M_j - S_{i,j}]$, followed by a constant part for $B_{i,j} \geq w_{i,j} \cdot M_j - S_{i,j}$

(the constant part corresponds to the maximum possible advertising opportunities that advertiser i may get); the utility function in region R_i is thus already concave so we do not need to focus more on it. Every other region R_k, $i<k\leq k^0$, will consist of the concave part $w_{i,j}M_j\frac{B_{i,j}}{\sum_{i'=1}^{k}B_{i',j}}$ for $B_{i,j}\in[w_{k+1,j}\cdot M_j-S_{k,j}-B_{k,j},\ w_{k,j}\cdot M_j-S_{k,j}-B_{k,j}]$, followed by the linear part $w_{i,j}\frac{B_{i,j}}{w_{k,j}}$ for $B_{i,j}\in[w_{k,j}\cdot M_j-S_{k,j}-B_{k,j},\ w_{k,j}\cdot M_j-S_{k,j}]$. Of course $B_{i,j}\geq0$, so if any of the endpoints of the aforementioned intervals is negative we simply replace it with 0. From Lemma 1, we can easily derive that $U_{i,j}(B_{i,j})$ is continuous in the domain $B_{i,j}\geq0$. It is also differentiable everywhere except for the points where the utility function transitions from the concave part to the linear part, and vice versa.

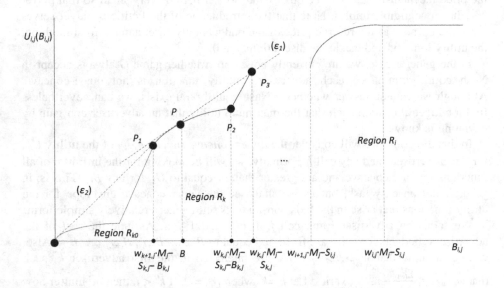

Fig. 4. Utility function $U_{i,j}$ when $k^0\geq i$ and its upper concave envelope

We now examine the derivatives in regions R_k, $i\leq k\leq k^0$. For region R_i, i.e., when $k=i$, the derivative in $(w_{i+1,j}\cdot M_j-S_{i,j}-B_{i,j},\ w_{i,j}\cdot M_j-S_{i,j})$ is $w_{i,j}M_j\frac{S_{i,j}}{(S_{i,j}+B_{i,j})^2}$, while it is 0 for $B_{i,j}>w_{i,j}\cdot M_j-S_{i,j}$. For region R_k, with $i\leq k\leq k^0$, the derivative in $(w_{k+1,j}\cdot M_j-S_{k,j}-B_{k,j},\ w_{k,j}\cdot M_j-S_{k,j}-B_{k,j})$ is $w_{i,j}M_j\frac{S_{k,j}+B_{k,j}}{(S_{k,j}+B_{k,j}+B_{i,j})^2}$, while the derivative in $(w_{k,j}\cdot M_j-S_{k,j}-B_{k,j},\ w_{k,j}\cdot M_j-S_{k,j})$ is $w_{i,j}/w_{k,j}$. Although $U_{i,j}$ is not differentiable at the transition points, the left $\partial_-U_{i,j}$ and right $\partial_+U_{i,j}$ derivatives obviously exist. We now state the following two results.

Lemma 3: $\partial_-U_{i,j}\big(w_{k+1,j}M_j-S_{k,j}-B_{k,j}\big) > \partial_+U_{i,j}\big(w_{k+1,j}M_j-S_{k,j}-B_{k,j}\big)$, for $i\leq k<k^0$.

Lemma 4: $\partial_-U_{i,j}\big(w_{k,j}M_j-S_{k,j}-B_{k,j}\big) < \partial_+U_{i,j}\big(w_{k,j}M_j-S_{k,j}-B_{k,j}\big)$, for $i<k\leq k^0$.

Lemma 3 implies that whenever we make a transition from the linear to the concave part (e.g., point P_1 in Figure 4) the first derivative gets lower, and concavity is maintained. In contrast, Lemma 4 suggests that when we move from the concave to the linear part (e.g., point P_2), the first derivative gets higher; this in turn violates concavity. We will show how to tackle this by considering region R_k, $i \leq k \leq k^0$, in Figure 4. The idea is to draw a line ε_1 from P_3 to the point P in the concave part of region R_k so that the line ε_1 is tangent to the curve. Based on our previous discussion, the derivative at P is $w_{i,j} M_j \frac{S_{k,j} + B_{k,j}}{(S_{k,j} + B_{k,j} + B)^2}$. On the other hand, the slope of ε_1 is

$$\frac{U_{i,j}(B_3) - U_{i,j}(B)}{B_3 - B} = w_{i,j} \frac{\frac{B_3}{w_{k,j}} - M_j \frac{B}{S_{k,j} + B_{k,j} + B}}{B_3 - B}.$$ Thus, we are looking for a B such that

$$w_{i,j} M_j \frac{S_{k,j} + B_{k,j}}{(S_{k,j} + B_{k,j} + B)^2} = w_{i,j} \frac{\frac{B_3}{w_{k,j}} - M_j \frac{B}{S_{k,j} + B_{k,j} + B}}{B_3 - B}.$$ But $B_3 = w_{k,j} \cdot M_j - S_{k,j}$, so the previous equation becomes after some algebraic manipulations:

$$S_{k,j} B^2 - 2(M_j w_{k,j} - S_{k,j})(S_{k,j} + B_{k,j}) B - (M_j w_{k,j} - S_{k,j})(S_{k,j} + B_{k,j})(S_{k,j} + B_{k,j} - M_j w_{k,j}) = 0 \quad (1)$$

Equation (1) is a quadratic equation, which accepts the two solutions $\frac{(M_j w_{k,j} - S_{k,j})(S_{k,j} + B_{k,j}) \pm \sqrt{(M_j w_{k,j} - S_{k,j})(S_{k,j} + B_{k,j}) S_{k,j} M_j w_{k,j}}}{S_{k,j}}$. First, note that $M_j \cdot w_{k,j} > S_{k,j}$ (since $B_4 > 0$), so the solutions are real numbers. Second, we keep the solutions with the minus because it is lower than $B_3 = M_j \cdot w_{k,j} - S_{k,j}$ and even $B_2 = M_j \cdot w_{k,j} - S_{k,j} - B_{k,j}$. Indeed, after performing some algebraic manipulations we get $\frac{(M_j w_{k,j} - S_{k,j})(S_{k,j} + B_{k,j}) - \sqrt{(M_j w_{k,j} - S_{k,j})(S_{k,j} + B_{k,j}) S_{k,j} M_j w_{k,j}}}{S_{k,j}} < M_j w_{k,j} - S_{k,j} - B_{k,j} \Leftrightarrow M_j w_{k,j} > S_{k,j} + B_{k,j}$, which is true. Now, there are 2 cases. If the solution is greater than $w_{k+1,j} M_j - S_{k,j} - B_{k,j}$ (see point P_1 in Figure 4), then we draw the line ε_1 from P to P_3 as we show in Figure 4. Else, we draw the line from P_1 to P_3 (we illustrate such a scenario with line ε_2 in region R_{k^0} in Figure 4). We summarize the two cases by writing

$$B = \max \left\{ \frac{(M_j w_{k,j} - S_{k,j})(S_{k,j} + B_{k,j}) - \sqrt{(M_j w_{k,j} - S_{k,j})(S_{k,j} + B_{k,j}) S_{k,j} M_j w_{k,j}}}{S_{k,j}}, w_{k+1,j} M_j - S_{k,j} - B_{k,j} \right\}.$$

We will now prove that the slope of ε_1 is greater than the right derivative at P_3. Indeed, the slope of ε_1 is $w_{i,j} M_j \frac{S_{k,j} + B_{k,j}}{(S_{k,j} + B_{k,j} + B)^2}$. The right derivative at P_3, on the other hand, is $w_{i,j} M_j \frac{S_{k,j}}{(w_{k,j} M_j)^2}$. But then $w_{i,j} M_j \frac{S_{k,j} + B_{k,j}}{(S_{k,j} + B_{k,j} + B)^2} > w_{i,j} M_j \frac{S_{k,j}}{(S_{k,j} + B_{k,j} + B)^2} > w_{i,j} M_j \frac{S_{k,j}}{(S_{k,j} + B_{k,j} + (M_j w_{k,j} - S_{k,j} - B_{k,j}))^2} = w_{i,j} M_j \frac{S_{k,j}}{(M_j w_{k,j})^2}$, which proves our claim. Moreover, it is easy to see that the slope of ε_1 is lower than the left derivative at P_1, since the opposite would imply that the line segment P_2-P_3 has a slope $w_{i,j}/w_{k,j}$ that is greater than the left derivative at P_1 $w_{i,j}/w_{k+1,j}$, which is untrue given that $w_{k,j} > w_{k+1,j}$.

We repeat the process described above in all regions. At the end of this process, we derive a utility function $\tilde{U}_{i,j}$ that is continuous everywhere, differentiable everywhere except for the points where it changes slope, and the left and right derivatives (which exist for all $B_{i,j} \geq 0$) are monotonically non-increasing in the allocated budget $B_{i,j}$. But then $\tilde{U}_{i,j}$ will be concave in terms of $B_{i,j}$. Now, recall that $U_i = \sum_{j=1}^{L} U_{i,j}$. If we repeat the above process for every $U_{i,j}$, $1 \leq j \leq L$, we can eventually form the function $\tilde{U}_i(\boldsymbol{B}_i; \boldsymbol{B}_{-i}) = \sum_{j=1}^{L} \tilde{U}_{i,j}(B_{i,j}; \boldsymbol{B}_{-i,j})$ (where $\boldsymbol{B}_{-i}$ denotes the vector of budget allocations of all advertisers but i). The function $\tilde{U}_i(\boldsymbol{B}_i; \boldsymbol{B}_{-i})$ is the sum of concave functions, so it is also concave in i's strategy $\boldsymbol{B}_i$. In the end, the new utility functions $\tilde{U}_i(\boldsymbol{B}_i; \boldsymbol{B}_{-i})$, $1 \leq i \leq N$, possess two important properties: (1) each $\tilde{U}_i(\boldsymbol{B}_i; \boldsymbol{B}_{-i})$ is continuous in $(\boldsymbol{B}_i; \boldsymbol{B}_{-i})$, and (2) each $\tilde{U}_i(\boldsymbol{B}_i; \boldsymbol{B}_{-i})$ is concave in $\boldsymbol{B}_i$ for any fixed value of $\boldsymbol{B}_{-i}$. Moreover, the strategy space of every advertiser is convex, closed and bounded. Consequently, based on Rosen's theorem [16] we can immediately derive that a Nash equilibrium exists. We denote that equilibrium by $\tilde{\boldsymbol{B}} = (\tilde{\boldsymbol{B}}_1, \dots, \tilde{\boldsymbol{B}}_N)$. Moreover, we call $\tilde{G}$ the new game when the utility functions are replaced by their upper-concave envelopes.

Note that $\tilde{\boldsymbol{B}}$ may not be an equilibrium of game G. This means that there may be players in game G who have an incentive to deviate if the strategy vector $\tilde{\boldsymbol{B}}$ is chosen. However, the following lemma shows that we can bound the maximum utility that a player can gain by deviating.

Lemma 5: Let the strategy vector $\tilde{\boldsymbol{B}}$ be a Nash equilibrium of game $\tilde{G}$. Then the maximum utility that player i can gain by deviating from $\tilde{\boldsymbol{B}}$ in game G is $\tilde{U}_i(\tilde{\boldsymbol{B}}_i; \tilde{\boldsymbol{B}}_{-i}) - U_i(\tilde{\boldsymbol{B}}_i; \tilde{\boldsymbol{B}}_{-i})$.

Essentially, the above result says that we can find a set of budget allocations such that we can know exactly the maximum utility that an advertiser may gain by deviating. Note that in the special case where the Nash equilibrium of game $\tilde{G}$ falls into the parts of $\tilde{U}_i$ that are equal to U_i, then the Nash equilibria of game $\tilde{G}$ are also Nash equilibria of game G.

7   Conclusion

The market for location-based advertising is set to witness an unprecedented growth over the next years. The massive proliferation of modern mobile phones with embedded geo-positioning functionality and the development of fast wireless communication protocols have created exciting opportunities for advertisers to reach the user base that is most relevant to them. On the other hand, sponsored search advertising has been a thriving market in the last decade for advertisers who want to advertise their product or service to online users posing relevant queries. Inspired by the enormous success of sponsored search and the immense potential for LBA, we address the market for location-based sponsored search advertising. We provide models that build on prior work in sponsored search advertising, but we additionally consider that advertisers are characterized by location-dependent valuations. We distinguish between three cases: (1) bids-only case, (2) budgets-only case, and (3) bids-and-budgets case, and analyzed the equilibrium strategies in the corresponding markets using game theoretical tools.

There are several research directions that we would like to pursue with regard to the market for location-based sponsored search advertising. First, we would like to extend our model so that it takes into account the more subtle issues that are involved in the sponsored search market such as the externalities between the displayed ads, or the more realistic scenario of advertisers who are interested in several keywords.

Second, our model assumed offline ad slot scheduling [9], where we estimate the number of queries in every cell, and then allocate to every advertiser a percentage of the ads in every cell. It would be interesting to deal with the more challenging problem of online ad slot scheduling, where the number of expected queries per cell is not available in advance. Finally, we would like to fully explore the equilibrium strategies in the bids-and-budgets case, as our current work provides equilibrium strategies only for the case where advertisers have sufficiently small or large budgets.

Acknowledgements. This work was supported by grant RPC11EG01 from HKUST. We thank the anonymous reviewers for their valuable comments.

References

1. Banerjee, S., Dholakia, R.: Does location-based advertising work? International Journal of Mobile Marketing 3(1) (2008)
2. Brainard, W., Scarf, H.: How to compute equilibrium prices in 1891. Cowles Foundation Discussion Paper 1270 (2000)
3. Bruner, G., Kumar, A.: Attitude toward location-based advertising. Journal of Interactive Advertising 7(2) (2007)
4. Clarke, E.H.: Multipart pricing of public goods. Public Choice 11(1), 17–33 (1971)
5. Dhar, S., Varshney, U.: Challenges and business models for mobile location-based services and advertising. Communications of the ACM 54(5), 121–129 (2011)
6. Debreu, G., Koopmans, T.C.: Additively decomposed quasiconvex functions. Mathematical Programming 24(1), 1–38 (1982)
7. Edelman, B., Ostrovsky, M., Schwarz, M.: Internet advertising and the generalized second price auction: Selling billions of dollars worth of keywords. American Economic Review 9(1), 242–259 (2007)
8. Feldman, J., Muthukrishnan, S.: Algorithmic methods for sponsored search advertising. In: Performance Modeling and Engineering, Springer, Heidelberg, pp. 91–124. Springer, Heidelberg (2008)
9. Feldman, J., Muthukrishnan, S., Nikolova, E., Pál, M.: A truthful mechanism for offline ad slot scheduling. In: Monien, B., Schroeder, U.-P. (eds.) SAGT 2008. LNCS, vol. 4997, pp. 182–193. Springer, Heidelberg (2008)
10. Feldman, M., Lai, K., Zhang, L.: The proportional-share allocation market for computational resources. IEEE Transactions on Parallel and Distributed Systems 20(8), 1075–1088 (2009)
11. Groves, T.: Incentives in teams. Econometrica 41, 617–631 (1973)
12. Jansen, B., Mullen, T.: Sponsored search: an overview of the concept, history, and technology. Int. J. Electronic Business 6(2), 114–131 (2008)
13. Johari, R., Tsitsiklis, J.N.: Efficiency loss in a network resource allocation game. Mathematics of Operations Research 29(3), 407–435 (2004)
14. Kelly, F.: Charging and rate control for elastic traffic. European Transactions on Telecommunications 8, 33–37 (1997)
15. Kempe, D., Mahdian, M.: A cascade model for externalities in sponsored search. In: 4th International Workshop on Internet and Network Economics, pp. 585–596 (2008)
16. Rosen, J.B.: Existence and uniqueness of equilibrium points for concave N-person games. Econometrica 33(3), 520–534 (1965)

17. Trimponias, G., Papadias, D., Bartolini, I.: Location-based sponsored search advertising. Technical Report,
    http://www.cse.ust.hk/~dimitris/PAPERS/SSTD13-TR.pdf
18. Unni, R., Harmon, R.: Perceived effectiveness of push vs. pull mobile location-based advertising. Journal of Interactive Advertising 7(2), 28–40 (2007)
19. Varian, H.: Position auctions. International Journal of Industrial Organization 25(6), 1163–1178 (2007)
20. Vickrey, W.: Counterspeculation, auctions, and competitive sealed tenders. Finance 16, 8–27 (1961)
21. http://advertise.bingads.microsoft.com/en-us/home
22. http://adwords.google.com/support/aw/?hl=en

A Group Based Approach for Path Queries
in Road Networks

Hossain Mahmud[1], Ashfaq Mahmood Amin[1], Mohammed Eunus Ali[1],
Tanzima Hashem[1], and Sarana Nutanong[2]

[1] Bangladesh University of Engineering and Technology, Dhaka, Bangladesh
{hossain.mahmud,ashfaq.m.amin}@gmail.com,
{eunus,tanzimahashem}@cse.buet.ac.bd
[2] Department of Computer Science, Johns Hopkins University, Maryland, USA
nutanong@cs.jhu.edu

Abstract. The advancement of mobile technologies and map-based applications enables a user to access a wide variety of location-based services that range from information queries to navigation systems. Due to the popularity of map-based applications among the users, the service provider often requires to answer a large number of simultaneous (or contemporary) queries. Thus, processing queries efficiently on spatial networks (i.e., road networks) have become an important research area in recent years. In this paper, we focus on path queries that find the shortest path between a source and a destination of the user. In particular, we address the problem of finding the shortest paths for a large number of simultaneous path queries in road networks. Traditional systems that consider one query at a time are not suitable for many applications due to high computational and service cost overhead. We propose an efficient group based approach that provides a practical solution with reduced cost. The key concept of our approach is to group queries that share a common travel path and then compute the shortest path for the group. Experimental results show the effectiveness and efficiency of our group based approach.

Keywords: Spatial query processing, Road networks, Clustering.

1   Introduction

With the proliferation of GPS-enabled mobile technologies, users access a wide variety of location-based services (LBSs) from different service providers. These LBSs range from simple information queries such as finding the nearest restaurant to navigational queries such as finding the shortest path to a destination. In this paper, we focus on the shortest path computation problem. In particular, we address the problem of finding the shortest paths for a large number of simultaneous (or contemporary) path queries in road networks. For example, tourists from nearby hotels may issue path queries to different sightseeing places at the same time. Traditional systems that evaluate *one query at a time* are not suitable for many applications as these systems cannot guarantee a cost-effective

M.A. Nascimento et al. (Eds.): SSTD 2013, LNCS 8098, pp. 367–385, 2013.
© Springer-Verlag Berlin Heidelberg 2013

and real-time response to the user in high load conditions [16,29]. We propose an efficient group based approach that provides a practical solution for path queries with reduced cost.

In a road network, users are often interested in a path to the destination that can be reached in minimum travel time. For such a scenario, travel times on road segments are considered as edge weights of the road network graph during the shortest (or fastest) path computation. Since travel time on a road segment is highly dynamic and depends on various real-time traffic conditions [8], it is not possible to accurately compute the travel time based on the network distance. To answer such user queries, an LBS provider needs to gather real time traffic conditions of the underlying road networks. However, it may not be possible for every LBS provider to have their own monitoring infrastructure for traffic updates due to high cost. Hence, to process queries on road networks, LBS providers subscribe to map services such as Google Maps [13], MapQuest [20], and Bing Maps [4] for traffic updates.

Due to huge client bases and popularity of map-based services, an LBS server often require to respond to a large number of simultaneous user queries. Thus, efficient processing of a large number of queries in road networks have become an important research topic. Specially, when an LBS server needs to call the map based services for every user request, it becomes a major bottleneck in providing a cost-effective and real-time response to the user [29] due to the following limitations. First, map based web services charge on per request basis (e.g., in Google Maps [13], an evaluation user can submit 2,500 requests per day and a licensed business user can submit 100,000 requests per day [1]) and thus the number of requests that one can make to obtain real-time traffic information is limited. Second, each map web service call incurs a significant delay in response, e.g., a web service call to fetch the travel time from the Microsoft MapPoint web service to a database engine takes 502 ms [16]. In addition to these limitations, well known approaches (e.g., Dijkstra [30] and A* Algorithm [21]) for the shortest path computation require expensive graph traversal operations, and thus incur huge computational overhead especially for a large number of queries.

In this paper, we propose a shared execution approach to process a large number of path queries simultaneously. The key idea of our shared execution approach to path queries comes from the path coherence properties of road networks. That is, two shortest paths originated at spatially close set of locations S and terminated at another spatially close set locations E are likely to share a common path of a significant length l and l tends to increase as the distances from S to E increase [24].

Based on the path coherence concept, we first group similar queries into clusters based on the similarities of the shortest path estimates called Q-lines. Specifically, a Q-line, is defined as the straight line vector connecting the source to the destination of a given path query. Each cluster, represented as a pair of source-destination regions, essentially is a group of path queries who have high probabilities of sharing a common travel path in their answers. Then, we execute each group as a single path query, that finds the best path from the source region to the destination region of the cluster.

Our group based heuristic to answer the shortest paths for a large number of simultaneous path queries significantly reduces the computational overhead. Note that, though the group based approach does not guarantee optimal shortest paths for all queries, the deviation from the optimal paths is found negligible. Extensive experimental studies in a real road network show that our group based approach is on average 9 times faster than the straightforward approach that evaluates each path query independently, in return of sacrificing the accuracy by 0.25%, which is acceptable for most of the users.

In summary, the contributions of this paper are as follows:

- We formulate the problem of group based path queries in road networks.
- We develop an efficient clustering technique to group path queries based on similarities of Q-lines that form the base of our efficient solution to process a large number simultaneous path queries in a road network.
- We conduct an extensive experimental study to show the efficiency and the effectiveness of our approach.

2 Related Works

To handle a large number of queries in modern database systems, shared execution of queries have recently received a lot of attention [2,9,28,29]. The core idea of all these approaches is to group similar queries (i.e., who share some common execution path) and then execute the group as a single query in the system. These approaches are found to be effective for many applications in handling high load conditions, whereas traditional systems that consider one query at a time fail to deliver the required performance for such applications. In this paper, we propose a shared execution approach for path queries on road networks, which is the first attempt of this kind.

The problem of finding the shortest path from a source to a destination on a graph (or spatial/road network) has been extensively studied in literature (e.g., [6,21,26,27,30]). Dijkstra's [30] algorithm is the most well-known approach for computing a single source shortest path with non-negative edge cost. Dijkstra [30] incrementally expands the search space, starting from the source, along network edges until the destination is reached. Hence, Dijkstra's algorithm requires to visit nodes and edges that are far away from the actual destination. Over the past two decades, a plethora of techniques have been proposed to address the limitations of Dijkstra's algorithm [3,11,12]. An alternative school of thought employs hill climbing algorithms, A* search [21] and RBFS [22], that use heuristics, e.g., Euclidian distances, to prune the search space. However, both Dijkstra's and hill climbing algorithms incur expensive graph traversal operations and require complete recomputation on every update. The above approaches assume that the road conditions will remain static. To address the dynamic load conditions of road networks, several approaches have been proposed [15,17].

Some techniques [3,11,12,23] rely on graph preprocessing under the assumption of static conditions (e.g., landmark) to accelerate query response times. Reach based routing [12] enhances responses by adding shortcut edges to reduced nodes' *reaches* during preprocessing. Landmark indexing [11] and transit node routing [3] boost run time query performance by using precomputed distances between certain set of landmarks chosen according to the algorithms. An adaptation of landmark based routing in dynamic scenarios [7] yields improved query times but requires a link's cost not to drop below its initial value. A dynamic variant of highway node routing [25] gives fast response times but can handle a very small number of edge weight changes. Several algorithms such as spatially induced linkage cognizance (SILC) [23], and path-coherent pairs decomposition (PCPD) [24] based on spatial path coherence have been proposed recently. These algorithms have high pre-computation overhead and space complexity.

An approach for all-time-shortest path queries have been proposed recently in [14]. For a *single* source and destination pair and a time interval, all-time-shortest path finds shortest paths for every start time in the given interval as for the same source-destination pair the shortest path may be different at different time of the day. Since two shortest paths for two different start times may share a common path, independent evaluation of a path query for each start time results in many redundant computation. To reduce this redundancy, [14] proposed a concept of critical time points, the time when two paths of two different start times get apart from each other.

A recent approach [19] continuously updates a user if the shortest path for a given source-destination pair changes due to dynamic traffic. Rather than recomputing on every update, this technique sends the user a new route only when delays change significantly.

All of the above approaches for path queries consider one query at a time, which is both costly and computationally expensive for an LBS server, especially in high load conditions. In this paper, we propose a novel clustering technique that finds similarities among path queries and executes similar queries as a group. There are several clustering algorithms [10], which use properties such as density, similarity, etc. to cluster nodes, lines and trajectories. However, the use of clustering techniques in shortest path calculation has not been addressed so far.

3   Group Based Path Queries (GBPQ)

We propose an efficient group based approach to compute path queries on road networks. A path query takes a source and a destination as inputs and returns a sequence of road segments that minimizes the total travel cost (e.g., travel time) from the source to the destination. Given a set of path queries, group based path queries ($GBPQ$) cluster n queries into m groups, where $n \gg m$, and evaluate each group of path queries collectively.

3.1   Intuition

The basic idea of our approach is developed based on a key observation of road networks, i.e., path coherence. Path coherence is a property where the shortest paths originated from spatially close set of locations S and terminated at another spatially close set of locations E are likely to share a common path of a significant length l and l has a tendency to increase as the distances from S to E increase. Thus, by grouping these source-destination pairs that share a common path for shared execution, reduce the computational overhead significantly. To group these source-destination pairs, we use the query line (Q-line) similarities based on perpendicular, parallel, and angular distances. A Q-line is a straight line vector connecting a source (e.g., s_1) to a destination (e.g., d_1). Based on the path coherence properties, it is highly likely that queries who have similar Q-lines will have a large portion of common travel path. Hence, we propose a group based approach that groups source and destination pairs based on their similarities of Q-lines and execute similar queries as a group in a road network.

3.2   Solution Overview

Our approach, GBPQ, groups source destination pairs of path queries into different clusters based on the similarities of Q-lines. We introduce the distance function and the areas of influence that form the base of our clustering algorithm. Fig. 1 illustrates our solution overview for eight path queries: (a) obtaining Q-lines from initial source and destination points of path queries, (b) clustering Q-lines and creating source-destination region pairs $\{R_s^1, R_d^1\}$ and $\{R_s^2, R_d^2\}$ based on the distance function and influence areas, (c) finding the shortest path between two regions, and (d) adding up internal paths inside a region to obtain the complete path.

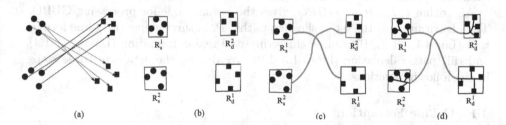

$$R_s^1 \quad R_d^2 \quad R_s^1 \quad R_d^2 \quad R_s^1 \quad R_d^2$$
$$R_s^2 \quad R_d^1 \quad R_s^2 \quad R_d^1 \quad R_s^2 \quad R_d^1$$

(a)   (b)   (c)   (d)

Fig. 1. An example of finding shortest paths in a group based framework

Note that since we have considered the shortest path between a source region and its corresponding destination region for a group, this path may not be the best path for every source-destination point pairs that belong to this group. Thus some path queries in GBPQ may result in a slightly larger path then the optimal shortest path. Our extensive evaluation shows the deviation of the path returned by GBPQ from the optimal path is only about 0.25% in the average case, which is within the acceptable limit of the users.

Algorithm 1. Evaluate_GBPQ(SD)

      Input : $SD = \{(s_1, d_1), (s_2, d_2), \ldots, (s_n, d_n)\}$
      Output: $P = \{p_1, p_2, \ldots, p_n\}$
1.1 **for** *each* $(s_i, d_i) \in SD$ **do**
1.2 | $l_i \leftarrow getStraightLineVector(s_i, d_i)$

1.3 $C \leftarrow$ Cluster_Queries $(l_1, l_2, \ldots, l_n)$
1.4 Compute $\{(R_s^1, R_d^1), (R_s^2, R_d^2), \ldots, (R_s^m, R_d^m)\}$
1.5 **for** *each* $(R_s^j, R_d^j) \in R$ **do**
1.6 | $SP_j \leftarrow ShortestPaths\ (R_s^j, R_d^j)$

1.7 **for** *each* $(s_i, d_i) \in SD$ **do**
1.8 | Find j such that $s_i \in R_s^j$ and $d_i \in R_d^j$
1.9 | $p_i \leftarrow$ Construct_Path (s_i, d_i, SP_j)

---

4 Algorithm

In this section, we present our algorithm for group-based path queries (GBPQ). The input to the algorithm is a set of n path queries, $SD = \{(s_1, d_1), (s_2, d_2), \ldots, (s_n, d_n)\}$, where (s_i, d_i) represents a path query from a source s_i to a destination d_i for $1 \leq i \leq n$, and the output of the algorithm is a set of approximate shortest paths, $P = \{p_1, p_2, \ldots, p_n\}$, where p_i represents the approximate shortest path for the path query (s_i, d_i). We propose a shared execution strategy that group similar queries using some key features of road networks. The algorithm first finds a common shortest path with respect to each group of path queries and then computes the approximate shortest path for each individual path query (s_i, d_i) based on the common shortest path of the group. Our approach sacrifices the accuracy of the query answers slightly, i.e., computes a slightly larger path than the optimal one, in turn for significant savings in computational overhead.

Algorithm 1, *Evaluate_GBPQ*, gives the pseudo code for processing GBPQ. The algorithm finds the set of shortest paths, P, in three steps: (i) Q-line formation (Lines 1.1–1.2), (ii) Q-line clustering and region formation (Lines 1.3–1.4) and (iii) path calculation (Lines 1.5–1.9). We discuss the details of three steps in the following sections.

4.1 Q-Line Formation

We define Q-line, l_i, as the straight line vector connecting the source s_i to the destination d_i of a path query. The concept of Q-line is used to predict the similarity of path queries, whose answers share common paths. The algorithm computes the set of Q-lines, $L = \{l_1, l_2, \ldots, l_n\}$ in Line 1.2, which is used for clustering the path queries in the second step of the algorithm.

4.2 Query Clustering and Region Formation

In this phase, we first cluster the path queries based on the similarities of Q-lines and then compute the *source and destination region pair* for each cluster.

A source and destination region pair consists of a source region and a destination region, where the source region (destination region) of a cluster is a minimum bounding rectangle (MBR) containing the locations of sources (destinations) of all Q-lines in the cluster.

Algorithm 1 finds a set of clusters $C = \{c_1, c_2, \ldots, c_m\}$ using the function *Cluster_Queries* (Line 1.3), where $m \le n$. After clustering the queries, Algorithm 1 (Line 1.4) computes the set of source and destination region pairs, $R = \{(R_s^1, R_d^1), (R_s^2, R_d^2), \ldots, (R_s^m, R_d^m)\}$, where R_s^j and R_d^j represent source region and destination region, respectively, of cluster c_j for $1 \le j \le m$.

Algorithm 2 shows the steps for *Cluster_Queries*. The detailed discussion of function *Cluster_Queries* is given at the end of this section. To measure the similarity among Q-lines, we define two metrics: (1) *distance function*, and (ii) *areas of influence*, which are used in Function *Cluster_Queries*. Thus, we next explain distance function and areas of influence.

Distance Function: As a similarity measure of two Q-lines, the distance between two Q-lines l_1 and l_2 is defined as an aggregate measure consisting of the following three distances.

1. Parallel distance $d_\parallel$: the length difference between the projection of l_1 onto l_2 and l_2 itself.
2. Perpendicular distance $d_\perp$: the mean length of two perpendicular distances measured from two corner points of l_1 onto l_2.
3. Angular distance d_θ: the angle between l_1 and l_2 weighted by the length of the longer Q-line.

We calculate the distance between two Q-lines by using the following formula:

$$distance = w_\perp d_\perp + w_\parallel d_\parallel + w_\theta d_\theta \tag{1}$$

The weight values $w_\perp$, $w_\parallel$ and w_θ are used to control the effective contribution of three components on the overall distance. For example, a larger $w_\parallel$ value reduces the length difference between a Q-line and its projection on the other Q-line. Similarly, a larger value of $w_\perp$ keeps the endpoints of two queries closer to each other. Likewise, a larger w_θ value gives more emphasis on the angular distance of two Q-lines. In our experiments, we keep all these weights to unity, so that all the three components of the distance function have equal effect on overall distance. Based on the distance function two path queries can be grouped together if their distance is less than a threshold value ψ. The formal definitions and impact of parallel, perpendicular and angular distances [10] are discussed below. Symbols used in these definitions are shown in Fig. 2a.

PARALLEL DISTANCE: Let s_j and d_j be the two endpoints of the Q-line l_j. If the projection of $s_j d_j$ over l_i is $p_s p_d$, then the parallel distance $d_\parallel$ is defined as maximum of the Euclidean distance of s_i to p_s and d_i to p_d, i.e., $d_\parallel = MAX(s_i p_s, p_d d_i)$.

PERPENDICULAR DISTANCE: Let $l_{\perp 1}$ and $l_{\perp 2}$ be the distance components of two Q-lines l_i and l_j as shown in Fig. 2a. Then the perpendicular distance $d_\perp$ of these two Q-lines is defined with second order Lehmer mean [5] of $l_{\perp 1}$ and $l_{\perp 2}$, i.e., $d_\perp = \frac{l_{\perp 1}^2 + l_{\perp 2}^2}{l_{\perp 1} + l_{\perp 2}}$.

ANGULAR DISTANCE: Let θ be the smaller intersecting angle between Q-lines l_i and l_j. Then their angular distance d_θ is defined as the product of sin component of the larger Q-line, i.e., $d_\theta = MAX\{l_i, l_j\} \times sin\theta$.

For clustering queries, similarities among Q-lines are measured using the distance function. However, computing the similarities for every pair of Q-lines will be prohibitively expensive, specially for a large number of Q-lines. We formulate a branch-and-bound approach that tremendously reduces the computational afford to identify clusters of similar Q-lines. Specifically, we introduce the concept of *areas of influence* to help prune a large number of Q-lines while forming clusters of queries and thus reduces the computational overhead significantly.

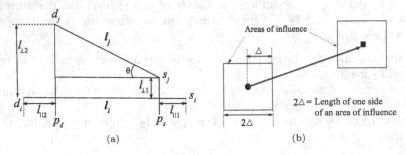

(a)   (b)

Fig. 2. (a) Components of the distance function, (b) Areas of influence of a Q-line

Areas of Influence: We define the areas of influence of a Q-line as a pair of regions, represented as two squares centering the two endpoints of the Q-line. If another Q-line has its source and destination inside these two squares, respectively, then we compute the distance between these two Q-lines to check whether they belong to the same cluster. This is because, there is a high probability that the distance between those two Q-lines is smaller than the threshold value ψ.

We define the side length of both squares as 2Δ, which denotes the size of the areas of influence. A larger value of Δ results in many unwanted Q-lines to be included for distance calculations and a smaller value results in a large number of small clusters. Since the appropriate value of Δ depends on the query set, we choose a suitable Δ through a empirical study in the experiment. Fig. 2b shows areas of influence using a pair of rectangles. Note that one can choose other types of shapes such as rectangle, circle or ellipse for its influence areas. Our approach is independent to the shape of the influence area.

Query Clustering: We use the concept of the distance function and the areas of influence to cluster similar queries in Algorithm 2. The input to algorithm is the set $L = \{l_1, l_2, \ldots, l_n\}$ of n Q-lines, and the output is a set $C = \{c_1, c_2, \ldots, c_m\}$ of m clusters. For each Q-line $l \in L$, we first find the set of Q-lines I that are inside

l's areas of influence (Line 2.4). Then we compute the initial *representative Q-line*, r, of the set I (Line 2.5). The representative Q-line of a given set of Q-lines is the average direction vector of those Q-lines. Average direction vector I of the set $I = \{l_1, l_2, \ldots, l_j\}$ is calculated with the following equation: $I = \frac{l_1+l_2+\ldots+l_j}{|I|}$, where $|I|$ is the cardinality of set I.

Algorithm 2. CLUSTER_QUERIES(L)

 Input : A set of Q-line $L = \{l_1, l_2, \ldots, l_n\}$
 Output: A set of clusters of queries $C = \{c_1, c_2, \ldots, c_m\}$
2.1 $L' \leftarrow Null$
2.2 **for** *each* $l \in L$ **do**
2.3 | $c_{new} \leftarrow Null$
2.4 | $I \leftarrow subsetWithinAreaOfInfluence(L, l)$
2.5 | $r \leftarrow representativeQ\text{-}line(I)$
2.6 | **for** *each* $i \in I$ **do**
2.7 | | **if** $distance(r, i) \leq \psi$ **then**
2.8 | | $c_{new} \leftarrow c_{new} \cup i$
2.9 | | $Update(r)$
2.10 | | Remove i from L
2.11 |
2.12 | **if** $size(c_{new}) \geq \mu$ **then**
2.13 | | $C \leftarrow C \cup c_{new}$
2.14 | **else**
2.15 | | $L' \leftarrow c_{new}$
2.16
2.17 **for** *each* $q \in L' \cup L$ **do**
2.18 | **for** *each* $c \in C$ **do**
2.19 | | **if** $distance(representativeQ\text{-}line(c), q) \leq \psi$ **and** q *is not classified*
      then
2.20 | | $c \leftarrow c \cup q$
2.21 | | $Update(r)$
2.22 | | Mark q as classified
2.23 |
2.24 | **if** q *is not classified* **then**
2.25 | | $C \leftarrow C \cup \{q\}$
2.26 | Remove q from L' or L

Now, for each element $i \in I$, we first calculate its distance from the current representative Q-line, r. If the value is less than or equal to ψ, i is added to a new cluster c_{new}. The representative Q-line r is then again updated to add the effect of newly added Q-line i (Lines 2.6 - 2.10) in c_{new}. We use the moving average process to update r, i.e., if r currently represents a cluster of j queries, when a new Q-line i is added into the cluster the value of r will be updated as $\frac{r*j+i}{j+1}$. The reason for updating r with the moving average is as follows.

Initial set I of Q-lines, formed based on the influence set of a Q-line $l \in L$, may contain queries that belong to more than one cluster. Since the initial representative Q-line r is the average of all Q-lines in the set, Q-lines of all probable clusters may have less distance than ψ and thus all Q-lines may have been wrongly clustered into a single cluster. Now, if we take a particular order (e.g., bottom-up or top-down) of Q-lines of I and update r for every $i \in I$ that is included in c_{new}, r is shifted towards the cluster c_{new}. Thus, other Q-lines in I that have higher distances from the updated r are excluded from c_{new}. The Q-lines included in c_{new} are removed from L. If c_{new} does not have enough number of Q-lines (i.e., $size(c_{new}) < \mu$), Q-lines of c_{new} are inserted into a new list L' (Lines 2.12 - 2.15).

After finishing the initial clustering process (Lines 2.6- 2.15), we have some Q-lines left which are not included in any cluster. Some Q-lines are not clustered because the representative Q-line was initially at a greater distance than ψ and later it did not fall into others' areas of influence, i.e., the remaining elements of the set L. The other set of Q-lines L' may be left unclustered because there are not sufficient amount of queries to form a new cluster (Line 2.15). For all these queries, we initially check them whether they fit into already created clusters. For a Q-line, if we find a cluster with a distance less than ψ, we add the query into that cluster and update the representative Q-line of the cluster. Otherwise, a new cluster is created and the Q-line is removed from L or L' (Lines 2.17- 2.26).

At this stage we have m clusters of Q-lines. Next for each cluster c_j, where $1 \leq j \leq m$, we compute a source-destination region pair (R_s^j, R_d^j), where R_s^j contains the source points and R_d^j contains the destination points of all path queries (or Q-lines) of that cluster. We represent R_s^j and R_d^j with two minimum bounding rectangles (MBRs), a source MBR and a destination MBR, respectively.

4.3   Path Calculation

The final step of Algorithm 1 is to compute the shortest path for every path query, which is done in two phases. The algorithm first computes the shortest path for each cluster for its source-destination region pairs as shown in Fig. 3a. Then the algorithm finds the approximate shortest path for each individual path query in the cluster using Function $Construct\_Path$ as shown in Fig. 3b. The details of $Construct\_Path$ is summarized in Algorithm 3.

The first phase, i.e., region to region shortest path is computed as follows. Essentially, as discussed earlier, a region pair consists of two MBRs, a source MBR and a destination MBR. For a source region, exit points of the region are identified by considering the outgoing edges of the region. On the other hand, for a destination region, entry points are identified by considering the incoming edges to the region. These regions act like virtual super-nodes, where exit (or entry) paths from the regions are the edges of those nodes. Then we apply a heuristic based approach, A* search algorithm, to find the k-shortest paths $\{SP_i^1, SP_i^2, ..., SP_i^k\}$ between a source region R_s^i to a destination region R_d^i.

The motivation of using k-shortest paths instead of a single shortest path is as follows. There can be scenarios, where a single shortest path between R_s^i and

R_d^i may result in a large path deviation for some path queries in this cluster. Thus, if we have multiple routes through different exit (entry) points of source (destination) regions, it is more likely that these k paths may serve well all path queries in the cluster. While computing k shortest paths, we take a different strategy than traditional k-shortest paths approach (i.e., k paths for a fixed source and destination pair). We choose k paths on the basis of different combinations of exit points of source and destination regions. For example, the first shortest path is essentially the actual shortest path between two regions (i.e., two super nodes). Thus for consecutive shortest paths, there should be at least one different exit point from source or destination region.

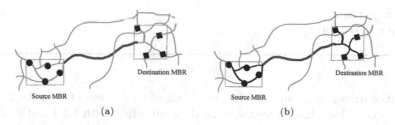

Fig. 3. (a) Finding the weighted shortest path, (b) Constructing shortest path by adding road segments within regions

The second phase of connecting source and destination points to the corresponding region's shortest path is computed as follows. For each query $s \in R_s^i, d \in R_d^i$ in a cluster c_i, we find a path $sp \in SP_i$ that gives the minimum overall shortest path for connecting s and d.

For some path queries, none of the k region to region shortest paths provide good answers. These queries need to be executed independently to ensure better paths for them. To identify such queries, we propose a heuristic based on the influence region, which can be defined as follows. If the path fragment from a source point s (destination point d) to the starting point (ending point) of all of the k paths is greater than $2\sqrt{2}\Delta$, we evaluate the query independently. Basically, $2\sqrt{2}\Delta$ is the maximum Euclidian distance of between two points of an influence region Δ (Section 4.2). The intuition behind this heuristic is that if it takes a long path to travel from a query source (destination) to region exit (entry) path, it is more likely that there may exist an alternative better path for such queries.

4.4   Maximum Error Bound

In this section, we derive the maximum error bound of the path returned by our approach for a given source s and destination d pair. Let us assume that s and d belongs to the cluster c_i represented as source destination region pair (R_s^i, R_d^i). Let p_i be the shortest path ($k = 1$) between R_s^i and R_d^i, where p_i^s and p_i^d is the starting and ending points of the path p_i, respectively. The minimum path that a source and destination pair belong to this cluster need to travel is p_i. Now, if

---

Algorithm 3. CONSTRUCT_PATH (s, d, SP_i)

    Input   : A source s, a destination d, weighted k-shortest paths
           $SP_i = \{SP_i^1, SP_i^2, ..., SP_i^k\}$
    Output: A path p

3.1 $p \leftarrow \infty$
3.2 **for** *each* $sp \in SP_i$ **do**
3.3 $f_1 \leftarrow$ shortest path between s and start point of sp
3.4 $f_2 \leftarrow$ shortest path between d and end point of sp
3.5   **if** $f_1 <= 2\sqrt{2}\Delta$ *and* $f_2 <= 2\sqrt{2}\Delta$ **then**
3.6 $p \leftarrow Min(p, f_1 + sp + f_2)$
3.7

3.8 **if** $p = \infty$ **then**
3.9 $p \leftarrow Dist(s, d)$
3.10 **return** p

---

the distance from s to p_i^s and the distance from p_i^d to d is less or equal to $2\sqrt{2}\Delta$ (Lines 3.5-3.6, Algorithm 3), we use p_i as the connecting path for s and d. Thus the maximum path that a source and destination pair need to travel through p_i is $p_i + 4\sqrt{2}\Delta$. Thus the maximum error for a path query can be bounded by $4\sqrt{2}\Delta$. Note that, queries that are evaluated independently return the actual shortest path and thus have zero error.

4.5 Discussion

In this paper, we have assumed that there are n submitted queries in the system for our clustering algorithm. We then apply our clustering technique to group these n queries into different subgroups. However, the value of n can be dynamically chosen based on the user given threshold, i.e., how long a user can wait for the query answer. The detailed study of finding a suitable n and other optimization such as re-using the cluster for future incoming queries in the system is the scope of future study.

    Note that, since our main focus in this paper is to show the effectiveness of group based approach compared to one query at a time approach, we limit our experiments using a well known shortest path algorithm A*. However, any other state-of-the-art shortest path algorithms (e.g., [3,27]) can be used.

5 Experimental Study

In this section, we evaluate the performance of our proposed algorithm by varying a wide range of parameters. We compare our group based path queries (GBPQ) approach with the naive approach that executes each query at a time using A* algorithm [21]. We have also used A* for our region to region shortest path in GBPQ. Since the improvement of our approach over the naive approach comes from clustering similar queries, the superiority of GBPQ still holds if we compare

GBPQ with any other shortest path algorithms. We simulate our experiment on a system with Intel core i5 2.86 GHz processor and 4 GB memory running Windows 7 ultimate. C++ is used to implement our algorithms.

A road network dataset of North America with 175,813 nodes, 179,179 edges, and a diameter of 18,579 units is used. At the beginning, the entire map data is loaded into the memory. For a single path query we need to select a source-destination pair on the map. However to simulate a group behavior we first partition the entire data space into a number of square windows. Then, we choose two random windows, one as a source region and the other as a destination region for a group of path queries, who have their source locations and destination locations inside the source and destination regions, respectively. Within a region, query points are generated using Gaussian and Zipf distributions.

Before moving on to the performance evaluation, we first discuss the performance of our clustering algorithm.

5.1 Cluster Generation

The performance of GBPQ largely depends on how accurately our clustering algorithm (Algorithm 2) can group similar queries. The performance of our clustering algorithm depends on a number of parameters. Thus, we have first conducted a wide range of experiments to determine suitable values of these parameters for our experiments.

Parameter Pruning: The efficiency of our clustering algorithm depends on three pruning parameters: i) half length of an areas of influence Δ, ii) minimum distance threshold ψ, iii) minimum number of queries μ in a cluster. Half length of an area of influence defines two surrounding regions around source and destination points. Queries in the same areas of influence have higher probabilities of belonging to the same cluster. Moreover, as per our distance measure, the maximum distance allowed between two Q-lines increases with the increase of Δ. The performance of GBPQ also depends on the distance threshold value ψ. The parameters ψ and Δ are also correlated as a higher value of ψ allows us to take a larger Δ. The other parameter μ determines the minimum cluster size, and thus has an impact on the number of clusters. However, choosing suitable values of these parameters is a challenge and may vary with the data sets. Thus, we resort to detailed experimental study [18] to choose default values of these parameters as $\Delta = 80$, $\psi = 160$ and $\mu = 30$ for experiments in Section 5.2.

Number of Clusters: We vary the number of queries n, the minimum query distance d_c, i.e., the distance between the source and the destination of each query, and the window size ω, and measure the number of clusters identified our algorithm. In our system, we aim to minimize the number of clusters, since the number of clusters negative affect the performance of the algorithm.

Figures 4a and 4b show the number of identified clusters using our algorithm for varying number of queries for Gaussian and Zipfian distribution, respectively. Figures show that the number of clusters increases as the number of queries

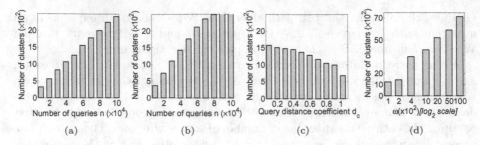

Fig. 4. Number of clusters for varying (a) number of queries (Gaussian) (b) number of queries (Zipfian)(c) query distance (d) window length

increases. In Fig. 4c, we can see that the number of clusters decreases as the query distances increase, which is expected.

Figure 4d shows the number of identified clusters for varying window length ω for a fixed number of 50,000 queries. As we increase ω, the number of identified clusters also increases. This is because for a larger window size the queries are randomly distributed in a wider space. For example, when there is only one window (i.e., the window size is same as the dataspace size), all queries are randomly distributed throughout the entire data space. However, even in this worst case, our algorithm identifies 7,216 clusters from 50000 queries generated in the entire data space. With the increased number of windows (i.e., for a smaller window size), the clusters become more apparent and the algorithm is able to identify them more efficiently.

Effectiveness of Clustering: The effectiveness of the clustering algorithm is measured as the percentage in which the paths generated from that cluster stay within the predefined error bound and need not be evaluated independently. In this set of experiments, we show the percentage of queries needed to be evaluated independently, i.e., the queries that cannot be served by the group's shortest path (Algorithm 3). Figures 5a and 5b show the percentage of query needs to be evaluated independently for different number of queries for Gaussian and Zipfian

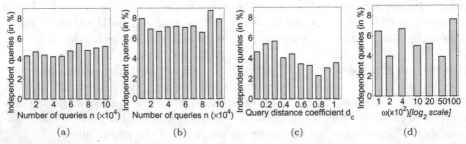

Fig. 5. Number of independent queries for varying (a) number of queries (Gaussian) (b) number of queries (Zipfian)(c) query distance (d) window length

distribution, respectively. Results show the percentage of independent evaluation ranges from 2% to 9% having an average of 5.44%. Figures 5c and 5d show the percentage of independent query evaluation for varying d_c and ω, respectively for Gaussian distribution. In summary, we can see that our clustering algorithm can always serve more than 90% of the queries using the corresponding group shortest paths.

5.2 Performance Evaluation

In this section, we show the performance of our algorithm based on the two performance metrics: processing time and the average percentage of deviation of the answer from actual shortest path. Processing time is the total query response time including the clustering time for a given number of queries. Clustering time is about 6 seconds on average for 50,000 queries and shows a linear relationship with number of queries for the selected parameter values. To compare our approach with the naive approach, We vary the following parameters: (i) the number n of queries; (ii) the minimum query distance d_c, i.e., the distance between the source point and destination point of each query; (iii) the window size ω; and (iv) the value of k. We calculate d_c as a function of ω.

Effect of Number of Queries: We vary the number n of queries in the range of 10,000 to 100,000 with a step size of 10,000 units. For both Gaussian and Zipf distributions we see that the processing time for GBPQ rises slightly with the increase of the values of n (Fig. 6a, 6c). Whereas, for the naive approach, the processing time increases significantly with the increase in the number of queries. When the value of n is 10,000, the processing times for GBPQ and the naive approach are approximately 100 and 1200 seconds, respectively. With an increased value n of 100000, the processing times for GBPQ and the naive approach are approximately 900 and 12600 seconds, respectively. GBPQ is on average 14 times faster than the naive approach, in our experiments. Moreover, our experimental results show that on an average the deviation of the path returned by GBPQ from the actual shortest path is only around 0.21% in case of Gaussian distribution (Fig. 6b) and around 0.27% in case of Zipf (Fig. 6d). Maximum deviation reported on about 75 run of different setup is 9.88% with the average of 7.43% and standard deviation of 1.78.

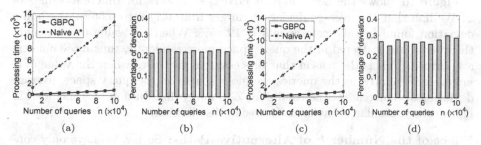

Fig. 6. Effect of number of queries (a-b) Gaussian and (c-d) Zipf distributions

Effect of Query Distance: In this set of experiments, we vary the minimum query distance coefficient d_c for generating queries. Figure 7a shows the results for Gaussian distribution of query points. We see from the figure that for GBPQ the processing time slowly increases as the value of d_c increase. On the other hand, the processing time increases significantly for the naive approach as the value of d_c increases. This is because two reasons, (i) a higher value of d_c corresponds to a longer distance between the source and destination; (ii) the number of nodes traversed for such a query is higher than that of the query which has a smaller d_c. The accuracy of our GBPQ increases as d_c increases. The percentage of average deviation of the answered path from the actual path reduces from 0.32% to 0.11% when the query distance increases from 1,000 units to 10,000 units (Fig. 7b). Maximum deviation found in these experiments is 9.22% where the set of maximum deviation has an average of 5.73% and standard deviation of 2.30. The results for Zipf query distribution (not included in figure) shows similar behavior as Gaussian distribution.

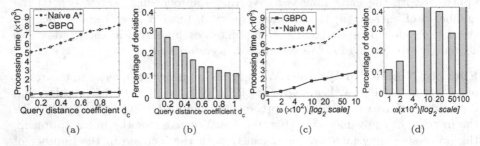

(a)   (b)   (c)   (d)

Fig. 7. Effect of the minimum query distance, (a) processing time (b) percentage of deviation; Effect of window size, (c) processing time (d) percentage of deviation

Effect of Window Size: In this set of experiments, we vary the window length ω and compare the performance of our approach with the naive approach. Figure 7c shows the processing time of GBPQ and the naive approach for Gaussian distribution of query points. The processing time increases as ω increases, e.g., when the window size is equal to the data space, the processing time of GBPQ is 62.3% lower then the naive approach. When the window size is 100×100, the processing time of GBPQ is 92% lower than the naive approach.

Figure 7d shows the deviations of GBPQ's answers for different values of ω. We find that the average deviation ranges from 0.11% to 0.56%. Maximum deviation found in this experiments is 12.85%. When the value of ω is high, the queries are scattered in the query space more randomly, thus more clusters are formed. This is the reason that processing is increases when the value of ω increases. Similarly, as the queries are more sparse on the query space, average deviation also increases. The results for Zipf query distribution (not included in figure) shows similar behavior as Gaussian distribution.

Effect of the Number k of Alternative Paths: So far, we have only considered a single path from a source region to a destination region (i.e., $k = 1$).

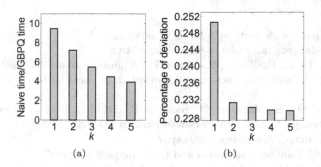

Fig. 8. Effect of k on (a) processing time (b) percentage of deviation

In this set of experiments, we vary the value of k as 1, 2, 3, 4, and 5, and see the effect on the performance of the system. As discussed in Section 4.3, a larger k can be used to provide alternative routes and hence to improve the accuracy of individual shortest path query. However, higher k means more path calculation, and thus requires more time to process the query. Figure 8 shows that with the increase of k, the gain in processing time by GBPQ to Naive A* decreases. Figure 8 also shows that the accuracy of path queries increases approximately 7.7% for an increase of k value from 1 to 2.

6   Conclusion

We have proposed a group based approach for processing a large number of simultaneous path queries in a road network. Our approach is based on a novel clustering technique that groups queries based on the similarities of their *Q-lines*. We introduce two concepts: the distance function and the areas of influence, that help us to effectively cluster similar queries and execute them as a group. Our group based approach to evaluate a large number of path queries provides a cost-effective solution with reduced computational overhead. Our experiments have shown that on an average our shared execution approach is 9 times faster than a traditional approach, where each path query is evaluated independently. Our approach achieves this significant improvement at the cost of sacrificing 0.25% accuracy in the average case. Besides query optimization for an LBS server, our proposed technique can be used in many applications such as car sharing and the support of intermodal transport that need clustering similar paths.

Acknowledgments. This research has been conducted at the department of Computer Science and Engineering (CSE), Bangladesh University of Engineering and Technology (BUET). This work is supported by BUET and CodeCrafters-Investortools Research Grant.

References

1. Google maps/google earth apis terms of service, http://code.google.com/apis/maps/terms.htm
2. Ali, M.E., Tanin, E., Zhang, R., Kulik, L.: A motion-aware approach for efficient evaluation of continuous queries on 3d object databases. VLDB J. 19(5), 603–632 (2010)
3. Bast, H., Funke, S., Matijevic, D., Sanders, P., Schultes, D.: In transit to constant time shortest-path queries in road networks. In: ALENEX (2007)
4. BingMaps, http://www.bing.com/maps/
5. Bullen, P.S.: Handbook of means and their inequalities (1987)
6. Delling, D., Goldberg, A.V., Werneck, R.F.F.: Faster batched shortest paths in road networks. In: ATMOS, pp. 52–63 (2011)
7. Delling, D., Wagner, D.: Landmark-based routing in dynamic graphs. In: Demetrescu, C. (ed.) WEA 2007. LNCS, vol. 4525, pp. 52–65. Springer, Heidelberg (2007)
8. Demiryurek, U., Banaei-Kashani, F., Shahabi, C.: Efficient K-nearest neighbor search in time-dependent spatial networks. In: Bringas, P.G., Hameurlain, A., Quirchmayr, G. (eds.) DEXA 2010, Part I. LNCS, vol. 6261, pp. 432–449. Springer, Heidelberg (2010)
9. Giannikis, G., Alonso, G., Kossmann, D.: Shareddb: Killing one thousand queries with one stone. PVLDB 5(6), 526–537 (2012)
10. Lee, J.G., Han, J.: Trajectory clustering: A partition-and-group framework. In: SIGMOD, pp. 593–604 (2007)
11. Goldberg, A.V., Harrelson, C.: Computing the shortest path: A search meets graph theory. In: SODA, pp. 156–165 (2005)
12. Goldberg, A.V., Kaplan, H., Werneck3, R.F.: Efficient point-to-point shortest path algorithms. Tech. Report (2005)
13. GoogleMaps, http://maps.google.com
14. Gunturi, V.M.V., Nunes, E., Yang, K., Shekhar, S.: A critical-time-point approach to all-start-time lagrangian shortest paths: A summary of results. In: Pfoser, D., Tao, Y., Mouratidis, K., Nascimento, M.A., Mokbel, M., Shekhar, S., Huang, Y. (eds.) SSTD 2011. LNCS, vol. 6849, pp. 74–91. Springer, Heidelberg (2011)
15. Koenig, S., Likhachev, M., Furcy, D.: Lifelong planning A*. Artificial Intelligence 155, 93–146 (1968)
16. Levandoski, J.J., Mokbel, M.F., Khalefa, M.E.: Preference query evaluation over expensive attributes. In: CIKM, pp. 319–328 (2010)
17. Likhachev, M., Ferguson, D., Gordon, G., Stentz, A., Thrun, S.: Anytime dynamic A*: An anytime, replanning algorithm. In: ICAPS (2005)
18. Mahmud, H., Amin, A.M., Ali, M.E., Hashem, T.: Shared execution of path queries on road networks. CoRR, abs/1210.6746 (2012)
19. Malviya, N., Madden, S., Bhattacharya, A.: A continuous query system for dynamic route planning. In: ICDE, pp. 792–803 (2011)
20. MapQuest, http://www.mapquest.com
21. Nilson, N.J., Hart, P.E.: A formal basis of the heuristic determination of minimum cost paths 4(2), 100–107 (1968)
22. Russell, S., Norvig, P.: Artificial Intelligence a modern approach, 2nd edn. (2006)
23. Samet, H., Sankaranarayanan, J., Alborzi, H.: Scalable network distance browsing in spatial databases. In: SIGMOD, pp. 43–54 (2008)

24. Sankaranarayanan, J., Samet, H., Alborzi, H.: Path oracles for spatial networks. PVLDB 2(1), 1210–1221 (2009)
25. Schultes, D., Sanders, P.: Dynamic highway-node routing. In: Demetrescu, C. (ed.) WEA 2007. LNCS, vol. 4525, pp. 66–79. Springer, Heidelberg (2007)
26. Terrovitis, M., Bakiras, S., Papadias, D., Mouratidis, K.: Constrained shortest path computation. In: Medeiros, C.B., Egenhofer, M., Bertino, E. (eds.) SSTD 2005. LNCS, vol. 3633, pp. 181–199. Springer, Heidelberg (2005)
27. Terrovitis, M., Bakiras, S., Papadia, D., Mouratidis, K.: Shortest path and distance queries on road networks: An experimental evaluation. In: PVLDB, pp. 406–417 (2012)
28. Thomsen, J.R., Yiu, M.L., Jensen, C.S.: Effective caching of shortest paths for location-based services. In: SIGMOD, pp. 313–324 (2012)
29. Zhang, D., Chow, C.-Y., Li, Q., Zhang, X., Xu, Y.: SMashQ: spatial mashup framework for k-nn queries in time-dependent road networks. Distributed and Parallel Databases, 259–287 (2013)
30. Zwick, U.: Exact and approximate distances in graphs - A survey. In: Meyer auf der Heide, F. (ed.) ESA 2001. LNCS, vol. 2161, pp. 33–48. Springer, Heidelberg (2001)

STEPQ: Spatio-Temporal Engine
for Complex Pattern Queries

Dongqing Xiao and Mohamed Eltabakh

Worcester Polytechnic Institute, MA 01604, USA
{dxiao,meltabakh}@cs.wpi.edu

Abstract. With the increasing complexity and wide diversity of spatio-temporal applications, the query processing requirements over spatio-temporal data go beyond the traditional query types, e.g., range, kNN, and aggregation queries along with their variants. Most applications require support for evaluating powerful spatio-temporal pattern queries (STPQs) that form higher-order correlations and compositions of sequences of events to infer real-world semantics of importance to the targeted application. STPQs can be supported by neither traditional spatio-temporal databases (STDBs) nor by modern complex-event-processing systems (CEP). While the former lack the expressiveness and processing capabilities for handling such complex sequence pattern queries, the later mostly focus on the *Time* dimension as the driving dimension, and hence lack the power of the special-purpose processing technologies established in STDBs over the past decades. In this paper, we propose an efficient and scalable spatio-temporal engine for complex pattern queries (*STEPQ*). STEPQ has several innovative features and ideas that will open the research in the area of integration between spatio-temporal databases and complex event processing.

1 Introduction

The recent advances and wide-spread popularity of mobile devices, wireless cellular phones, and Global Positioning Systems (GPS) have enabled spatio-temporal applications in various domains to continuously monitor and track all objects of interest. Thus with the increasing complexity and wide diversity of spatio-temporal applications, the query and data exploration requirements go beyond the traditional spatio-temporal query types, e.g., *range, k — nearest-neighbor (kNN)*, and *aggregation* queries [3,4], to more expressive and semantics-rich spatio-temporal pattern queries (or STPQ) that require higher-order correlation among events. In Table 1, we illustrate several of STPQs from different applications. Evidently, STPQs are prevalent in many applications as they capture real-world semantics that otherwise would have been lost or delegated to the application layer for ad-hoc and inefficient processing. It is not meaningful to assume that a suspicious criminal activity in Q1 or the alert condition for a patient in Q3 depend solely on a single data instance (or even snapshot) of the data stream - rather separate snapshots of instances in the high-speed stream must be trapped at the right moments of time and synchronized to determine the correct match of such a complex STPQ query.

In this paper, we envision the STEPQ system—Spatio-Temporal Engine for complex Pattern Queries— that addresses the unique challenges of handling

M.A. Nascimento et al. (Eds.): SSTD 2013, LNCS 8098, pp. 386–390, 2013.
© Springer-Verlag Berlin Heidelberg 2013

Table 1. Examples of spatio-temporal pattern queries (STPQs)

Q1	Report child-abuse criminals who stay in a school area A1 for more than x minutes and then move to a suspicious area A2 within one hour (E.g.,suspicious criminal activity).
Q2	Report cars that stay in my *kNN* over interval T and continuously are getting closer to my moving car.
Q3	Send alert to patient *P*, if she stays in contact (within distance D for at least interval T) with a patient having a transferable disease (E.g., health threat).
Q4	For consecutive areas A1, A2, and A3, report speeding cars (over the speed limit for at least x mins) in A1 and A3 but not in A2 (E.g., testing effect of radar signs over A2 on drivers' behavior).
Q5	Report restaurants located in kNN of two moving cars and getting closer to both cars over interval T (E.g., find common nearby restaurants in direction of moving cars).

STPQs, including: (1) They embed powerful semantics not captured by current spatio-temporal query types, (2) Unlike traditional query types that can be evaluated on each instance of the database in isolation, STPQs require correlation among spatio-temporal events (both in time and in space) over multiple instances of the database, (3) They require a full-fledged query engine equipped not only with efficient event-processing techniques but also with effective spatio-temporal processing capabilities, and (4) Unlike state-of-art event-processing techniques (CEP) that have no control over the input stream of events, the STPQs generate these streams of events, and hence crucial optimization strategies can be deployed to control which higher-order events to generate and when. These challenges combined make the state-of-art in complex event processing (CEP), e.g., [8], not applicable since CEP techniques cannot process traditional *range* or *kNN* queries efficiently, also state-of-art in spatio-temporal databases (STDBs), e.g., [2,7], fall short since they lack the expressiveness power and processing capabilities of handling complex pattern queries.

2 Limitations of State-of-Art Techniques

Conceptually, STPQs can be viewed as two-layered queries where the first layer runs traditional spatio-temporal queries, e.g., *range* and *kNN*, on top of the raw input stream coming from moving objects (we refer to these queries as *base* queries). The second layer runs complex pattern-matching queries on top of the results generated from the base queries. Thus, with the state-of-art technology, there are two possible approaches to support STPQs, namely **application-level** and **middleware-level** as depicted in Figures 1(a) and (b). In the application-level approach, all of the pattern matching and event correlation is done at the application level to impose the query semantics, which is clearly an ad-hoc and inefficient solution since (1) each application applies its own semantics independently, (2) mobile devices usually have limited power and processing capabilities, and (3) STDBs may send streams of unnecessary results plus the lack of many possible optimizations that could have been performed by the execution engine. The middleware-level approach, which consists of loosely coupled CEP systems, e.g., [8,1,5] and STDBs is a more feasible approach. However, it has serious drawbacks and limitations including:

(1) Coupling Hurdles: There are several linking problems that emerge between the STDB and CEP layers such as: (a) STDBs deploy incremental

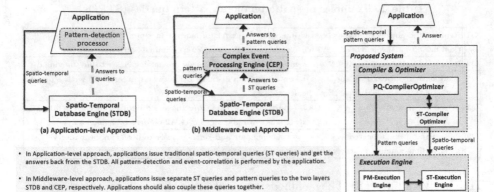

Fig. 1. Possible architectures for supporting spatio-temporal pattern queries

evaluation techniques for purposes of efficiency and scalability whereas CEP systems do not handle incremental updates, and (b) The base queries can themselves be moving objects, e.g., Queries Q2, Q3, and Q5 in Table 1, and hence CEP systems need to get as input not only the query answer, but also the query points.

(2) Optimization Hurdles: Since STPQs generate both the base queries and the pattern-matching queries, then several optimization opportunities arise that cannot be leveraged with loosely coupled STDB and CEP layers. For example, in Query Q4, the three *range* queries over areas *A1*, *A2*, and *A3* will be concurrently running, although queries over *A2* and *A3* should run only if there is a match in the previous areas.

(3) Synchronization and Transformation Hurdles: A STPQ may require not only executing multiple base queries to generate events, but also synchronizing their execution. For example, Query Q5 in Table 1 requires synchronizing the execution of two moving *kNN* queries and then intersecting their results. Such synchronization and transformation over the event streams are not feasible in the middleware-level approach and not even supported by current STDBs.

3 STEPQ System: Vision and Challenges

Given the above limitations, it is clear that engineering existing systems to handle STPQs is not the right approach. In the following we envision the architecture of the proposed STEPQ system and the involved challenges. The system consists of two standard layers; compilation/optimization and execution layers as illustrated in Figures 1(c). In the compilation/optimization layer, the *pattern-query compiler & optimizer* (*PQ-CompilerOptimizer*) component, which is the central component of the system, is responsible for compiling and optimizing the entire query. Given a spatio-temporal pattern query, *PQ-CompilerOptimizer* decomposes it into one or more traditional queries (the *base* queries) and pattern-matching queries. The individual base queries are compiled and optimized using an extended *spatio-temporal compiler & optimizer (ST-CompilerOptimizer)* that works under the control of the *PQ-CompilerOptimizer*. In contrast, pattern-matching queries are fully compiled by *PQ-CompilerOptimizer*. The base queries will be executed by the extended spatio-temporal execution engine (*ST-ExecutionEngine*), while the pattern-matching queries will be executed by the *pattern-matching execution engine (PM-ExecutionEngine)*. The continuously generated results from the base queries will drive the progress of the pattern-matching queries. The key characteristics and challenges in STEPQ are (More details and examples can be found in [6]):

- **Leveraging & Extending State-of-Art in STDBs:** It is crucial to leverage the existing technology in STDBs. This is achieved by the *ST-CompilerOptimizer* and *ST-ExecutionEngine* components that retain all the innovations in STDBs such as continues and incremental evaluation, spatial-aware operators and access methods, and scalable execution. Moreover, base queries will be subject to new optimizations triggered by *PQ-CompilerOptimizer*.

- **Coherent Integration between Spatio-Temporal and Pattern-Matching Techniques:** This is achieved by having a single system with interacting components orchestrated by the *PQ-CompilerOptimizer*. Such integration allows the sharing of base queries across multiple pattern-matching queries, activating/suppressing base queries when needed, and seamless flow between the generated streams from the base queries to the pattern-matching queries.

- **Cross-Cutting Optimizations:** This is achieved by the *PQ-CompilerOptimizer* component that enables *PM-ExecutionEngine* to provide feedback information to *ST-ExecutionEngine* to control the execution of the base queries depending on the progress of the pattern queries. Cross-cutting optimizations require new communication mechanisms (and feedback loop) between the pattern-matching and base queries to control what events to generate and when.

- **Synchronized Query Processing:** STPQs may require not only executing multiple base queries, but also synchronizing their execution and jointly processing their results. Hence, new execution plans and synchronization strategies need to be integrated in the evaluation of both spatio-temporal and pattern-matching queries.

- **Event Model and Query Language:** New—possibly extensible—query languages and event models are needed to meet the diverse requirements of STPQs. For example, new concepts such as *event sets* need to be introduced to provide logical grouping of events produced from the base queries. The significance of event sets is two-fold. First, each answer set produced from a base query can be pipelined and processed as one unit, and hence further operations, e.g., synchronization and transformation, can be applied on the event sets. Second, event sets provide an efficient mechanism for anticipating when events should occur in the future, and hence they enable continuity/persistency operations.

References

1. Adaikkalavan, R., Chakravarthy, S.: SnoopIB: Interval-based event specification and detection for active databases. TKDE 59(1), 139–165 (2006)
2. Behr, T., Guting, R.H.: Fuzzy Spatial Objects: An Algebra Implementation in SEC-ONDO. In: Proceedings of the International Conference on Data Engineering, ICDE, pp. 1137–1139 (2005)
3. Benetis, R., Jensen, C.S., Karciauskas, G., Saltenis, S.: Nearest Neighbor and Reverse Nearest Neighbor Queries for Moving Objects. In: Proceedings of the International Database Engineering and Applications Symposium, IDEAS, pp. 44–53 (2002)
4. Cai, Y., Hua, K.A., Cao, G.: Processing Range-Monitoring Queries on Heterogeneous Mobile Objects. In: Proceedings of the International Conference on Mobile Data Management, MDM (2004)
5. Demers, A., Gehrke, J., Panda, B.: Cayuga: A general purpose event monitoring system. In: CIDR, pp. 412–422 (2007)
6. Eltabakh, M.: STEPQ: Extensible Spatio-Temporal Engine for Complex Pattern Queries. Technical Report WPI-CS-TR-13-02
7. Gedik, B., Liu, L.: MobiEyes: Distributed Processing of Continuously Moving Queries on Moving Objects in a Mobile System. In: Bertino, E., Christodoulakis, S., Plexousakis, D., Christophides, V., Koubarakis, M., Böhm, K. (eds.) EDBT 2004. LNCS, vol. 2992, pp. 67–87. Springer, Heidelberg (2004)
8. Wu, E., Diao, Y., Rizvi, S.: High-performance complex event processing over streams. In: Proceedings of the ACM SIGMOD International Conference on Management of Data, pp. 407–418 (2006)

Cost Models for Nearest Neighbor Query Processing over Existentially Uncertain Spatial Data

Elias Frentzos[1], Nikos Pelekis[2], Nikos Giatrakos[3,*], and Yannis Theodoridis[1]

[1] Department of Informatics, University of Piraeus, Piraeus, Greece
[2] Department of Statistics & Insurance Science, University of Piraeus, Piraeus, Greece
`{efrentzo,npelekis,ytheod}@unipi.gr`
[3] Dept. of Electronics & Computer Engineering, Technical University of Crete, Crete, Greece
`ngiatrakos@softnet.tuc.gr`

Abstract. A major challenge posed by real-world applications involving spatial information deals with the uncertainty inherent in the data. One type of uncertainty in spatial objects may come from their existence, which is expressed by a probability accompanying the spatial value of an object reflecting the confidence of the object's existence. A challenging query type over existentially uncertain data is the search of the Nearest Neighbour (NN), as the likelihood of an object to be the NN of the query object does not only depend on its distances from other objects, but also from their existence. In this paper, we present exact and approximate statistical methodologies for supporting cost models for Probabilistic Thresholding NN (PTNN) queries that deal with arbitrarily distributed data points and existential uncertainty, with the aid of appropriate novel histograms, sampling and statistical approximations. Our cost model can be also modified in order to support Probabilistic Ranking NN (PRNN) queries with the aid of sampling. The accuracy of our approaches is exhibited through extensive experimentation on synthetic and real datasets.

Keywords: Spatial Databases, Existential Uncertain Data, Nearest Neighbor Query Processing.

1 Introduction

In the literature, two types of uncertainty have gained the interest of the research community, namely the *locational* and the *existential* uncertainty. *Locationally* uncertain are the objects that do exist but their location is uncertain. This kind of uncertainty is described by a probability density function. On the other hand, *existentially* uncertain objects are those that their uncertainty emanates from their existence, and this is expressed by a probability E_x accompanying the spatial value of an object x reflecting the confidence of x's existence. As a motivating example, consider the case where an image processing tool extracts some interesting formations of pixels that may or may not correspond to a predefined type of objects due to low image resolution. Another example involves semantically-enriched representations of

---

* Work done during the author's PhD studies at the Dept. of Informatics, University of Piraeus.

M.A. Nascimento et al. (Eds.): SSTD 2013, LNCS 8098, pp. 391–409, 2013.

trajectories of moving objects [8], where a point of interest may be part of a semantic trajectory of a user if the latter has been predicted to perform an activity at that place. Existential uncertainty is also natural in the case of fuzzy classification [3], [13].

The related work on existentially uncertain data [3], [13] focuses on two probabilistic versions of several spatial queries. A *thresholding* query returns the objects that satisfy some spatial condition with probability more than a given threshold *t*, while a *ranking* query returns the objects that satisfy a spatial condition in order of their confidence. Dai et al. [3] proposed search algorithms for the above two types of spatial range and NN queries, where the existentially uncertain data are indexed by 2-dimensional R-trees [7] or appropriate augmented variants of them. In [13] authors also present appropriate algorithms for *Spatial Skyline* [9], and *Reverse Nearest Neighbor* [10] queries, based on the idea of incremental NN search.

In this paper, we focus on the *probabilistic thresholding* (PTNN) and *probabilistic ranking nearest neighbor* (PRNN) queries on existentially uncertain data. In a nutshell, a PTNN query seeks for spatial objects whose probability of being the NN of a query object exceeds a given threshold *t*, while a PRNN query returns only the *m* most probable NNs. The motivation is that, this type of query presents a quite involved search complexity, as the probability of an object to be the NN depends not only on the location, but also on the existential probability of other objects. Moreover, compared to the other operators presented in [13], they are more popular with broader applicability. In [4] we utilized a statistical model in order to estimate the number *f* of NNs that are to be retrieved from the database so as to be at least *CI* % confident (i.e. *CI* is a user-defined confidence, e.g. 99%) that the PTNN search will end without the need to retrieve $n > f$ NNs. The concept is to provide efficient search algorithms, with predetermined cost, and with custom defined certainty (as high as required) of resolution. On the other hand, this is a case which can be only applied to uniform data and existential uncertainty distribution.

We are aware that PTNN2D and PRNN2D are overwhelmed in terms of efficient query processing by the other schemes proposed in [3], which employ augmented versions of R-trees and 3D R-trees. However, experience has shown that it is very difficult for commercial Spatial Database Management Systems (SDBMS) to support novel proposals, especially when they require altering the data structures used on their engines. Then again, PTNN2D and PRNN2D while not optimal, they can be directly employed with conventional 2D R-trees already implemented in commercial SDBMS. Moreover, the analysis provided in this paper can be easily modified in order to provide similar results that support all schemes of [3].

Outlining the major issues addressed in this paper, our main contributions are:

- Following the assumption of uniformity regarding the existential uncertainty distribution, we present an exact statistical-based analysis for the determination of the *discrete distribution probability density function (dpdf)*, that a PTNN query terminates after having retrieved exactly n objects; exploiting this analysis, we present a cost model for the forecasting of the number of disk page accesses required to process a PTNN query, given that the dataset is indexed by R-trees [7], as well as it is uniformly distributed in the data space. We further exploit well-known properties of distribution expected values in order to provide an approximate model for PTNN and PRNN queries.

- We show how to utilize histograms in order to relax the assumption of uniformly distributed data points and existential uncertainty and provide an efficient cost model that predicts the number of disk page accesses required to process PTNN, over arbitrarily distributed data and existential uncertainty. We also utilize random sampling so as to achieve better forecasts, as well as, overpass the problem that is faced regarding an analytical PRNN cost model calculation. Specifically, we alternately apply the results of our statistical analysis and the sampling method, over augmented versions of well-known histograms [1], together with the approach of [11].
- Finally, we report the results of a comprehensive set of experiments, which demonstrates the correctness and accuracy of our analysis.

To the best of our knowledge, our work is the first on these topics. The rest of the paper is structured as follows: Section 2 overviews the background work. Section 3 describes the statistical analysis of PTNN queries based on the assumption of uniformly distributed data and existential uncertainty. In Section 4, we present the details of an efficient cost model for PTNN and PRNN queries that supports arbitrary distributions regarding the problem parameters, Section 5 evaluates the accuracy of our model through an extensive experimental study over several datasets, while, Section 6 provides conclusions and interesting research directions.

2   Background

2.1   Probabilistic NN Search over Spatial Data with Existential Uncertainty

Formally, a PTNN query takes as input a query object q and a probability threshold t, while the data are represented as tuples of the form (x, E_x). As proposed by Dai et al. [3], the 2-dimensional PTNN (PTNN2D) algorithm, illustrated in Figure 1, iteratively retrieves spatially nearest objects in a Best-First (BF) mode [6], and terminates only after the value of P^{first} becomes smaller than the given threshold t. The PTNN2D algorithm iteratively calculates the value of P^{first}, which is the probability that no object retrieved before the current object x is the actual NN, according to [3]:

$$P_x^{first} = \prod_{i=1}^{n-1}(1-E_i),$$ (1)

where n-1 is the number of objects that are closer to the query object than the current object x, i.e., the number of objects retrieved from the BF algorithm before x, and E_i their existential uncertainty. Then, the probability that x is the actual NN, is [3]:

$$P_x = E_x \cdot P_x^{first}$$ (2)

The intuition behind the PTNN2D algorithm is that once $P^{first} < t$, we are sure that the subsequent nearest objects, even if they exist with 100% probability, they cannot be the NN of q, so the algorithm can safely terminate. Note also that PTNN2D algorithm utilizes R-tree indexes so as to incrementally retrieve the k-th NN; as such, the R-tree can be replaced by other access method supporting incremental NN search.

```
        Algorithm PTNN2D(q, 2D R-tree on S, t)
1.          P^first=1;
2.          While P^first ≥ t and more objects in S do
3.              x:=next NN of q in S (use BF [7]);
4.              P_x:= P^first.E_x;
5.              If P_x ≥ t then output (x, P_x);
6.              P^first= P^first.(1-E_x);
```

Fig. 1. Probabilistic NN on a 2D R-tree with thresholding

```
        Algorithm PRNN2D(q, 2D R-tree on S, m)
1.          P^first:=1;
2.          H = ∅; /*Heap of m objects with highest P_x*/
3.          P^m:=0; /* P_x of m-th object in H*/
4.          While P^first ≥ P^m and more objects in S do
5.              x:=next NN of q in S (use BF [7]);
6.              P_x:= P^first · E_x;
7.              If P_x ≥ P^m then
8.                  Update H to include x;
9.                  P^m:= m-th probability in H;
10.             P^first:= P^first · (1-E_x);
```

Fig. 2. Probabilistic NN on a 2D R-tree with ranking

Similarly, a ranking spatial query returns the objects which qualify the spatial predicates of the query, in order of their confidence. Ranking queries can be also thresholded by a parameter m returning thus the m most confident objects. Therefore, a *probability ranking NN* (PRNN) query takes as input a query object q and the number of objects required with the highest probability, over data of the same form, i.e., (x, E_x). Dai et al. [3], also propose the 2-dimensional PRNN (PRNN2D) algorithm, illustrated in Figure 2, which iteratively retrieves spatially nearest objects x in a Best-First (BF) mode iteratively calculating P_x and P^{first} using Eq.1 and Eq.2 respectively. The difference here is that the output is a heap H containing the m most probable NN objects. Therefore the threshold used to terminate is based on P^m which is the P_x of the m-th object in the heap H and the algorithm terminates only after the value of P^{first} becomes smaller than P^m.

2.2   Cost Models for NN Search over Conventional Spatial Data

Tao et al. [11] present an efficient cost model for the optimization of NN queries in low and medium-dimensional spaces. They provide a closed formula for the estimation of (a) the average nearest distance D_k from the query point q to its k-th NN and (b) the number of tree nodes whose MBRs intersect the vicinity circle $\Theta(q, D_k)$ with center q and radius D_k, which is equal with the average number of node accesses $NA(k)$ required by an R-tree to retrieve the k-th NN. Specifically, the analysis of [6] shows that the average nearest distance D_k is estimated by:

$$D_k \approx \frac{2}{C_V}\left[1-\sqrt{1-\left(k/N\right)^{1/d}}\right]$$

(3)

$$C_v = \frac{\sqrt{\pi}}{\left[\Gamma(d/2+1)\right]^{1/d}} \tag{4}$$

where d is the dimensionality and N denotes the cardinality.

These formulas work only with uniformly distributed data in the search space. On the other hand, real-world data employ arbitrary distributions; as such Tao et al. [11], provide an extension of the model presented above, by using MinSkew histograms.

Specifically, the *MinSkew* technique proposed by Acharya et al. [1], is a binary space partitioning (BSP) technique employing the *spatial skew* definition provided in [1]. Each MinSkew Histogram HS can be seen as a set of spatial disjoint buckets B_i that cover the whole data space: $HS = \left\{ B_i : \bigcup(B_i) = S \wedge \bigcap(B_i) = \varnothing \right\}$ and $B_i = \left\{ \left[x_{i,L}, x_{i,U} \right], \left[y_{i,L}, y_{i,U} \right] \right\}$. The main advantage of this technique is that the area grouped together within the same bucket has small spatial skew, i.e., objects are almost uniformly distributed inside it; as a result, it is usually assumed that the data distribution inside each bucket B_i is uniform.

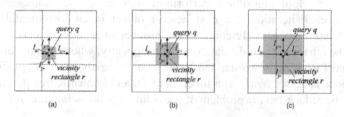

Fig. 3. Estimating the "radius" of the vicinity rectangle L_r [11]

[11] provides an algorithm that works over an input histogram HS and a query point q. The algorithm employs the notion of the *vicinity rectangle* that approximates the *vicinity circle* so as to minimize the number of complex (vicinity) circle-(histogram) rectangle intersection discoveries, and reduces them to less expensive rectangle – rectangle inspections. The algorithm initially determines the distances that q needs to travel along each dimension so as to reach the boundaries of each histogram bucket (cf. Figure 3), and stores them in a heap. Then, utilizing the histogram, the algorithm iterates by computing the expected number of points En found inside the vicinity rectangle formed by the next distance in the heap; if En is smaller than k, i.e., the number of requested nearest neighbors, the appropriate vicinity radius is calculated (reduced) according to the formula:

$$L_r \approx \left(\frac{L_{old}^{d}\left(k - En\right) - L^d\left(k - En_{old}\right)}{En_{old} - En} \right)^{1/d} \tag{5}$$

where L is the "diameter" (i.e., side length) of the current vicinity rectangle, while L_{old} and En_{old} are the respective diameter and expected number of points found inside the vicinity rectangle in the previous iteration, respectively. In the case En is smaller than

k, the algorithm proceeds with the next distance in the heap until En becomes greater than k. Finally, D_k is obtained by $D_k = L_r/C_v$.

After obtaining D_k the cost model developed for uniform data is applied. Specifically, the query cost in terms of node accesses $NA(k)$ is provided by the following equation:

$$NA(k) = \sum_{i=0}^{\log_f \frac{N}{f}} \left[\frac{N}{f^{i+1}} \cdot \left(\frac{L_i - (L_i/2 + s_i/2)^2}{1 - s_i} \right)^d \right]$$ (6)

where N is the cardinality of the dataset, f is the average node fanout, s_i the extent of a level-i node and L_i calculated as a function of D_k and the respective s_i. We have also to note that N is determined based on the *local density* provided by the histogram in the area "near" the query point. The interested reader is cited to [11] for more details.

In our approach, we make use of the techniques proposed in [11], so as to estimate the radius of the *vicinity circle* D_k required to be browsed in order to process PTNN and PRNN queries. Specifically, both PTNN2D and PRNN2D browse the database according to the distance of the query to the dataset objects until a probabilistic criterion is met. Both algorithms perform a number of iterations, continuously requesting, in each iteration, the next nearest object in an incremental way. The number of iterations is actually equal to the number of nearest objects to the query that have to be retrieved from the database. Consequently, when utilizing an R-tree, as PTNN2D and PRNN2D suggests, and given that the analysis of [11] estimates the number of node accesses $NA(k)$ as a function of D_k and known R-tree parameters, our problem can be reduced to the problem of providing a good estimation of D_k.

Table 1. Table of notations

Notation	Description
x, E_x	A data point and its existential probability
S	A dataset of tuples (x, E_x)
q, t, m	The query object, threshold probability of a PTNN query and number of requested objects of a PRNN query
P^{first}	The probability that no object retrieved before the current object x is the actual NN
P_n^{first}	The probability that no object retrieved before the n-th iteration is the actual NN
P_x	The probability that an object x is the actual NN
$P_{exact}(n)$	The probability that the PTNN algorithm terminates after having retrieved exactly n objects
H	A heap used in the PRNN algorithm
P^m	the P_x of the m-th object in the heap H
$EV(u)$	The expected (average) value of a given variable u
D_k	The nearest distance from the query point q to its k-th NN

3   Statistical Analysis of PTNN Queries

In this section, aiming at a statistical analysis of probabilistic thresholding NN queries, we initially calculate the *expected number* of iterations $EV(n)$ needed for the PTNN2D algorithm to terminate, and then we make use of existing work on cost models so as to determine the average number of node accesses $NA(EV(n))$ needed in order to process such queries over conventional R-trees. In the sequel, due to the difficulty of extending the exact solution to support such queries over arbitrary distributed data, we present an approximate solution regarding PTNN queries. We close the section by discussing the extension of this model in the case of PRNN queries. In this first approach, we make two assumptions regarding the data distribution:

- *data uniformity assumption*: points x_i are uniformly distributed in the data space,
- *uncertainty uniformity assumption*: the existential uncertainty E_x of all objects in S is uniformly distributed inside the unit interval $[0,1]$.

Both assumptions are relaxed in the subsequent section where an efficient cost model is presented. Table 1 summarizes the notation used in the rest of the paper.

3.1   Exact Statistical Analysis of PTNN Queries

To start with, we provide a lemma from which a cost model for PTNN queries is straightforwardly devised in the case of uniformly distributed data and existential uncertainty. More specifically, the first step towards a cost model for the PTNN2D Algorithm 3, is to determine the *dpdf* that the algorithm terminates after exactly n iterations, i.e., the distribution of the number of objects retrieved before P^{first} becomes less than the given threshold t. Formally, we provide the following lemma:

Lemma 1: *The dpdf that the PTNN2D algorithm terminates after exactly n iterations, under the uncertainty uniformity assumption, is given by the following formula:*

$$P_{exact}(n) = \frac{(-1)^{n-1} t \ln(t)^{n-1}}{(n-1)!} \qquad (7)$$

where t is the algorithm threshold.

Proof: Our goal is to determine the *dpdf* $P_{exact}(n)$, such that, the algorithm terminates after having retrieved exactly n objects. For this we distinguish between two cases, namely $n = 1$ and $n > 1$. In the first case, the algorithm terminates with a single iteration iff the value of $P_2^{first} = \prod_{i=1}^{1}(1-E_i) = (1-E_1)$ calculated at the end of the first iteration (i.e., line 7 in Figure 1) is less than the given threshold t. Thus, from the uncertainty uniformity assumption, it holds that $P_{exact}(1) = P(1-E_1 \le t) = P(E_1 \ge 1-t) = t$. Given that $-1^0 = (\ln(t))^0 = 0! = 1$ we have proved Lemma 1 in the case where $n = 1$.

In the second case, i.e., $n > 1$, the algorithm terminates iff P_{n+1}^{first}, which is calculated at the end of the nth iteration (i.e., line 7 in Figure 1), becomes less than t after *exactly* n iterations. In other words, we must first determine the conditional probability that P^{first} becomes less than t after n iterations, given also that it must not terminate before reaching n iterations:

$$P_{cond}(n) = P\left(\prod_{i=1}^{n}(1-E_i) \le t \mid \prod_{i=1}^{m}(1-E_i) > t, \forall m \le n-1\right) \tag{8}$$

Then, the total probability that the algorithm terminates after having retrieved exactly n objects can be obtained by multiplying P_{cond} with the probability that the algorithm has not terminated until reaching n iterations:

$$P_{exact}(n) = P_{cond}(n) \cdot P\left(\prod_{i=1}^{m}(1-E_i) > t, \forall m \le n-1\right) \tag{9}$$

Moreover, since $0 \le E_1 \le 1 \Leftrightarrow 0 \le 1 - E_1 \le 1$, it also holds that $(1-E_1) \ge .. \ge \prod_{i=1}^{n-2}(1-E_i) \ge \prod_{i=1}^{n-1}(1-E_i)$. Therefore, given that $\prod_{i=1}^{n-1}(1-E_i) > t$, it stands that $\prod_{i=1}^{m}(1-E_i) > t, \forall m \le n-2$; then, (8) and (9) can be rewritten as follows:

$$P_{cond}(n) = P\left(\prod_{i=1}^{n}(1-E_i) \le t \mid \prod_{i=1}^{n-1}(1-E_i) > t\right) \tag{10}$$

$$P_{exact}(n) = P_{cond}(n) \cdot P\left(\prod_{i=1}^{n-1}(1-E_i) > t\right) \tag{11}$$

Since the values of E_x follow the uniform distribution, the same also stands for $1-E_x$; as such the product of the n uniformly distributed values of $1-E_x$ should follow the *uniform product distribution*, i.e., the distribution of the product of n uniformly distributed uncorrelated variables $x_1, x_2, .. x_n$, with *pdf* given by [12]:

$$P_{x_1 x_2 .. x_n}(u) = \frac{(-1)^{n-1}}{(n-1)!}(\ln u)^{n-1} \tag{12}$$

where u is the product $\prod x_i$.

In our case, we first set as u the product $\prod_{i=1}^{n-1}(1-E_i)$, and then determine the amount of objects $X \in S$, such that $\prod_{i=1}^{n}(1-E_i) = (1-E_n) \cdot u \le t$ which leads to:

$$(1-E_n) \le t/u \tag{13}$$

Given that $(1-E_n)$ is also uniformly distributed, it should hold that the amount of objects fulfilling the above expression V_n is

$$V_n = t/u \tag{14}$$

Known the above, we can calculate the probability $P_{cond}(n)$ by summing up (i.e., integrating) the amount of objects V_n for each value of u, weighted by the value of the distribution of u, and divided by the respective sum (i.e., integral) of the distribution of u. Moreover, since it is known that $u = \prod_{i=1}^{n-1}(1-E_i) \geq t$, the above integrals should involve only the values of u between t and 1. Summarizing:

$$P_{cond}(n) = \frac{\int_t^1 \frac{t}{u} P_{n-1}(u)\,du}{\int_t^1 P_{n-1}(u)\,du} \tag{15}$$

Moreover, the total probability that the algorithm has not been terminated until reaching n iterations (i.e, $\prod_{i=1}^{n-1}(1-E_i) > t$), can be easily calculated, using the pdf of the product of $n-1$ uniformly distributed variables:

$$P\left(\prod_{i=1}^{n-1}(1-E_i) > t\right) = \int_t^1 P_{n-1}(u)\,du \tag{16}$$

Finally, by substituting (15) and (16) into (11) and performing the necessary calculations, we have proved Lemma 1 in the case where $n > 1$ ∎

Lemma 1 provides us with the dpdf that the algorithm terminates after exactly n iterations. The dpdf expressed by (7) is a closed formula, since it involves only the logarithm of the threshold t and the factorial of n. Obviously, the density of the probability obtained from (7) for several values of n, is dominated by the factorial of $n-1$; as such, it is expected that as the number of iterations grows, the respective probability density will tend to zero very fast. In the sequel we present a corollary derived from Lemma 1, which helps us to determine the cost model for PTNN queries over existentially uncertain data that follow the uncertainty uniformity assumption.

Corollary 1: *The expected number of iterations in the execution of the PTNN2D algorithm, under the uncertainty uniformity assumption, is:*

$$EV(n) = 1 - \ln(t) \tag{17}$$

Proof: The expected number of iterations needed from the PTNN2D algorithm to terminate is actually the mean value of (7) for each $n \in \mathbb{N}$. As such, $EV(n)$ can be calculated by averaging the dpdf $P_{cond}(n)$ over all possible values of n.

$$EV(n) = \sum_{i=1}^{\infty} \frac{(-1)^{i-1} t \ln(t)^{i-1}}{(i-1)!} \cdot i \tag{18}$$

Equation (18) cannot be straightforwardly evaluated since it involves infinity; however, we may calculate its limit:

$$\sum_{i=1}^{\infty} \frac{(-1)^{i-1} t \ln(t)^{i-1}}{(i-1)!} \cdot i = \lim_{n \to \infty} \sum_{i=1}^{n} \frac{(-1)^{i-1} t \ln(t)^{i-1}}{(i-1)!} \cdot i \tag{19}$$

which after the necessary calculations turns into:

$$\sum_{i=1}^{\infty} \frac{(-1)^{i-1} t \ln(t)^{i-1}}{(i-1)!} \cdot i = 1 - \ln(t) \tag{20}$$

Finally, by substituting (20) into (18) we have proved Corollary 1   ■

Obviously, the expected number of iterations $EV(n)$ needed from the PTNN2D in order to terminate, is equal with the number of NNs needed to be retrieved from an existentially uncertain spatial database queried with a query point and a given threshold t. Thus, we may employ the analysis presented in [11], so as to estimate the average radius D_k on which the $EV(n)$-th NN will be found, under the data uniformity assumption. Apparently, this model can be applied in our case where the dimensionality d is 2 and the value of $\Gamma(d/2+1)$ is $\Gamma(2/2+1)=1$; then, by substituting the expected number of n produced by (17) into the number of k NNs requested, (3) can be rewritten as follows:

$$D_k \approx \frac{2}{\sqrt{\pi}} \left[1 - \sqrt{1 - \sqrt{\frac{1 - \ln(t)}{N}}} \right] \tag{21}$$

From this point on, the analysis of [6] that estimates the number of node accesses $NA(EV(n))$ in the case of uniform data distribution (which is identical with our *data uniformity assumption*) remains unaffected; the single modification to be made is to calculate D_k using (21) instead of (3), and then apply Eq.(6) accordingly. Concluding, the cost model for PTNN queries over existentially uncertain data that follow both the uncertainty uniformity and the data uniformity assumptions is based on (21), which estimates the distance from the query point that has to be browsed from the database so as to answer such a query; then, the required node accesses $NA(EV(n))$ can be straightforwardly estimated by replacing the D_k into the analysis of [11].

3.2   Approximate Statistical Analysis of PTNN Queries

Unfortunately, the extension of the above-described theoretical model in the case of arbitrarily distributed data is not straightforward at all. Histograms widely used in order to provide statistical estimations in DBMS, pose insuperable problems to this extension due to their discrete nature. Specifically, given the simplest case where a

1-dimensional histogram $HS = \left\{ \left[0, E_1 \right], \left[E_1, E_2 \right], ..., \left[E_{m-1}, 1 \right] \right\}$ is used to describe the existential uncertainty distribution in a given point in space, the distribution of the exact number of iterations following the methodology of Lemma 1 would be given as a function defined in m^n parts. This is due to the fact that the resulted distribution would be calculated as the product of n sets containing m spaces each. Obviously, such an approach is not practical. On the other hand, we may provide an approximate solution which utilizes the notion of the expected value of the probability of a random object retrieved in the n-th iteration to be included in the query results. More formally, we provide the following Lemma 2.

Lemma 2: *The number of iterations n required for the expected value P^{first} of the PTNN2D algorithm to become equal to the threshold t, is given by:*

$$EV\left(P_n^{first} \right) = t \Rightarrow n = 1 + Log_{1-EV(E_x)}(t) \tag{22}$$

where $EV(E_x)$ is the expected value of existential uncertainty E_x of a random x in S.

Proof: Our main objective is to express $EV(P^{first})$ as a function of known values. Towards this goal, we know that the expected (mean) value of a random variable produced as the product of two other random variables, is equal to the product of the expected value of the two variables. Formally, given two random variables u and v the following stands:

$$EV(u \cdot v) = EV(u) \cdot EV(v) \tag{23}$$

From the definition of P^{first} (1) and from (23), we have that the expected value $EV(P^{first})$ after n iterations is:

$$EV\left(P_n^{first} \right) = \left(1 - EV\left(E_x \right) \right)^{n-1} \tag{24}$$

Now, in order for (24) to become equal to t we have:

$$\left(1 - EV\left(E_x \right) \right)^{n-1} = t \Rightarrow n - 1 = Log_{1-EV(E_x)}(t)$$

which proves Lemma 2.   ∎

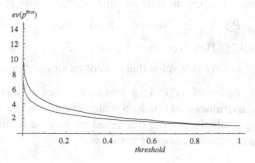

Fig. 4. Estimating the number of iterations of PTNN2D over uniform data by exact (solid line) and approximate solutions (doted line)

It is clear that Lemma 2 can be utilized in order to provide an approximate estimation for the number of iterations needed by the PTNN2D algorithm in order to terminate. It certainly does not provide the exact value of $EV(n)$ as Corollary 1 does, however, it provides strong evidence that the algorithm may terminate when n becomes greater than the value provided. What is more, Lemma 2 does not utilize the uncertainty uniformity assumption; as such it can be applied over data with arbitrary distributed existential uncertainty, relaxing therefore our uncertainty uniformity assumption. Also interestingly when employing Lemma 2 under the uncertainty uniformity assumption, where $EV(E_x)=0.5$, (22) results in $n = 1 + Log_{0.5}(t)$. A comparison between this result and (17) is given in Figure 4. It is clear that the approximate solution always overestimates n, with its difference from the exact solution increasing when the value of t becomes less than 0.2 (and the number of iterations increases above 3).

3.3   Discussion on PRNN Queries

One may suggest that Lemma 1 and its Corollary 1 can be easily extended to cover the case of PRNN queries, processed by the PRNN2D algorithm, since their main difference is on their termination condition, i.e., the continuously evolving value of P^m employed instead of the constant value of t. Towards this goal, we could utilize the fact that the expected (mean) value of a random variable produced as the product of two independent random variables, is equal with the product of the expected value of the two variables. However, the calculation of a theoretical value for P^m is a very hard task which involves order statistics [2]. Specifically, even in the – simplest – case of $m=2$, the expected value of the m – th P_x inside H, is determined by distinguishing between two cases regarding the order of values in H, i.e., $\{P_1, P_2\}$, $\{P_2, P_1\}$:

- In the case where $E_1 \geq 0.5$, since $P_1^{first} = 1$, it follows that $P_1 \geq 0.5$, $P_2^{first} = (1 - E_1) < 0.5$ and $P_2 < 0.5$. Therefore the order of P_i inside H will be $\{P_1, P_2\}$. Now, given from the uncertainty uniformity assumption that $EV(P_1)=EV(E_1)=0.5$ and $EV(P_2)=EV\left(P_2^{first}E_2\right) = EV\left(P_2^{first}\right)EV(E_2) = 0.25 \cdot 0.5 = 0.125$, $H=\{0.5, 0.125\}$ and $EV(P^m)=0.125$.

- In the case where $E_1 < 0.5$, it follows that $P_1 < 0.5$ and $P_2^{first} = (1 - E_1) > 0.5$; therefore for $P_2 > P_1$ it should hold that $E_2 > E_1/(1 - E_1)$, and $EV(P^m)=EV(P_1)=0.25$. However, in the case of $E_2 < E_1/(1 - E_1)$ it follows that $P_1 > P_2$, and $EV(P^m)=EV(P_2)$, a value that cannot be straightforwardly calculated.

It is clear that the calculation of $EV(P^m)$ for arbitrary values of m is a very demanding task. However, the usefulness of such a calculation can be argued, since by approximate sampling methods as those described in the next section, we may obtain good estimates of the expected number of iterations.

4   A Cost Model for PTNN and PRNN Queries

In the exact analysis of Section 3, we assumed that both data points and their existential uncertainty are uniformly distributed in their space. In this section, we relax both assumptions and apply our approach to arbitrarily distributed data with the employment of augmented *histograms*. The presence of the histogram is to provide (a) local estimations regarding the density of existentially uncertain spatial objects in the neighborhood of the query point, and, (b) statistics that can be used in order to estimate the number of iterations needed from the PTNN2D algorithm to terminate.

4.1   Augmented Histograms

The proposal of Acharya et al. [1], may be easily extended in order to support our scenario of existentially uncertain spatial objects, by augmenting it in a third dimension describing the existential uncertainty. Formally, the proposed histogram is

$$HS = \left\{ B_i : \cup (B_i) = S \times [0,1] \wedge \cap (B_i) = \varnothing \right\} \quad \text{and} \quad B_i = \left\{ \left[x_{i,L}, x_{i,U} \right], \left[y_{i,L}, y_{i,U} \right], \left[E_{i,L}, E_{i,U} \right] \right\},$$

and the data distribution inside each 3D bucket B_i is considered as uniform. The histogram is created using the methodology of [1] by simply treating the existential uncertainty dimension as an additional spatial dimension.

4.2   A Sampling-Based Approximation Method

The above proposed histogram, besides its conventional use, i.e., to estimate the local density of data, it can be used in order to produce a 1D histogram of the data points' existential uncertainty distribution in the area "near" the query point. Subsequently, random values of existential uncertainty can be produced following the local distribution provided by the 1D histogram, and then, used to simulate the behavior of the PTNN2D algorithm. The basic dilemma that is posed towards a good estimation following such a technique is to provide an efficient termination condition for the sampling process. This condition can be provided by computing the standard deviation of the sampled mean value:

$$\sigma_{mean} = \frac{\sigma}{\sqrt{N}} \tag{25}$$

where σ is the sample standard deviation and N the sample size. Then, by using the hypothesis that n follows the normal distribution, and a confidence interval $CI=95\%$, the expected number of iterations $EV(n)$ is:

$$\bar{n} - 1.96 \frac{\sigma}{\sqrt{N}} \le EV(n) \le \bar{n} + 1.96 \frac{\sigma}{\sqrt{N}} \tag{26}$$

where N is the number of observations (number of PTNN2D simulation runs), σ_n the (computed so far) standard deviation of n and 1.96 is the approximate value of the 97.5 percentile point of the normal distribution, used in the construction of approximate 95% confidence interval.

Figure 5 illustrates the algorithm *SamplePTNN2D* which summarizes the proposed methodology regarding the estimation of the number of iterations of the PTNN2D algorithm using sampling. The algorithm utilizes a 1D histogram *HS* describing the existential uncertainty distribution in the local query area, the algorithm's threshold t, and the precision p (e.g., 5%) of the expected value of n. The precision is used instead of an absolute value of standard deviation in order to compute it as a percentage of the calculated mean value. The algorithm begins by instantiating *km*, i.e. the calculated mean of the number of iterations needed by the PTNN2D algorithm to terminate, and *kt*, which is the standard deviation of the calculated mean. After that (lines 4-6), the algorithm instantiates P^{first}, N (i.e. the number of PTNN2D simulations) and n (i.e., the number of iterations of the PTNN2D algorithm in the current run). In lines 7-11, the PTNN2D algorithm is simulated and the number of iterations n in its current run is determined. The histogram *HS* is used in line 8 in order to produce random values based on the local area's existential uncertainty distribution. After simulation, the algorithm calculates the new mean value of the number of iterations required in every run, as well as the mean's standard deviation (line 13). The algorithm eventually terminates and returns the calculated mean value of iterations when there is 95% probability (which is included in the area $1.96 \times kt$) that the mean differs by at most p regarding its accurate value.

```
       Algorithm SamplePTNN2D(HS 1D Histogram, threshold t, precision p)
1.        km:=0; //calculated mean iterations
2.        kt:=+∞; // calculated stdev of mean iterations
3.     While p·km<1.96·kt do
4.        N:=N+1; //num of runs
5.        P^first:=1;
6.        n:=0; //run's iterations
7.        While P^first≥t do // simulate PTNN2D
8.           n:=n+1;
9.           E_x:=ProduceRandomValue(HS);
10.          P^first:= P^first·(1-E_x);
11.       End While;
12.       km:=Mean(n);
13.       kt:=Stdev(n)/Sqrt(N);
14.    End While;
15.    Return km;
```

Fig. 5. Sampling algorithm for estimating the number of iterations of PTNN2D

Interestingly, the method of sampling can be directly applied with limited only modifications in the case of PRNN queries. The respective *SamplePRNN2D* algorithm is illustrated in Figure 6. The algorithm utilizes the same ideas as *SamplePTNN2D* with the only difference that PRNN2D is simulated (instead of PTNN2D) between lines 9-17, with P^m calculated and eventually tested so as to be used as a termination condition. This observation enables us to introduce a cost model for PRNN queries as well, as described in the following section.

```
      Algorithm SamplePRNN2D(HS 1D Histogram,# objects m, precision p)
  1.      km:=0; //calculated mean iterations
  2.      kt:=+∞; // calculated stdev of mean iterations
  3.      While p·km<1.96·kt do
  4.          P^first:=1;
  5.          N:=N+1; //num of runs
  6.          H:=∅;
  7.          P^m:=+∞;
  8.          n:=0; //run's iterations
  9.          While P^first≥P^m do //simulate PRNN2D
 10.              n:=n+1;
 11.              E_x:=ProduceRandomValue(HS);
 12.              Px:=P^first·E_x;
 13.              If P_x≥ P^m then
 14.                  Update H to include x;
 15.                  P^m:= m-th probability in H;
 16.              P^first:= P^first·(1-E_x);
 17.          End While;
 18.          km:=Mean(k);
 19.          kt:=Stdev(k)/Sqrt(k);
 20.      End While;
 21.      Return km;
```

Fig. 6. Sampling algorithm for estimating the number of iterations of PRNN2D

4.3   An Effective Cost Model for PTNN and PRNN Queries

In this section we present an effective cost model for PTNN queries that works over arbitrarily distributed spatial data with existential uncertainty. The proposed cost model is calculated using the algorithm presented in Figure 7, which employs several ideas presented in [12]. In particular, algorithm *EstimateThresholdDk* takes as input a simple spatial histogram, an augmented histogram, a query point q and a threshold t, and estimates the radius D_k of the vicinity circle that has to be browsed by the PTNN2D algorithm. The radius D_k is then applied over Eq.(6) so as to estimate the number of node accesses NA that are needed in order to answer the query. The algorithm initially (lines 2-4) determines the critical vicinity rectangle "radiuses", i.e., the rectangle's half-side, on which the object's density changes. These radiuses

```
Algorithm EstimateThresholdDk(Histogram HS, Augmented Histogram AHS, point q, threshold t)
  1.  HP = new min-Heap
  2.  for each bucket B in HS;
  3.      Determine the radius that is needed for a rectangle
          with center q to reach B and add it to HP
  4.  end for
  5.  EnOld=:0; lOld:=0;
  6.  While true do // algorithm eventually terminates at line 13
  7.      l=:HP.pop;
  8.      En=:HS.Density(q,l)*(4*l*l);//calculate # objects inside rec
  9.      m=:AHS.MeanValue(q,l)
 10.      k=:Log(t)/Log(1-m)+1
 11.      If k<En then
 12.          Compute Lr by equation (5)
 13.          Return Lr/Sqrt(PI)
 14.      Else
 15.          lOld=l;EnOld=En;
 16.      End if
 17.  End while
```

Fig. 7. Algorithm *EstimatedThresholdDk* for computing D_k

are determined by simply calculating the distance that q needs to travel along each axis so as to reach each bucket's boundaries. After their calculation, these values are inserted into a min-heap so as to be used in incremental order.

Then, the algorithm iteratively retrieves candidate critical distances l on which the vicinity rectangle's density is changed (via the min heap), and calculates (line 8) the expected number of objects En found inside it, by simply multiplying the local density produced by HS by the area of the respective vicinity rectangle. It also determines in line 9 via the augmented histogram, the mean value m of the existential uncertainty of objects found inside the vicinity rectangle, using as input the query point q as well as the radius l of the vicinity rectangle. The value m is eventually used in line 10 to calculate the (approximated) number of nearest neighbors k that must be retrieved in order for the PTNN2D algorithm to terminate. Then, in line 11, the values of k and En are compared, in order to determine whether the number of required nearest neighbors k is less than the objects contained inside the (so far calculated) vicinity rectangle. If it is not so, the algorithm stores l in $lOld$ and En in $EnOld$ to be used by Eq.5 in a subsequent iteration, and performs another iteration, so as to use a greater critical radius l (which are stored in the minheap). Eventually, the algorithm terminates by computing Lr via Eq.(5), and returning Dk (lines 12-13) when the iteratively increasing radius of the vicinity rectangle, produces an approximate number of objects contained inside the respective vicinity rectangle, greater than k.

The previously presented algorithm provides a good approximation of the number of objects that have to be retrieved from the database in order for the PTNN2D to terminate. However, this can be also achieved via sampling, as described in the previous section. Specifically, lines 9-10 of the *EstimatedThresholdDk* can be replaced with (a) the calculation of a 1-dimensional histogram, *AHS*, and (b) algorithm *SamplePTNN* (cf. Figure 5) that estimated k based on a 1-dimensional histogram of existential uncertainty. Similarly, by replacing lines 9-10 with the calculation of the 1D histogram and the algorithm *SampleRTNN* used to estimate the number of iterations of PRNN2D, algorithm *EstimatedThresholdDk* may be also used as a cost model for PRNN search.

Summarizing, the proposed cost model based on the *EstimatedThresholdDk* algorithm, can be used for estimating the radius of the vicinity circle, used for both PTNN and PRNN queries.

5   Experimental Study

Our experimental study is based on real point datasets. In particular, as in [13], we used the San Francisco roads' dataset (SF) dataset. Due to the lack of a real spatial dataset with objects having existential probabilities, we generated probabilities for the objects, using the following methodology. As [13] suggests, we first generated K = 10 anchor points on the map in positions of high data density. These points model locations around which there is large certainty for the existence of data. For each point x of the dataset, we find the closest anchor and we assign an existential probability inversely proportional to its distance from it. Thus, the distribution of probabilities around the anchors is a Zipfian one. The probabilities are normalized w.r.t. the maximum probability.

We conducted our experiments on a Windows XP workstation with AMD Athlon II X4 640 3GHz processor CPU. All evaluated methods were implemented using the .NET framework. Two statistical measures were used so as to demonstrate the behavior of our model. The *average radius of the vicinity circle* $\overline{D}$, the *average estimated radius of the vicinity circle* $\overline{D_e}$, and the *average error in the estimation of the vicinity circle* $\overline{DS}$. Formally, these measures are defined as:

$$\overline{D} = \frac{1}{n}\sum_{i=1..n} D_i, \; \overline{D_e} = \frac{1}{n}\sum_{i=1..n} D_i^e, \text{ and, } \overline{DS} = \frac{1}{n}\sum_{i=1..n} \left| D_i - D_i^e \right|$$

where n is the number of executed queries, D_i the actual distance of the vicinity circle from the *i-th* query, and D_i^e the estimated radius of the vicinity circle via the respective cost model. We distinguish between, $\overline{D}$ and $\overline{DS}$, in order to disclose the details of the behavior of our model, as will be shown in the following experiments. In order to test the accuracy of the proposed model, we performed 500 PTNN queries in locations selected driven by the dataset density, under various threshold values and counted the actual number of iterations that the algorithm performed. We also compared the values gathered from the experiment with the one calculated using our model. The corresponding results are illustrated in Figure 8(a) and (b), regarding the PTNN2D algorithm with estimates gathered via Lemma 2 and sampling, respectively. It is clear that the values $\overline{D}$ and D_i^e displayed in both bars (actual and estimated vicinity circle radiuses) are almost identical, meaning that the estimation gathered by our model is very accurate, with an error that never exceeds 12%, regarding the average number of iterations for all 500 queries. Moreover, the mean deviation $\overline{DS}$ (i.e., the average unsigned error of the estimation in each individual query), illustrated by the error bars, is between 20% and 50% in all experimental settings, increasing with the threshold. This is actually an expected result since the increase of threshold

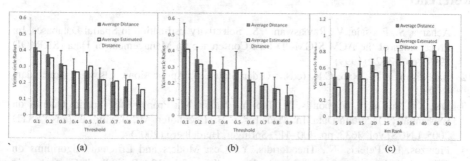

(a)   (b)   (c)

Fig. 8. Average actual and estimated search radius of the PTNN2D algorithm scaling the threshold using (a) mean probability, (b) Sampling, and (c) the PRNN2D algorithm scaling the number of objects requested

results in decreasing the number of iterations of the PTNN2D algorithm, which leads to the deviation growth. A comparison between the two alternative ways of estimation, i.e., Lemma 2 and sampling, results that the latter performs slightly better.

Similar results are exposed regarding the PRNN2D algorithm, illustrated in Figure 8(c), where our estimations are once again accurate, with an error that never exceeds 15%, except of the case where small cardinalities of objects are requested, i.e., smaller than 10, where the error reaches 25%.

6   Conclusions and Future Work

In this paper, we have worked on the problem of performing *probabilistic thresholding nearest neighbor* and *probabilistic ranking nearest neighbor* queries over existentially uncertain spatial point datasets [3],[13]. Following a statistical approach, we estimate the average number of the nearest neighbors required for processing PTNN queries as a function of the threshold t and then, utilizing existing approaches [6], we propose a cost model for such queries. We have also provided approximate solutions for the same problem, which turn out to be applicable over arbitrarily distributed data. Our experimental study proves the effectiveness and efficiency of the proposed techniques. There are numerous interesting research directions arising from this work, including the application of our model in data spaces of higher dimensionality, its extension in order to support reverse nearest neighbor, and spatial skyline queries according to [13], as well as objects with time-varying existential uncertainty.

Acknowledgements. Elias Frentzos was supported by the Greek States Scholarships foundation. Nikos Pelekis and Yannis Theodoridis were supported by the European Union Seventh Framework Programme (FP7/2007-2013) under grant agreement n°270833, ICT project DATASIM (www.datasim-fp7.eu).

References

1. Acharya, S., Poosala, V., Ramaswamy, S.: Selectivity Estimation in Spatial Databases. In: Proceedings of the ACM SIGMOD Int'l Conference on Management of Data (SIGMOD 1999), pp. 13–24 (1999)
2. Balakrishnan, N., Rao, C.R. (eds.): Order Statistics: Applications. Elsevier, Amsterdam (1998)
3. Dai, X., Yiu, M.L., Mamoulis, N., Tao, Y., Vaitis, M.: Probabilistic Spatial Queries on Existentially Uncertain Data. In: Medeiros, C.B., Egenhofer, M., Bertino, E. (eds.) SSTD 2005. LNCS, vol. 3633, pp. 400–417. Springer, Heidelberg (2005)
4. Frentzos, E., Pelekis, N., Theodoridis, Y.: Cost Models and Efficient Algorithms on Existentially Uncertain Spatial Data. In: Proceedings of the 12th Panhellenic Conference in Informatics (PCI 2008), Samos, Greece (2008)
5. Frentzos, E., Gratsias, K., Theodoridis, Y.: On the Effect of Location Uncertainty in Spatial Querying. IEEE Trans. Knowl. Data Eng. 21(3), 366–383 (2009)

6. Hjaltason, G., Samet, H.: Distance Browsing in Spatial Databases. ACM Transactions in Database Systems 24(2), 265–318 (1999)
7. Manolopoulos, Y., Nanopoulos, A., Papadopoulos, A.N., Theodoridis, Y.: Rtrees: Theory and Applications. Springer (2005)
8. Parent, C., Spaccapietra, S., Renso, C., Andrienko, G., Andrienko, N., Bogorny, V., Damiani, M.L., Gkoulalas-Divanis, A., Macedo, J., Pelekis, N., Theodoridis, Y., Yan, Z.: Semantic Trajectories Modeling and Analysis. ACM Computing Surveys (2013)
9. Sharifzadeh, M., Shahabi, C.: The Spatial Skyline Queries. In: Proceedings of the 32nd International Conference on Very Large Data Bases (VLDB), Seoul, Korea (2006)
10. Stanoi, I., Agrawal, D., Abbadi, A.: Reverse Nearest Neighbor Queries for Dynamic Databases. In: Proceedings of the SIGMOD Workshop on Research Issues in Data Mining and Knowledge Discovery (2000)
11. Tao, Y., Zhang, J., Papadias, D., Mamoulis, N.: An Efficient Cost Model for Optimization of Nearest Neighbor Search in Low and Medium Dimensional Spaces. IEEE Trans. Knowledge and Data Eng. 16(10), 1169–1184 (2004)
12. Weisstein, E.W.: Uniform Product Distribution. From MathWorld. A Wolfram Web Resource
13. Yiu, M., Mamoulis, N., Dai, X., Tao, Y., Vaitis, M.: Efficient Evaluation of Probabilistic Advanced Spatial Queries on Existentially Uncertain Data. IEEE Trans. Knowledge and Data Eng. 21(1) (2009)

Processing Probabilistic Range Queries over Gaussian-Based Uncertain Data

Tingting Dong[1], Chuan Xiao[1], Xi Guo[2], and Yoshiharu Ishikawa[1]

[1] Nagoya University, Japan
{dongtt,chuanx,y-ishikawa}@nagoya-u.jp
[2] The Chinese University of Hong Kong, China
guoxi@se.cuhk.edu.hk

Abstract. Probabilistic range query is an important type of query in the area of uncertain data management. A probabilistic range query returns all the objects within a specific range from the query object with a probability no less than a given threshold. In this paper we assume that each uncertain object stored in the databases is associated with a multi-dimensional Gaussian distribution, which describes the probability distribution that the object appears in the multi-dimensional space. A query object is either a certain object or an uncertain object modeled by a Gaussian distribution. We propose several filtering techniques and an R-tree-based index to efficiently support probabilistic range queries over Gaussian objects. Extensive experiments on real data demonstrate the efficiency of our proposed approach.

1 Introduction

In recent years, uncertain data management has received considerable attention in the database community. It involves a large variety of real-world applications, ranging from mobile robotics, sensor networks to location-based services.

Among all the problems in the area of uncertain data management, *probabilistic range query* is an important one for processing uncertain data in real-world applications. A probabilistic range query returns all the data objects that appear within the given search region with probabilities no less than a given probability threshold.

For instance, consider a self-navigated mobile robot moving in a wireless environment. The robot builds a map of the environment by observing nearby landmarks through devices such as sonar and laser range finders. Due to the inherent limitation brought about by sensor accuracy and signal noises, the location information acquired from measuring devices is not always precise. At the same time, the robot also conducts probabilistic localization [19] to estimate its own location autonomously by integrating its movement history and the landmark information. This can cause impreciseness in the location of the robot, too. In consequence, probability queries have evolved to tackle such impreciseness; e.g., "find landmarks lying within 5 meters from my current location with a probability at least 80%".

M.A. Nascimento et al. (Eds.): SSTD 2013, LNCS 8098, pp. 410–428, 2013.

Typically for such applications, uncertain objects are stored in the databases and associated with probability distributions. A commonly used distribution for such a purpose is a multi-dimensional *Gaussian distribution* which is widely adopted in statistics, pattern recognition [7] and localization in robotics [19].

In this paper we study the case where the locations of data objects are uncertain, whereas the location of the query object is either exact or uncertain. Specifically, data object are described by Gaussian distributions with different parameters to indicate their differences in uncertainty. A query object can be either a certain point in the multi-dimensional space or an uncertain location represented by a multi-dimensional Gaussian distribution. We solve the probabilistic range query problem according to the above setup.

A straightforward approach to this problem is to compute the *appearance probability* [18] for each data object and output it if this probability is no less than the threshold. However, the probability computation usually requires costly numerical integration for accurate result [16], rendering it prohibitively expensive to compute for all the data objects and check if the query constraint is satisfied. Thus, such computations should be reduced as much as possible.

There have been solutions to probabilistic range queries that can handle Gaussian-based uncertain data, yet based on specific assumptions. For example, U-tree [16] assumes that each uncertain object is located within a *pre-defined uncertainty region*. It constructs an index for all objects based on this region to reduce the number of candidates that require the expensive numerical integration. Besides, Gauss-tree [3] is proposed for probabilistic identification queries, but the Gaussian distributions they follow must be *independent* in each dimension. When these assumptions are violated, these solutions no longer work. One problem of U-tree is that it is not easy to decide a suitable extent of the uncertainty region for a real world object. In this paper we solve these problems with generic Gaussian distributions without any of these assumptions; i.e., the objects can locate in an infinite space as opposed to U-tree, or have correlations between dimensions as opposed to Gauss-tree.

Furthermore, we propose filtering techniques to generate candidate Gaussian objects and only compute probability integration for these candidates. Equipped with the filtering techniques, an R-tree-based indexing method is proposed to accelerate query processing. The index structure is inspired by the idea of TPR-tree [21], of which the (Minimum Bounding Box) MBBs vary with time. The difference is that in our index, a parent MBB not only varies with the probability threshold but also tightly encloses all the child MBBs.

In our preliminary work [10], we propose query processing algorithms for probabilistic range queries, assuming that *only the location of the query object* is uncertain and modeled by a Gaussian distribution, but data objects are certain multi-dimensional points. An R-tree can be used to manage these certain data points and process queries, which is different from the situation here. In this paper, we extend the uncertainty to data objects and propose novel solutions. A precedent report of this work has appeared in [11]. The approach proposed in [11] approximates the Gaussian distribution by an upper-bounding function which is in a simple exponential form. An R-tree-like hierarchical index structure is proposed and an exponential summary function is defined to cover multiple upper-bounding functions or summary functions. Nevertheless, the summary function is so sensitive to child functions that it will become dramatically large if child Gaussians are sparsely distributed in the space or one of them has big variances, leading to loose index structure and weak filtering power.

Our contributions are summarized as follows:

1. We formalize two types of probabilistic range queries with respect to the query object: a certain point and an uncertain location represented by a Gaussian distribution, while data objects are represented by Gaussian distributions with different parameters.
2. For the two types of queries, we propose several effective filtering techniques to prune unpromising objects.
3. We design a novel R-tree-based index structure to support probabilistic range queries on Gaussian objects.
4. We demonstrate the efficiency of our approach through comprehensive experimental performance study.

The rest of the paper is organized as follows. Section 2 defines our problem. We present our filtering strategies in Section 3. Section 4 describes our index structure. We discuss the extension of our approach to support other models and queries in Section 5. Experiment results and analyses are covered by Section 6. Section 7 reviews related work. Section 8 concludes the paper.

2   Problem Definition

In this section, we first define Gaussian objects, and then define probabilistic range queries on two types of query objects: point objects and Gaussian objects.

2.1   Gaussian Objects

The Gaussian distribution, also known as the normal distribution, is a continuous probability distribution defined by a bell-shaped probability density function described with a mean value and a standard deviation. In this paper, we assume that data objects are modeled by Gaussian distributions in a d-dimensional space. A point x referred in the paper, by default, is in a d-dimensional numerical space, namely, $x = (x^1, .., x^d)^t$.

Definition 1 (Gaussian objects). *A Gaussian object o is represented by its possible locations (points) and the probability density it appears at each location. Formally, the probability density that o is located at x_o is captured by a d-dimensional Gaussian probability density function*

$$p_o(x_o) = \frac{1}{(2\pi)^{\frac{d}{2}} |\Sigma_o|^{\frac{1}{2}}} \exp\left[-\frac{1}{2}(x_o - \mu_o)^t \Sigma_o^{-1}(x_o - \mu_o) \right]. \tag{1}$$

μ_o is the mean location (center) of o. Σ_o is a $d \times d$ covariance matrix. $|\Sigma_o|$ (resp. Σ_o^{-1}) is the determinant (resp. inverse matrix) of Σ_o.

2.2   Probabilistic Range Queries on Gaussian Objects

Given a dataset of Gaussian objects $\mathcal{D}$, a query object q, a distance threshold δ, and a probability threshold θ, a *probabilistic range query (PRQ) on Gaussian objects* retrieves all the data objects $o \in \mathcal{D}$ such that the distance between o and q is no more than δ with a probability no less than θ.

In this paper, we consider two types of query objects for q:

1. The query object is a point, namely, $q = (x_q^1, x_q^2, \dots, x_q^d)^t$.
2. The query object is a Gaussian object, namely,

$$p_q(\boldsymbol{x}_q) = \frac{1}{(2\pi)^{\frac{d}{2}}|\boldsymbol{\Sigma}_q|^{\frac{1}{2}}} \exp\left[-\frac{1}{2}(\boldsymbol{x}_q - \boldsymbol{\mu}_q)^t \boldsymbol{\Sigma}_q^{-1}(\boldsymbol{x}_q - \boldsymbol{\mu}_q)\right].$$

The *probabilistic range query with point query object* (PRQ-P) is formally defined as

$$\text{PRQ-P}(q, \mathcal{D}, \delta, \theta) = \{o \mid o \in \mathcal{D}, \Pr(\|\boldsymbol{x}_o - \boldsymbol{q}\| \le \delta) \ge \theta\},$$

where $\|\boldsymbol{x}_o - \boldsymbol{q}\|$ represents the Euclidean distance between $\boldsymbol{x}_o$ and $\boldsymbol{q}$. We call the region consisting of the points with distance no more than δ from the query object *the query region*. $\Pr(\|\boldsymbol{x}_o - \boldsymbol{q}\|)$, the probability integration within the δ range of the query, is computed by

$$\Pr(\|\boldsymbol{x}_o - \boldsymbol{q}\| \le \delta) = \int \chi_\delta(\boldsymbol{x}_o, \boldsymbol{q}) \cdot p_o(\boldsymbol{x}_o) d\boldsymbol{x}_o, \tag{2}$$

where

$$\chi_\delta(\boldsymbol{x}_o, \boldsymbol{q}) = \begin{cases} 1, & \|\boldsymbol{x}_o - \boldsymbol{q}\| \le \delta; \\ 0, & \text{otherwise.} \end{cases} \tag{3}$$

The integration in Eq. 2 is not in a closed-form and hence cannot be computed directly. Numerical solutions such as Monte Carlo methods can be employed to evaluate the probability. We use the *importance sampling* [15] in this paper. Specifically, we generate $\boldsymbol{x}_o$ as per the probability function $p_o(\boldsymbol{x}_o)$, and increment the count when Eq. 3 is satisfied. Finally, the value of the integration can be obtained by dividing the count by the number of samples generated. Generally speaking, however, Monte Carlo methods are only accurate only if the number of samples is sufficiently large (at the order of 10^6) [16]. Therefore, integral computation induces expensive cost.

Fig. 1 illustrates the PRQ-P query in a 2-dimensional space. The Gaussian object o exists in the space with decreasing probability density as it spreads from the center $\boldsymbol{\mu}_o$. We project the probability surface of o to a plane and show the diminishing trend with gradient colors. A PRQ-P query finds the Gaussian objects located in the proximity of the query point with a required probability. Computing the probability using Eq. 2 corresponds to integrating the probability density function of o within the shaded area around q.

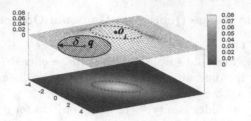

Fig. 1. PRQ-P Query

Similar to PRQ-P, the *probabilistic range query with Gaussian query object* (PRQ-G) is defined as

$$\text{PRQ-G}(q, \mathcal{D}, \delta, \theta) = \{o \mid o \in \mathcal{D}, \Pr(\|\boldsymbol{x}_o - \boldsymbol{x}_q\| \leq \delta) \geq \theta\},$$

where $\Pr(\|\boldsymbol{x}_o - \boldsymbol{x}_q\| \leq \delta)$ is computed by

$$\Pr(\|\boldsymbol{x}_o - \boldsymbol{x}_q\| \leq \delta) = \iint \chi_\delta(\boldsymbol{x}_o, \boldsymbol{x}_q) \cdot p_o(\boldsymbol{x}_o) \cdot p_q(\boldsymbol{x}_q) d\boldsymbol{x}_o d\boldsymbol{x}_q, \tag{4}$$

where

$$\chi_\delta(\boldsymbol{x}_o, \boldsymbol{x}_q) = \begin{cases} 1, & \|\boldsymbol{x}_o - \boldsymbol{x}_q\| \leq \delta; \\ 0, & \text{otherwise.} \end{cases}$$

To compute the integration in Eq. 4, although we can simply generate random numbers for two Gaussian distributions $p_o(\boldsymbol{x}_o)$ and $p_q(\boldsymbol{x}_q)$ respectively, a more efficient method is shown in [11]. It constructs a $2d$-dimensional Gaussian distribution by combining the two d-dimensional Gaussian distributions.

3   Filtering Based on Approximated Region

A naïve algorithm to answer PRP-P or PRP-G queries is to pair the query object with every data object and perform integration check with either Eq. 2 or Eq. 4. The algorithm becomes prohibitively expensive for large datasets. So we develop our approach based on a filter-and-refine paradigm; i.e., to obtain a set of candidate objects and then compute the integration for the candidates only.

In this section, we first introduce the notion of ρ-region that leverages the two thresholds δ and θ, and then propose the ρ-region-based filtering techniques to handle PRP-P and PRP-G queries.

3.1 ρ-Region

Definition 2 (ρ-region). *Consider a Gaussian object o and the integration of its probability density function $p_o(\boldsymbol{x}_o)$ over an ellipsoidal region*

$(\boldsymbol{x}_o - \boldsymbol{\mu}_o)^t \boldsymbol{\Sigma}_o^{-1} (\boldsymbol{x}_o - \boldsymbol{\mu}_o) \leq r^2$. *Let r_ρ be the value of r within which the result of the integration equals ρ:*

$$\int_{(\boldsymbol{x}_o - \boldsymbol{\mu}_i)^t \boldsymbol{\Sigma}_o^{-1} (\boldsymbol{x}_o - \boldsymbol{\mu}_o) \leq r_\rho^2} p_o(\boldsymbol{x}_o) d\boldsymbol{x}_o = \rho.$$

We call the ellipsoidal region

$$(\boldsymbol{x}_o - \boldsymbol{\mu}_o)^t \boldsymbol{\Sigma}_o^{-1} (\boldsymbol{x}_o - \boldsymbol{\mu}_o) \leq r_\rho^2$$

the ρ-region of o.

In Fig. 1, the dotted ellipsoidal curve illustrates a ρ-region. Because the probability density of a Gaussian distribution decreases as we move away from the center of the object, if the query object is distant enough from the center, the probability integration within the query region will not reach the probability threshold θ. In other words, it is possible to determine whether a data object can satisfy the query condition by deriving a suitable ρ-region with the threshold θ (will be introduced in Section 3.3 and Section 3.4) and examining whether the ρ-region intersects the query region.

To compute r_ρ with a given ρ, we borrow the approach proposed in our previous work [10]. It transforms the integration over an ellipsoidal region to an integration over a d-dimensional sphere region. By assigning $\boldsymbol{\mu}_o = \boldsymbol{0}$ and $\boldsymbol{\Sigma}_o = \boldsymbol{I}$ in Eq. 1, we have the *normalized Gaussian distribution*

$$p_{\text{norm}}(\boldsymbol{x}) = \mathcal{N}(\boldsymbol{0}, \boldsymbol{I}) = \frac{1}{(2\pi)^{d/2}} \exp\left[-\frac{1}{2}\|\boldsymbol{x}\|^2\right].$$

Based on this probability density function, the following property can be derived.

Property 1. [10] Consider integration of $p_{\text{norm}}(\boldsymbol{x})$ over $\|\boldsymbol{x}\|^2 \leq r^2$. For a given ρ $(0 < \rho < 1)$, let $\tilde{r}_\rho$ be the radius within which the integration becomes ρ:

$$\int_{\|\boldsymbol{x}\|^2 \leq \tilde{r}_\rho^2} p_{\text{norm}}(\boldsymbol{x}) d\boldsymbol{x} = \rho.$$

Then $r_\rho = \tilde{r}_\rho$ holds.

The preceding property indicates that we can compute $\tilde{r}_\rho$ and hence r_ρ for a given ρ value. Therefore, we can construct a (ρ, r_ρ)-table offline (numerical integration is necessary) and obtain the ρ-region by looking up the corresponding r_ρ from this table. If there is no matched entry for a given ρ, we conservatively return the corresponding r_ρ of the smallest value greater than ρ, so correctness of the result can be guaranteed.

The ellipsoidal shape of a ρ-region renders it difficult to quickly examine whether the ρ-region intersects the query region as well as develop an indexing scheme based on prevalent spatial indexes such as R-tree. Hence we will study the minimum bounding box (MBB) which tightly bounds the ρ-region.

3.2   Minimum Bounding Box of ρ-Region

Fig. 2 shows the MBB of a ρ-region of a 2-dimensional Gaussian object o. Let w_j denote the length of its edge along the j-th dimension. The following property holds [10].

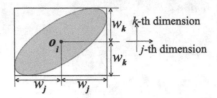

Fig. 2. MBB of ρ-Region

Property 2. The value of w_j $(j = 1, \ldots, d)$ is given as

$$w_j = \sigma_j r_\rho \tag{5}$$

where σ_j corresponds to the standard deviation for the j-th dimension

$$\sigma_j = \sqrt{(\Sigma_o)_{jj}}$$

where $(\Sigma_o)_{jj}$ represents the (j, j)-th element of the matrix Σ_o.

For a data object o, since σ_j can be calculated from the covariance matrix Σ_o, the scale of the MBB is determined uniquely by r_ρ, and hence ρ. Consequently, in order to establish the filtering conditions utilizing the MBBs, it is essential to explore the relation between ρ and the probability threshold θ. Next we will present our filtering techniques for PRQ-P and PRP-G, respectively.

3.3   Filtering Policies for PRQ-P Queries

Our filtering policies to process PRP-P queries are divided into two cases: $\theta < 0.5$ and $\theta \geq 0.5$.

Case 1: $\theta < 0.5$. Consider the four data objects o_1, o_2, o_3, o_4 shown in Fig. 3(a). $bb_i(\rho)$ denotes the MBB of o_i's ρ-region.

First, let's consider o_4. Since the probability that o_4 is located inside its ρ-region is ρ, the probability of being outside the ρ-region's MBB, is definitely less than $1 - \rho$. Furthermore, given the line symmetry of the Gaussian distribution, the probability that o_4 exists inside the query region is at most $(1 - \rho)/2$. Hence, if $\rho = 1 - 2\theta$, and $bb_4(\rho)$ and the query region do not overlap, the probability that o_4 lies in query region must be less than θ. Second, for objects o_1 and o_3, since their mean locations are inside the query region, it is obvious that their MBBs intersect the query region. Therefore, we include them into the candidate set without examining their MBBs. Third, for object o_2, we check and find its MBB intersects the query region, and then it becomes a candidate.

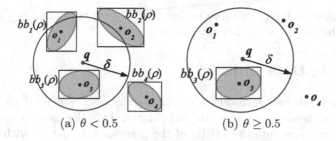

(a) $\theta < 0.5$ (b) $\theta \geq 0.5$

Fig. 3. Filtering for PRQ-P Queries

In summary, when $\theta < 0.5$, a data object is a candidate only if its $bb_i(1-2\theta)$ intersects the query region.

Case 2: $\theta \geq 0.5$. We show our idea in Fig. 3(b). If the probability that a data object exists in the query region reaches a θ no less than 0.5, it is necessary that its mean location lies inside the query region. In this way, o_2 and o_4 can be pruned, whereas o_1 and o_3 are considered as candidates.

Moreover, for all candidates, let $\rho = \theta$ and compute their $bb_i(\theta)$s. If the query region fully contains $bb_i(\theta)$; e.g., o_3, the probability that this object lies within the query region is definitely greater than θ. We validate it as a result without computing the numerical integration.

3.4 Filtering Policies for PRQ-G Queries

For PRQ-G queries, since both the query object q and the data object o are in Gaussian distributions, we obtain both of their MBBs of ρ-regions.

As shown in Fig. 4(a), assume the distance between the two MBBs is exactly δ. The probability that q and o_i are both located inside their ρ-regions at the same time is ρ^2, assuming q and o_i are independent. Therefore, the probability that the distance between q and o_i is no more than δ is at most $1 - \rho^2$. For a given probability threshold θ, let $1 - \rho^2 = \theta$; i.e., $\rho = \sqrt{1-\theta}$. We exclude o_i from the candidate set if the minimum distance between $bb_i(\rho)$ and $bb_q(\rho)$ is more than δ.

Moreover, let $\rho^2 = \theta$; i.e., $\rho = \sqrt{\theta}$. If the maximum distance of $bb_i(\rho)$ and $bb_q(\rho)$ is less than δ, as shown in Fig. 4(b), o_i is guaranteed to be located inside

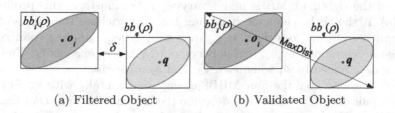

(a) Filtered Object   (b) Validated Object

Fig. 4. Filtering for PRQ-G Queries

the query region with a probability no less than θ, and becomes a result without computing the exact probability integration.

4   Indexing Data Objects

The filtering conditions introduced in the previous section need to know the value of θ and hence ρ to generate candidates. In order not to scan all the data objects and compute the MBBs of the ρ-regions on the fly with the given θ, an immediate solution is to index the MBBs for a sufficiently large $\rho_{\max}$. Because the MBB with a larger ρ always consumes the one with a smaller ρ, it can support all the queries such that the ρ values computed from θ satisfy the condition $\rho \leq \rho_{\max}$. However, the efficiency of the index is compromised for small ρ values. This method serves as a baseline algorithm (we use an R-tree to index MBBs and name it FR-tree), and will be compared in the experiment with the indexing technique we are going to present.

Inspired by the TPR-tree [21], we propose an R-tree-based index structure which stores the MBBs in a parametric fashion. It works for arbitrary probability thresholds and range thresholds, and there is no no need to assume the two thresholds are given prior to index construction. The MBBs can be dynamically computed as we traverse the index. Furthermore, the bounding box of a node (at both leaf and non-leaf levels) tightly encloses all its children's bounding boxes regardless of the θ value, as opposed to the TPR-tree within which all child bounding boxes are bounded in a loose manner.

Our index is a balanced, multi-way tree with the structure of an R-tree. Each entry in leaf nodes contains a data object in the form of $o_i = (id_i, \boldsymbol{\mu}_i, \boldsymbol{\Sigma}_i)$, where id_i is the object id, $\boldsymbol{\mu}_i, \boldsymbol{\Sigma}_i$ are the mean location and the covariance matrix of the Gaussian distribution. In a non-leaf node, each entry has a pointer to a child node and the bounding box enclosing all the bounding boxes from the child node.

Consider an object o_i with mean location $(x_i^1, .., x_i^d)^t$. Its MBB is a rectangle parameterized with r_ρ. From Eq. 5, the extent of the MBB in the j-th dimension can be represented by

$$[x_i^j - w_i^j, x_i^j + w_i^j] = [x_i^j - \sigma_i^j r_\rho, x_i^j + \sigma_i^j r_\rho]. \tag{6}$$

Seeing the MBBs grow with r_ρ, in order to tightly bound the MBBs (or bounding boxes) in child nodes, it is necessary to search each dimension for the leftmost and the rightmost MBBs under varying ρ. We illustrate this problem in Fig. 5(a). It shows the changing bounding box that encloses the MBBs of four 2-dimensional objects' ρ-regions as r_ρ increases. When r_ρ is less than r_1, the left edge is determined by o_1, and it becomes o_3 when r_ρ exceeds r_1. The right bound is determined by o_4 when $r_\rho < r_2$, and o_2 otherwise.

Fig. 5(b) shows how the four MBBs changes horizontally with r_ρ. For each object, a pair of symmetrical lines describe the left and the right coordinations of the MBB. The lines have different slopes due to the difference in the standard deviations of the objects. The bold polylines illustrate the left and right

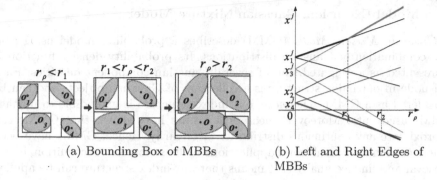

(a) Bounding Box of MBBs   (b) Left and Right Edges of
                                            MBBs

Fig. 5. Bounding Box with Varying ρ

coordinations of the bounding box. Therefore the problem becomes how to find
the bold polylines. To this end, a bounding box can be represented by several
segments with respect to r_ρ.

We store in the index the j-th dimension of a bounding box in the form of

$$(\langle x_1^j, \sigma_1^j, r_1\rangle, .., \langle x_k^j, \sigma_k^j, +\infty\rangle).$$

For example, for the four objects in Fig. 5(a), the left and the
right coordinations of the bounding box are $(\langle x_1^j, \sigma_1^j, r_1\rangle, \langle x_3^j, \sigma_3^j, +\infty\rangle)$ and
$(\langle x_4^j, \sigma_4^j, r_2\rangle, \langle x_2^j, \sigma_2^j, +\infty\rangle)$, respectively.

We can find all the segments in the j-th dimension through a sort on the coordinations first and then a linear scan from the object whose standard deviation
has the value on the j-th dimension. The time complexity is $O(n \log n)$, where
n is the number of its child nodes. The number of segments in a bounding box
is at most n.

To process a query, θ is converted to ρ, and then r_ρ with the pre-computed
(ρ, r_ρ) table. Starting with the root node, for PRQ-P we compare the query
region (MBB of the query object for PRQ-G) with the bounding box, and check
the filtering condition. Given an r_ρ, we scan the stored jth-dimension of the
bounding box and find α such that $r_{\alpha-1} \leq r_\rho < r_\alpha$. Then the extent of the
bounding box on the j-th dimension can be computed through Eq. 6 using
values x_α, σ_α, and ρ.

Note that our index is different from TPR-tree: (1) The bounding boxes of
TPR-tree change towards one direction in a rate (velocity), while our bounding
boxes change towards two opposite directions symmetrically with r_ρ. (2) The
bounding boxes of TPR-tree are tight only when an update occurs, while our
bounding boxes are always tight.

5 Discussion

In this section, we discuss the extension of our approach to other types of
uncertainty models and queries.

5.1   Model Extension: Gaussian Mixture Model

A *Gaussian Mixture Model* (GMM) describes a probability model using a finite combination of Gaussian distributions. Its probability density function is represented as a weighted sum of several Gaussian component densities. Each leaf node in our index structure is a collection of Gaussian objects and can be considered as a GMM. Therefore, our index structure is extendible to manage GMM-based data. Moreover, theoretical results have shown that GMMs can approximate any continuous distribution arbitrarily. Hence, GMMs have been widely used in many real-world applications such as biometric recognition, image retrieval and finance analysis. It means that our index structure can be applied to applications with various types of data.

5.2   Probabilistic Nearest Neighbor Queries

Although we focus on probabilistic range queries in this paper, our index structure can also support other types of queries such as probabilistic nearest neighbor queries. A common approach to answer a probabilistic nearest neighbor query is based on the *minimal maximum distance*. It states that if the minimum distance of the bounding box of a data object or a set of bounding boxes from the query object is greater than the minimal maximum distance of all current bounding boxes from the query object, this data object or set of bounding boxes can be excluded from the searching list of this query. By defining appropriate query processing strategies, our index structure can answer probabilistic nearest neighbor queries as well.

6   Experiments

We report our experiment results and analyses in this section.

6.1   Experimental Setup

We design two baseline approaches for experimental evaluation. One baseline approach is to sequentially scan the dataset and compute probability integration with the query. We name it Scan and evaluate our filtering techniques by comparing candidate number and query processing time with it. The other baseline approach indexes the MBBs of the ρ-region with $\rho_{\max} = 0.99$. Because the MBB with a larger ρ always consumes the one with a smaller ρ, it can support all the queries such that the ρ values computed from θ satisfy $\rho \leq \rho_{\max}$. We equip this index with our filtering techniques and name it FR-tree, and evaluate our index structure by comparing filtering time and IO access with it. Our proposed index is referred to as G-tree.

Three real datasets are used in our experiments. MG and LB are two 2-dimensional datasets of Montgomery and Long Beach road networks (39K and

52K respectively)[1]. Airport is a 3-dimensional dataset containing latitudes, longitudes and elevations of 41K airports in the world [2]. All datasets are normalized to $[0, 1000]^d$. LB is used by default.

We randomly generate PRQ-P and PRQ-G queries. The probability threshold θ lies within $[0.01, 0.99]$, and the query range δ is randomly chosen from $[10, 100]$ for MG and LB, and $[100, 200]$ for Airport. We randomly generate covariance matrices for both data Gaussian objects and query Gaussian objects.

We implemented the index structure by extending the spatial index library SaiL [3] [9]. It is a generic framework that integrates spatial and spatio-temporal index structures and supports user-defined datatypes and customizable spatial queries. We conducted experiments using a PC with Intel Core 2 Duo CPU E8500 (3.16GHz), RAM 4GB, running Fedora 12. We construct an index of all data objects for both PRQ-P and PRQ-G, and store it in the secondary memory.

6.2   Query Performance Evaluation

The average query response time of 200 PRQ-P (resp. PRQ-G) queries (10K samples are used for numerical integration) is 0.242 seconds (1.250 seconds resp. PRQ-G) for G-tree, and 120.764 seconds (236.725 seconds resp. PRQ-G) for Scan, almost 500 (190 resp. PRQ-G) times that of G-tree. Among the overall response time, the integral computation takes up 0.237 seconds (1.246 seconds resp. PRQ-G) for G-tree, and 120.692 seconds (236.577 seconds resp. PRQ-G) for Scan. This indicates that probability integration dominates the overall query processing and is computationally expensive. Consequently, it is important to reduce candidate objects which need to perform integration as much as possible.

Among 50,747 objects in LB, the average candidate number of G-tree is 93 for PRQ-P (335 for PRQ-G). The number of validated objects by integration is 65 for PRQ-P (156 for PRQ-G). So for PRQ-P 69.9% (46.6% for PRQ-G) of the candidates identified by our approach are real results. This demonstrates the effectiveness of our proposed filtering techniques.

In the sequel, we exclude the integral part from query processing and focus on evaluating the filtering and indexing performance of FR-tree and G-tree.

We run the two algorithms to process 10K queries on the three datasets and show the average filtering time and IO access of PRQ-P (resp. PRQ-G) in Fig. 6(a) – 6(b) (resp. Fig. 6(d) – 6(e)). For PRQ-P, the filtering time of G-tree is half of that of FR-tree, because the IO access of G-tree is 90% less than that of FR-tree, though the segmented bounding boxes in G-tree are more complex to process than those in FR-tree. The reduction on PRQ-G is more substantial. The filtering time of G-tree on MG and LB is 71% less than that of FR-tree, and 61% on Airport. The IO access of G-tree of three datasets is 6% that of FR-tree.

As a ρ_{max} is adopted to process queries with any θ, the bounding boxes in FR-tree are very loose. This causes more IO accesses and increases filtering

---

[1] http://www.census.gov/geo/www/tiger

[2] http://www.ourairports.com/data

[3] http://libspatialindex.github.com/

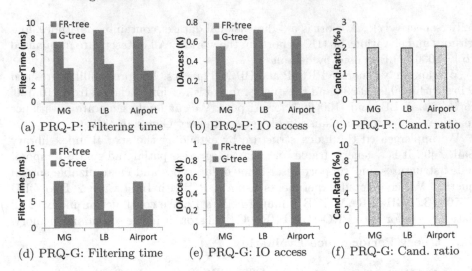

Fig. 6. Performance of PRQ-P and PRQ-G Queries

time. On the other hand, since the bounding boxes in G-tree are constructed in a parametric fashion, they can be calculated dynamically for arbitrary θ and hence are compact. Another interesting observation is that the IO access almost resembles the candidate number, indicating most IOs are spent on retrieving data objects.

Fig. 6(c) and Fig. 6(f) shows the candidate ratio of PRQ-P and PRQ-G, which is calculated by dividing the candidate number by the total number of objects. The candidate number of FR-tree and G-tree is the same since we equip FR-tree with our filtering techniques.

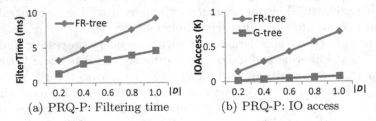

Fig. 7. Varying $|\mathcal{D}|$: Filtering time and IO access (PRQ-P)

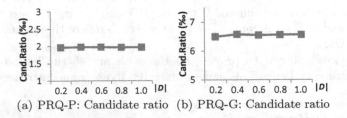

Fig. 8. Varying $|\mathcal{D}|$: Candidate ratio

The candidate ratio is around 2‰ for PRQ-P and 6‰ for PRQ-G on the three datasets. This reveals that only a very small percentage of data objects will become candidates owing to our filtering techniques.

Varying Dataset Size. To evaluate the scalability of our approach, we randomly extract 20%, 40%, 60%, 80% and 100% of LB dataset and show the filtering time and IO access of two methods in Fig. 7(a) – 7(b) on PRQ-P queries. The performance on PRQ-G queries reveals a similar trend and hence is omitted here due to space limit. As the dataset size becomes larger, the filtering time and IO access of FR-tree almost increase linearly. G-tree displays a steady increasing trend and always outperforms FR-tree.

As shown in Fig. 8(a) – 8(b), the candidate ratio of PRQ-P retains 2‰ when varying the dataset size $|\mathcal{D}|$, and 6.5‰ for PRQ-G, demonstrating the steadiness and scalability of our approach with respect to the dataset size.

Varying Query Range. We vary the query range δ from 10 to 100 by 10 and show the performance on PRQ-P queries in Fig. 9(a) – 9(b). The performance on PRQ-G queries is similar and hence omitted. As δ increases, FR-tree consumes much more time and more IO accesses on filtering processing. In contrast, G-tree exhibits much slower increasing trends. Fig. 10(a) – 10(b) shows that the candidate ratio of both PRQ-P and PRQ-G also increases with δ, but for PRQ-P it is only 3.4‰ (11.6‰ for PRQ-G) even if δ achieves 100.

Varying Probability Threshold. We vary θ from 0.1 to 0.9 and show the performance in Fig. 11(a) – 12(b) for both PRQ-P and PRQ-G queries. For PRQ-P, the filtering time and IO access of both FR-tree and G-tree decreases gradually with θ when it is less than 0.5. When θ exceeds 0.5, the filtering time

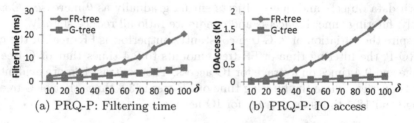

(a) PRQ-P: Filtering time   (b) PRQ-P: IO access

Fig. 9. Varying δ: Filtering time and IO access (PRQ-P)

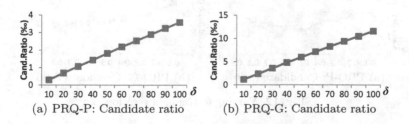

(a) PRQ-P: Candidate ratio   (b) PRQ-G: Candidate ratio

Fig. 10. Varying δ: Candidate ratio

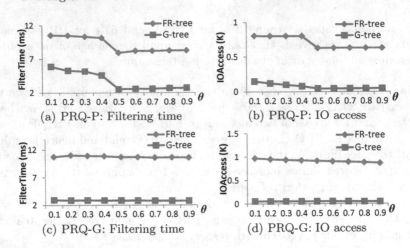

Fig. 11. Varying θ: Filtering time and IO access

slightly rebounds. This is consistent with our filtering condition which assigns $\rho = 1 - 2\theta$ if $\theta < 0.5$ and $\rho = \theta$ if $\theta \geq 0.5$. Because when $\theta \leq 0.5$, ρ decreases when θ moves towards larger values, and bounding boxes shrink. So most of non-candidates can be filtered quickly and and less IO accesses are needed, and hence it accelerates filtering.

On the contrary, when $\theta \geq 0.5$, ρ increases with θ, each bounding box enlarges and consequently the filtering time and IO access rises. However, an object needs to satisfy the constraint that the center must be located within the query region, and thus the increase in filtering time is not obvious in this case. The reason also accounts for the trend of G-tree on candidate ratio in Fig. 12(a).

For PRQ-G queries, as we set $\rho = \sqrt{1 - \theta}$ for filtering, the bounding boxes of both data objects and query objects shrink gradually as θ increases. Consequently, filtering time, IO access and candidate ratio all reduce slightly.

Despite the variation of θ, G-tree constantly outperforms FR-tree. In the case of PRQ-P, the filtering time of FR-tree amounts to 2.4 times that of G-tree on average and 9.4 times on average for IO access. This contrast is more evident on PRQ-G queries, where the filtering time of FR-tree is 3.8 times that of G-tree on average and 18.4 times on average for IO access.

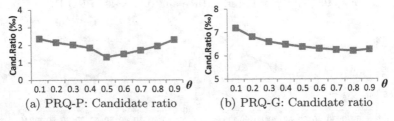

Fig. 12. Varying θ: Candidate ratio

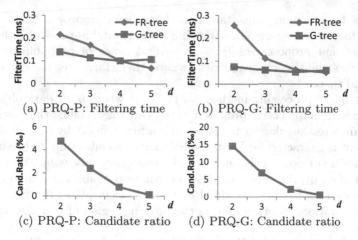

(a) PRQ-P: Filtering time   (b) PRQ-G: Filtering time

(c) PRQ-P: Candidate ratio   (d) PRQ-G: Candidate ratio

Fig. 13. Varying d: Filtering time and Candidate ratio

Varying Dimensionality. We also study the impact of dimensionality using randomly generated synthetic datasets with the size 20K and the query range within $[100, 200]$. Fig. 13 shows the scalability of FR-tree and G-tree against dimensionality.

As shown in Fig. 13(a) and Fig. 13(b), in both cases of PRQP and PRQG, the filtering time of two trees reduces with increasing d because the object density decreases with d. This can be confirmed by the decreasing trend of candidate ratio of both PRQP and PRQG in Fig. 13(c) and Fig. 13(d).

It is also notable that the filtering performance of FR-tree begin to exceed that of G-tree at $d = 5$. The explanation is that candidate retrieval becomes less frequent as the object density decreases, and hence the operation of comparing the query region with node MBBs dominates the filtering procedure. While FR-tree's MBBs can be obtained directly from the index structure, G-tree needs to compute the exact MBBs from scratch for all nodes.

Index Construction. We evaluate the index construction on the Airport dataset. The node capacities of the indexes are selected to optimize query performance for both FR-tree and G-tree.

The index size of FR-tree is 10.0MB, and the construction time is 5 seconds on average. G-tree has a size of 10.7MB, slightly larger than FR-tree due to the segmented bounding boxes in its entries. It takes 60 seconds on average to build. Although G-tree needs more construction time, considering the superior query performance and the index construction can be done offline, the index construction is in an affordable manner.

7   Related Work

Uncertain Data Management. We focus on research work in the area of unertain data management that is closely related to our work. A number of approaches for managing uncertain data have been proposed. Early research primarily focuses on queries in a moving object database model [6,14,20,22]. [5] classifies and proposes solutions to several types of probabilistic queries including probabilistic range queries, where their target is merely the one-dimensional space.

A range query processing method for the case where both data objects and query object are imprecise is proposed in [4]. But they assume that each object exists within a rectangular area. [24] models a fuzzy object by a fuzzy set where each element is characterized by its probability of membership. (The sum of all probabilities is not necessarily one.) For efficient query processing, they propose the notion of α-cut, the subset of elements whose probabilities are no less than a user-specified probability threshold α, to filter elements of the fuzzy object. However, the rationales of computing the filtering region of two algorithms are different.

As an index structure for Gaussian distributions, Gauss-tree [3] is proposed for probabilistic identification query. It assumes all Gaussian distributions are probabilistically *independent* in each dimension. This imposes heavy restriction on the generality of the approach and the overall accuracy of the query result is limited. In [12], Lian et al. propose a generic framework to tackle the local correlations among uncertain data.

Indexing Uncertain Data for Range Queries. [1] presents various structures on uncertain data that support range queries in the one-dimensional case. In terms of probabilistic range queries in a multi-dimensional space, Tao et al. propose U-tree [16]. Uncertain objects are assumed to follow arbitrary probability distributions within uncertainty regions. Zhang et al. propose a quadtree-based index structure U-Quadtree [23] for range searching on multi-dimensional uncertain data. They mainly focus on representing uncertainty by discrete instances inside a minimal bounding box. The difference lies in that we take advantage of specific properties of Gaussian distribution and index uncertain objects distributed in an infinite space.

Spatial Data Indexing. The traditional spatial database has been well studied and many indexing methods have been proposed [2,8,13] to support spatial query processing. The well-known ones are R-tree [8] and its extension R*-tree [2], which index objects by deriving their minimum bounding rectangle (MBR). TPR-tree [21] and TPR*-tree [17] are proposed to index moving objects. But none of them can be applied directly to index Gaussian objects for our problem.

8   Conclusion

In this paper, we study the probabilistic range queries over uncertain data. We assume that the location of the query object is either fixed or follows a multi-dimensional Gaussian distribution. The locations of data objects are represented by Gaussian distributions. Given these assumptions, we define two types of probabilistic range queries with respect to the query object. To expedite query processing, we propose several filtering techniques to effectively reduce non-candidate objects. We further propose a novel R-tree-based index structure to efficiently process queries. We conduct experiments on real datasets to evaluate our proposed approach.

Acknowledgement. This research is supported by the FIRST program, Japan.

References

1. Agarwal, P.K., Cheng, S.-W., Yi, K.: Range searching on uncertain data. ACM Trans. Algorithms 8(4), 43:1–43:17 (2012)
2. Beckmann, N., Kriegel, H.-P., Schneider, R., Seeger, B.: The R*-tree: an efficient and robust access method for points and rectangles. In: Proc. ACM SIGMOD (1990)
3. Böhm, C., Pryakhin, A., Schubert, M.: The Gauss-tree: Efficient object identification in databases of probabilistic feature vectors. In: Proc. ICDE (2006)
4. Chen, J., Cheng, R.: Efficient evaluation of imprecise location-dependent queries. In: Proc. ICDE (2007)
5. Cheng, R., Kalashnikov, D.V., Prabhakar, S.: Evaluating probabilistic queries over imprecise data. In: Proc. ACM SIGMOD (2003)
6. Cheng, R., Kalashnikov, D.V., Prabhakar, S.: Querying imprecise data in moving object environments. IEEE TKDE 16(9), 1112–1127 (2004)
7. Duda, R.O., Hart, P.E., Stork, D.G.: Pattern Classification, 2nd edn. Wiley (2000)
8. Guttman, A.: R-trees: A dynamic index structure for spatial searching. In: Proc. ACM SIGMOD, pp. 47–57 (1984)
9. Hadjieleftheriou, M., Hoel, E., Tsotras, V.J.: Sail: A spatial index library for efficient application integration. GeoInformatica 9, 367–389 (2005)
10. Ishikawa, Y., Iijima, Y., Yu, J.X.: Spatial range querying for Gaussian-based imprecise query objects. In: Proc. ICDE, pp. 676–687 (2009)
11. Kodama, K., Dong, T., Ishikawa, Y.: An index structure for spatial range querying on Gaussian distributions. In: Proc. Fifth International Workshop on Management of Uncertain Data (MUD 2011), pp. 1–7 (2011)
12. Lian, X., Chen, L.: A generic framework for handling uncertain data with local correlations. PVLDB 4(1), 12–21 (2010)
13. Manolopoulos, Y., Nanopoulos, A., Papadopoulos, A.N., Theodoridis, Y.: R-Trees: Theory and Applications. Springer (2005)
14. Pfoser, D., Jensen, C.S.: Capturing the uncertainty of moving-object representations. In: Güting, R.H., Papadias, D., Lochovsky, F.H. (eds.) SSD 1999. LNCS, vol. 1651, pp. 111–131. Springer, Heidelberg (1999)
15. Press, W.H., Teukolsky, S.A., Vetterling, W.T., Flannery, B.P.: Numerical Recipies: The Art of Scientific Computing, 3rd edn. Cambridge University Press (2007)
16. Tao, Y., Cheng, R., Xiao, X., Ngai, W.K., Kao, B., Prabhakar, S.: Indexing multidimensional uncertain data with arbitrary probability density functions. In: Proc. VLDB (2005)
17. Tao, Y., Papadias, D., Sun, J.: The TPR*-tree: An optimized spatio-temporal access method for predictive queries. In: Proc. VLDB, pp. 790–801 (2003)
18. Tao, Y., Xiao, X., Cheng, R.: Range search on multidimensional uncertain data. ACM TODS 32(3) (2007)

19. Thrun, S., Burgard, W., Fox, D.: Probabilistic Robotics. The MIT Press (2005)
20. Trajcevski, G., Wolfson, O., Hinrichs, K., Chamberlain, S.: Managing uncertainty in moving objects databases. ACM TODS 29(3), 463–507 (2004)
21. Šaltenis, S., Jensen, C.S., Leutenegger, S.T., Lopez, M.A.: Indexing the positions of continuously moving objects. In: Proc. ACM SIGMOD, pp. 331–342 (2000)
22. Wolfson, O., Sistla, A.P., Chamberlain, S., Yesha, Y.: Updating and querying databases that track mobile units. Distributed and Parallel Databases 7(3), 257–287 (1999)
23. Zhang, Y., Zhang, W., Lin, Q., Lin, X.: Effectively indexing the multi-dimensional uncertain objects for range searching. In: EDBT, pp. 504–515 (2012)
24. Zheng, K., Zhou, X., Fung, P.C., Xie, K.: Spatial query processing for fuzzy objects. VLDB Journal 21(5), 729–751 (2012)

Mining Co-locations under Uncertainty

Zhi Liu and Yan Huang

Computer Science and Engineering
University of North Texas
zhiliu@my.unt.edu, huangyan@unt.edu

Abstract. A co-location pattern represents a subset of spatial features whose events tend to locate together in spatial proximity. The certain case of the co-location pattern has been investigated. However, location information of spatial features is often imprecise, aggregated, or error prone. Because of the continuity nature of space, over-counting is a major problem. In the uncertain case, the problem becomes more challenging. In this paper, we propose a probabilistic participation index to measure co-location patterns based on the well-known possible world model. To avoid the exponential cost of calculating participation index from all possible worlds, we prove a lemma that allows for instance centric counting, avoids over-counting, and produces the same results as using possible world based counting. We use this property to develop efficient mining algorithms. We observed through both algebraic analysis and extensive experiments that the feature tree based algorithm outperforms uncertain Apriori algorithm by an order of magnitude not only for co-locations of large sizes but also for datasets with high level of uncertainty. This is an important insight in mining uncertainty co-locations.

1 Introduction

A co-location represents a subset of spatial features whose events frequently appear together in spatial proximity. In Epidemiology, incidents of different but related diseases occur in different places. These diseases may exhibit co-location patterns where some types of diseases tend to occur in spatial proximity. In Ecology, different types of animals can be observed in different locations. There exist patterns such as symbiotic relationship and predator-prey relationship. Different types of crimes committed and different types of road accidents may also exhibit co-location. Many spatial data are uncertain with approximations and errors in real world. In Epidemiology, the occurrence of a disease may not be geo-located precisely and may be often associated with several locations, e.g. home and work place. In Ecology, observation of spices is often imprecise. Finding co-locations under uncertainty is useful for these domains.

In the problem of mining certain co-locations, a set S of spatial features is given and each spatial feature s is associated with a set of events $s.E$. The spatial feature of a given event e is denoted as $s(e)$. A set of events E is supporting a subset of spatial feature $S' \subseteq S$ if: (1) E forms a clique using a user given distance threshold; (2) for any $e_1 \in E, e_2 \in E$ and $e_1 \neq e_2$, we

M.A. Nascimento et al. (Eds.): SSTD 2013, LNCS 8098, pp. 429–446, 2013.

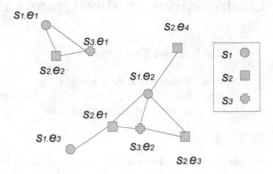

Fig. 1. Example Co-location. Three spatial features: $\{s_1, s_2, s_3\}$. s_1 has three events, s_2 has four events, and s_3 has two events.

have $s(e_1) \neq s(e_2)$; (3) $\cup_{e \in E}\{s(e)\} = S'$. The participation ratio $PR(s, S')$ of a spatial feature s in a subset of spatial feature $S' \subseteq S$ is the probability of an event of s participating in a supporting clique of S'. For example, in Figure 1, there are three spatial features $S = \{s_1, s_2, s_3\}$. Two events of different spatial features are connected if their distance is less than a user specified threshold. Feature s_2 has four events and three of them participate in a clique supporting $\{s_1, s_2, s_3\}$. So $PR(s_2, \{s_1, s_2, s_3\})$ is $3/4$. Then participation index $PI(S')$ is defined as $PI(S') = \min_{s \in S'}\{PR(s, S')\}$. In Figure 1, $PI(s_1, s_2, s_3) = \min\{2/3, 3/4, 2/2\} = 2/3$. The problem of co-location mining is to find all subsets of spatial features with participation indices above a user defined threshold.

When an event is uncertain, there are two main challenges: (1) How to define participation ratio and participation index in a probabilistic manner? (2) How to efficiently find co-locations when the number of uncertain events is large? Let us assume a simple dataset of 5 spatial features. Let each spatial feature have 4 events and each event has 3 possible locations. Then we will be $3^{4^5} = 3.49 \times 10^9$ possible worlds.

Contributions

- We formulate a framework for mining uncertainty co-locations. The framework is based on possible world model. The participation index to measure co-location with uncertainty is defined in a probabilistic manner;
- We prove a lemma that allows an instance centric algorithm to be developed which is significantly faster than using all possible worlds to compute the participation index;
- We propose an Uncertain Apriori co-location mining algorithm (UApriori) and an Uncertain Feature Tree based algorithm (UFTree) to efficiently mine co-locations under uncertainty; we further propose event-based optimizations and table search techniques for optimizing these algorithms;
- Algebraic analysis and experiments on a large Shanghai taxi trajectory dataset and synthetic datasets show the effectiveness and efficiency of our

proposed model and algorithms. We observed that the UFTree algorithm outperforms UApriori by an order of magnitude not only for co-locations of large sizes but also for datasets with high level of uncertainty. This is an important insight in mining uncertainty co-locations.

2   Problem Definition

We follow the commonly used uncertainty model Block-Independent Disjoint Scheme [3] to define the problem of uncertainty co-location. A probabilistic spatial feature s is given by a set of uncertain events $s = \{e_1, e_2, \ldots\}$. An uncertain event e_i is represented by a set of d-dimensional points $u_1, u_2, \ldots$ reflecting all possible instances of e_i. Each instance u_k is assigned with a probability $P(u_k)$ denoting the probability that e_i appears at u_k. Let h be a user given distance threshold. In Figure 2(a), there are three spatial features. Spatial feature s_1 has three uncertain events, spatial feature s_2 has four uncertain events, and spatial feature s_3 has two uncertain events. Each instance of an uncertain event of s_1, s_2, and s_3 is represented by a circle, a square, and a plus sign respectively. Probabilities are assigned to instances of an uncertain event that sums up to 1 within the event. Two instances of different spatial features that are within a user given threshold h are connected by a line. In this example, s_1, s_2, s_3 may be three animal species, e.g. $s_1 =$ Egyptian plover, $s_2 =$ Nile crocodile, and $s_3 =$ monitor lizard. An event of s_1 represents a particular Egyptian plover, e.g. plover Smith. And the instance of the plover Smith is an location where Smith is spotted.

A possible world $w_s = u_1^s, u_2^s, \ldots$ of a spatial feature s is a set of instances containing one instance from each event of s and occurring with a probability of $P(w_s) = \prod_k P(u_k^s)$. Let W_s be the set of all possible worlds for s, then $\sum_{w \in W_s} P(w) = 1$. We are given a set n of probabilistic spatial features $S = \{s_1, s_2, \ldots, s_n\}$, a possible world w_S is given by the combination of a possible world of each spatial feature, i.e. $w_{s_1}, w_{s_2}, \ldots, w_{s_n}$ with a probability of $P(w_S) = \prod_i P(w_{s_i})$. Obviously, let W_S be the set of all possible world for S, then $\sum_{w \in W_S} P(w) = 1$ as well.

Definition[Complete Possible World] Given a subset of probabilistic spatial features S, W_S^c includes all the instances of all the events of S and is called the complete possible world with respect to S. For any instance, we use $e(u)$ to denote the event associated with u and use $s(u)$ to denote the spatial feature of u.

All of the instances in Figure 2(a) is an example of complete possible world of $\{s_1, s_2, s_3\}$.

Definition[Supporting Clique] A set of instances c in the complete world $w_{S'}^c$ is supporting a subset of spatial feature $S' \subseteq S$ if: (1) c forms a clique using a user given distance threshold; (2) $s(u_1) \neq s(u_2)$ (which also implies $e(u_1) \neq e(u_2)$) for any $u_1 \in c, u_2 \in c$ and $u_1 \neq u_2$; (3) $\cup_{u \in U} \{s(u)\} = S'$. c is called a supporting clique of S'.

In Figure 2(a), $\{s_1, s_2, s_3\}$ has 6 supporting cliques. However, not all of them can appear together in the same possible worlds as explained later.

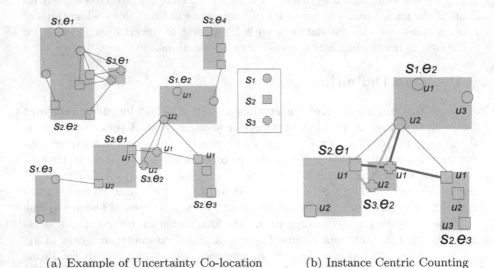

(a) Example of Uncertainty Co-location   (b) Instance Centric Counting

Fig. 2. In 2(a), feature s_1 has three uncertain events, i.e. $\{s_1.e_1, s_1.e_2, s_1.e_3\}$. Event $s_1.e_1$ has three uncertain instances.

The participation ratio and index are used to prune the co-locations with low prevalence. Different from certain cases, the participation ratio can not be directly calculated by counting the instances participating in a supporting clique. In an uncertain case, the participation ratio of an instance will be determined by both the probability of possible worlds and the supporting cliques it participates.

Definition[Participation Ratio of an Instance] For a subset S' of S and a user given distance threshold, the participation ratio of an instance u of spatial feature s_i in S' is defined as the sum of probabilities of all the possible worlds that instance $s_i.u$ participates in a supporting clique of S' and this can be calculated as:

$$PR(u, S') = \sum_{\{w | \exists c, u \in c, c \subseteq w, w \in W_{S'}, c \text{ supports } S'\}} p(w) \qquad (1)$$

In Figure 2(b), $PR(s_1.e_2.u_2, \{s_1, s_2, s_3\})$ is the sum of the probabilities of all the possible worlds that contain at least one supporting clique of $\{s_1, s_2, s_3\}$ with $s_1.e_2.u_2$ participating. In this example, it includes all of the worlds that contain one or more of the three supporting cliques that $s_1.e_2.u_2$ participates in the figure. Since $s_1.e_2, s_2.e_1, s_3.e_2$ and $s_2.e_3$ have 3, 2, 2, 3 possible instances respectively, the total number of possible world should be 36. Among those, 3 possible worlds have a supporting clique of $\{s_1.e_2.u_2, s_2.e_1.u_1, s_3.e_2.u_1\}$, 3 contain that of $\{s_1.e_2.u_2, s_2.e_1.u_1, s_3.e_2.u_2\}$, and 2 contain that of $\{s_1.e_2.u_2, s_2.e_3.u_1, s_3.e_2.u_1\}$. Given equal probability of instances, all of the possible worlds have the same probability of 1/36. However, one possible world contains both of the cliques

$\{s_1.e_2.u_2, s_2.e_1.u_1, s_3.e_2.u_1\}$ and $\{s_1.e_2.u_2, s_2.e_3.u_1, s_3.e_2.u_1\}$. So there are totally $3 + 3 + 2 - 1 = 7$ possible worlds that $s_1.e_2.u_2$ participates in a supporting clique of $\{s_1, s_2, s_3\}$. The participation ratio of $s_1.e_2.u_2$ should be $PR(s_1.e_2.u_2) = 7 \times (1/36)$.

Definition[Probabilistic Participation Ratio] Let $s_i.U$ be all possible instances of s_i where $s_i \in S'$ and $|s_i|$ be the number of events of s_i. The probabilistic participation ratio (short as participation ratio or PR here after) of s_i is:

$$PR(s_i, S') = \frac{1}{|s_i|} \sum_{u \in s_i.U} PR(u, S') \qquad (2)$$

Definition[Probabilistic Participation Index] For a subset $S' \subseteq S$, the probabilistic participation index (short as participation index or PI) $PI(S')$ of S' is defined as follows:

$$PI(S') = \min PR(s_i, S') \qquad (3)$$

The problem of mining co-location under uncertainty is to find all subsets of spatial features with participation indexes above a user given threshold θ.

3   Instance Centric Counting

The naive way to calculate the participation ratio of an instance in S' is to enumerate all the possible worlds and then sum up the probabilities of the worlds where the instance participates in a supporting clique of S'. However this method is very expensive as the number of possible worlds is very large. We propose an instance centric calculation of the participation ratio, taking supporting cliques involved into consideration. However, in this method, avoiding over-counting is a challenge. We prove a Lemma to allow instance centric counting which will enable efficient algorithms to be developed. Using this method, we can get the same result as summing up possible worlds above. We first define the relationship between a possible world and a clique.

Definition[Clique Probability] For a supporting clique c of S', the probability $P(c)$ is the sum of the probability of all of the worlds which contains the clique c. And it can be represented as:

$$P(c, S') = \sum_{\{w | w \in W_{S'}, c \in w\}} P(w) \qquad (4)$$

Finding all the possible worlds that contain c is very expensive. However, it is easy to prove the lemma 1. The proof is similar of that of lemma 2 and is omitted due to space constraint.

Lemma 1. *For a supporting clique c of S', the probability $P(c, S')$ is equivalent to:*

$$P(c, S') = \prod_{u \in c} P(u) \qquad (5)$$

If an instance u only participates in one supporting clique, then the participation ratio of u is the probability of the clique. However, an instance can participate in multiple cliques in the same world, resulting in over counting if we simply sum up the probabilities of all the supporting cliques that u participates (naive PR calculation). For example, Figure 2(b) is a subset of Figure 2(a) which includes all events and cliques that relate to $s_1.e_2.u_2$. The two instances of $s_3.e_2$ cannot happen at the same time (yellow lines and green lines could not happen in the same world). So not all three cliques can happen in the same world. However, clique $\{s_1.e_2.u_2, s_2.e_1.u_1, s_3.e_2.u_1\}$ and $\{s_1.e_2.u_2, s_2.e_3.u_1, s_3.e_2.u_1\}$ can co-exist, resulting in over counting of the worlds that contain both if we use the naive PR calculation.

Definition[Coexistence] Two supporting cliques c_1 and c_2 of S' can coexist, denoted as $\odot(c_1, c_2) = 1$, in the same possible world under the following condition: $\forall u_1 \in c_1, u_2 \in c_2$, if $e(u_1) = e(u_2)$, then $u_1 = u_2$, where $e(u)$ denotes the event associated with the instance u.

Definition[Coexistence of a Set of Supporting Cliques] A set of supporting cliques C can coexist and is denoted as $\odot(C) = 1$ if every pair of the cliques can coexist.

Definition[Probability of Coexisting Clique Set] The probability of a set of coexisting supporting cliques C of S' is the sum of the probabilities of those possible worlds in which C happens:

$$P(C, S') = \sum_{\{w | \forall c \in C, c \subseteq w, w \in W_{S'}\}} P(w) \tag{6}$$

Lemma 2. *The probability of a set of coexisting supporting cliques C of S' can be calculated as:*

$$P(C, S') = \prod_{u \in I} P(u) \tag{7}$$

where I is the union of all instances of the events of the cliques in C.

Proof: Let E be the set of all events. For any $E' \subseteq E$, it is easy to see that the sum of the probabilities of all possible worlds $W_{E'}$ that include an instance from each event in E' is 1. The probability of coexisting clique set is the sum of all possible worlds which contain the coexisting supporting clique set C. We can divide E into two parts: E_1 is the set of events having an instance in the coexisting cliques and E_2 is the complementary set. So a possible world that contains C can also contain any other possible worlds of E_2. So the summation can be calculated as $\prod_{u \in I} P(u) \times (\sum_{w \in W_{E_2}} P(w))$. We know that $\sum_{w \in W_{E_2}} P(w) = 1$, so the probability of a set of coexisting supporting cliques is $\prod_{u \in I} P(u)$.

From lemma 1 we know that the probability of a clique c_1 is the sum of all of the possible worlds that contain the clique. However, those possible worlds can contain both cliques c_1 and c_2. So the probabilities of the possible worlds which contain both cliques will be counted two times if they can coexist when counting the participation ratio of an instance. To summarize, when there are coexisting cliques and they contribute to the participation of the same instance, over counting will happen. We define such cliques as star supporting cliques.

Definition[Star Supporting Clique Set] All the supporting cliques of S' that an instance u participates is called the star supporting clique set of u in S' and is denoted as $\star(u, S')$.

Lemma 3. *To an instance $s_i.u_k$ in a complete possible world, the star supporting clique set $\star(s_i.u_k, S')$ of S' can be used to calculate the participation ratio of the instance as follows where $l = |\star(s_i.u_k, S')|$:*

$$
\begin{aligned}
PR(s_i.u_k, S') = &\sum_{c_j \in \star(s_i.u_k, S')} P(c_j, S') \\
&- \sum_{c_1 \in \star(s_i.u_k,S'), c_2 \in \star(s_i.u_k,S'), \odot(c_1,c_2)} P(c_1 \cup c_2, S') + \ldots \\
&+ (-1)^{l+1} \sum_{c_1 \in \star(s_i.u_k,S'), \ldots c_l \in \star(s_i.u_k,S'), \odot(c_1, \ldots c_l)} P(c_1.. \cup c_l, S')
\end{aligned}
\tag{8}
$$

Proof: From the definition of participation ratio, $PR(u_k) = \sum_{w_i \in W_{S'}} P(w_i)$, where $W_{S'}$ is the set of possible worlds which instance u_k participates in forming a supporting cliques of S'. So we can prove lemma 3 by checking whether the probability of each possible world in $W_{S'}$ has been counted and only counted once. For each w_i in $W_{S'}$, let us assume that w_i has n coexisting cliques $\{c_1, c_2 \ldots c_n\}$. From the definition of the probability of coexisting clique set, we know that every time we calculate $P(C, S')$, we count the probability of w_i once. Here C represents any possible combinations of coexisting cliques in w_i. So when we calculate $P(c_1)$ as well as $P(c_1 \cup c_2)$ and any other combinations, we count $P(w_i)$ once. In lemma 3, for w_i, each of $P(c_1) \ldots P(c_n)$ is added to the probability of w_i once. Each of $P(c_1 \cup c_2) \ldots P(c_{n-1} \cup c_n)$ subtracts $P(w_i)$ once. So $P(w_i)$ has been counted $C_n^1 - C_n^2 + \ldots (-1)^{n+1} C_n^n = 1$ times.

Lemma 4. *(Anti-monotone Property) The participation index is anti-monotone with respect to the number of features in the co-location.*

Proof: Let l, l' be two co-locations and $l' \subseteq l$. We use W_l and $W_{l'}$ to represent the set of possible worlds which have supporting cliques of these two co-locations. For each possible world w in W_l, w contains supporting cliques of l. Because $l' \subseteq l$, w will also has supporting cliques of l'. So if $w \in W_l$, then $w_i \in W_{l'}$. Therefore, $W_l \subseteq W_{l'}$. From the definition of participation index, $PI(l) = \sum_{w \in W_l} P(w)$ and $PI(l') = \sum_{w \in W_{l'}} P(w)$, we can conclude that $PI(l') \geq PI(l)$.

4   Mining Co-location from Uncertain Data

In this section, we describe the framework of mining uncertain co-locations. While participation ratio can be calculated by counting supporting cliques [10] in the certain case, the participation ratio of uncertain co-locations is calculated by the probabilities of instances. Thus we divide the process of mining co-location in two steps, (1) finding uncertainty co-locations and their supporting cliques; (2) calculating the participation ratios. In section 4.1, we propose an Apriori based algorithm to generate supporting cliques under uncertainty. Section 4.2 will present the feature tree driven method combined with maximal clique method. We propose event-based pruning and clique-feature table searching to reduce the computational cost. In section 4.3, we will present the algorithm for calculating participation ratios and section 4.4 presents algebraic analysis of the computational complexity.

In the following part, we use C_k to represent a set of size k cliques, use L_k to represent a set of size k co-locations and use C_M to represent a set of maximal cliques.

4.1   Uncertain Apriori (UApriori) Co-location Miner

We first propose an Apriori like algorithm in the instance level and will present event level pruning in section 4.1. We process in the uncertain instance level to generate the supporting cliques for each co-location. Algorithm 1 shows the process of how to generate uncertainty co-location by Apriori algorithm. Here we use an example to illustrate. Figure 3(a) is the Apriori-gen process for the example in figure 2(b). We use plane sweep algorithm to get the neighbor relation of between the instances of S_1, S_2, S_1, S_3 and S_2, S_3 (the first three tables). They are also the supporting cliques C_2 of size-2 co-locations L_2. Assuming all of them have a participation index greater than the threshold, we can generate size-3 co-locations by them. First we generate the supporting cliques for the co-location $\{s_1, s_2, s_3\}$ from C_2 through joins. For example, we use the $size$-2 cliques $\{s_1.e_2.u_2, s_2.e_1.u_1\}$ and $\{s_1.e_2.u_2, s_3.e_2.u_1\}$ in C_2 to generate the $size$-3 supporting clique $\{s_1.e_2.u_2, s_2.e_1.u_1, s_3.e_2.u_1\}$. When generating the new supporting clique, we calculate the distance between $s_2.e_1.u_1$ and $s_3.e_2.u_1$ by their coordinates instead of searching the c_2 table to find the record of clique $\{s_2.e_1.u_1, s_3.e_2.u_1\}$.

Algorithm 1. Generating co-locations from searching table

Apply plane sweep algorithm to generate instance neighbor relation C_2
Generate size 2 co-locations L_2 by the neighbor relation
$k = 2$
while $L_k \neq \emptyset$ **do**
   $k = k + 1$
   Generate supporting cliques C_k from C_{k-1} for co-location L_k
   Calculate the participation index of each size k co-location
   Generate size k co-location set L_k
end while
Return $L_2 \cup L_3 ... \cup L_{k-1}$

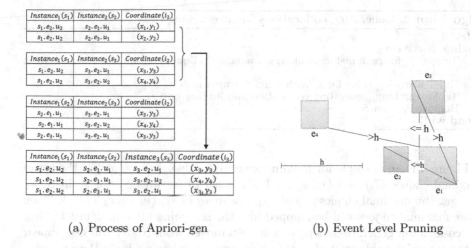

$Instance_1(s_1)$	$Instance_2(s_2)$	$Coordinate(i_2)$
$s_1.e_2.u_2$	$s_2.e_1.u_1$	(x_1,y_1)
$s_1.e_2.u_2$	$s_2.e_3.u_1$	(x_2,y_2)

$Instance_1(s_1)$	$Instance_2(s_3)$	$Coordinate(i_2)$
$s_1.e_2.u_2$	$s_3.e_2.u_1$	(x_3,y_3)
$s_1.e_2.u_2$	$s_3.e_2.u_2$	(x_4,y_4)

$Instance_1(s_2)$	$Instance_2(s_3)$	$Coordinate(i_2)$
$s_2.e_1.u_1$	$s_3.e_2.u_1$	(x_3,y_3)
$s_2.e_1.u_1$	$s_3.e_2.u_2$	(x_4,y_4)
$s_2.e_3.u_1$	$s_3.e_2.u_1$	(x_3,y_3)

$Instance_1(s_1)$	$Instance_2(s_2)$	$Instance_3(s_3)$	$Coordinate(i_3)$
$s_1.e_2.u_2$	$s_2.e_1.u_1$	$s_3.e_2.u_1$	(x_3,y_3)
$s_1.e_2.u_2$	$s_2.e_1.u_1$	$s_3.e_2.u_2$	(x_4,y_4)
$s_1.e_2.u_2$	$s_2.e_3.u_1$	$s_3.e_2.u_1$	(x_3,y_3)

(a) Process of Apriori-gen   (b) Event Level Pruning

Fig. 3. Example for the process of UApriori

Event Level Pruning (UApriori-E). However in uncertainty, each event of a feature may have many uncertain instances. Both algebraic and experiments show that with large number of instances, the join process to generate cliques become inefficient. To optimize this process, we propose an event level pruning approach to apply event neighbor relation checking first and only proceed to instance level join when the minimal distance between two events are within user given distance.

There are three relations between two events: the minimum distance larger than threshold h, the maximal distance less than h, and $Distance_{Max} > h \cap Distance_{Min} < h$. For the first case, such as e_1 and e_4 in figure 3(b), we can prune e_4 directly. For the second, such as e_1 and e_2, all instances in e_2 can form neighbor relation with instances in e_1 without calculating the distances between them. We only need to calculate the distance between instances in the third case.

4.2   Uncertain Feature Tree Co-location Miner (UFTree)

Feature tree has been used in mining long pattern association rules. However, the main challenge in using the same structure in co-location mining is the lack of given set of transactions. In this paper, we propose feature tree based methods in mining uncertain co-locations that deal with the lack of transactions. In the first step, we use the plane sweep algorithm and the Bron-Kerbosch algorithm [4] to find all maximal cliques in the complete possible world (optimization will be discussed shortly) and add them into C_M. A naive way to generate the supporting cliques is to enumerate all sub-cliques of every maximal clique and map them into different co-locations. But without a pruning process, the cost is prohibitive especially when some large size cliques exist. Here we build a searching table to help generating the supporting cliques and the co-location set. Algorithm 2 describes the framework of this process.

Algorithm 2. Generating co-locations from searching table

$L = \emptyset$
while $S \neq \emptyset$ **do**
    To $s_i \in S$, for each instances of s_i, get maximal cliques contain this instance from C_M
    Build searching table for s_i with other features in S
    Table searching, generating co-location and adding into L
    Remove s_i from S
end while

Figure 4 is an example for building searching table. $\{c_1, c_2, ..c_6\}$ is the set of maximal cliques C_M and $\{s_1, s_2, ..s_6\}$ is the set of features S. For s_1 in S, we first get the maximal cliques containing instances of s_1: $\{c_1, c_2, c_3, c_4, c_5\}$. Then those maximal cliques will be mapped into the searching table in figure 4. After the co-location generating step, s_1 will be removed from S and will not appear in the searching table built after this. For example, when we build the searching table for s_2, instances from s_1 will not be included in the table.

Maximum Cliques:

$c_1: \{s_1.u_1, s_2.u_1, s_3.u_1, s_5.u_1\}$
$c_2: \{s_1.u_1, s_2.u_2, s_4.u_1\}$
$c_3: \{s_1.u_2, s_3.u_1, s_4.u_2\}$
$c_4: \{s_1.u_3, s_2.u_3, s_4.u_2\}$
$c_5: \{s_1.u_4, s_3.u_2, s_5.u_2\}$
$c_6: \{s_2.u_1, s_3.u_2, s_4.u_3, s_6.u_1\}$

Searching Table

	c_1	c_2	c_3	c_4	c_5
s_1	$s_1.u_1$	$s_1.u_1$	$s_1.u_2$	$s_1.u_3$	$s_1.u_4$
s_2	$s_2.u_1$	$s_2.u_2$		$s_2.u_3$	
s_3	$s_3.u_1$		$s_3.u_1$		$s_3.u_2$
s_4		$s_4.u_1$	$s_4.u_2$	$s_4.u_2$	
s_5	$s_5.u_1$				$s_5.u_2$

Fig. 4. Searching Table

To search this table and generate co-locations, we introduce the searching strategy on a tree-based structure. The breadth first searching with follow set pruning approach.

Uncertainty Feature Tree Searching with Follow Set Pruning (UFTree-FP Algorithm). We introduce the follow set searching strategy to generate co-locations on the table. We first present key definitions for this strategy.

Definition[Pre-Node Set] Node $\mathcal{N}$ is a node in the feature tree. The pre-node set $Pre\{\mathcal{N}\}$ of $\mathcal{N}$ is the set of nodes which have the same father node as $\mathcal{N}$ and at the left side of $\mathcal{N}$ in the feature tree.

Definition[Follow set] The follow set of node $\mathcal{N}$ is a set of features which may form co-locations with the features in node $\mathcal{N}$.

For example, for the root node of the tree in figure 5(a), the set of $\{s_2, s_3, s_4, s_5\}$ is the follow set of this root node. Feature s_1 will be combined with each of them to generate new co-locations in the next level of this tree.

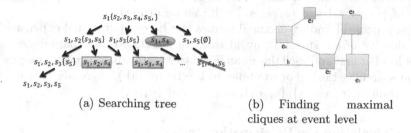

(a) Searching tree

(b) Finding   maximal
cliques at event level

Fig. 5. Strategie on searching the table

Algorithm 3 describes the process of breadth-first search to generate co-locations related to s_i. When searching the table, we will generate the follow set for each node. First we set the follow set of the root node as $S_{s_i} - s_i$ and add the root node to the node list. Then we process each node in the node list. For every node, we combine the feature set of the node with each feature S_{Fi} in the follow set and get a new co-location. If the participation index of the new co-location is greater than the threshold, this node will be added into the node list and the feature S_{Fi} will be added into the follow sets of every pre-node of this node.

Algorithm 3. The breadth-first searching algorithm

Set co-location set $L = \emptyset$
Node list $NL = \emptyset$
Add root node to NL
Set the follow set of root node as $\mathcal{S}_{s_i} - s_i$
$i = 0$
while $i \leq NL.size$ **do**
　$\mathcal{N} = NL[i]$
　Combine the feature set of $\mathcal{N}$ with each feature s_{Fi} in the follow set and generate
　a new co-location l_n
　Calculate the participation index PI
　if $PI \geq$ threshold **then**
　　Add $\mathcal{N}'$ to the node list, the feature set of $\mathcal{N}'$ composed by the features in l_n
　　and the follow set is node $\mathcal{N}$'s follow set besides s_{Fi}
　　Add s_{Fi} to the follow sets of each pre-node of $\mathcal{N}'$
　　$L = L \cup l_n$
　end if
end while
Return L

Figure 5 (b) is an example for breadth-first search. During the search for the node of $\{s_1, s_3\}$, if the participation ratio is greater than the threshold, s_3 will be added into the follow set of node $\{s_1, s_2\}$. The participation ratio of $\{s_1, s_4\}$ is less than the threshold, so the sub-tree of this node $\{s_1, s_4, s_5\}$ will be pruned and s_4 will not be added into the follow sets of the pre-nodes of $\{s_1, s_4\}$. So the node of $\{s_1, s_2, s_4\}$ and $\{s_1, s_3, s_4\}$ will also be pruned.

The process of finding maximal cliques can be computational expensive with high density of instances. To avoid this, we generated maximal cliques in the event level at first and used this as an index to generate instance level cliques used in the searching table. For example, in the figure 5(b), there are four maximal event cliques: $\{e_1, e_2, e_3\}, \{e_2, e_4\}, \{e_3, e_5\}$ and $\{e_4, e_5\}$.

4.3   Calculating the Participation Index

We have proposed lemma 3 to calculate the participation ratio of each possible instance. If there was no coexisting cliques, lemma 3 can be represented as: $PR(s_i.u_k, S') = \sum_{c_j \in \star(s_i.u_k, S')} P(c_j, S')$. So we can calculate all the participation ratios of possible instances by calculating the probability of each supporting clique and add the probability to the ratio of the instances belonging to this clique. If c_i and c_j are two coexisting cliques and $\{u_1, u_2...u_n\}$ are their common possible instances, to these instances, their participation ratios $PR(u)$ will be deducted by $P(c_i \cup c_j)$ according to the lemma 3. So we need to calculate the probabilities of different combinations of coexisting cliques and add or subtract the results to the ratios of instances shared by those cliques according to lemma 3. After getting all of the participation ratios, we can calculate the participation indexes of co-locations by the method described in the definition of probabilistic participation ratio and participation index.

4.4   Computational Complexity

Table 1 summarizes the parameters used in complexity analysis as well as in the experiment section. In the UApriori algorithm, when generating $size(k+1)$ co-locations from $size\ k$, the worst case is: for each supporting clique, we have to see if they have $k-1$ instances in common with other cliques. If we define $N_{C(k)}$ as the number of supporting cliques of size k co-locations, the cost of generating supporting cliques is: $O(\sum_{k=2}^{w-1}(k-1) \mid N_{C(k)} \mid)$. For the uncertain feature tree algorithm, the cost of generating step can be divided into three parts: finding out all maximal cliques, building searching table, and generating supporting cliques. Finding maximal cliques is a NP problem [4]. Thus we propose to do this step on event level in our approach to avoid high computational cost. For each feature, we use N_{CL} as the average number of supporting cliques of each co-location and N_{F_i} as the number of cliques related to each feature. So the cost of building the searching table is $\mid F \mid N_{F_i} N_{CL}$. To generate supporting cliques for each co-location set, we need to search a line of the searching table and the worst case is $O(N_{F_i})$. So the cost of UFTree algorithm is $O(\mid F \mid N_{F_i} N_{CL} + N_{F_i} \sum_{k=2}^{w} \mid C(k) \mid)$.

Table 1. Parameters

N_s	Number of Features
O_L	The overlapping ratio of co-locations
$\overline{I}$	Average number of instances of an event
N_L	Number of co-locations
$\overline{L}$	Average length of co-locations
$\overline{E}$	Average number of events of each feature
$\overline{C}$	Average number of supporting cliques of a co-location in event level

5   Experiments

In this section, we will first compare the algorithms on different synthetic datasets to show the influence of parameters. Since there were no algorithms solving the same problem with us, here we only compared our own method and analysis the results. Then we will use the Shanghai taxi trip dataset to illustrate the application of our algorithms in real world.

5.1   Synthetic Data Generation

Our model of the synthetic data is similar of that of paper [10]. Parameters in table 1 are used to generate synthetic data. We generated uncertainty datasets with a broad range of values for the chosen parameters to evaluate the performance of those algorithms. The length of co-location, the average number of supporting cliques, and the average number of instances in an event will be picked from three Poisson distributions $P(\overline{L})$, $P(\overline{C})$ and $P(\overline{I})$. The generating step will be divided into three steps: 1. Generating a co-location; 2. Generating event level supporting cliques for this co-location; 3. Transforming each event into possible instances using a Poisson distribution in an area of given size. We generated the datasets by setting different values for $\overline{L}$ (from 2 to 11), $\overline{E}$ (from 3 to 16), $\overline{I}$ (from 2 to 8), and setting 5 values for O_L : $0, 20\%, 40\%, 60\%, 80\%$.

5.2   Performance on Synthetic Data

Figure 6 presents the running time of the algorithms under different values of those parameters. We first compare those algorithms by changing the value of $\overline{E}$ and $\overline{I}$. The results are shown in figure 6(a) and 6(b). Figure 6(a) shows the influence of $\overline{E}$. $\overline{E}$ has little influence on the UFTree algorithm. However, with the increasing value of $\overline{E}$, the advantage of event pruning process becomes significant between UApriori and UApriori-E algorithms. In figure 6(b), we can see that with the increase of the average number of instance, the running times of both UApriori-E and UApriori algorithm increase quickly. But the UFTree algorithm still use a very short time. This phenomenon illustrates that generating cliques in the searching table performs much better than the Apriori-gen process.

From figure 6(c), we can conclude that the overlapping ratio has litter influence on those algorithms. In figure 6(d), we tested the influence of the threshold of participation ratio on running time.

The UFTree algorithm performs much better due to its fast matching process and searching strategy. To illustrate this phenomenon, we can simplify the expression of the computational complexity and just concentrate on the generating process to analyze the computational cost. For UApriori, UApriori-E and UFTree algorithms, the worst case of generating the supporting cliques can be represented by $O((k-1)+(\overline{E}\times\overline{I})^2)$, $O((k-1)+\overline{E}^2+\overline{E}\times(\overline{I})^2)$ and $O((\overline{E}\times\overline{I})^2)$ in generating the supporting cliques of a size k co-locations. But with different values of parameters, the costs of UApriori and UFTree algorithm are typically far less than that. For example, if event from one feature can only form one event pair with another event and there is only one instance pair between them, the cost of UApriori and UFTree will become $O((k-1)+\overline{E}^2+\overline{E}\times\overline{I}^2)$ and $O(\overline{E})$. So with the increase of $\overline{L}$, the UApriori Algorithm becomes much more slower and the UFTree algorithms provides a much better performance. We can see that in the results of figure 6(e).

Tree based algorithm has been used in long pattern association rule mining. A significant advantage of this algorithm is its subset pruning strategy. For example, when we get $\{S_1, S_2, S_3\}$, we need not to test $\{S_2, S_3\}$ due to the anti-monotone property. In our experiment mentioned above (figure 6(a) to 6(e)), we didn't use this strategy. In figure 6(f), we compared the UFTree algorithm with and without the subset pruning: UFTree-Subset Pruning and UFTree algorithm. The results show that subset pruning can improve the performance especially in mining co-locations of large sizes. The reason is larger sized co-locations contain more subsets (equal to the number of power set). From the results, we can see that when the size of co-locations is larger than 8, the UTFree-Subset Pruning method made a significant improvement.

5.3   Experiments on Real Data

Data for the framework is based on trips of 17,000 Shanghai taxis for one day (May 29, 2009); the dataset contains 432,327 trips. Each trip includes the starting and destination coordinates and the start/end time. Our real data experiment is based on the premise that if people from two or more areas often take taxi to some other common areas, these two or more areas may have some potential characteristics in common. For example, they may be all university areas or residential areas. Of course, people from adjacent areas also tend to have the same destinations because adjacent areas tend to have similar characteristics. But the taxi trips start and end coordinates can be different even when they have the same start and destination areas. For example, a university may have several entrances and people can get on or off taxis at any entrance. So it becomes an uncertainty co-location problem. The university area is an event and each entrance can be a possible instance in our uncertainty co-location model.

We generate the dataset according to our uncertainty model from the real world taxi trips. Firstly, we divided the region of Shanghai into $500 \times 300m$

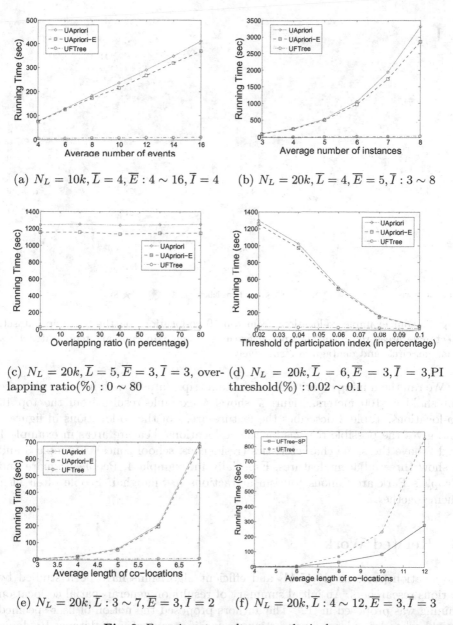

(a) $N_L = 10k, \overline{L} = 4, \overline{E} : 4 \sim 16, \overline{I} = 4$ (b) $N_L = 20k, \overline{L} = 4, \overline{E} = 5, \overline{I} : 3 \sim 8$

(c) $N_L = 20k, \overline{L} = 5, \overline{E} = 3, \overline{I} = 3$, over- (d) $N_L = 20k, \overline{L} = 6, \overline{E} = 3, \overline{I} = 3$,PI lapping ratio(%) : $0 \sim 80$ threshold(%) : $0.02 \sim 0.1$

(e) $N_L = 20k, \overline{L} : 3 \sim 7, \overline{E} = 3, \overline{I} = 2$ (f) $N_L = 20k, \overline{L} : 4 \sim 12, \overline{E} = 3, \overline{I} = 3$

Fig. 6. Experiment results on synthetic data

rectangles. Top 680 areas with more than 300 trip starting points are chosen as feature areas. Each area is a spatial feature. For each feature s, an area e with a trip starting in s and ending in e is an event of s. The trip ending points in an event e of s is then clustered into small groups, each representing an instance with probability as the ratio of the number of trip ending points in the small group over that in e.

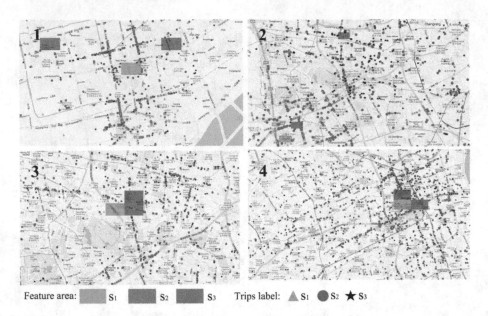

Feature area: ▮ S₁ ▮ S₂ ▮ S₃   Trips label: ▲ S₁ ● S₂ ★ S₃

Fig. 7. Four example co-locations from top 10 co-locations on the taxi trip dataset. Rectangle areas represent the spatial features. Their trip ends are represented by a triangle, circle and pentagram respectively.

We ran the mining algorithms on the taxi trips dataset with instance distance threshold as 100 meters. Figure 7 shows 4 example results from the top 10 co-locations. Table 2 describes the feature areas of the co-locations in figure 7 and gives the possible reasons for the co-locations. Feature areas in example 1 and 2 have the same characteristics (residences, school zones). Example 3 and 4 show three adjacent features. Especially in example 4, People's Square and People's Park are famous tourism attractions in Shanghai. People often visit them together.

6   Related Work

Co-location patterns [10,15,11] and efficient algorithms have been studied by various researchers. An initial summary of results on general spatial co-location mining was proposed in [12]. The authors proposed the notion of user-specified neighborhoods in place of transactions to specify group of spatial items. By doing so, they can adopt traditional association rule mining algorithms, i.e., Apriori [2] to find spatial co-location rules. An extended version of their approaches was presented in [10]. Fast mining algorithms are proposed in [15]. The algorithms combine the discovery of spatial neighborhoods with the mining process. The algorithms work on a given pattern, star, clique, or generic, and calculate the participation index using an extension of a spatial join algorithm. Others have considered complex co-location patterns including negative co-locations [11] and

Table 2. Examples of real data co-location

Co-location Samples with Top Rank Participation Index
1. {Chengxi Flower Garden }, { Shanxin Apartment}, {Zongtongwan Garden}
Explanation: All of those three areas are large and up-scale neighborhoods with many community facilities and amenities.
2. {Shanghai Longbai Middle School, Longbai Rainforest Preschool}, {Longbai No.1 Elementary School, Xijiao Apartment}, {Shanghai Tianshan Middle School}
Explanation: These three areas largely overlap with a school zone.
3. {Hongxian Unit, Xinle Unit}, {Hongxian Unit}, {Xianxia Villa}
Explanation: Three adjacent residential areas.
4. {People's Square, Shanghai Grand Theatre},{People's Square}, {People's Park}
Explanation: Three famous adjacent entertainment and tourism areas in Shanghai.

zonal co-locations [6] with dynamic parameters, i.e., repeated specification of zone and interest measure values according to user preferences. The problem of mining co-location patterns with rare spatial features has been studied in [9]. The authors applied a new measure which considers the maximum participation ratio of the co-location patterns.

In [1], the authors study the problem of frequent pattern mining with uncertain data. They extend the Apriori-based, hyper-structure based, and pattern growth approaches and conclude that experimental behaviour of different classes of algorithms is very different in the uncertain case as compared to the deterministic case. In [13], the authors studied on the probabilistic spatially co-locations and proposed a dynamic programming algorithm which is suitable for parallel computation. They proposed the uncertainty model by introducing the concept of existential probability of an instance. The probability of a possible world can be calculated as the product of the probabilities of presence or absence of all uncertainty instances. By multiplying the probability of possible worlds and the participation index under certain case together, they can get the finial participation index. This model is only considers the probability of existence while our model considers the possible locations of all instances.

The lexicographic tree (enumeration tree) has been used in mining maximal or long pattern association rules. The counting [14] and pruning [5,16,8] methods in the search tree to generate association rules are not applicable to our problem. In [14], the authors presented the maximal clique generation method. They modified the Bierstone's algorithm [7] and used the concept of α-related to weaken the clique generation process to avoid the edge density is too high.

7   Conclusion

In this paper, we formulated the framework for mining co-locations in uncertain database and defined the notion of participation index in the uncertain case. We presented the UApriori algorithm to mining co-locations in uncertain data and proposed event level pruning (UApriori-E) to make it more efficient. We presented the uncertain feature tree co-location algorithms and introduced different searching and pruning methods. Our experiment showed that in mining large size co-locations or dealing with datasets with high level of uncertainty, the feature tree algorithm with follow set pruning provided the best performance. In future, we will generalize our model to account for distributions of instances instead of the possible instance model used in this paper.

References

1. Aggarwal, C.C., Li, Y., Wang, J., Wang, J.: Frequent pattern mining with uncertain data. In: Proceedings of the 15th ACM SIGKDD (2009)
2. Agrawal, R., Srikant, R.: Fast algorithms for mining association rules in large databases. In: Proc. VLDB (1994)
3. Bernecker, T., Emrich, T., Kriegel, H.-P., Renz, M., Zankl, S., Züfle, A.: Efficient probabilistic reverse nearest neighbor query processing on uncertain data. In: Proc. VLDB Endow. (2011)
4. Bron, C., Kerbosch, J.: Algorithm 457: finding all cliques of an undirected graph. Commun. ACM (1973)
5. Burdick, D., Calimlim, M., Flannick, J., Gehrke, J., Yiu, T.: Mafia: a maximal frequent itemset algorithm. In: TKDE (2005)
6. Celik, M., Kang, J.M., Shekhar, S.: Zonal co-location pattern discovery with dynamic parameters. In: Proceedings of the Seventh IEEE ICDM (2007)
7. Corneil, D.G., Mulligan, G.D.: Corrections to bierstone's algorithm for generating cliques. Commun. ACM (1972)
8. Gouda, K., Zaki, M.J.: Efficiently mining maximal frequent itemsets. In: ICDM (2001)
9. Huang, Y., Pei, J., Xiong, H.: Mining co-location patterns with rare events from spatial data sets. Geoinformatica (2006)
10. Huang, Y., Shekhar, S., Xiong, H.: Discovering colocation patterns from spatial data sets: A general approach. In: TKDE (2004)
11. Munro, R., Chawla, S., Sun, P.: Complex spatial relationships. In: Proceedings of the Third IEEE ICDM (2003)
12. Shekhar, S., Huang, Y.: Discovering spatial co-location patterns: A summary of results. In: Jensen, C.S., Schneider, M., Seeger, B., Tsotras, V.J. (eds.) SSTD 2001. LNCS, vol. 2121, pp. 236–256. Springer, Heidelberg (2001)
13. Wang, L., Wu, P., Chen, H.: Finding probabilistic prevalent colocations in spatially uncertain data sets. In: IEEE TKDE (2013)
14. Zaki, M.J.: Scalable algorithms for association mining. In: TKDE (2000)
15. Zhang, X., Mamoulis, N., Cheung, D.W., Shou, Y.: Fast mining of spatial collocations. In: ACM SIGKDD (2004)
16. Zou, Q., Chu, W.W., Lu, B.: Smartminer: a depth first algorithm guided by tail information for mining maximal frequent itemsets. In: ICDM (2002)

Querying Incomplete Geospatial Information in RDF*

Charalampos Nikolaou and Manolis Koubarakis

National and Kapodistrian University of Athens, Greece
{charnik,koubarak}@di.uoa.gr

1 Introduction

Incomplete information has been studied in-depth in relational databases and knowledge representation. It is also an important issue in Semantic Web frameworks such as RDF, description logics, and OWL 2. In [6], we introduced RDF^i, an extension of RDF for representing incomplete information using constraints. We defined a semantics for RDF^i and studied SPARQL query evaluation in this framework. Given the current interest in publishing geospatial datasets as linked data (e.g., by Ordnance Survey in the UK), RDF^i is an excellent framework for encoding, possibly incomplete, qualitative and quantitative geospatial information which is found in these published datasets. RDF^i is also interesting because when the constraint language used can express the topological relations of RCC-8 [8], the recent OGC standard GeoSPARQL [7] for querying geospatial information expressed in RDF, becomes a special case of RDF^i.

In this paper, we propose the problem of *implementing an efficient query processing system for incomplete temporal and geospatial information in* RDF^i as a challenge to the SSTD community. For the case of incomplete temporal information in relational databases, two such systems have been implemented in the past [1,3], but their query languages are rather limited. There is also the paper [2] which studies a closely related problem, temporal relationships in databases, but which has no related implementations. Finally, the knowledge representation language Telos [5] allows the modelling of incomplete temporal knowledge but well-known implementations such as ConceptBase have not implemented this functionality. To the best of our knowledge, no such relational database system or RDF store exists for the geospatial case, although there are some description logic reasoners that come close in terms of functionality [11]. In the rest of the paper, we present the RDF^i framework and outline the hard problems that have to be solved if such a query processing system is to become a reality. As in the theoretical foundation of [6], we concentrate on geospatial information only.

2 RDF^i by Example

RDF^i [6] is a framework that extends RDF in a general way with the ability to represent and query incomplete information. Incomplete information often

---

* This work was supported by the European FP7 project TELEIOS (257662) and the Greek NSRF project SWeFS (180).

M.A. Nascimento et al. (Eds.): SSTD 2013, LNCS 8098, pp. 447–450, 2013.

```
gag:Region   rdfs:subClassOf geo:Feature.     gag:WestGreece  rdf:type gag:Region.
gag:Municipality rdfs:subClassOf geo:Feature. gag:OlympiaMuni rdf:type gag:Municipality.
noa:Hotspot rdfs:subClassOf geo:Feature.      noa:hotspot     rdf:type noa:Hospot.
noa:Fire rdfs:subClassOf geo:Feature.         noa:fire        rdf:type noa:Fire.
gag:OlympiaMuni geo:hasGeometry ex:oGeo.   ex:oGeo rdf:type sf:Polygon.
ex:oGeo geo:asWKT "POLYGON((..))"^^geo:wktLiteral.
noa:hotspot geo:hasGeometry ex:rec.        ex:rec geo:asWKT "POLYGON((..))"^^geo:wktLiteral.
gag:WestGreece geo:sfContains gag:OlympiaMuni.      noa:hotspot geo:sfContains noa:fire.
```

Fig. 1. RDF^i database using the vocabulary of GeoSPARQL

arises when sensing the real-world due to the inherent imprecision of measuring devices. For example, in wild-fire monitoring applications, satellite images are analyzed to detect hotspots (i.e., pixels of the image corresponding to geographic regions that are probably on fire). Due to the medium resolution of the satellite images, these hotspots correspond to rectangles that are 3km by 3km wide. Thus, a useful representation of the real world situation of a hotspot is to state that there is a geographic region with unknown exact coordinates where a fire is taking place, and that region is included in the known geometry for the hotspot. Such information is usually combined with other relevant sources, such as administrative boundaries, to aid decision makers in managing the fire.

This scenario is captured by the RDF^i database of Fig. 1 using the vocabulary offered by GeoSPARQL (namespace `geo`). This database contains data about the hotspot, the fire, the administrative region of West Greece, and the municipality of Olympia in which the hotspot has been detected. To make the example more interesting, we have assumed that the exact administrative boundary for West Greece is unknown. The second set of triples in Fig. 1 encode the geometries of Olympia and the hotspot using the Well-Known Text standard format, while the last two triples state the containment relations that hold between West Greece and the municipality of Olympia, and the hotspot and fire.

RDF^i uses the concept of e-literals to represent property values that exist but are unknown or partially known. Partial knowledge about property values is expressed by a quantifier-free formula of a first-order *constraint language*. For simplicity, we have not used the more powerful syntax of RDF^i to capture our partial knowledge about West Greece and the hotspot. Instead we have expressed it as triples as in GeoSPARQL (last two triples of Fig. 1). The following is a SPARQL query that uses the topology vocabulary extension of GeoSPARQL to query the database of Fig. 1 for fires that are inside the region of West Greece.

`SELECT ?f WHERE {?f rdf:type noa:Fire. gag:WestGreece geo:sfContains ?f.}`

The specification of GeoSPARQL does not propose a semantics or algorithm for computing the answer to such a query, although the answer is entailed by the triples of Fig. 1. The answer can be computed by computing the entailment of relation `geo:sfContains` between `gag:WestGreece` and `noa:hotspot` from the fact that the geometry of the hotspot is contained in the geometry of Olympia, and then include it in the computation of the transitive closure for relation `geo:sfContains` to derive the triple `gag:WestGreece geo:sfContains noa:fire`. In contrast, SPARQL query evaluation over RDF^i databases as studied in [6] gives an algorithm for computing such entailments.

3   Query Processing Challenges

As the example of the previous section shows, and as we have shown more generally in [6, Theorem 2], the challenge with which a system is faced for answering queries such as the above is the efficient computation of the entailment relation $\Phi \models \Theta$ where Φ, Θ are quantifier-free first-order formulae of a constraint language that is capable of expressing the topological relations of various frameworks such as RCC-8, DE-9IM, OGC Simple Features, etc. Computing such an entailment usually reduces to checking the consistency of constraint networks that involve qualitative spatial relations among regions identified by a URI and constant ones. This combination of qualitative and quantitative constraints has been studied in detail for temporal constraints but similar results do not exist for spatial constraints. Only recently there has been some work on topological relations among polygonal regions, but is limited to atomic and complete constraint networks, which are far away from real datasets.

RDF stores supporting linked geospatial data are expected to scale to *billions* of triples like their non-spatial counterparts and recent work in this area is encouraging [4]. Can this level of scalability be achieved when qualitative spatial relations come into play? A good approach here might be to start with algorithms with low polynomial complexity and try to implement them as efficiently as possible. In the temporal case, this approach has been followed successfully by temporal reasoners. In addition, there might be cases where network structure can be exploited (e.g., hierarchical organization of geographical regions).

To answer the above question, we have compared the performance of checking the consistency of tractable RCC-8 constraint networks using the well-known Path Consistency (PC) algorithm as implemented in the state of the art qualitative spatial reasoners Renz Solver [9], PyRCC-8 and PyRCC-8 ∇ [10], and a relational counterpart of the PC algorithm of [9] as a SQL program in PostgreSQL only to reach the same conclusion. Our findings were that none of the implementations scale so as to be qualified for use in implementations of query processing algorithms for the entailment problem described above.

Fig. 2a shows the performance of these implementations using real-world linked geospatial datasets containing only qualitative spatial relations from RCC-8 (this is a much simpler problem than the one considered in Fig. 1). Each dataset is presented in increasing order of its size, while the sizes range from $1500/2000$ nodes/edges to $276728/590443$ nodes/edges. For the last dataset there are no measurements for any of the implementations, because they all crass either immediately (reasoners) or after some amount of time (PostgreSQL) due to memory allocation errors[1]. For the first case, this is because they allocate a two-dimensional array to represent the input constraint network. This array is of size N^2 where N is the number of nodes. Thus, even for medium-sized graphs, these implementations fail to run, a drawback that is not present in the relational-based implementation which can complete one or two iterations of PC.

The most interesting observation for such graphs is that they are sparse, hence representations of the input graph that are based on two-dimensional arrays are

---

[1] Setup: Intel Xeon E5620, 2.4 GHz, 12MB L3, 48GB RAM, RAID 5, Ubuntu 12.04.

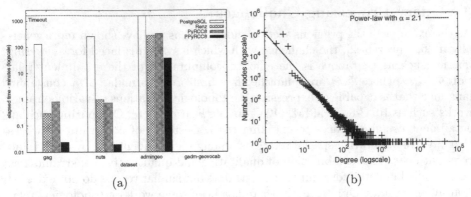

(a)   (b)

Fig. 2. (a) Performance of PC for various datasets (b) Degree distribution for gadm-geovocab

not appropriate. This fact is depicted in Fig. 2b where the degree distribution among the nodes of the largest dataset is shown. Fig. 2b shows that the input RCC-8 constraint network is sparse following a power-law distribution. The primary characteristic of such graphs is that most of the nodes have very low degree, while only a small number of the nodes have high degree number leading to star-shaped graphs. Another observation is that real datasets comprise constraint graphs with edges of three kinds: (non-)tangential proper part, externally connects, or equals. Such kind of edges reflect networks that are composed of components with a hierarchical structure which calls for further optimization.

References

1. Brusoni, V., Console, L., Terenziani, P., Pernici, B.: Later: Managing temporal information efficiently. IEEE Expert 12(4), 56–64 (1997)
2. Chaudhuri, S.: Temporal relationships in databases. In: VLDB, pp. 160–170 (1988)
3. Griffiths, A., Theodoulidis, B.: SQL+i: Adding temporal indeterminancy to the database language SQL. In: Morrison, R., Kennedy, J. (eds.) BNCOD 1996. LNCS, vol. 1094, pp. 204–221. Springer, Heidelberg (1996)
4. Kyzirakos, K., Karpathiotakis, M., Koubarakis, M.: Strabon: A Semantic Geospatial DBMS. In: Cudré-Mauroux, P., et al. (eds.) ISWC 2012, Part I. LNCS, vol. 7649, pp. 295–311. Springer, Heidelberg (2012)
5. Mylopoulos, J., Borgida, A., Jarke, M., Koubarakis, M.: Telos: Representing knowledge about information systems. ACM Trans. Inf. Syst. 8(4), 325–362 (1990)
6. Nikolaou, C., Koubarakis, M.: Incomplete information in RDF. In: RR (2013)
7. Open Geospatial Consortium: OGC GeoSPARQL - A geographic query language for RDF data. OGC® Implementation Standard (2012)
8. Randell, D.A., Cui, Z., Cohn, A.G.: A spatial logic based on regions and connection. In: KR (1992)
9. Renz, J., Nebel, B.: Efficient methods for qualitative spatial reasoning. Journal of Artificial Intelligence Research (JAIR) 15, 289–318 (2001)
10. Sioutis, M., Koubarakis, M.: Consistency of Chordal RCC-8 Networks. In: ICTAI (2012)
11. Stocker, M., Sirin, E.: PelletSpatial: A hybrid RCC-8 and RDF/OWL reasoning and query engine. In: OWLED (2009)

Link My Data: Community-Based Curation of Environmental Sensor Data

Heiko Müller, Chris Peters, Peter Taylor, and Andrew Terhorst

Intelligent Sensing and Systems Laboratory, CSIRO, Hobart, Australia
{heiko.mueller,chris.peters,peter.taylor,andrew.terhorst}@csiro.au

Abstract. One of the biggest obstacles to reuse of third-party sensor data is a lack of knowledge about data properties (e. g., provenance and quality) leading to a lack of trust in the data. LINK MY DATA (LMD) is a first step towards overcoming this problem. LMD provides a platform for data curation that allows users to share knowledge about individual sensors and sensor observations. The system supports annotation and transformation of sensor data on the Web to improve data quality and (re-)usability. Transformations, for example, allow to remove gaps or change the temporal resolution for time series of sensor observations. Transformation results are stored as views on the original data and made available for other users. Within this demonstration we present the main features of LMD. We show how LMD makes it easy to curate other people's data. The audience is welcome to interact with LMD's web-based user interface to share their knowledge about sensor data on the Web.

1 Motivation

Sensors play an important role in environmental monitoring and modelling. The recent years have seen a vast increase in the amount of environmental sensor data being published on the Web. Data (re-)use, however, is cumbersome for various reasons: First, sensor data comes in different types (e. g., aggregated, instantaneous, cumulative) and often has different temporal and spatial scales requiring transformation to homogenise the data. Second, data quality is often varying and a lack of knowledge about data properties (e. g., the environment in which a sensor is deployed, or awareness of events like flooding that temporarily deteriorate the quality of observations) hampers our ability to decide whether a given piece of data is fit for its purpose.

LINK MY DATA (LMD) is a first step towards overcoming this problem by creating a common platform for sharing knowledge about sensor data on the Web. The goal is to leverage the collective knowledge of data users to improve data quality and (re-)usability. LMD achieves this through use of a flexible annotation system. Annotations are either free form text, elements from controlled vocabularies, or references to other objects or resources on the Web. Temporal annotations highlight events that influence a sensor and the quality of its observations. Examples of annotations include descriptions of time series interpolation types or highlighting regions of missing data. Annotations help making data properties explicit and are a first step towards improving data (re-)usability.

M.A. Nascimento et al. (Eds.): SSTD 2013, LNCS 8098, pp. 451–455, 2013.

LMD has a strong focus on historic time series data from terrestrial sensors, like stream gauges or automatic weather stations, used for example in data-driven modelling [7]. Users often fix quality flaws in historic data individually, e. g., they eliminate gaps in the data when using the data. Currently, such modifications are not shared with other users. LMD addresses this problem by allowing users to define and share modified, and annotated, versions of online data. We provide a set of transformation operators that allow users to define transformation programs for time series data on the Web. Transformation results are stored as virtual or materialised views on the original data and are published on the Web. We thereby ensure that users can benefit from data cleaning efforts of previous users and avoid redoing modifications over and over again.

LMD is among the first approaches towards active user participation in sensor data curation. It is inspired by current trends towards semantically-enriched sensor data [6], crowdsourcing [3], and collaborative data management [4,2]. We describe the overall system architecture and main features of LMD in Section 2. The proposed demonstration is outlined in Section 3.

2   System Overview

The overall architecture of LMD is shown in Figure 1. The system is implemented in Java and run as a web service using Apache Tomcat. Data is stored in a dedicated data store. There are four main components in the architecture: The *Data Access Layer* provides data access and abstracts from the underlying data store. We currently use MongoDB[1], mainly for its scalability and its flexibility in storing objects with similar but varying schema. The *Query Engine* allows searching for objects with specific annotations (e.g., sensors in a region that observe rainfall in a specified data format). The *Time Series Engine* is responsible for executing data transformation programs. The system is accessible via a *Web API*. All data is represented in JSON.

LMD has several features that distinguish it from other systems for annotation (see [5,8] for surveys) and collaborative data curation (e. g., [1,4,2]). In LMD, any resource with a uniform resource identifier (URI) can be annotated. The system itself provides resolvable resource locators (using HTTP URLs) for all system generated objects. In LMD, there are three types of annotations: literals, links and temporal annotations. Literals and links are similar to triples in RDF, i. e., labeled edges in a graph connecting resources and values. Values, however, can be structured, i. e., JSON documents. Temporal annotations support definition of intervals and points in time that are of importance to a resource (e. g., the period during which a sensor was out of order or calibration events). Given that all annotations have URLs, we can annotate them as well, e. g., to provide references to further information about an event or to maintain provenance information.

Time series in LMD are sequences of (time, value)-pairs that are published as (part of) structured documents on the Web. Each document may contain multiple time series. LMD allows users to define *source descriptions* that specify how

---

[1] http://www.mongodb.org

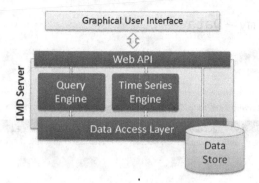

Fig. 1. Architecture of the Link My Data platform

to extract time series from documents. Source descriptions are format-specific. We currently support JSON, XML, and CSV as document formats. For XML and JSON, we use path expressions to specify the elements that belong to individual time series in the document. LMD represents time series as sequences of JSON objects and assigns unique identifiers to them, thus allows annotation.

LMD is equipped with a set of operators for transforming time series data. This allows users to manipulate existing data and share the results. Given that we cannot update existing data directly (it may be stored anywhere on the Web), we store transformation results as virtual or materialised views in our data store. The current set of operators supports temporal aggregation, update of observed values, automatic annotation generation (e. g., gap detection), as well as concatenation and merging of time series. Operators take one or more time series as input and produce a time series as output. Operators are chained together to form a data transformation pipeline. These pipelines are specified either programmatically or using a simple scripting language (see Figure 2).

A special operator allows generation of transformation pipelines on-the-fly from time series annotations. This operator can be used to define generic transformation pipelines that are applicable to different time series. We then associate such transformers with certain annotations. For example, if a time series is annotated as of type cumulative rainfall then a script is applicable that transforms cumulative rainfall into hourly proceeding total. Generic transformers follow the idea of 'active objects', i. e., objects that present users with a set of actions that are applicable to them.

3 Demonstration

LMD comes with a Java library for application development. We used this library to implement a graphical user interface (see Figure 2). A current prototype is available at http://lmd.it.csiro.au.

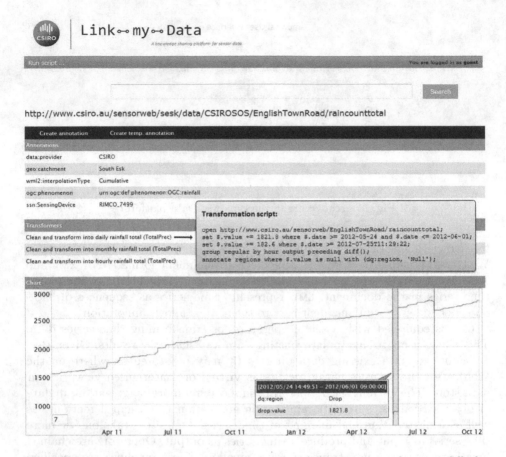

Fig. 2. Screenshot showing annotations and a transformation script for a rainfall observation time series

In our demonstration we show how LMD makes it easy to create, update, and delete annotations. We demonstrate LMD's search and query capability that enables searching for annotated resources based on keywords and conditions over annotations. LMD's ability to annotate annotations is another powerful feature. We show how provenance information about annotations can be represented as annotations and used to validate an annotation, i.e., detect changes to the original data that cause an existing annotation to become obsolete.

To demonstrate LMD's capability to manipulate existing sensor data, we use sensor data collected by the South Esk Hydrological Sensor Web (SESK)[2]. Annotations are used to highlight quality flaws in the data. We show how to derive transformation scripts from these annotations to provide clean views on the data. A particular example is shown in Figure 2. Here, we first update existing values

---

[2] http://www.csiro.au/sensorweb/sesk

in a time series of cumulative rainfall observations. We then aggregate the data to produce a time series representing hourly rainfall totals.

Annotation scripts can be defined manually for individual time series. The full power of LMD's transformation capability, however, is shown when defining generic scripts that apply to different time series and are parameterised by existing annotations. We give examples of such generic scripts in our demonstration, e. g., to transform cumulative rainfall into hourly, daily, or monthly totals.

We use sensor observations from several web data services such as SESK and the Australian Cosmic Ray Sensor Network[3] in our demonstration. We demonstrate how to add new source descriptions to extract time series data. Participants are encouraged to interact with the system at any time to provide annotations or define data transformations. We are interested in setting up source descriptions for data services that are of interest to participants to allow them to make the data available for annotation and transformation and thereby enhance data quality and (re-)usability by others.

Acknowledgements. The Intelligent Sensing and Systems Laboratory and the Tasmanian node of the Australian Centre for Broadband Innovation is assisted by a grant from the Tasmanian Government which is administered by the Tasmanian Department of Economic Development, Tourism and the Arts.

References

1. Auer, S., Dietzold, S., Lehmann, J., Riechert, T.: Ontowiki: A tool for social, semantic collaboration. In: Workshop on Social and Collaborative Construction of Structured Knowledge (CKC) (2007)
2. Buneman, P., Cheney, J., Lindley, S., Müller, H.: The database wiki project: a general-purpose platform for data curation and collaboration. SIGMOD Record 40(3), 15–20 (2011)
3. Doan, A., Ramakrishnan, R., Halevy, A.Y.: Crowdsourcing systems on the worldwide web. Commun. ACM 54(4), 86–96 (2011)
4. Liang, S., Chen, S., Huang, C.Y., Li, R.Y., Chang, D.Y., Badger, J., Rezel, R.: Geocens: Geospatial cyberinfrastructure for environmental sensing. In: GIScience 2010, Zurich, Switzerland (2010)
5. Reeve, L., Han, H.: Survey of semantic annotation platforms. In: Proceedings of the 2005 ACM symposium on Applied computing, SAC 2005, pp. 1634–1638 (2005)
6. Sheth, A., Henson, C., Sahoo, S.S.: Semantic sensor web. IEEE Internet Computing 12(4), 78–83 (2008)
7. Solomatine, D., See, L., Abrahart, R.: Data-driven modelling: Concepts, approaches and experiences. Practical Hydroinformatics 68, 17–30 (2008)
8. Uren, V., Cimiano, P., Iria, J., Handschuh, S., Vargas-Vera, M., Motta, E., Ciravegna, F.: Semantic annotation for knowledge management: Requirements and a survey of the state of the art. Web Semant 4(1), 14–28 (2006)

---

[3] http://waterml2.csiro.au/cosmoz

CrowdPath: A Framework for Next Generation Routing Services Using Volunteered Geographic Information

Abdeltawab M. Hendawi, Eugene Sturm, Dev Oliver, and Shashi Shekhar

Department of Computer Science and Engineering,
University of Minnesota, Minneapolis, MN 55455, USA
{hendawi,oliver,shekhar}@cs.umn.edu, sturm049@umn.edu

Abstract. Our proposed system *CrowdPath* is based on the hypothesis that people know their commute area better than conventional routing services that use traditional digital roadmaps and shortest path algorithms. The knowledge and experiences of drivers reflected in volunteered commute routes may provide better routes. By leveraging such available volunteered geographic information (VGI), our goal is to investigate next-generation routing services to further reduce travel time, fuel consumption, and improve navigation. Previous related work summarizes GPS tracks into a landmark graph which is used for answering routing queries. In contrast, *CrowdPath* directly queries a collection of map-matched GPS tracks to recommend paths from a source location to a destination. Our evaluation using real GPS tracks illustrates the promise of *CrowdPath* in significantly reducing travel time compared to routes from common routing providers. In the future, *CrowdPath* may be extended to adapt route recommendations by start time and provide safe paths using volunteered crime and accident reports.

1 Introduction

Given a source location, a destination, and a user preference function (e.g., minimize travel time, fuel consumption), routing services provide a set of paths from source to destination that optimizes the user preference function. Examples of routing services include in-car GPS devices, web-based applications, cellphones, etc.

The proliferation of volunteered geographic information (VGI) such as GPS tracks donated by individuals via forums such as OpenStreetMap [3] has created an opportunity for providing next generation routing services. Next-generation routing is important for critical societal applications such as further reducing fuel consumption, commute time, and traffic congestion, as well as improving navigation. Preliminary evidence for the transformative potential of next-generation routing includes the experience of UPS, which saves millions of gallons of fuel by simply avoiding left turns [2]. Such turn modeling information (as well as traffic light synchronization, slowdowns at curves) may be gleaned from the volunteered commute routes of individuals.

This problem is challenging because accurate modeling of driving conditions such as traffic delays, potholes, traffic light synchronization, and weather conditions, is very difficult. Accurate modeling of user behavior and recent changes in maps is also challenging.

M.A. Nascimento et al. (Eds.): SSTD 2013, LNCS 8098, pp. 456–461, 2013.

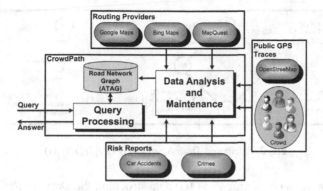

Fig. 1. The *CrowdPath* System Architecture

Previous related work summarizes GPS tracks into a landmark graph which is used for answering routing queries. For example, T-Drive [4] considers GPS tracks for taxis where the tracks are sampled every five minutes. This approach is able to offer rush hour vs non rush hour routing services for high cab-traffic areas. Edges in the landmark graph correspond to a sequence of segments in the roadmap used to pose routing queries.

In contrast, the proposed *CrowdPath* system directly queries a collection of map-matched GPS tracks to recommend paths from a source location to a destination. *CrowdPath* is based on the hypothesis that people know their commute area better than conventional routing services that use traditional digital roadmaps and shortest path algorithms. In fact, individuals examine many paths using alternate preference functions (e.g., minimize travel time, minimize fuel consumption, avoid potholes, scenic routes) from their home to work until they find their favorite ones. This may be seen in the way some people do not completely follow directions given by routing providers in certain nearby areas where they have more experience. Based on this observation, *CrowdPath* collects volunteered GPS tracks for people's daily trips from different sources such as OpenStreetMap [3]. These tracks are analyzed to extract only the valid ones that do not contain unreasonable behavior such as speed limit violations, e.g., speeds above 90 miles per hour on highways. The travel time and distance costs of all possible sub-trip combinations inferred from the valid tracks are compared to the ones recommended by routing providers.

Scope and Outline: A comparison with T-Drive and shortest path algorithms is limited to a conceptual discussion. A detailed comparison is outside the scope of this work. We assume that GPS commute routes are donated and issues such as privacy are beyond the scope of the present research. These limitations may be investigated in future work.

2   System Overview

System Architecture: The architecture of the *CrowdPath* system is given in Figure 1. *CrowdPath* has three main types of data sources, namely, *public GPS traces*, *routing providers*, and *risk reports*. The system has three main components, namely, the

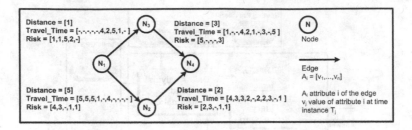

Fig. 2. Attributed Time Aggregated Graph (ATAG)

attributed time aggregated graph (ATAG) data structure, the *Data Analysis and Main-tenance* module, and the *Query Processing* module. This section discusses the data sources and system components.

Data Sources: The proposed framework relies on data extracted from three main sources. (1) *Public GPS Traces.* Our main source of volunteered data is OpenStreeMap (OSM) which allows the download of volunteered GPS traces filtered by areas of interest. We use a tool provided on the OSM wiki called JOSM which allows us to view OSM data and collect the GPS traces for a specified area. For example, in Minneapolis, Minnesota, USA, we found about 326 GPS tracks consisting of hundreds of thousands of GPS points. (2) *Routing Services.* This data source accesses three major mapping and routing services, namely, Google Maps, Bing Maps, and MapQuest to extract direction, distance cost, and travel time cost for their recommended paths between the start and end points of a user trip. The data is accessed by sending *http* requests to each provider to get the recommended path between start and end points. The time and distance costs are then obtained from the returned path. (3) *Risk Reports.* The final data source contains data about car accidents and crime linked to their locations and times during the day. This source will be included in future versions with more routing preference functions.

ATAG Data Structure: To save all the data in a way that allows us to compute the shortest path from a given source location to a destination, we introduce the *attribute time aggregated graph* (ATAG). Potentially ATAG may be used to support other preference routing functions, e.g., less fuel consumption, less car accidents, less pollution, or paths with more services. This graph data structure is an adapted version of the existing time aggregated graph (TAG) [1] in which intersections are represented as nodes and roads as edges, and each edge can have multiple weights. The existing TAG graph carries values for only one feature or attribute such as travel time cost. By contrast, each edge in the ATAG data structure can have more than one attribute, e.g., distance, time, risk; for each edge, we store multiple weights in different time slots. Figure 2 illustrates the idea of storing multiple attributes for each single edge in the ATAG data structure. As can be seen in the figure, each edge has three different attributes (i.e., distance, travel time, and risk) and each attribute can have either one value like the distance attribute or many values like the travel time attribute which has values in ten different time instances. The edge between (N_1, N_3) has a weight of one distance unit for all times of the day, four units for travel time starting at the sixth time slot (weights for travel time do not exist during the first five time slots), and four units for risk starting on the first time slot of the day. The main difference between ATAG and the landmark graph [4] is that the latter summarizes GPS tracks into the main road segments. In contrast, ATAG directly maps the GPS tracks to their equivalent road edges in the underlying road network graph. Also, the advantage of introducing ATAG over the traditional time-series graph [1] is that the former does not need to replicate the data which in turns saves storage and reduces I/O cost.

Analysis and Maintenance Module: The main function of this module is to extract and store valid GPS tracks from the set of all tracks obtained. These valid tracks are made available to the query processor for answering routing queries. The analysis and maintenance module only extracts the tracks that do not violate certain validation checks (e.g., speed limit). The module then map matches the points inside each track to their corresponding nodes in ATAG. After that, the costs, e.g., travel time or distance, between each pair of points are used to update the weights of the equivalent edges at the same time slot of the day. Another potential technique is to store valid tracks as a whole and retrieve them without requiring further computation. This can be done using a three dimensional matrix (start, destination, time slot), however, the matrix will be sparse.

Query Processing Module: The dynamic nature of the underlying road network is captured through ATAG where each edge could have multiple weights at different time slots of the day. Traditional shortest path algorithms that assume one weight for each edge are not suitable for answering routing queries in our dynamic graph. The main challenge here is to design efficient, valid, and expandable query processing algorithms to take into consideration the spatio-temporal aspects of the network while evaluating routing queries. An efficient algorithm reduces computation cost, a correct algorithm returns a valid path, and an expandable routing algorithm answers potential queries using the multiple attributes on edges in just one traversal, e.g., finding shortest paths with less fuel consumption. To this end, we leverage the SP-TAG algorithm [1] to find the optimal path between two nodes given a graph structure with multi-weight edges.

3   Demonstration

The goal of this demonstration is to solicit audience feedback on how to improve the *CrowdPath* system to include other routing preference functions (e.g., fuel consumption, pollution, services on roads, potential coupons and sales, safety, etc.). We plan to prioritize these preference functions based on the feedback we get.

Our demonstration will include the following. First, the user can interact with the *CrowdPath* system to browse the available GPS-tracks and examine the structure of associated .gpx files (basic format for storing GPS tracks). The user is also able to examine the validation techniques we applied to the volunteered GPS-tracks to eliminate the tracks with issues (e.g., speed limit violations and topologically unreasonable tracks). The audience will see a sample of the valid tracks versus the invalid ones. In addition, we will present some of the GPS-tracks that are similar to the routes given by the conventional routing services and show how we do the analysis and comparison. Furthermore, we will present a coverage map to illustrate the distribution of the obtained

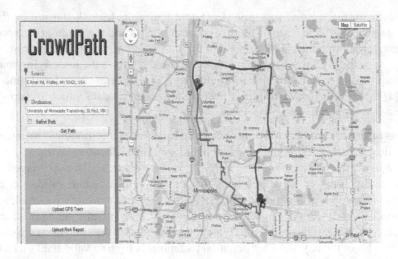

Fig. 3. The *CrowdPath* Main GUI (Best in color)

volunteered GPS-tracks in different areas on the map. Moreover, the user is able to upload new GPS tracks from which the system extracts valid tracks and sub-tracks. Finally, the user can examine the main GUI of the *CrowdPath* system, Figure 3, to ask for a routing service by entering the start and destination locations, specified as latitude and longitude or as an exact address. The user will be able to compare the routes returned by conventional routing providers (blue line) and the route recommended based on the volunteered GPS tracks (red line) in terms of the total travel time and distance. For example, consider the start location *3308 California St NE, Minneapolis, MN 55418, USA* and the destination *University of Minnesota Transit way, St Paul, MN 55114, USA.* A traditional routing service recommends a route that has a travel time of 19 minutes whereas the route recommended by *CrowdPath* has a travel time of 14 minutes. In this test case, the *CrowdPath* route is faster and saves about 26% of the total driving time, which may reduce fuel consumption and the impact on both the environment and traffic.

4   Conclusion

In this demonstration paper we investigated the problem of next generation routing services using volunteered commute routes. This problem is important for critical societal applications such as further reducing travel time, and fuel consumption. We presented the *CrowdPath* system that is based on the hypothesis that people know their commute area better than conventional routing services that use traditional digital roadmaps and shortest path algorithms. *CrowdPath* directly queries a collection of map-matched GPS tracks to recommend paths from a source location to a destination. We presented an evaluation using real GPS tracks that demonstrates the promise of *CrowdPath* in significantly reducing travel time compared to routes from common routing providers.

In future work, we plan to elaborate further on the system details (e.g., data structures and algorithms). We also plan to extend *CrowdPath* to adapt route recommendations by start time and provide safe paths using volunteered crime and accident reports.

References

1. George, B., Kim, S., Shekhar, S.: Spatio-temporal network databases and routing algorithms: A summary of results. In: Papadias, D., Zhang, D., Kollios, G. (eds.) SSTD 2007. LNCS, vol. 4605, pp. 460–477. Springer, Heidelberg (2007)
2. Lovell, J.: Left-hand-turn elimination (December 2007), http://goo.gl/3bkPb
3. OSM. Public GPS traces (January 2013), http://www.openstreetmap.org/traces
4. Yuan, J., Zheng, Y., Zhang, C., Xie, W., Xie, X., Sun, G., Huang, Y.: T-drive: driving directions based on taxi trajectories. In: ACM SIGSPATIAL GIS, California, USA (November 2010)

Interactive Toolbox for Spatial-Textual Preference Queries

Florian Wenzel, Dominik Köppl, and Werner Kießling

Department of Computer Science, University of Augsburg
D-86135 Augsburg, Germany
{wenzel,koeppl,kiessling}@informatik.uni-augsburg.de

Abstract. Spatial-textual data is ubiquitous on websites such as
OpenStreetMap or Wikipedia and is published progressively by Open
Government Data Initiatives. Those data sources provide the base for
novel mashup applications for Location-Based Services. We present a
mobile prototype for the San Francisco area based on *Preference SQL*
that allows an intuitive expression of spatial keyword queries. Search
can further be extended towards temporal, categorical, and numerical
attributes. Our demo provides a fully flexible toolbox for generating extended spatial-textual queries in non-metric spaces which are evaluated
using a *Best-Matches-Only* query model. Spatial relevance is defined by
an asymmetric routing distance supporting complex geometries. Textual
relevance is determined using the Apache Lucene library.

1 Introduction

Websites such as Tripadvisor or Wikipedia provide spatial-textual data in the
form of geo-tagged content. Additionally, Open Government Data Initiatives
lead to the publication of large amounts of public spatial data by cities such as
San Francisco or Berlin. All these data sources are a base for up-and-coming
mobile mashup applications that combine base information to provide novel
Location-Based Services (LBS). These integrated applications require a dynamic
combination of different data sources in form of temporary relations. Additionally, numerical, categorical, and temporal attributes are of importance. Spatial
Keyword (SK) Queries have to be extended towards this end to combine all
these attributes in a semantically intuitive fashion in order to provide high quality personalized search results. These preconditions render the use of current
index-based approaches inapplicable. Furthermore, the heterogeneity of spatial
and textual data among data sources demands a flexible query and data model.
The representation of spatial resources as a mix of points and complex geometries asks for a flexible spatial relevance definition. As asymmetric distances
occur through one-way streets and terrain topology, distance axioms have to
be relaxed towards non-metric spaces to allow queries based on net distances.
Semi-structured and unstructured text pose additional demands on textual relevance determination. The Preference SQL System [2] provides a rich toolbox
for developers that meets these new challenges. Base preference constructors can

M.A. Nascimento et al. (Eds.): SSTD 2013, LNCS 8098, pp. 462–466, 2013.

be combined in a flexible fashion via complex preference constructors to form personalized Preference SQL queries. Spatial relevance is evaluated using an asymmetric routing distance. Textual relevance is determined using the Apache Lucene library. We illustrate how developers can employ Preference SQL for innovative LBS by presenting a mobile LBS application for the San Francisco area. This research prototype illustrates how base preferences can be defined and combined without knowledge of any query syntax based on three provided use cases. Movie scenes, restaurants, or landmarks can be retrieved according to their spatial, textual, temporal, numerical, and categorical attributes. Furthermore, the individual importance of base preferences can be modified to form complex Preference SQL queries. To the best of our knowledge, there is no other framework to treat this kind of extended SK queries on non-metric spaces.

2   Preference SQL Overview

The presented demo application relies on the Preference SQL system as underlying representation of preferences as strict partial orders [1]. Preference SQL enhances the SQL standard by a **PREFERRING** clause that specifies preferences by means of preference constructors given in [2]. These preferences are evaluated as *soft constraints* on base of the results generated by SQL hard constraints. The syntax allows for further post-filtering.

Preference evaluation follows a ***Best-Matches-Only query model***, that defines for a preference $P = (A, <_P)$ on a relation $R = (A_1, \cdots, A_n) \supseteq A$ the preference selection operator by

$$\sigma[P](R) := \{t \in R \mid \nexists t' \in R : t.A <_P t'.A\}.$$

The Preference SQL system is implemented as Java-middleware on top of conventional database systems such as Oracle or Postgres and provides a JDBC driver for seamless integration into applications. The system implements a parser, heuristic and cost based optimizer, and efficient evaluation algorithms such as BNL, BNL++, LESS, SFS, and Hexagon [2].

2.1   Constructor-Based Approach

The system follows a constructor based approach by dividing preferences into base preferences operating on attributes, and complex preferences that combine multiple preferences. For an overview of the corresponding query syntax we refer to [2]. Base preference constructors displayed in a 'is-a'-hierarchical view in Figure 1 provide a flexible toolbox for the expression of base preferences on spatial, textual, temporal, numerical, and categorical domains. All displayed base preferences are sub-constructors of the $SCORE_d$ preference, which generates a preference order using a scoring function $f(x)$ as follows:

$$x <_P y \text{ iff } f_d(x) > f_d(y) \text{ with } f_d(x) = \begin{cases} \left\lceil \frac{f(x)}{d} \right\rceil & \text{if } d > 0 \\ f(x) & \text{else} \end{cases}$$

The optional d-parameter can be used to form equivalence classes of equally important $f(x)$ values. Numerical preferences define $f(x)$ based on deviation from a desired input value. In the case of BETWEEN, numerical values within a user-defined interval are preferred. Categorical preferences such as LAYERED define $f(x)$ based on sets of preferred domain values.

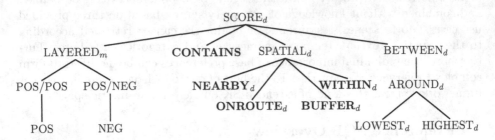

Fig. 1. Taxonomy of base preference constructors

Complex preference constructors provide the ability to either combine multiple preferences or to fine-tune the behavior of a preference. The dual operator δ is such a complex preference. It reverses a preference order, i.e. $a <_{P\delta} b :\Leftrightarrow b <_P a$. Equal importance of base preferences is achieved by applying a Pareto constructor that performs Skyline evaluation. The Prioritization constructor determines the first mentioned preference as more important. Only in cases of equality or indifference considering the first preference, the second preference is evaluated. Ranked importance can be implied by using the RANK_F constructor which performs ranking according to a ranking function F.

2.2 Toolbox of Spatial-Textual Preference Constructors

The presented system provides full flexibility for developers to generate intuitive Preference SQL queries. Within this toolbox, spatial and textual constructors are key components and are thus highlighted consecutively. In contrast to query engines using an exact match query model, spatial-textual preference queries are evaluated as soft constraints following the Best-Matches-Only query model.

▷ **Spatial Preferences:** Given a query geometry g_q preferred by the user, spatial preferences determine those data geometries d_i of a database relation that are best matches according to the spatial relevance defined by the preference constructor. The query model uses Keyhole Markup Language (KML) to define g_q. The data model supports geometry types of underlying PostGIS or Oracle Spatial database extensions. As shown in Figure 1, NEARBY, WITHIN, ONROUTE and BUFFER can be used to express spatial preferences which define relevance based on distance. WITHIN describes a preference on an attribute $A \ni d_i$ favoring geometric objects that are within or close to a region g_q. A geometric object d_i is better than d_j if $dist(g_q, d_i) < dist(g_q, d_j)$. Applicable distances are described below. In the given query model, relevance can be expressed by using a point, a region, or a line as query geometries g_q. Hence, the preferences NEARBY and ONROUTE follow the concept of WITHIN, but differ in the fact that NEARBY accepts a point and ONROUTE a linestring instead of a region for g_q. A more comprehensive intention can be expressed with the BUFFER constructor which also accepts a region and treats geometries closer to g_q as more favorable, but geometries within g_q are considered as least favorable.

As sub-constructors of SCORE, the preference order is induced by a scoring function $f_{dist}(x) := dist(g_q, x)$. $dist$ can be substituted with ST_MaxDistance, ST_Distance and net_dist. ST_Distance calculates the minimal distance between two geometries, whereas ST_MaxDistance computes the maximal distance:

$$ST\_Distance(A, B) := \min_{a \in A, b \in B} \|a - b\|_2$$

$$ST\_MaxDistance(A, B) := \max_{a \in A, b \in B} \|a - b\|_2$$

Here, $\|\cdot\|_2$ is the classic Euclidean norm and A, B are geometric objects regarded as a set of points. Both functions correspond to the SQL/MM standard and are executed by the database system. Alternatively, net_dist is a distance proprietary to Preference SQL as it employs a distance calculated using PgRouting, based on a road network from OpenStreetMap. In contrast to Euclidean distance, the routing result is inherently asymmetric. In urban environments, one-way streets lead to asymmetric results. Using costs such as duration, asymmetry gets even more apparent in mountainous areas where terrain topology becomes of importance. Preference SQL further accounts for transport modality by letting users define routing to be performed for cars or pedestrians.

▷ **Textual Preferences:** The CONTAINS constructor is provided for textual domains. Apache Lucene is used for evaluation by using Lucene's Score as scoring function, consequently the query model provides full Lucene functionality with respect to search keywords, including wildcards and fuzzy search. These scores can be customized by using implemented similarity distances like tf–idf or by defining new ones. Considering the data model, language stemmer and tokenizer for semi-structured data like Wikipedia entries can be specified. This functionality provides a powerful means to state preferences on unstructured or semi-structured text presented on websites in the form of reviews or descriptions. Furthermore, text-search functionality can be combined with any other kind of base preference constructors with the help of complex preference constructors.

3 Showcase Application

We present a dynamic HTML5 application based on the jQuery Mobile framework which communicates with Preference SQL via JDBC. Based on use cases in the San Francisco area, we demonstrate how Preference SQL can be used by developers to provide users with an intuitive preference based LBS. Spatial preferences can be defined by drawing query geometries on a map. Additional icons allow to express those preferences with the dual operator applied. In a dropdown list, preferred city districts can be selected which are also visualized on a map. The top of the screen defines the three consecutively described operations.

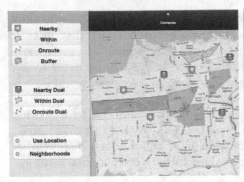
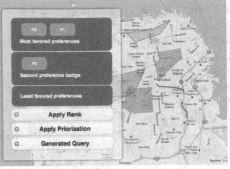

(a) Spatial Preference Selection   (b) Complex Preference Composition

▷ **Define:** A pop-up lets users select one of the following use cases: (U1) combines movie locations listed by the SF Data Project[1] with movie data from the IMDB database[2]. (U2) combines restaurant inspection results from the SF Data Project with reviews from Tripadvisor[3]. (U3) joins geometry data and tags of Points and Regions of Interest (POI/ROI) from OpenStreetMap[4] with geo-tagged content from Wikipedia[5]. For each use case, a form pop-up provides input for individual base preferences on attributes of the joined data sources. A text-search functionality further allows input of complex search terms. Aliases are assigned to each base preference which are displayed in an overview overlay.

▷ **Compose:** A pop-up allows users to compose complex preferences. Initially, three levels are displayed in which base preferences can be dragged from the overview. Levels are arranged in order of importance, with the most important level at the top of the list. As soon as a preference is placed in the lowest level, an additional level is created underneath. Levels are combined using the Prioritization constructor. Preferences within a level are interpreted as equally important by the use of the Pareto constructor. Ranking and further Prioritization can be applied to each level. The generated Preference SQL query is displayed instantly.

▷ **Search:** After clicking the search button, the generated query is evaluated by the Preference SQL system and best-matching results are shown.

References

[1] Kießling, W.: Foundations of Preferences in Database Systems. In: Proceedings of 28th Int. VLDB Conference, pp. 311–322. Morgan Kaufmann (2002)
[2] Kießling, W., Endres, M., Wenzel, F.: The Preference SQL System - An Overview. IEEE Data Engineering Bulletin 34(2), 11–18 (2011)

---

[1] http://www.datasf.org
[2] http://www.imdb.com/interfaces
[3] http://www.tripadvisor.com
[4] http://www.openstreetmap.org
[5] http://www.wikipedia.org

Where Have You Been Today?
Annotating Trajectories with DayTag

Salvatore Rinzivillo[1], Fernando de Lucca Siqueira[2], Lorenzo Gabrielli[1],
Chiara Renso[1], and Vania Bogorny[2]

[1] ISTI - CNR, Pisa, Italy
[2] Universidade Federal de Santa Catarina, Florianopolis, Brazil

Abstract. Traditionally, the information about human mobility behavior, called *diary*, is acquired from volunteers by means of paper-and-pencil surveys. These diaries, representing the mobile activities of individuals, are semantically rich, but lack in spatial and temporal precision. An alternative way is collecting diaries by annotating with activities the GPS tracks of individuals. This is more accurate from a spatio-temporal point of view, but the manual annotation becomes a burdensome work for the user. The tool we propose, called DayTag, is designed as a personal assistant to help an individual to reconstruct her/his diary from the GPS tracks collected by a smartphone. The user interacts through the software to visualize and annotate the trajectories, thus resulting in a simple way to get user diaries.

1 Introduction

The study of the mobility behavior of people in urban areas is essential in any transportation management and planning scenario. For this reason, traditionally this information is collected by means of paper-and-pencil surveys, that are filled in by a limited number of selected volunteers. These surveys detail a typical day of a citizen moving in a city, thus reporting the main trips including the location and time of the daily activities (go to work, go shopping, etc).

Diaries manually collected are semantically very rich and can be very useful for mobility data analysis, since the user may express specific activities he/she performed at the stopped locations. However, this kind of collection lacks in spatial and temporal accuracy, relying only on the user memory. Also, people tend not to report small movements like stopping at the ATM to get cash or at a coffe shop for a coffee, just limiting the reporting to the main activities. Furthermore, these diaries usually represent one specific day and not a typical behavior of the individual over a longer period. Moreover, they miss several spatio-temporal essential information such as the georeferenced location where the user stopped, the route covered during the movement, the duration of the single trips and the single stops. On the other hand, collecting diaries by semantically annotating the traces of individuals automatically gathered by GPS-enabled devices offers a cheap and easy way to collect accurate spatio-temporal information. Almost any modern smartphone can be used as a GPS tracker, thus getting a precise

M.A. Nascimento et al. (Eds.): SSTD 2013, LNCS 8098, pp. 467–471, 2013.

location of the movements. However, the downside is that the manual annotation of activities from a GPS track is a burdensome work and it can cause errors because of user mistaking the annotations.

Since the GPS track collections are becoming more and more common and useful in several application domains, we propose a tool, called *DayTag*, designed as a personal assistant to help an individual to annotate GPS trajectories with activities, thus reconstructing her/his daily diary. The GPS logs are analyzed to extract the visited location (the stops) together with the background geographical information. The system offers a graphical interface by automatically highlighting the relevant visited locations and the time spent during the visit. The user interacts with the software to visualize, correct, complete and annotate her/his trajectories. This results in a simple and quick way to get reliable diaries. We provide a demo of the software using traces of volunteers collected in GPX format with commercial smartphones.

Other tools for diaries collection are available in the literature as applications to be installed on the smartphone and thus the annotation results "online". For example, Easytracker [3] allows the manual real time annotation providing a first phase of automatic segmentation of the trajectory and a second phase to manually annotate each segment with the transportation mode. Another example of smartphone application is TripZoom [2] for monitoring the mobile behavior of the user sensing not only his/her movements, but also taking into account the behavior of the social network of the user. Both applications offer an annotation task in real time at the smartphone level, while the approach of DayTag is to offer an a-posteriori semi-automatic annotation at desktop level. To the best of our knowledge, DayTag is the first tool enabling the user to annotate his/her own trajectory in a a-posteriori fashion using a desktop application. The advantages are manifold: during the movement the user may find the annotation difficult (e.g. while driving a car) or forget to annotate the activity on the smartphone in real time; or the annotation capabilities are limited by the smartphone interface. A desktop application highlighting the trajectories and the main stops allows the user to be more accurate in the description of the context like the place, the motivation of the movement, the transportation mode and other contextual information like the weather.

2   The DayTag Tool

User diaries are essentially annotated trajectories. The annotation can be done at several levels like the places that people visit, the activity or purpose of the visit (e.g. go to shopping, go to work), the transportation means used during the movement, but also other contextual information like the weather, the temperature, events. The choice of the annotation dimensions is inspired from a conceptual model for semantic trajectories called COnSTANT [1], in turn based on the move (the segment of a trajectory where the user is changing the position) and stop (the part of a trajectory where the individual does no change the spatial position) model [6]. In DayTag we support the annotation of both stops and moves: the user can annotate each segment specifying the following *start and end date and time* of the stop (or the move), the *purpose* for the movement - or *activity*, the *transportation mean* and the *weather* conditions.

The trajectory annotation using a state-of-the-art GPS data viewer is not a simple task for a user. Consider the traces of a volunteer collected during a day in Figure 1. We can see that while the spatial component is quite clear the other annotation attributes of the diary are still difficult to be grasped. For example, how much time does it take for the user to move from one stop to the successive one? A manual annotation procedure may include the collection of the user annotations in a separate file and then manually join these annotation with the spatio-temporal data coming from the GPS device.

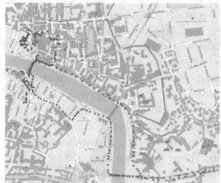

start	end	weather	activity	transp.	place	stop/move
17.49	17.56	sunny	goto shopping	car	via san martino	move
17.56	18.16	sunny	shopping	walking	via san martino	stop
18.16	19.30	sunny	shopping	walking	borgo stretto	move
19.30	19.32	sunny	meet friends	walking	piazza vettovaglie	move
19.32	19.33	sunny	meet friends	walking	piazza vettovaglie	stop
19.33	21.00	sunny	meet for aperitivo	walking	piazza san omobono	move
21.00	21.02	sunny	walking around	walking	Orzo Bruno Pub	move
21.02	21.30	sunny	have beer	walking	Orzo bruno Pub	stop
21.30	21.48	sunny	back home	walking	Via San Martino	move
21.48	21.51	sunny	back home	car	home	move

Fig. 1. Visualization of a raw trajectory collected by a smartphone in the area of Pisa, with no annotations. Screenshot of the annotation file to be filled in manually by the user. The link between the annotation information and the GPS track have to be done manually.

With the purpose of supporting the user in annotating her/his own diary from the GPS tracks, we developed DayTag offering a visual interface and trajectory mining algorithms to compute stop places. DayTag is developed in Java and it exploits available Open Gis Consortium standards to represent and store spatial data on a DBMS. Indeed, the persistence layer is based on the PostgreSQL with PostGIS extension [4]. The graphical interface is developed using Swing libraries. The tools allows the user to load the raw GPS tracks and to preprocess them by removing noise points and by automatically detect stops and moves. Once the mobility episodes are extracted, the user can annotate her activities. The annotated trajectories are then stored in a DB. DayTag graphical user interface is composed of two linked displays: the *Time Display* and the *Map Display* (see Figure 2). The Time Display shows the temporal evolution of the movement, while The Map Display renders the raw GPS points and the visited locations on a map. The Time Display shows the user position with respect to a conceptual reference system of visited locations. To detect such places, the system analyzes the GPS trace of the user to automatically split the sequence of points into moves and stops [5, 6]. When two stops are geographically *close* to each other they are represented by a unique location. This generalization enables the user to abstract from the actual GPS coordinates and to focus only on the relevant visited places. The Time Display shows a time line oriented horizontally and, for each location, it shows a distinct axis. The position of the user along time is presented with a linestring along the same location axis, i.e. the user stays in the corresponding location, or with a line connecting two location axes. By default, the moves (oblique segments) and stops (horizontal segments) are rendered as gray lines as they are not annotated. When a user annotates a move (or stop) with an activity the line color changes based on the selected activity type. In Figure 2 (left), the first movement goes from location 0 to location 1 and it is annotated with the activity *work* (code 1 in our internal representation). The movement starts at 7:45:05 and stops at 08:15:25. Then the user stops in location 1 for almost four hours and then moves to location 2 with activity type *Social* (code 3). After a hour and a half he/she goes back to work to location 1. At this stage the locations are labeled by default with progressive numbers, however the user can specify mnemonic names for future references.

The Map Display depicts the geographic context of the user movements. The map is browsable by the user with operations like pan and zoom. The stop locations drawn on this panel are linked with the locations represented in the Time Panel. When the user selects the location on the Time Panel the extent of the Map Display is adjusted in order to show the selected location. Similarly, when the user click on a move (stop) on the Time Display, the corresponding trajectory (location) are zoomed in the Map Display.

3   The SSTD 2013 Demo

During the SSTD 2013 conference we intend to show the whole process of annotating personal trajectories first showing the difficulties of manual annotations and, in second step, we will demonstrate how our tool can significantly simplify this task.

We present two screenshots of the annotation process in Figure 2. From the Time Display it is easy to identify the moves (respectively stops) that are not yet annotated, since they are rendered with gray color. When a move (resp. stop) is selected with the mouse, a dialog box appears to select the context information. The dialog box is shown in Figure 2. We can notice that some of the attributes of the selected move (resp. stop) are already filled: the start and end time, the origin location and the destination location can be derived from the GPS data preprocessed during the loading phase of the data. Other information, like weather condition can be inserted by the user or are proposed by default from the attributes specified in the previous activities. The *goal* attribute specifies which activity - selected from a predefined menu - is performed with the selected move (resp. stop). During the conference, we will provide a set of GPS traces already collected by our volunteers that will be annotated and analysed on site. The attendees of the conference may also voluntary provide their own GPS traces

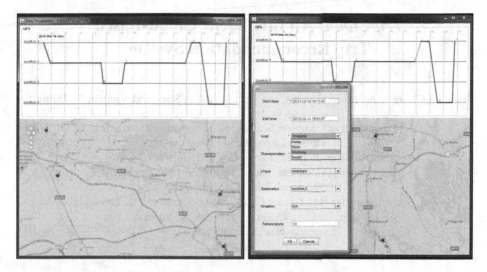

Fig. 2. DayTag interface (left) and how to annotate a trajectory (right)

to visualize and annotate their own movements. DayTag will be available for downloading after the conference so each attendee can later exploit the system to annotate his/her movements in other contexts. The attendee participating in the demo will be asked to fill-in a questionnaire to collect user's feedback on tool usage and interface comprehension.

Acknowledgments. The work was mainly supported by EU project FP7-PEOPLE-SEEK (No. 295179), FP7-FET-DATASIM (270833) and CNR-CNPQ Bilateral Project 2012.

References

1. Bogorny, V., Renso, C., de Aquino, A.R., de Lucca Siqueira, F., Alvares, L.O.: Constant – a conceptual data model for semantic trajectories of moving objects. Transactions in GIS (2013)
2. Broll, G., Cao, H., Ebben, P., Holleis, P., Jacobs, K., Koolwaaij, J., Luther, M., Souville, B.: Tripzoom: an app to improve your mobility behavior. In: Proc. of the 11th Int. Conf. on Mobile and Ubiquitous Multimedia, pp. 57:1–57:4 (2012)
3. Doulamis, A., Pelekis, N., Theodoridis, Y.: Easytracker: An android application for capturing mobility behavior. In: Panhellenic Conference on Informatics (2012)
4. The Open Source Geospatial Foundation. PostGIS, http://postgis.net/
5. Tietbohl Palma, A., Bogorny, V., Kuijpers, B., Alvares, L.O.: A clustering-based approach for discovering interesting places in trajectories. In: SAC (2008)
6. Spaccapietra, S., Parent, C., Damiani, M.L., de Macêdo, J., Porto, F., Vangenot, C.: A conceptual view on trajectories. Data Knowl. Eng. 65(1), 126–146 (2008)

TripCloud: An Intelligent Cloud-Based Trip Recommendation System

Josh Jia-Ching Ying[1], Eric Hsueh-Chan Lu[2], Bo-Nian Shi[1], and Vincent S. Tseng[1,*]

[1] Department of Computer Science and Information Engineering
National Cheng Kung University
No.1, University Road, Tainan City 701, Taiwan (R.O.C.)
[2] Department of Computer Science and Information Engineering
National Taitung University
No.684, Sec. 1, Zhonghua Rd., Taitung City, Taitung County 95002, Taiwan (R.O.C.)
{jashying,ericlu416,bernie_0914}@gmail.com,
tsengsm@mail.ncku.edu.tw

Abstract. With the advance of Location-Based Services (LBS), researches on trip recommendation have attracted extensive attentions. Among them, one active topic is trip planning. In the previous studies on trip planning, various user constraints such as travel time, travel budget, attraction categories, etc., have been considered and users' past travel logs were analyzed for travel recommendation. However, such kind of trip planning approaches cause the computational complexity to increase significantly. Hence, in this paper, we demonstrate a cloud-based travel recommendation system named *TripCloud*, which is built by extending our previous work, *Personalized Trip Recommendation (PTR)*, for meeting user's multiple constraints with efficient trip planning. *TripCloud* encapsulates several data mining techniques and a cloud-based trip planning model to rate the interestingness of each attraction and plan an interesting trip, respectively. Visualization interface is also provided to exhibit the recommended trips based on the characteristics of user constraints.

Keywords: Trip Planning, Recommendation Techniques, Cloud Computing, Location-Based Social Network, Data Mining.

1 Introduction

Traveling is one of the most important entertainments in a modern society. Traditionally, before traveling to an unfamiliar city, one of possible ways of trip planning for tourists is to ask travel agencies to schedule a trip or directly buy a tour package. For example, the trip planned by travel agencies usually includes some famous attraction such as Stature of Liberty and Time Square when the targeted city is New York City. However, such popular trip may not be satisfied by everyone. With the advances of intelligent mobile devices and Web 2.0 techniques, many kinds of

---

* Corresponding author.

M.A. Nascimento et al. (Eds.): SSTD 2013, LNCS 8098, pp. 472–477, 2013.

applications on web services and Location-Based Services (LBSs) such as Gowalla, FourSquare and Facebook have been developed. Based on these approaches, users can easily record and share their daily lives and travel experiences via their mobile devices. Hence, another possible way of trip planning is to search the travel information from websites and plan the trip. This information benefits tourists to add or remove some attractions for planning a personalized trip. Take a scenario as an example. Suppose a tourist wants to travel to New York City. He/She may not know which attractions are worth visiting because this is the first time he visits New York City. He may search for the attractions in New York City by the travel guide websites such as Lonely Planet and Yahoo Travel and schedule a travel trip by some trail recommenders [4]. However, the whole procedure takes lots of time for planning a personalized trip since the amount of travel information is very huge. Although the user can additionally search some travel blogs and check the comments about the attractions, it takes more time to search the information and to put them together for trip planning.

In our previous work, we have developed a framework named *Personalized Trip Recommendation* (*PTR*) [2] that can meet multiple user requirements for travel recommendation. However, the computation cost is quite high in real applications. Although *PTR* have modified the trip planning algorithm Trip-Mine [3] by panel computing architecture, the algorithm cannot satisfy real applications still when the number of attractions is very large. To efficiently plan a personalized trip with user's multiple requirements, in this paper, we demonstrate a novel system named *TripCloud*, which extends *PTR* to a cloud-based architecture for efficiently making travel recommendation. The core idea here is to view trip planning as a combination of different queries, where each query performs an attraction retrieval task. These queries are done separately and then the answer is obtained by combining all the results of queries. Finally, an efficient algorithm named Trip-Mine [3] is adopted to plan the optimal trip based on the combined results of queries. Inherently, this search-and-combine process could be realized by *MapReduce* techniques [1] of cloud computing.

2 Personalized Trip Recommendation

In this section, we briefly introduce our previous work, named *Personalized Trip Recommendation* (*PTR*). As mentioned earlier, tourists may ask travel agencies to plan a trip or directly buy a tour package before traveling to an unfamiliar city. However, popular trip may not be satisfied by everyone. With the advance of Location-Based Social Network (LBSN) such as Gowalla, FourSquare and Facebook, users can record and share their travel experiences via such social media. Hence, another possible way of trip planning is to search the travel information from websites and plan the trip. However, it is hard to distinguish that which attractions are suitable to visit since the travel constraints specified by tourists are different. Take a scenario as an example, suppose that Tom has only 8 hours and 100 dollars to travel New York. The intuitive idea is to choose the interesting attractions, e.g. Metropolitan Museum of Art and Times Square, from social websites and arrange them to the most interesting trip that satisfies the multiple user-specific constraints, e.g., travel time and travel budget. The temporal properties of attractions need to be considered. For example, the opening time of Metropolitan Museum of Art is only in the morning and afternoon. Thus it is incorrect to arrange this attraction in the evening. On the contrary, Times Square is more suitable to be arranged in the evening. However, such idea is inefficient since there are thousands of attractions in large cities such as New York. Therefore, it is essential to develop a personalized travel recommendation system which can automatically recommend the suitable attractions at suitable time for tourists and the most interesting trip that satisfies the multiple user constraints.

To provide an efficient and personalized travel recommendation system with multi-constraints, we have proposed the *Personalized Trip Recommendation* (*PTR*) framework to plan a personalized trip that satisfies multiple user-specific constraints. In *PTR*, we design an attraction scoring component to evaluate the personalized score of attraction by considering user preferences. The proposed attraction scoring component consists of two aspects: 1) User-based attraction score for measuring how interesting the attraction is for a specific user. For example, although *Times Square* and *Metropolitan Museum of Art* are very famous in *New York*, the score of *Times Square* may be higher than that of *Metropolitan Museum of Art* for the tourist who is interested in fashion or shopping. 2) Temporal-based attraction score for measuring how suitable users visit the attraction at a specific time. Different attractions may have different suitable time periods. For example, *Times Square* in *New York* is more suitable to be visited in the evening. Finally, two kinds of scores are fused by a user-specific weight parameter as the final score. Besides, the personalized score is evaluated by using check-in logs and attraction information.

In *PTR*, we extract valuable attraction information from LBSNs as knowledge bases to support an efficient and personalized travel recommendation system that considers multi-constraints at the same time. The key contributions are summarized as follows: 1) We propose the *Personalized Trip Recommendation* (*PTR*) framework, a new approach for trip planning which considers multi-constraints, user preferences and temporal properties, simultaneously. The problems and ideas have not been well explored in the research community. 2) We propose an attraction scoring component for automatically estimating the interesting score of attraction by considering user preferences and temporal properties.

3 Cloud-Based Trip Planning

In this section, we continue to descript the system architecture and major components of *TripCloud*. In order to efficiently response users' queries, the *TripCloud* system employs *MapReduce* techniques [1] of cloud computing to process trip planning task. As Fig. 1 shows, the system architecture consists of three phases: user interface, cloud-based trip planning and attraction interestingness learning. In the user interface, we utilize the webpage for intermedia between users and our system. A user can submit his/her requirements (i.e., queries) to our system and receive the result from our system. The attraction interestingness learning phase follows our previous work, *PTR* [2], to estimate user-based and temporal-based scores of each attraction for each user from the Gowalla data.

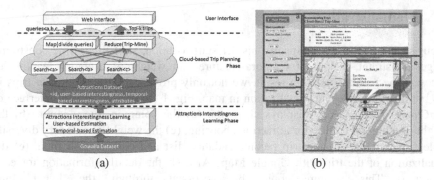

(a)   (b)

Fig. 1. System Architecture and User Interface of *TripCloud*

As mentioned earlier, adopting cloud computing architecture could enable our travel recommendation to achieve a real application. Thus, the main contribution of this demonstration is extending our previous work, *PTR* [2], in cloud computing architecture. To do so, in Trip Planning Phase, we can view trip planning problem as a problem of combination of different queries, each query is a search processes of interesting attractions. Therefore, we could divide the trip planning problem into several parallel search processes of interesting attractions and adopt our previous work Trip-Mine [3] to select and merge these interesting attractions.

To perform the cloud-based search, we use Hadoop as a basic framework to construct the *TripCloud* system. Hadoop is a software framework that supports distributed computing, it enables applications to use thousands of computers to process a big data and achieve high performance. HDFS (Hadoop Distributed File System) is a distributed and scalable file system for the Hadoop framework. HDFS will split a file into several blocks and store these blocks on different nodes, it is more efficient when we want to read a file on HDFS. HBase is a non-relational distributed database, and provides all functions of Google BigTable [9][10] based on the framework of HDFS. To realize cloud computing, a distributed computing framework, named MapReduce, in Hadoop could utilize HBase and process dataset on a lot of computers. MapReduce consists of map and reduce:

- "Map" step: The master node divides the input data into smaller data, and assigns them to nodes. The worker nodes process these data and send the result of processing back to the master node. For our work, the input data of search process of interesting attraction contains all the attractions in the attraction database, and then the data is split by the different query terms into many partitions and loaded by mappers. For each map task, it will find the top k attractions based on the attraction interestingness that is the weighted average of user-based and Temporal-based interestingness. Here, the weight is specified by users.

- "Reduce" step: The master node collects the results from workers and combines them in pre-defined way to generate the answer to the original problem. For our work, the input data are the <key, value> pairs obtained from the map function, key is attraction id and value is the weighted average of user-based and temporal-based interestingness. Reduce task collects all the attractions into a set, form trips with the top k interestingness by performing our previous work, Trip-Mine [3], and output the trips, which are the recommended trips.

4   Demonstration

In our system, an user can specify his/her start location, start time, time constraint, budget constraint, weight of temporal feature and weight of category feature. To represent the recommendation result, we not only provide the description of trips in text but also illustrate the visualization in map. Fig. 1 (b) shows the web interface of *TripCloud*. The interface displays six parts: (a) the common query input, (b) the weight of temporal feature for attraction choosing, (c) the weight of category diversity while trip ranking, (d) the trip recommendation list and computation time, (e) the visualization of the trip with Google Maps API, (f) the detail information for each attraction. This system could be accessed through the hyper link, **http://140.116.247.182:9999/tripPlan/**. We can observe that our system not only allows users to specify multiple requirements but also computes the recommendation result efficiently.

To build our *TripCloud*, we use a cloud-based framework which provides efficient computation. Fig. 2 shows the extra power brought by the cloud can do to trip recommendation. As shown in Fig. 2, we query the system 100 times user different number of nodes of cloud server. Indeed, we can observe that more data nodes bring more computation efficiency in cloud computing. This result strongly suggest that cloud-based framework is one of possible way to solve trip planning problem, especially, such trip planning problem has been theoretically categorized a kind of NP hard problem.

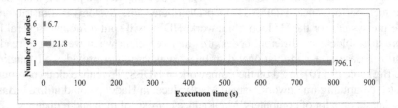

Fig. 2. System Architecture and User Interface of *TripCloud*

5   Conclusion

This paper describes *TripCloud*, a system for travel recommendation meeting user's multiple requirements. Through extending PTR to cloud computing architecture, our *TripCloud* system not only allows users to specify multiple requirements but also computes the recommendation result efficiently. *TripCloud* thus contributes towards better travel recommendation satisfied users' multiple requirements.

References

1. Dean, J., Ghemawat, S.: MapReduce: Simplified data processing on large clusters. In: Sixth Symposium on Operating System Design and Implementation (OSDI 2004) (2004)
2. Lu, E.H.-C., Chen, C.-Y., Tseng, V.S.: Personalized Trip Recommendation with Multiple Constraints by Mining User Check-in Behaviors. In: Proceedings of The 20th ACM SIGSPATIAL International Conference on Advances in Geographic Information Systems (GIS), Redondo Beach, California, November 6-9 (2012)
3. Lu, E.H.-C., Lin, C.-Y., Tseng, V.S.: Trip-Mine: An Efficient Trip Planning Approach with Travel Time Constraints. In: Proceedings of IEEE International Conference on Mobile Data Management (MDM), Lulea, Sweden, June 6-9 (2011)
4. Yoon, H., Zheng, Y., Xie, X., Woo, W.: Social Itinerary Recommendation from User-generated Digital Trails. In: Personal and Ubiquitous Computing. Springer (2011)

The Array Database That Is Not a Database: File Based Array Query Answering in Rasdaman

Peter Baumann, Alex Mircea Dumitru, and Vlad Merticariu

Jacobs University, Bremen 28759, Germany
`firstNameInitial.lastName@jacobs-university.de`

Abstract. Array DBMSs extend the set of supported data structures in databases with (potentially large) multi-dimensional arrays. This information category actually comprises a core data structure in many scientific applications.

When it comes to Petabyte archives, storage costs prohibit importing (i.e., copying) such data into a database. Therefore, *in-situ* processing of database queries is required, that is: evaluating queries on the original files, without previous insertion into the database. We have implemented such an *in-situ* feature for the rasdaman Array DBMS. In this demonstration, we show with rasdaman how query processing in array databases can simultaneously rely on arrays stored in the database — as usual — and in operating system files, like preexisting archives.

1 Introduction

While scientific data convey a significant diversity of data structures, the high volume parts often consist of n-D spatio-temporal or statistical arrays; examples include confocal microscopy images in the Life Sciences, satellite imagery in the Earth Sciences, climate simulation output, and telescope and simulation data in Space science. Due to the lack of array support by standard databases, scientists tend to employ proprietary file-based implementations for serving and analyzing such data.

Array DBMSs like rasdaman [6], SciDB [5], and SciQL [11] close this gap by extending the set of supported data structures in databases with unlimited-size multi-dimensional arrays. In our work with large-scale data centers we use rasdaman, the historically first and "comprehensively implemented" [12] Array DBMS. As it turns out, the massive amounts of such data often cannot be imported into a database due to storage costs. Therefore, *in-situ* processing of database queries is required, that is: evaluating queries on the original files, without previous insertion into the database.

We have implemented such an *in-situ* feature for the rasdaman Array DBMS. In this demonstration, we show how query processing in array databases can simultaneously rely on arrays stored in the database — as usual — and out of the array in operating system files. Examples are taken from real-life large-scale data center offerings and use cases.

M.A. Nascimento et al. (Eds.): SSTD 2013, LNCS 8098, pp. 478–483, 2013.

2   In-situ Query Processing in rasdaman

In rasdaman [6][13], arrays are modeled as functions mapping coordinates from an n-dimensional interval to some value set, which can consist of atomic values (such as greyscale pixels) or composite values (such as RGB or horizontal / vertical windspeed). The conceptual model of rasdaman introduces a new column type, array, which is parameterized with array cell ("pixel", "voxel") type and array extent. The rasdaman query language, *rasql*, allows the composition of expressions on arrays embedded into the SELECT/FROM/WHERE style of SQL. Simplified, rasql adds n-dimensional signal and image processing operators to SQL while remaining set-oriented.

In the server, arrays get partitioned into so-called tiles [6] which form the unit of storage. In the original implementation, a tile is mapped to a relational BLOB and also forms the unit of access for the query engine — in other words, a BLOB tile is always read as a whole. The rasdaman engine, during piecewise ingest or update, automatically splits data into the previously defined tiling structure. This target structure can be preset through a storage layout sub-language which extends the INSERT statement [7].

The *in-situ* mechanism allows to register and de-register external files containing array data. A core design decision was to reference files on the level of tiles so that a file resembles a tile; not only is the resulting solution more scalable, it also allows to use the preexisting tile access and processing methods. Additionally, this allows to mix file and database tiles to accommodate "hot spots" through dedicated database structures.

2.1   Conceptual Extensions

The rasdaman query language offers fully fledged SELECT, INSERT, UPDATE, and DELETE statements in the tradition of SQL.

For registering files, the VALUES clause in the INSERT statement is substituted by a REFERENCING clause which we introduce by way of example. To this end, we assume grayscale TIFF images g1.tiff, g2.tiff, g3.tiff, and g4.tiff, each of extent 100*100 and all sitting in directory /my/tiff/images. The following query imports them as component tiles of one array in table MyImages:

```
INSERT INTO MyImages REFERENCING (GreyImage)
"file://my/tiff/images/g1.tiff" [0:99, 0:99],
"file://my/tiff/images/g2.tiff" [100:199, 0:99],
"file://my/tiff/images/g3.tiff" [0:99, 100:199],
"file://my/tiff/images/g4.tiff" [100:199, 100:199]
```

The cast operation, (GreyImage), tells the system about the pixel type of the images. Four images are registered, each one sitting in one corner of a square. The resulting object covers the whole square, which has an extent of { 0, ..., 199 } × { 0, ..., 199 }. Input files may sit on the local file system or may be fetched via some other supported protocol, such as http. This way, files can even sit on a remote server allowing a greater level of flexibility in the partitioning of data.

De-registering is done during DELETE of an array. When affecting collections that contain referenced files, the operation only removes the metadata associated with the files leaving the data on the file system untouched. This prevents accidental deletion of archive data and is consistent with the *in-situ* way of delegating the archive maintenance to system administrators.

2.2   Implementation

In addition to the pre-existing database access classes reading database BLOBs, an access variant has been added which accesses the previously registered files. File format independence is achieved by using the Geospatial Data Abstraction Library (GDAL) [2].

Relying on files accessed by processes outside the DBMS's control is inherently dangerous, as the database engine cannot expect that the file contents at query time is the same as has been at registration time. As the files are completely outside DBMS control, they might change their contents, structure, or even might be removed at all. This has been addressed by adding plausibility checks at access time. An initial check is done at insert time to verify that the file exists and that it can be read properly by the rasdaman engine. Upon each read a check is performed whether the file still exists and is coherent with the array type; verifications include in particular the array cell type and the number of cells in each dimension. This allows a greater degree of freedom as other tools can perform modifications on the file, either at metadata level (think of adding a tag in a tiff file) or at data level (think value correction) as long as the structure remains coherent.

All further processing inside the server is identical. Thereby, all features of rasdaman remain available without any restriction, including query optimization and parallel tile evaluation.

While thorough performance evaluation is a next step to be undertaken, preliminary observations reveal that there is no significant difference between accessing files or database tiles. Processing in the engine is the same anyway, so the only effect remaining is from potentially inadequate tiling structures – such as defining horizontal slices and performing vertical access.

It depends on the data format chosen whether such effects can be remedied by an appropriate tiling structure. Consider a 3-D x/y/t image timeseries. While TIFF, as a 2-D format, would result in a slice-by-slice structure which is good for extracting timeslices but disastrous for cuts along time, a genuinely 3-D format like NetCDF can indeed store 3-D cubes, which would yield overall good access performance in all directions. This effect will also be shown in the demonstration.

3   Related Work

A standard describing database integration of external files is SQL/MED (Management of External Data) [3]. Applications are enabled to access tuple data in both databases and DBMS-external files. Several commercial systems support storage of data in external files. For example, Oracle offers the BFILE concept [4] where a pointer to a file is stored. Like with rasdaman, access is read-only reflecting the fact that these files are not under DBMS control. However, there are two distinct differences: BFILEs constitute a specific column type visible to the query writing user; moreover, BFILEs cannot be involved in regular tuple set operations — in plain words, a BFILE is not a replacement for a table, but a semantic-less BLOB without regular query processing support.

Semantic support is introduced by Alagiannis et al. [9] who have coined the term *in-situ* for database processing directly on the file system, without previous database import. They call such a DBMS a "NoDB" system. Their own NoDB implementation, a modified version of PostgreSQL 9.0 called PostgresRaw, is able combine tuples stored in a file with those stored in the database. While the approach is similar to ours in that file-based query evaluation takes place, the data structure setup is rather different as rasdaman operates on arrays rather than on tuple sets. As Array DBMS implementation shows [5,6] array query processing is quite distinct from tuple processing.

Array DBMSs are emerging; while rasdaman has been the only such system since 1994 [6], recently several projects have been started, such as SciDB [5] and SciQL [11]. While SciDB does not support *in-situ* features, this issue is being addressed in SciQL; Ivanova et al. [10] describe a method to answer array queries on image files. At query time they load these files into the database so that, with the next incoming query addressing the same file, data, indexes, etc. are already available. Obviously, their approach differs fundamentally from ours in that they still rely on database loading whereas rasdaman circumvents this step completely. According to users the main shortcoming of such an approach is not so much the overhead of the first loading, but the duplication of storage when it comes to realistically large archive sizes.

4 Demonstration

In the demonstration we will provide a laptop with a set of data files obtained from Earth Science data centers participating in the EarthServer initiative [1].

A first set of files will consist of satellite scenes which together from one mosaic — that is, they contain the same spectral channels and together cover some area.

A second scenario consists of 2-D and 3-D data files over the same region. The 2-D files will be satellite imagery again, the 3-D file contains rainfall and temperature data. We will show how a join combines both objects while reading from files. Some queries convey a comparable performance, for others a difference in performance behavior is directly observable. An example for the former case is a trim operation: reading a tile vs reading a file does not make a significant difference.

Fig. 1. Four-layer map produced by a rasdaman query

An example for the latter case is timeseries analysis where file-based access requires opening every timeslice while the database can establish tiles as time sequences, controlled by the rasdaman storage layout language.

A third demonstration combines, in a join operation, objects sitting in the database with *in-situ* objects and applies additional operations, thereby underlining that the *in-situ* objects are fully integrated into query processing. Such a use case is given by map browsing in the style of the OGC Web Map Service (WMS) standard [8]. Figure 1 shows a map overlay where four layers are generated coming from four different database objects[1].

At the bottom, there is an airborne image (in greyscale), it is copied as is. Two layers constitute water areas and water lines; they come from binary maps where the zero values are mapped to transparent and the ones are colored according to an RGB value provided in the query. The top layer is derived from a Digital Elevation Model (DEM); in this example, a traffic light classification has been performed to highlight areas endangered by a potential flooding. Altogether, this constitutes a nontrivial query with high practical relevance.

Participants can try out these and other queries themselves in an ad hoc fashion on these and other data available.

5   Conclusion

We introduce the implementation of *in-situ* data processing through a database query engine, with the specific focus on Array Databases. Data centers in the Earth Sciences require such capabilities to avoid copying massive amounts of data. Our approach of adding references to the files instead of on-demand importing fits well with their requirements.

For the future we plan to extend the *in-situ* capabilities with more flexibility and specific optimizations; in particular, we want to allow array objects to be composed of both database tiles and files. Further, we will perform large-scale evaluations, as several superscale data centers have expressed interest in offering query-based services while not having to change their backends.

References

1. EarthServer - Big Earth Data Analytics (seen on March 8, 2013)
2. GDAL - Geospatial Data Abstraction Library (seen on March 8, 2013)
3. ISO/IEC 9075-9:2008 Information technology - Database languages - SQL - Part 9: Management of External Data (SQL/MED)
4. Oracle bfile (seen on March 8, 2013)
5. Scidb (seen on March 8, 2013)
6. Baumann, P.: On the management of multi-dimensional discrete data. VLDB Journal 4(3), 401–444 (1994)
7. Baumann, P., Feyzabadi, S., Jucovschi, C.: Putting pixels in place: A storage layout language for scientific data. In: Proc. 10th ICDM, December 14, pp. 194–201 (2010)

---

[1] Data courtesy Geoinformation Brandenburg.

8. Baumann, P., Jucovschi, C., Stancu-Mara, S.: Efficient map portrayal using a general-purpose query language. In: Bhowmick, S.S., Küng, J., Wagner, R. (eds.) DEXA 2009. LNCS, vol. 5690, pp. 153–163. Springer, Heidelberg (2009)
9. Alagiannis, I., et al.: NoDB: efficient query execution on raw data files. In: Proc. ACM SIGMOD, May 20, pp. 241–252. ACM (2012)
10. Ivanova, M., Kersten, M., Manegold, S.: Data vaults: A symbiosis between database technology and scientific file repositories. In: Ailamaki, A., Bowers, S. (eds.) SSDBM 2012. LNCS, vol. 7338, pp. 485–494. Springer, Heidelberg (2012)
11. Kersten, M., Zhang, Y., Ivanova, M., Nes, N.: SciQL, a query language for science applications. In: Proc. Workshop on Array Databases, March 25 (2011)
12. Machlin, R.: Index-based multidimensional array queries: safety and equivalence. In: Proc. ACM PoDS, pp. 175–184 (2007)
13. Widmann, N., Baumann, P.: Performance evaluation of multidimensional array storage techniques in databases. In: Proc. IDEAS, pp. 408–413 (1999)

Reliable Spatio-temporal Signal Extraction and Exploration from Human Activity Records

Christian Sengstock, Michael Gertz, Hamed Abdelhaq, and Florian Flatow

Database Systems Research Group, Heidelberg University
{sengstock,gertz,abdelhaq,flatow}@informatik.uni-heidelberg.de

Abstract. Shared multimedia, microblogs, search engine queries, user comments, and location check-ins, among others, generate an enormous stream of human activity records. Such records consist of information in the form of text, images, or videos, and can often be traced in time and space using associated time/location information. Over the past years such spatio-temporal activity streams have been heavily studied with the aim to extract and explore spatio-temporal phenomena, like events, place descriptions, and geographical topics. Despite the clear intuition and often simple techniques to extract such knowledge, the amount of noise, sparsity, and heterogeneity in the data makes such tasks non-trivial and erroneous. This demonstration offers a visual interface to compare, combine, and evaluate spatio-temporal signal extraction and exploration approaches from large-scale sets of human activity records.

1 Introduction

Today huge and steadily increasing streams of human activity records can be accessed, such as shared multimedia, microblogs, search engine queries, user comments, and location check-ins. Besides a timestamp, a growing number of these records contain information about where the activity happened. Such spatio-temporal activity streams are becoming a valuable resource to extract and explore spatio-temporal phenomena, such as the identification of events and place descriptions [1, 2], hurricane trajectories [3], and geographical topics [4–6]. Those works are motivated by treating the information contained in user activity records as proxy observations of the real world. A record at a particular location and time can hence be seen as an implicit vote for or against a phenomenon [7]. Extracted spatio-temporal phenomena from user activities are particularly important as covariates for predictive models in political and social sciences, market research, or ecology [7, 8]. In turn, modeling the spatio-temporal distribution of human activities can also be used to predict the spatio-temporal context of activities when context information is missing [9, 10].

Inherent issues in processing such data are the huge amount of spatial heterogeneity (some regions have large amounts of observations while large parts have very few or none), spatio-temporal noise (location and time information is imprecise or incorrect), and feature sparsity (records are represented by only a small subset of a potentially large set of noisy and redundant features) [2, 3, 6].

M.A. Nascimento et al. (Eds.): SSTD 2013, LNCS 8098, pp. 484–489, 2013.

In fact, by neglecting statistical significance in the context of those issues, one easily draws wrong conclusions from the activity distributions.

In this work we present a framework that allows to implement approaches to extract and explore spatio-temporal phenomena from huge, noisy and heterogeneous activity sources in a unifying way. The framework is designed to process millions of spatio-temporal records and can easily be used in a streaming environment. We demonstrate a visual interface built on top of the framework that allows to compare, combine, and evaluate different signal extraction and exploration approaches. Known phenomena distributions can be loaded into the system as phenomena layers to evaluate extracted signals against ground truth knowledge. We use large-scale Twitter and Flickr data sets consisting of several millions of activity records to demonstrate the results of different approaches.

2   Definitions and Methodology

We generally treat geo-located Twitter posts and Flickr images as sets of *human activity records*. A source of human activity records is represented by a set of observations $\mathcal{O} = \{o_1, \ldots, o_n\}$. Each observation $o = (u, X, l, t)$ is performed by a user $u \in \mathcal{U}$, consists of a set of discrete features $X \subseteq \mathcal{X}$, has an associated point location $l \in \mathcal{L}$, and a timestamp $t \in \mathcal{T}$. We collapse l and t into a spatio-temporal context variable $c = (l, t)$ describing the observations in context space $\mathcal{C}$ and use C either to denote a subset or a random variable of that space. For Twitter, we have observations consisting of a set of extracted textual features (terms). For Flickr, we have observations consisting of a set of image tags. The features might also be weak labels obtained from a classifier predicting if an observation contains evidence for a particular phenomenon or not, as in [3]. This includes heuristic classifiers indicating evidence or non-evidence of a phenomenon by checking if an observation contains one or several features, as in [2] and [8].

A *spatio-temporal phenomenon* can be any social, cultural, or physical entity or process distributed in geographic space and/or time. For example, a festival occurring at a location during a certain time interval, the likelihood of road accidents, or a hurricane. The strength of the phenomenon E in a context subset C can be modeled by a probability distribution $P(C|E)$. By this, any phenomenon is represented as a spatio-temporal signal spreading its mass in context space, and the characteristics of the underlying distribution indicates the type of the phenomenon. For instance, we might think of a bridge as a distribution with equal mass at all locations and time points where the bridge exists, and of a hurricane trajectory as a path in spatio-temporal context space with equal mass at all points of the path.

The aim of *signal extraction* is to estimate a distribution $P(C|\mathcal{O}_E)$, describing the signal of phenomenon E given the positive and negative proxy observations $\mathcal{O}_E$. Recent work in signal extraction include the extraction of road accidents [2], and the extraction of hurricane trajectories in [3], both using Twitter data and appropriate classifiers to obtain the weak labels.

Table 1. Implemented Extraction, Exploration, and Signal Processing Routines

Extraction	Lit.	Exploration	Lit	Signal Processing
Raw Counts	[1]	Bump/Burst Detection	[1]	Gaussian Smoothing
Count Transformations	[6]	LATM	[4]	Baseline Extraction
Binom Model (BM)	[9, 10]	Geo-temporal Clusters	[5]	Baseline-aware Smoothing
BM / Uniform Prior	own	Lat Geo Feature Extraction	[6]	
BM / Adaptive Beta Prior	own			

The aim of *signal exploration* is to find meaningful signals $P(C|E_{1,...,k})$ in observations $\mathcal{O}$. Interesting signals might appear in the distribution of a single feature $x \in \mathcal{X}$, described by $P(C|X = x)$. E.g., in [1], several techniques to extract representative event and place tags are described. This task can be formulated as finding features whose distribution $P(C|X = x)$ has peaks in context space $\mathcal{C}$. Due to feature sparsity, meaningful phenomena signals might only be found by looking at distributions formed by a combination of features $X = x_1, \ldots, x_p$. This can be formalized by introducing a set of latent variables, $Z = Z_{1,...,k}, k < p$, representing both, distributions over features $P(X|Z_{1,...,k})$ and distributions in context space $P(C|Z_{1,...,k})$. The variables are assumed to represent underlying phenomena that generate the data $P(\mathcal{O}|Z)$, reducing the problem to estimate the latent variables Z. The discovery of spatio-temporal phenomena based on feature combinations using latent variable models, dimensionality reduction, and clustering has been studied in [4, 6, 5].

3   System Overview

Our system consists of (1) a data layer, (2) a functionality layer for signal processing, extraction, and exploration routines, and (3) a user interface. In the following we shortly describe the three components.

The *data layer* holds the spatio-temporal feature counts in a sparse 2D matrix, called sparse feature field $S^{|\mathcal{C}| \times |\mathcal{X}|}$. The spatial and temporal dimensions are discretized (see [6, 9, 10]) and projected into a single context dimension $\mathcal{C}$ spanning the rows, while the features $\mathcal{X}$ span the columns. The discretization bin-width of the spatial and the temporal domain is treated as a resolution parameter. The data layer provides a set of low level routines to manage the data: Spatial and temporal projection, aggregation, and context/feature slicing. Because of the sparse nature of the activity counts in context space, we are able to process datasets with millions of records and hundred thousands of features in high resolution context spaces efficiently in main memory. Starting with a high resolution grid, we can lower the resolution (re-parametrize) using fast convolution-based smoothing or matrix multiplication-based cell merging operations.

Signal processing, extraction and exploration routines are implemented in the *functionality layer*. The routines are functions $f : S \mapsto S'$, taking a sparse feature field as argument and returning a transformed instance. Hence, extracted phenomena signals are instances of a sparse feature field themselves and the routines are easily re-useable for different tasks. For our demonstration, we implemented

Extraction / Exploration   Geo-temporal View   Geo View
Feature Statistics Table   Spatio-temporal Signal

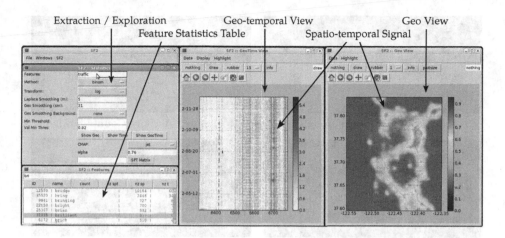

Fig. 1. Visual User Interface: Selected views to analyze and explore extracted signals

existing and novel extraction and exploration approaches, among basic signal processing routines (see Table 1).

The *visual user interface* allows to perform extraction and exploration tasks. The resulting signals and standard errors can be analyzed and visualized in different ways: Geographic distribution on a map, geo-temporal distribution on a geo-temporal grid, temporal distribution, and feature statistics tables (see Figure 1). Moreover, phenomena distributions can be loaded as sparse feature field layers allowing to evaluate extracted signals against ground truth using statistical hypothesis/model tests (Likelihood-ratio, Perplexity) and similarity measures (Mean Squared Error, Kullback-Leiber Divergence, Cosine Similarity).

4   Demonstration

We demonstrate the significance of our system by three tasks. (1) *Record Feature Engineering and Parametrization Impact*: We extract different types of textual record features from the Twitter and Flickr data (raw terms, stemmed terms, term types, weak labels) and analyze their respective signals. We demonstrate the importance of appropriate parametrization by using different strengths of Gaussian smoothing (corresponding to different resolutions) and demonstrate the impact of minimum activity count thresholds. (2) *Extraction and Exploration Comparison*: We compare the geographic and geo-temporal distributions of extracted signals using various approaches (see Table 1 for different techniques and Figure 2 for an example of different signal extraction results). We demonstrate that raw counts and the binomial model fail in finding reliable signals due to the spatio-temporal trend and the spatial heterogeneity by analyzing their respective standard errors. We also compare the results of four implemented exploration routines and show how resolution and normalization can be used as user-based parameters in the analysis. (3) *Ground Truth Evaluation*: We compare extracted

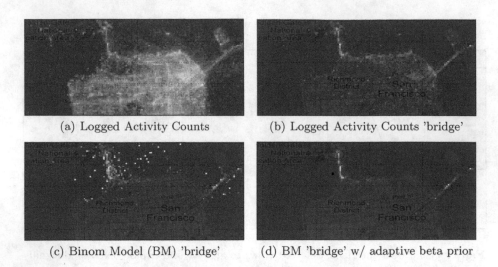

(a) Logged Activity Counts   (b) Logged Activity Counts 'bridge'

(c) Binom Model (BM) 'bridge'   (d) BM 'bridge' w/ adaptive beta prior

Fig. 2. Extracted Twitter signals from 5 million tweets covering the San Francisco area. Plot (a) shows the overall distribution of activity and (b),(c), and (d) show extracted signals for tweets containing the term 'bridge' (heuristic classifier).

signals with ground truth layers (buildings, regions, festivals, etc.) using different metrics. We show ranked lists of extraction and exploration routines for given phenomena on the basis of those metrics.

5   Conclusions

In this demonstration we present a visual user interface to perform spatio-temporal signal extraction and exploration tasks on large-scale sets of noisy and heterogeneous human activity records. The system allows to study the pros and cons of various approaches by user-based exploratory analysis and quantitative evaluation against ground truth knowledge. The demonstration provides insights in the particular problems of noise and sparsity within spatio-temporal human activity data sets, and the impact of preprocessing and parametrization to various extraction and exploration approaches. The system is built on top of a flexible data layer that allows to implement and re-use various approaches in a unifying way and to process millions of activity records efficiently.

References

1. Rattenbury, T., Naaman, M.: Methods for Extracting Place Semantics from Flickr Tags. ACM Transactions on the Web 3(1), 1–30 (2009)
2. Xu, J.-M., Bhargava, A., Nowak, R., Zhu, X.: Socioscope: Spatio-temporal Signal Recovery from Social Media. In: Flach, P.A., De Bie, T., Cristianini, N. (eds.) ECML PKDD 2012, Part II. LNCS, vol. 7524, pp. 644–659. Springer, Heidelberg (2012)

3. Sakaki, T., Okazaki, M., Matsuo, Y.: Earthquake Shakes Twitter Users: Real-time Event Detection by Social Sensors. In: Proc. of WWW 2010, pp. 851–860 (2010)
4. Yin, Z., Cao, L., Han, J., Zhai, C., Huang, T.: Geographical Topic Discovery and Comparison. In: Proc. of WWW 2011, pp. 247–256 (2011)
5. Zhang, H., Korayem, M., You, E., Crandall, D.J.: Beyond Co-occurrence: Discovering and Visualizing Tag Relationships from Geo-spatial and Temporal Similarities. In: Proc. of WSDM 2012, pp. 33–42 (2012)
6. Sengstock, C., Gertz, M.: Latent Geographic Feature Extraction from Social Media. In: Proc. of GIS 2012, pp. 149–158 (2012)
7. Jin, X., Gallagher, A., Cao, L., Luo, L., Han, J.: The Wisdom of Social Multimedia: Using Flickr For Prediction and Forecast. In: Proc. of MM 2010, pp. 1235–1244 (2010)
8. Zhang, H., Korayem, M., Crandall, D.J., Lebuhn, G.: Mining Photo-sharing Websites to Study Ecological Phenomena. In: Proc. of WWW 2012, pp. 749–758 (2012)
9. Wing, B.P., Baldridge, J.: Simple Supervised Document Geolocation with Geodesic Grids. In: Proc. of ACL 2011, pp. 955–964 (2011)
10. OHare, N., Murdock, V.: Modeling Locations with Social Media. Journal of Information Retrieval 16(1), 30–62 (2012)

UniModeling: A Tool for the Unified Modeling and Reasoning in Outdoor and Indoor Spaces*

Sari Haj Hussein**, Hua Lu, and Torben Bach Pedersen

Department of Computer Science, Aalborg University, Denmark
{sari,luhua,tbp}@cs.aau.dk

Abstract. This paper demonstrates UniModeling; a tool for the unified modeling and reasoning in outdoor and indoor spaces. UniModeling supports constructing unified graph models of outdoor and indoor spaces and RFID deployments in these spaces. It enables probabilistic incorporation of RFID data that facilitates the tracking of moving objects and enables the search for them to be optimized. Furthermore, UniModeling is empowered with three reasoning applications that pertain to the positioning of RFID readers in outdoor and indoor spaces and the points of potential traffic (over)load in these spaces. The utility of UniModeling is demonstrated in concrete steps through applying it to the modeling and reasoning about RFID deployment and baggage handling in an example airport.

1 Introduction

Ubiquitous receptor devices (e.g., RFID readers, and wireless sensor networks) are increasingly deployed in outdoor and indoor spaces (OI-spaces [1]) to enable new classes of so-called receptor-based applications. These applications need to span seamlessly both outdoor (O-) and indoor (I-) spaces in order to deliver their functionality. However, related work has mostly focused on the modeling of indoor spaces [2–5]. Related work that proposes ontologies for representing and reasoning about outdoor and indoor environments is either designed for navigation, or is ill-suited for reasoning about RFID-tracked moving objects [6, 7]. To compensate for the aforementioned lack, this paper demonstrates UniModeling; a tool for creating unified graph models of OI-spaces and RFID deployments in these spaces. This tool is empowered with a probabilistic translator that enhances and complement the knowledge about the locations of RFID-tracked moving objects in OI-spaces. UniModeling can also perform high-level reasoning about the positioning of RFID readers in OI-spaces and the locations of potential traffic load in these spaces, so-called bottleneck points (BPs).

The remainder of this paper is organized as follows. Section 2 presents the UniModeling system architecture. Section 3 concretizes the implementation and application details of UniModeling. The demonstration scenario is lastly given in Section 4. For a full description of the modeling foundations, translation, and reasoning applications in UniModeling, the reader is referred to [8, 9].

---

* This work was supported by the BagTrack project sponsored by the Danish National Advanced Technology Foundation under grant 010-2011-1.
** Corresponding author.

M.A. Nascimento et al. (Eds.): SSTD 2013, LNCS 8098, pp. 490–495, 2013.

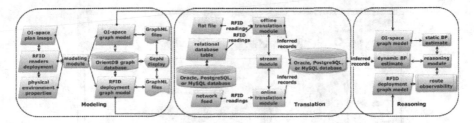

Fig. 1. The UniModeling system architecture

2 System Architecture

The UniModeling system encompasses four modules (Figure 1). *The modeling module* creates the OI-space and RFID deployment graphs. These graphs are stored in an OrientDB graph database (http://www.orientdb.org/), and they can be visualized in UniModeling and/or Gephi (http://gephi.org/). *The stream module* registers RFID data streams whose source is a flat file, a database table, or a network connection. *The offline and online translation modules* incorporate RFID data streams. The former module reads from a flat file or a database table, whereas the latter one reads from network feeds. *The reasoning module* delivers three reasoning applications; (1) route observability: A measure of the extent to which a given route[1] is covered by RFID readers. (2) static BP estimate: It determines the static, time-independent likelihood that a semantic location[2] is a BP. (3) dynamic BP estimate: It has the same role as the static BP estimate; however, it achieves this role by incorporating timestamped RFID data streams.

3 System Implementation

The real-world baggage handling plan in Aalborg Airport is used as a running example. This plan comprises two sub-plans; the I-space and O-space plans in Aalborg Airport hall (Figure 2) and apron[3] respectively. UniModeling is nonetheless applicable to any OI-space in which motion is partially constrained (e.g., due to the presence of obstacles in O-spaces and floor plans in I-spaces).

Modeling an OI-Space and an RFID Deployment: Relying on the space plan (Figure 2), the OI-space graph is created. A modeler delineates *semantic locations* (using geometric shapes) and specify *binary sub-routes* (using arrow shapes). In Figure 2, the check-in desks (CD) and check-in conveyor (CC) are locations, and (CD, CC) is a binary sub-route indicating motion from CD to CC. UniModeling automatically recognizes (CD|CC) as a *connection point*. Next, UniModeling converts the locations into vertices and sub-routes into edges (an edge direction matches the motion direction and the order of the sub-route). Furthermore, UniModeling labels the edges using sets taken

---

[1] A particular sequence of locations followed by a moving object in an OI-space.

[2] A location that has a meaningful interpretation to the RFID-based application.

[3] The open part of an airport in which airplanes are parked, fueled, boarded by passengers, and loaded with baggage.

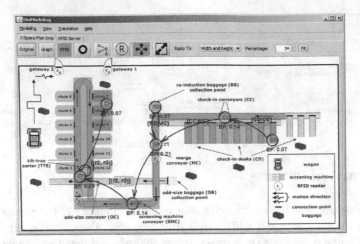

Fig. 2. The I-space plan in Aalborg Airport hall overlaid with the graph models

Fig. 3. An ongoing translation of Aalborg Airport RFID data

from the power set of the connection points. In Figure 2, CD and CC are converted into vertices, and (CD, CC) is converted into an edge connecting between these two vertices. The edge (CD, CC) is directed from CD to CC and labeled CD|CC.

Next the OI-space graph is transformed into an RFID deployment graph. A modeler places the RFID readers in their positions on the space plan (Figure 2). Based on this placement, UniModeling labels the vertices and edges in the RFID graph. In Figure 2, r_1 is positioned inside MC away from any connection point. Therefore, r_1 is added to $label(MC)$. However, r_2 and r_3 are adjacently positioned at SMC|TTS, and r_2 reads before r_3 when moving from SMC to TTS across SMC|TTS. Thus, (r_2, r_3) is added to $label(SMC, TTS)$. UniModeling can extend the RFID graph into a property graph [10] by allowing the vertices and edges to have various properties (key/value pairs) from the physical hall and apron environments. An important property to the route

Table 1. Tables (a) and (b) show the appearance and intermediate records of bag_1 during $[t_1, t_8]$ respectively (s-time and e-time are the start and end time of an appearance)

(a) Appearance records						(b) Intermediate records				
ar-id	obj-id	reader-id	s-time	e-time		ar-id	obj-id	loc	s-time	e-time
ar_1	bag_1	r_1	t_1	t_2		ar_1	bag_1	MC	t_1	t_2
ar_2	bag_1	r_2	t_5	t_6		ar_2	bag_1	SMC	t_5	t_6
ar_3	bag_1	r_3	t_7	t_8		ar_2	bag_1	TTS	t_5	t_6
						ar_2	bag_1	TTS	t_5	t_6
						ar_2	bag_1	TTS	t_5	t_6
						ar_3	bag_1	SMC	t_7	t_8
						ar_3	bag_1	TTS	t_7	t_8
						ar_3	bag_1	TTS	t_7	t_8
						ar_3	bag_1	TTS	t_7	t_8

Table 2. Tables (a) and (b) show the probabilistic and inferred records of bag_1 during $[t_1, t_8]$ respectively (prob-loc and infer-loc are probability distributions on the set of semantic locations)

(a) Probabilistic records				(b) Inferred records			
obj-id	prob-loc	s-time	e-time	obj-id	infer-loc	s-time	e-time
bag_1	[MC : 1]	t_1	t_2	bag_1	[MC : 1]	t_1	t_2
bag_1	[SMC : .25, TTS : .75]	t_5	t_6	bag_1	[MC : .39, SMC : .40, TTS : .21]	t_3	t_4
bag_1	[SMC : .25, TTS : .75]	t_7	t_8	bag_1	[SMC : .30, TTS : .70]	t_5	t_6
				bag_1	[SMC : .14, TTS : .86]	t_7	t_8

observability reasoning application (Section 2) is the coverage weight of RFID readers among locations. This vertex property specifies the quotient of the overlap between a reader reading zone and a location area. For a location l and a reader r, this means: $cw(l) = \frac{ZONE(r) \cap AREA(l)}{ZONE(r)} : ZONE(r) \cap AREA(l) \neq \emptyset$.

Incorporating RFID Data: The translator in the offline and online translation modules incorporates uncleansed RFID data (Figure 3) using three essential steps:

1) *Condensing:* UniModeling condenses RFID readings into appearance records[4] by employing a pre-processing module [11]. See Table 1(a) for instance.

2) *Probabilistic translation:* UniModeling translates the appearance into probabilistic records using the vertex and edge labels in the RFID graph. In Table 1(a) and Figure 2, $ar_1.reader\text{-}id = r_1 \in label(MC)$, which yields record 1 in Table 1(b). However, $ar_2.reader\text{-}id = r_2 \in label(SMC, TTS)$ & $label(TTS, TTS)$, which yields records 2-5 in Table 1(b). UniModeling executes an SQL query on Table 1(b) to obtain the probabilistic records in Table 2(a). To show the effect of this query, TTS appears three times in records 3-5 in Table 1(b), therefore TTS probability is .75 in record 2 of Table 2(a).

3) *Inferring the information gaps:* UniModeling infers the information gap $[t_3, t_4]$ in Table 2(a) via a dynamic Bayesian network (DBN) [12] whose beliefs are updated using the Estimated Posterior Importance Sampling algorithm for Bayesian Networks (EPIS-BN) [13]. Applying EPIS-BN to Table 2(a) yields the inferred records in Table 2(b). Note that the knowledge obtained from the translator is both (1) complete and (2) more informative about the locations of baggage in transit. To clarify (1), Table 2(b) communicates full observability of bag_1 during $[t_1, t_8]$, whereas the observability delivered is only partial in Table 1(a) during the same period (note the information gap $[t_3, t_4]$). To give an example on (2), ar_3 in Table 1(a) tells that bag_1 passed under r_3 during $[t_7, t_8]$.

---

[4] A record that stores the first and last detection of an RFID tag by a reader.

Due to the adjacent positioning of r_2 and r_3 (Figure 2), this information piece is deficient and possibly inaccurate. Contrary to this, record 4 in Table 2(b) tells that bag_1 is highly likely to be at TTS and less likely to be at SMC during $[t_7, t_8]$. All in all, the translator better facilitates baggage tracking and enables the search for lost baggage to be optimized.

Reasoning about RFID Deployment and Baggage Handling: The route observability is tested using the example route MC $\rightarrow$ SMC $\rightarrow$ TTS in Figure 2. The coverage weights along this route are $cw(\text{MC}) = \{r_1 \rightarrow 1\}$, $cw(\text{SMC}) = \{r_2 \rightarrow 0.8, r_3 \rightarrow 0.2\}$, and $cw(\text{TTS}) = \{r_2 \rightarrow 0.2, r_3 \rightarrow 0.8\}$ respectively. UniModeling determines the observability (3.2221) using[5]: $obs(\text{MC} \ldots \text{TTS}) = \sum_{l \in (\text{MC} \ldots \text{TTS})} \sum_{cw(l)} \log(cw(l) + 1)$. UniModeling also indicates that 3.2221 is less than the maximum, attainable observability which is 4.1699 (the obs bounds are derived in [9]). This suggests a possibility to adjust Aalborg Airport RFID deployment in order to improve the reading environment and thereby reduce or even eliminate the occurrence of RFID anomalies. Next, the static BP estimate is tested using the location TTS in Figure 2. TTS's degree[6] is 4, and double the number of edges in the OI-space graph is 14. UniModeling takes the ratio of these two quantities and reports the static BP estimate 0.29 in Figure 2. Note in Figure 2 that TTS estimate is higher than that for the rest of the locations. This necessitates careful planning for RFID deployment at TTS. Finally, the dynamic BP estimate is tested. The translated Aalborg Airport RFID data (Table 2(b)) is used as an input to an algorithm (see [9]) that determines the dynamic BP estimates over different monitoring periods in different days. These estimates constitute a good model that highlights congestion by baggage and ranks the standard of each responsibility area in Aalborg Airport.

4   Demonstration Scenario

The running example in this paper (the real-world baggage handling plan in Aalborg Airport) will be used as a demonstration case. First, the OI-space and RFID graphs of Aalborg Airport will be created. Then, a few routes used for transporting baggage will be marked on the graphs. The observabilities of these routes will be determined. This is followed by varying the placement of readers and monitoring the change in the observabilities. Next, the statis BP estimate will be computed for the airport locations (the graph vertices). Aalborg Aiport RFID data will then be translated into inferred records. Congestion problems will next be monitored through plotting the dynamic BP estimates over different monitoring periods in different days. Important messages about RFID deployment and baggage handling quality in Aalborg Airport will last be given.

References

1. Worboys, M.: Modeling indoor space. In: ISA (2011)
2. Hagedorn, B., Trapp, M., Glander, T., Döllner, J.: Towards an indoor level-of-detail model for route visualization. In: MDM (2009)

---

[5] Logarithms are to the base 2.
[6] The number of edges whose head or tail is TTS; the loop is counted twice.

3. Li, D., Lee, D.L.: A lattice-based semantic location model for indoor navigation. In: MDM (2008)
4. Li, X., Claramunt, C., Ray, C.: A grid graph-based model for the analysis of 2d indoor spaces. Computers, Environment and Urban Systems 34(6) (2010)
5. Lu, H., Cao, X., Jensen, C.S.: A foundation for efficient indoor distance-aware query processing. In: ICDE (2012)
6. Yang, L., Worboys, M.: A navigation ontology for outdoor-indoor space (work-in-progress). In: ISA (2011)
7. Niu, W.T., Kay, J.: Personaf: framework for personalised ontological reasoning in pervasive computing. User Modeling and User-Adapted Interaction 20(1), 1–40 (2010)
8. Hussein, S.H., Lu, H., Pedersen, T.B.: Towards a unified model of outdoor and indoor spaces. In: GIS (2012)
9. Hussein, S.H., Lu, H., Pedersen, T.B.: Reasoning about rfid-tracked moving objects in symbolic indoor spaces. In: SSDBM (2013)
10. Rodriguez, M.A., Neubauer, P.: Constructions from dots and lines. Bulletin of ASIS&T 36(6) (2010)
11. Jensen, C.S., Lu, H., Yang, B.: Graph model based indoor tracking. In: MDM (2009)
12. Koller, D., Friedman, N.: Probabilistic graphical models: principles and techniques. MIT Press (2009)
13. Yuan, C., Druzdzel, M.J.: An importance sampling algorithm based on evidence pre-propagation. In: UAI (2003)

The Spatiotemporal RDF Store Strabon

Kostis Kyzirakos, Manos Karpathiotakis, Konstantina Bereta, George Garbis,
Charalampos Nikolaou, Panayiotis Smeros, Stella Giannakopoulou,
Kallirroi Dogani, and Manolis Koubarakis

National and Kapodistrian University of Athens, Greece
{kkyzir,mk,Konstantina.Bereta,ggarbis}@di.uoa.gr,
{charnik,psmeros,sgian,kallirroi,koubarak}@di.uoa.gr

Abstract. Strabon is a very scalable and efficient RDF store for storing
and querying geospatial data that changes over time. We present the
geospatial and temporal features of Strabon and we demonstrate their
utilization in the fire monitoring and the burn scar mapping applications
of the National Observatory of Athens.

1 Introduction

In this paper we present the spatiotemporal features of the system Strabon[1] and
their utilization in the fire monitoring and the burn scar mapping applications
of the National Observatory of Athens (NOA), which have been developed in
the context of the project TELEIOS[2].

Strabon is a semantic geospatial DBMS for storing and querying geospatial
data that changes over time. It implements the data model stRDF, the query
language stSPARQL and the respective part of the OGC standard GeoSPARQL
[1]. The data model stRDF is an extension of the W3C standard RDF for representing time-varying geospatial data. The query language stSPARQL is an
extension of the query language SPARQL 1.1 and it has been implemented in
Strabon offering scalability to billions of stRDF triples. The initial versions of the
data model stRDF and the query language stSPARQL have been described in
[5]. The geospatial features of the most recent versions of stRDF and stSPARQL,
and their implementation in Strabon have been presented in [7]. This work also
presents the architecture of the system, the optimizations followed for storing
and querying geospatial data, and an experimental evaluation which shows that
Strabon in most cases performs better than any other geospatial DBMS that
has been competed with. The valid time dimension of the data model stRDF
and the query language stSPARQL has been described in detail in [2].

We were motivated by the fire monitoring application of NOA to extend RDF
and SPARQL 1.1 with the ability to store and query geospatial information that
changes over time. In the fire monitoring service stRDF is used to represent satellite image metadata (e.g., time of acquisition, geographical coverage), knowledge

---

[1] http://strabon.di.uoa.gr/
[2] http://www.earthobservatory.eu/

M.A. Nascimento et al. (Eds.): SSTD 2013, LNCS 8098, pp. 496–500, 2013.

extracted from satellite images (e.g., a certain image region is a hotspot) and auxil-iary geospatial datasets encoded as linked data. The hotspot products are encoded to stRDF, so that they can be combined with auxiliary linked geospatial data. In this application, the user-defined time dimension is used as the *detection time* of hotspots.

The burn scar mapping (BSM) application of NOA motivated us to design the valid time dimension of stRDF and stSPARQL and extend Strabon with valid time support, in order be able to retrieve the evolution of the land cover of areas though time. Strabon is one of the very few RDF stores that supports storing and quering the valid time of triples.

This document is structured as follows. In Section 2 we describe the function-alities and the architecture of the system Strabon. In Section 3 we present the use of Strabon in the fire monitoring application of NOA and in Section 4 we present the use of Strabon in the burn scar mapping application of NOA. In Section 5 we describe the scenarios that will be used to demonstrate the spatiotemporal functionalities of the system.

2   The Spatiotemporal Features of Strabon

The system Strabon implements the most recent versions of stRDF and stSPARQL. In the new version of stRDF, we use the widely adopted OGC stan-dards Well Known Text (WKT) and Geography Markup Language (GML) to represent geospatial data as literals of datatype `strdf:geometry`[3]. The new ver-sion of stSPARQL extends SPARQL 1.1 with the machinery of the OGC-SFA standard. We achieve this by defining one URI for each of the SQL functions defined in the standard and use them in SPARQL queries. Similarly, we have de-fined a Boolean SPARQL extension function for each topological relation defined in OGC-SFA (topological relations for simple features), [4] (Egenhofer relations) and [3] (RCC-8 relations). In this way stSPARQL supports multiple families of topological relations our users might be familiar with. Using these functions stSPARQL can express topological relations between geometry objects in the select or in the filter part of the query.

To be able to represent periods, we have also introduced the `strdf:period` datatype. The valid time annotation of a triple can be added at the end of the triple and it can be either a literal of the `strdf:period` or the `xsd:dateTime` datatypes. The temporal constants NOW and UC (i.e, "Until Changed") are also introduced. The first denotes the current timestamp and the second the time persistence of a triple when placed as the ending time of a period. The query language stSPARQL also defines a wide variety of temporal functions to express relations between intervals, or instants and intervals. Period constructors and temporal aggregates are also defined. A full reference of the spatial and temporal functions provided in stSPARQL can be found online[4].

Strabon 3.0 is a fully-implemented, open-source, storage and query evalua-tion system for stRDF/stSPARQL and the corresponding subset of GeoSPARQL.

---

[3] http://strdf.di.uoa.gr/ontology
[4] http://www.strabon.di.uoa.gr/stSPARQL

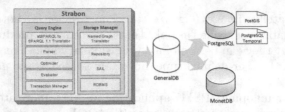

Fig. 1. Architecture of the system Strabon

Strabon has been implemented by extending the well-known RDF store Sesame transparently so that it can be compatible with most recent versions of Sesame. It also uses a DBMS as a backend, which can be a spatially and temporally enabled PostgreSQL database. MonetDB can also serve as backend for the geospatial features only. As Figure 1 shows, Strabon also consists of the *storage manager* and the *query engine.*

The storage manager stores stRDF triples using the per predicate scheme of Sesame and dictionary encoding. For each predicate table, two B+ tree two-column indices are created. For each dictionary table a B+ tree index on the id column is created. The geometries of spatial literals are stored as values of the `geometry` datatype in a separate table that uses an R-tree-over-GiST spatial index on the respective geometry column. Similarly, literals of the `strdf:period` datatype are stored in a separate table with a GiST index on values of the `PERIOD` datatype, which is a primitive provided by PostgreSQL Temporal. The *Named Graph Translator* translates the temporal triples of stRDF to standard RDF triples following the named graphs approach as discussed in [2].

In the query engine, the optimizer and the evaluator have been implemented by modifying the corresponding components of Sesame to be able to evaluate spatial and temporal functions efficiently. The main idea is to push the evaluation of spatial and temporal extension function to the database level, so that the respective indices should be used to increase performance. The *stSPARQL to SPARQL 1.1 Translator* translates the temporal triple patterns to triple patterns using the named graph approach.

3   The Fire Monitoring Application

NOA operates an MSG/SEVIRI satellite acquisition station, and has developed a real-time fire hotspot detection service for effectively monitoring a fire-front. As soon as the images are acquired (every 5 or 15 minutes), they are stored as arrays in MonetDB. Then they are cropped and georeferenced using the SciQL query language and hotspots are detected producing shapefiles that are then translated into the stRDF data model. A number of refinement steps as stSPARQL update operations are then performed to increase the accuracy of the resulting products, i.e., false alarms and omission errors are detected and proper corrections are made, producing the final products of the processing chain. The stSPARQL update operations that take place in the fire monitoring processing chain are described in [6]. In the fire monitoring application described in [6], only the refined results are eventually presented to the user. In the application that we present here, the user can also execute the refinement steps using the graphical user interface. Finally, we enrich the dataset that derives from the processing chain by combining it with the following two datasets (both compiled in the context of TELEIOS): (*i*) the dataset describing the coastline of Greece[5], and (*ii*) the dataset describing the Greek environmental landscape[6].

Strabon can also expose results in KML or GeoJSON and the results of each query posed can be depicted in a different layer of a map and overlay the retrieved data. This feature is very important for Earth Observation (EO) experts, as they can execute queries that retrieve EO data enriched with additional geospatial information from the Linked Open Data Cloud and visualize them in a map.

4   The Burn Scar Mapping Application

The Burn Scar Mapping (BSM) is another application of the National Observatory of Athens. It involves the damage assessment using Landsat images, i.e. the estimation of the burned areas after wildfires. The products of the BSM service are in the form of shapefiles that contain information about the burned areas, i.e., their id, geometry, etc. We translate the shapefiles into stRDF graphs. The valid time dimension of the data model stRDF is also used to annotate triples to encode that an area has a respective land cover in a specific time period. The geospatial features of Strabon allow us to generate burned area maps automatically and enrich the displayed geometries with auxiliary information from other geospatial datasets or from the linked data cloud. The temporal features of Strabon allow us to express queries regarding the evolution of the land cover of an area that got burned at one or more time points of the time line. The final products of the BSM processing chain can be exploited into this direction, as they allow us to perform a time-series analysis since 1984.

5   Demonstration

The demonstration of the system is divided into two parts. The first part shows the fire monitoring application of NOA that uses the geospatial features of Strabon combined with the user-defined time dimension. Firstly, the service is initialized and the user can see a map depicting the current hotspots and then she can navigate into archive data and select hotspots from previous fire seasons and varying time ranges and display them in the map, as in Figure 2(a). Then she will execute the respective stSPARQL query and update operations that take part in the processing chain of the fire monitoring application.

The second part of the demonstration shows the burn scar mapping application of NOA that uses the geospatial features combined with the valid time dimension of Strabon. Similar to the fire monitoring application, the user will be able to execute queries against Strabon asking for burned areas and then these

---

[5] http://geo.linkedopendata.gr/coastline_gr/
[6] http://geo.linkedopendata.gr/corine/

(a)   (b)

Fig. 2. (a) Fire monitoring application of NOA (b) Strabon endpoint

areas will be displayed at a separate layer on the map. The dataset that will be used in this demo will contain the burned areas from the shapefiles provided by NOA combined with the CORINE land cover dataset. By this way, the user will be able to execute a query to retrieve the evolution of the land cover of an area that got burned at one or more points in the timeline. The user will be provided with a Strabon endpoint where she will be able to see and execute some pre-defined queries, as in Figure 2(b) and then write similar queries in the textarea of the endpoint and execute them against the Strabon backend. For example, a user could pose a query to retrieve the previous land cover of burned area or the geometries of areas that were initially coniferous forests and then got burned.

References

1. Open Geospatial Consortium. OGC GeoSPARQL - A geographic query language for RDF data. OGC Candidate Implementation Standard (2012)
2. Bereta, K., Smeros, P., Koubarakis, M.: Representation and querying of valid time of triples in linked geospatial data. In: Cimiano, P., Corcho, O., Presutti, V., Hollink, L., Rudolph, S. (eds.) ESWC 2013. LNCS, vol. 7882, pp. 259–274. Springer, Heidelberg (2013)
3. Cohn, A., Bennett, B., Gooday, J., Gotts, N.: Qualitative Spatial Representation and Reasoning with the Region Connection Calculus. Geoinformatica 1(3), 275–316 (1997)
4. Egenhofer, M.J.: A Formal Definition of Binary Topological Relationships. In: Litwin, W., Schek, H.-J. (eds.) FODO 1989. LNCS, vol. 367, pp. 457–472. Springer, Heidelberg (1989)
5. Koubarakis, M., Kyzirakos, K.: Modeling and Querying Metadata in the Semantic Sensor Web: The Model stRDF and the Query Language stSPARQL. In: Aroyo, L., Antoniou, G., Hyvönen, E., ten Teije, A., Stuckenschmidt, H., Cabral, L., Tudorache, T. (eds.) ESWC 2010, Part I. LNCS, vol. 6088, pp. 425–439. Springer, Heidelberg (2010)
6. Kyzirakos, K., Karpathiotakis, M., Garbis, G., Nikolaou, C., Bereta, K., Sioutis, M., Papoutsis, I., Herekakis, T., Mihail, D., Koubarakis, M., Kontoes, C.: Real Time Fire Monitoring Using Semantic Web and Linked Data Technologies. In: 11th International Semantic Web Conference, Boston, USA (2012)
7. Kyzirakos, K., Karpathiotakis, M., Koubarakis, M.: Strabon: A Semantic Geospatial DBMS. In: Cudré-Mauroux, P., et al. (eds.) ISWC 2012, Part I. LNCS, vol. 7649, pp. 295–311. Springer, Heidelberg (2012)

Author Index